T0292752

Cooperation and Its Evolution

Life and Mind: Philosophical Issues in Biology and Psychology
Kim Sterelny and Robert A. Wilson, Series Editors

Cooperation and Its Evolution, Kim Sterelny, Richard Joyce, Brett Calcott, and Ben Fraser, editors, 2012

Ingenious Genes: How Gene Regulation Networks Evolve to Control Development, Roger Sansom, 2011

Yuck! The Nature and Moral Significance of Disgust, Daniel Kelly, 2011

Laws, Mind, and Free Will, Steven Horst, 2011

Perplexities of Consciousness, Eric Switzgebel, 2011

Humanity's End: Why We Should Reject Radical Enhancement, Nicholas Agar, 2010

Color Ontology and Color Science, Jonathan Cohen and Mohan Matthen, editors, 2010

The Extended Mind, Richard Menary, editor, 2010

The Native Mind and the Cultural Construction of Nature, Scott Atran and Douglas Medin, 2008

Describing Inner Experience? Proponent Meets Skeptic, Russell T. Hurlburt and Eric Schwitzgebel, 2007

Evolutionary Psychology as Maladapted Psychology, Robert C. Richardson, 2007

The Evolution of Morality, Richard Joyce, 2006

Evolution in Four Dimensions: Genetic, Epigenetic, Behavioral, and Symbolic Variation in the History of Life, Eva Jablonka and Marion J. Lamb, 2005

Molecular Models of Life: Philosophical Papers on Molecular Biology, Sahotra Sarkar, 2005

The Mind Incarnate, Lawrence A. Shapiro, 2004

Organisms and Artifacts: Design in Nature and Elsewhere, Tim Lewens, 2004

Seeing and Visualizing: It's Not What You Think, Zenon W. Pylyshyn, 2003

Evolution and Learning: The Baldwin Effect Reconsidered, Bruce H. Weber and David J. Depew, editors, 2003

The New Phrenology: The Limits of Localizing Cognitive Processes in the Brain, William R. Uttal, 2001

Cycles of Contingency: Developmental Systems and Evolution, Susan Oyama, Paul E. Griffiths, and Russell D. Gray, editors, 2001

Coherence in Thought and Action, Paul Thagard, 2000

Cooperation and Its Evolution

edited by Kim Sterelny, Richard Joyce, Brett Calcott, and Ben Fraser

A Bradford Book
The MIT Press
Cambridge, Massachusetts
London, England

This book was set in Stone Sans and Stone Serif by Toppan Best-set Premedia Limited, Hong Kong.

Library of Congress Cataloging-in-Publication Data

Cooperation and its evolution / edited by Kim Sterelny, Richard Joyce, Brett Calcott, and Ben Fraser.
p. cm.—(Life and mind: philosophical issues in biology and psychology)
"A Bradford Book."
Includes bibliographical references and index.
ISBN 978-0-262-01853-1 (hardcover : alk. paper), 978-0-262-55278-3 (paperback)
1. Evolution (Biology)—Philosophy. 2. Evolutionary psychology. 3. Cooperation. I. Sterelny, Kim. II. Joyce, Richard, 1966–. III. Calcott, Brett. IV. Fraser, Ben.
QH366.2.C657 2013
576.8–dc23
2012025348

Contents

Introduction: The Ubiquity, Complexity, and Diversity of Cooperation

Kim Sterelny, Richard Joyce, Brett Calcott, and Ben Fraser

The Structure of the Collection

Cooperation and its evolution has become an increasingly pivotal issue for evolutionary biology as the modern synthesis has developed. The result has been rich and varied, and a glance at this book's table of contents will show the diversity of the work on cooperation represented here. The chapters in this collection sample richly from the tree of life, though with some concentration on bacteria, social insects, and, especially, hominins. The contributions depend on a wide range of research tools: empirical survey, formal theory and simulation, conceptual clarification and modeling, and some shameless speculation (a prerogative of philosophy through the ages). They explore a variety of themes and mechanisms. We begin here with an eagle's-eye view of the overall organization of the collection, and then identify five more specific themes. Most of these themes arise in multiple places throughout this volume, and hence do not map in any straightforward manner onto the internal structure of the collection. Thus our goal in what follows is to emphasize the themes connecting the chapters, rather than to introduce the chapters sequentially.

The volume is divided into two parts. Part I, "Agents and Environments," explores a set of ideas connecting cooperation in social organization, especially complex social organization, to the conditions that make cooperation profitable and stable. Although the stability conditions of cooperation have been the focus of much insightful research, that research has been somewhat one-sided. (See Calcott, 2008.) In our opinion, stability—especially stability in the face of the threat of cheating—has been explored at the expense of investigating the environmental conditions and phenotypic possibilities that make cooperative interaction pay in the first place. In particular, much of the formal and experimental work on cooperation and its stability has explored the ways that social partners give one another incentives to cooperate via the threat of punishment or the withdrawal of cooperative relations (Trivers, 2002; Fehr & Fischbacher, 2003); rather less work has explored the conditions that make cooperation potentially profitable. We do not neglect destabilizing threats to cooperation, of course, but one aim of the collection is to redress this imbalance. With some exceptions, then, the focus of Part I is on the interactions between agent, population, and environment.

Part II, "Agents and Mechanisms," focuses (though not exclusively) on how proximate mechanisms emerge and operate in the evolutionary process, and how they shape evolutionary trajectories. The interaction between proximate and ultimate factors has recently regained its status as a controversial topic. It was once received wisdom—stemming notably from Ernst Mayr's influence—that there is a good heuristic division of labor between proximate and evolutionary biology. In particular, it was supposed that evolutionary biologists could assume, to a first approximation and within reasonable limits, that phenotypic variation is largely unconstrained, and that selection operates against a background of a stable environment and population structure. (See Orr, 2005.) Notoriously, evolutionary developmental biology and niche construction theory called this division of labor into question, arguing that the range of phenotypic possibility is constrained in various ways and that environments are labile and malleable in ways crucial to evolutionary trajectories (Odling-Smee et al., 2003; West-Eberhard, 2003; Arthur, 2004). Very recently, Mayr's picture has been vigorously restated. A group of evolutionary theorists (many connected to Stuart West's group at Oxford) has argued that much work on the evolution of cooperation conflates proximate and ultimate factors. (See Scott-Phillips et al., 2011; West et al., 2011.) Brett Calcott's chapter takes up these issues explicitly, but in general the chapters in Part II are skeptical about Mayr's division of labor (explicitly so in the case of the chapter by Joseph Henrich and coauthors), and they freely explore the interactions between proximate biology (especially psychology), social organization, and selection.

Five Themes

As we see it, the centrality of cooperation to evolutionary biology derives from at least five factors.

1 The Generation and Division of Profit

Within a Darwinian framework, cooperation is prima facie puzzling because it often seems risky. It is true that agents that act together can often secure resources or ward off dangers that they could not accomplish as lone individuals; that is, there is often a potential profit to cooperation. But often, (i) these cooperative acts have costs; (ii) the potential profit does not depend on every agent that stands to benefit paying those full costs; and (iii) there is no automatic, costless mechanism that distributes the profits created in proportion to costs paid. Thus arises the notorious free-rider threat to cooperation: Even when cooperation is potentially profitable it seems likely to decay, because those that cheat by parasitizing on the efforts of cooperators are even fitter than the cooperators themselves (Williams, 1966). An enormous amount of the cooperation literature has focused on this issue. (See Hammerstein, 2003.) As noted above, one distinctive feature of this collection is that considerable space is devoted to exploring the mechanisms that make cooperation profitable, and not just mechanisms that ensure that the division of profit stabilizes rather than undermines cooperation.

The collection includes a substantial set of chapters exploring various aspects of the profitability and stability problem. In recent work, Ronald Noë has developed an important distinction between partner control and partner choice models of cooperation. Models of contingent cooperation ("I will cooperate if you cooperate") are partner control models: Cooperative behavior depends on the threat of adverse consequences. The chapter by Noë and Bernhard Voelkl points out that in natural systems agents can often choose their social partners, and that these choices often determine both the profitability and stability of cooperation. They develop a theory of biological markets to show how the need to choose well, and to be chosen, can stabilize cooperative interactions. Ben Fraser's chapter complements that of Noë and Voelkl on choice and markets, exploring the problems of honest signals in such markets.

The chapters by Deborah Gordon, Adam Hart, and Andrew Cockburn form a linked and contrasting set. For there to be selection for cooperation, a group of agents acting together must have higher per capita expected productivity than any of them acting alone. Those researching social insects (and humans) typically see this profit of cooperation as the synergy of collective action: Collectives can operate powerfully in their world, exploiting coordination, teamwork, division of labor, and sheer numbers of individuals. Social insects cooperate because cooperation can achieve so much. Gordon and Hart exemplify this approach; Hart's chapter, for example, is on teamwork and the conditions that allow tasks to be partitioned among agents who individually complete subtasks. Cockburn's chapter (which synthesizes an enormous literature on cooperative breeding in birds) contrasts sharply: Cooperation is more profitable than going it alone, not because of the power of collective action, but because individual action is so profitless. In different ways, birds that help others at the nest do so because they have very poor prospects of individual success. Cooperative breeding may be a miserable prospect, but in certain circumstances trying to find a mate and breeding as a pair is even worse.

The collection does not focus on defection management, but it does not neglect problems of stability and defection. For example, Simon Huttegger and Brian Skyrms's chapter in Part II takes up the problem of the stability of the informational environment. A ring network is an efficient form of informational exchange, as information is shared cheaply and fully. Huttegger and Skyrms argue (on the basis of a family of models) that agents with relatively simple learning strategies can often find their way to ringlike structures, and can usually stay there: This is an aspect of Skyrms's general argument that the evolution of cooperative signaling systems is not dependent on the prior evolution of cognitive complexity (Skyrms, 2010).

Kim Sterelny and Paul Seabright also address these problems in their respective chapters. Sterelny argues that the Pleistocene–Holocene transition undermines the conditions that made hominin cooperation stable in the Pleistocene, and that as a result the continuing stability of cooperation (indeed, its expansion) is an unsolved problem. In response, Seabright takes up this challenge and suggests some solutions. One aspect of this puzzle is the emergence and stability of property rights, for these were an essential part of the agricultural revolution in the Holocene, as humans switched from foraging to farming. The stability of these rights is problematic, since respect for property becomes less and less in the interests of most of the individuals in social collectives. While both Sterelny and Seabright touch on this problem, it is Herbert Gintis's chapter that focuses most directly on property rights, arguing that they emerged from territoriality in animals.

The chapter by Hanna Kokko and Katja Heubel likewise explores the problem of stability through a wonderful case study of the Amazon molly, *Poecilia formosa*. As with many organisms in cooperation dilemmas, this fish faces the problem of the tragedy of the commons—in their case, with respect to an utterly essential resource: mates. The molly reproduces asexually, in the sense that it needs no male genes, but it does need the sperm of a closely related male species to induce parthenogenetic reproduction. Yet the more successfully the molly reproduces, exploiting the advantage of asexual reproduction, the less they have access to these males. Asexual reproduction is so efficient that the mollies drive their sexual relatives extinct, followed of course by their own extinction. Stability here depends on a metapopulation structure: Local extinctions of the related males and then of the mollies is typically followed by colonization from elsewhere.

Finally, one of the dilemmas of cooperation is the problem of reconciling trust and complementarity. One way of ensuring trust is to interact with partners whose fitness interests are aligned with one's own, as in cooperation among close relatives. But that has a cost: Cooperation is most profitable if it involves division of labor and complementary specialization, and that is less likely the more genetically similar partners are (Queller, 2000). Haim Ofek's chapter explores this problem and the unique hominin solution to it: trade. According to Ofek, humans are uniquely cooperative among vertebrates, as their complex division of labor owes its existence to the free market as a central organizing institution.

2 Transitions in Individuality

Reflection on the costs of cooperation, and the potential impact of those costs on the stability of otherwise profitable cooperative interchange, has a long history in evolutionary theory. (See Williams, 1966; Trivers, 1985; Hamilton, 1996.) Initially, the problem seemed to have a restricted scope, but David Buss (1987) and (especially) John Maynard Smith and Eörs Szathmary (1995) showed that cooperation is not a peripheral feature of the history of life—not a form of behavior found in just a few species of social animals—but is, rather, ubiquitous. One overarching trend in the history of life has been an increase in complexity: from prebiotic and marginally biotic systems of various kinds, to prokaryotes, eukaryotes, multicellular organisms, and social collectives. Buss, Maynard Smith, and Szathmary all pointed out that this macroevolutionary pattern of increasing maximal complexity depended on a series of revolutions in cooperation, as more complex evolutionary agents (metazoans, eusocial insect colonies) emerged out of cooperatively interacting simpler ones. Groups become new evolutionary individuals as the members of those groups go through an evolutionary transition from independence through contingent cooperation to mandatory cooperation. Transitions in individuality, then, seem to imply a shift in the unit of selection: Composite agents evolve from collectives, but composites themselves are differentially fit, not merely groups of individuals with competing fitness interests. Since this process is not typically instantaneous, it seems to follow that there must be periods during which both the composite and the individuals that form the composite are properly seen as Darwinian individuals with fitness interests of their own. (See Okasha, 2006; Godfrey-Smith, 2009; Calcott & Sterelny, 2011.) Hence there is an intimate connection between transitions in individuality and multilevel selection.

3 Levels of Selection

The neo-Darwinian synthesis took the organism to be the focal level of selection: Evolution is tracked by changes in gene frequencies, but gene frequencies change because genes have differential effects on the fitness of the organisms that carry them. Selection builds adaptations, and adaptations are traits of organisms that (on average) benefit those organisms. There were always a few puzzling cases that were hard to fit into this framework. Some genes increase their relative frequency by biasing gene replication mechanisms rather than by adapting the organism of which they are parts. For example, genes that are copied into only one sex are advantaged by biasing the sex ratio toward that sex (Burt & Trivers, 2006). And sexual reproduction itself is problematic (Williams, 1975). But cooperation was the basis of the most widely canvassed challenge to idea that selection acts only on individual organisms (especially once the lessons of Buss and Maynard Smith and Szathmary were absorbed), for many cases of cooperation seem to be altruistic. Organisms do not just engage in collective activities that are mutually profitable (e.g., naked mole rats constructing and maintaining a shared system of burrows); they sometimes act in ways that are individually costly but beneficial to others. Warning calls, for example, can attract predators' attention; and sometimes individuals forgo their own individual reproduction to enhance the reproductive success of another. Selection on individuals cannot favor traits that systematically cause individuals to sacrifice fitness benefits without compensation. One response has been to identify cryptic individual fitness benefits, leading not just to a more subtle appreciation of cost-benefit trade-offs (see, e.g., Zahavi & Zahavi, 1997), but to a revolutionary change in how fitness is conceptualized (Gardner, 2009). But the other response is to suppose that apparent altruism is real (Sober & Wilson, 1998; Nowak et al., 2010); it evolves because selection acts on collectives as well as the individuals that compose them (Okasha, 2006; Godfrey-Smith, 2009).

Unsurprisingly, the two chapters of this collection focused on social insects—those by Gordon and Hart—discuss the relations between collectives and the individuals from which they are composed. Gordon, for example, discusses the ways in which ant nests act as cohesive ecological units. But the two most ambitious chapters on this theme are those of Don Ross and of Jessica Flack and her coauthors. The latter proposes a framework that generalizes across not only the transitions identified by Maynard-Smith and Szathmary, but across ecological systems as well. The core idea is intriguing: that new forms of integration emerge in response to so-called *slow variables*, providing constraints that stabilize a preexisting range of possibilities. Stability is important: If one can anticipate stable conditions, one can invest in size and life span. The challenge to individuals (and collectives) is to detect and exploit these relative invariances. Flack et al. use examples from social organization (persisting power inequalities) and reef ecology (persisting physical substrates) to illustrate the power of their ideas. One important theme in Ross's chapter is the relationship between individuals and the collectives they form. Instead of seeing the properties of social collectives as explained, bottom up, from the intrinsic characters of their individual members, Ross argues that (in the human case) the explanatory arrow is from collective to individual—that this is crucial to explaining the extraordinary levels of human cooperation, and thus that cooperation depends more on social mechanisms like markets and norms than on developmentally entrenched features of human psychology. His chapter thus links neatly to Ofek's chapter, which emphasizes the importance of trade as an antidote to the disruptive power of sexual recombination.

In this collection, the issues surrounding the debate about units of selection are discussed largely via hominin evolution, rather than the usual suspects of eusocial insects or volvocine algae (Holldobler & Wilson, 2007; Michod, 2007). None of the chapters focuses exclusively on group selection in hominin evolution, though Sterelny, Seabright, and Henrich et al. all discuss its potential role. But many of them explore the role of culture and cultural inheritance in setting up the conditions that make selection on groups powerful. Cultural inheritance tends to reduce phenotypic variance within the group and increase variance between groups, and these conditions make selection on groups effective. Nicholas Shea's chapter, in particular, is entirely centered around culture as an inheritance system. Shea seeks to establish the fact that information transmission in human cultural traditions is usefully illuminated by attending to the distinction between detection-based and selection-based modes of passing on adaptively significant information.

4 Externalism and Beyond

As Peter Godfrey-Smith (1996) notes, the early foundational versions of the neo-Darwinian synthesis were externalist: conceiving of selection as adapting lineages of agents to an independent environment. For example, Mayr, in his classic early exposition of the division of labor between evolutionary and proximate biology (Mayr 1961), chose bird migration as his illustrative example. This picture goes along with seeing organisms as autonomous robots who self-assemble and operate on the world on the basis of their inherited design. Bird migration fits this model very well, as many birds migrate independently of their parents and migration is a response to an independent and very stable feature of the environment (the effect of axial tilt on seasonality), a feature not changed by the phenomenon of migration. We can thus see migration as an evolved, adapted, genetically canalized response to an autonomous environment. The evolution of cooperation differs from migration in part because the selective environment is dynamic rather than fixed, and it coevolves with the changing phenotype of the individual agents (see Sterelny, forthcoming). Moreover, in many lineages, social behavior is contingent not just on the long-term developmental history of the agent but on features of the immediate environment. Cooperation depends, often, on capacities to assess and coordinate. A migrating bird does not have to signal to, or negotiate with, the climate. It does have a decision problem (when should it fly?), but its informational environment is simple and transparent: It can rely on one or a few stable, simple cues (e.g., changes in day length). In view of the extra demands on cooperative behavior, constraints on the supply of phenotypic variation are likely to play a critical role in the evolutionary trajectory (Stevens & Hauser, 2004). Finally, the dynamic nature of selective environments implies that an evolutionary account of the origins of cooperation does not automatically become an explanation of its persistence. Cooperation has been an important issue in part because it moves evolutionary theory beyond its early simplification of selection on canalized phenotypes in relation to a fixed environment. To understand the evolution of cooperation one needs to understand interactions between agents, and so new formal tools have been developed, such as frequency-dependent selection and evolutionary game theory. But one needs more than these formal tools: One needs to represent the effects of agents on their selective environment and how those effects change over time, and the ways the inherited proximate biology of agents influence their evolutionary possibilities.

These issues are relevant to virtually every chapter in the collection, for the collection as a whole views cooperation as the result of coevolutionary interactions between agents and their social environment. Many of the chapters emphasize that agents intervene in their social environments in ways that amplify cooperative possibilities; see especially the chapters by Ofek, Ross, and Dan Fessler and Katinka Quintelier. Many discuss the ways that the proximate biology of agents constrain or amplify social possibilities. Fiery Cushman, for example, argues that punishment is critical in enforcing human prosociality, but that it is an effective mechanism in leveraging prosocial action only because of the prior evolution of sophisticated learning capacities. Without these capacities, agents could not learn from the negative consequences of antisocial acts, for often these are displaced in space and time. Felix Warneken compares the proximate psychology of cooperation in young humans, bonobos, and chimps. Young great apes have prosocial motivations, but human children are more than simply motivated to help; they are also much more aware of when and how help is needed—especially the need for informational help, to which humans are very attuned and to which young bonobos and chimps are largely blind. Humans are poised to share information in a way that other great apes are not. No doubt this is both the consequence and partial cause of the unique hominin evolutionary trajectory.

These themes are especially prominent in Part II, where they are viewed largely, but not exclusively, through the lens of hominin evolution. June Tangney and her coauthors, for example, discuss the moral emotions of guilt and shame as signaling systems, arguing that guilt, in particular, often plays an important signaling role in social reintegration. But Part II is not only about hominins: Livio Riboli-Sasco, Francois Taddei, and Sam Brown view signals through the lens of bacterial evolution, pointing out that cooperation is an information-hungry strategy. The conditions that make it profitable are specific and labile, not features of the environment that the bacterium can assume as a stable default. (For example, how closely the bacterium is related to others in the local neighborhood often matters.) So signaling and detection systems are not optional extras for the cooperative prokaryote.

5 Humans as a Model System

Recent work on human evolution has increasingly appreciated the fact that humans are much more cooperative than other great apes, and that one of the key challenges in understanding human evolution is to understand the human cooperation explosion. (See Hrdy, 2009; Tomasello, 2009; Bowles & Gintis, 2011.) This is obviously an important explanatory project in itself, but equally obviously, one would expect there to be reciprocal illumination between general accounts of the evolution of cooperation and specific hypotheses about the evolution of human cooperation. This collection devotes considerable space to the special case of human cooperation, in part motivated by the idea of encouraging this mutual illumination. Many of these chapters are in complementary linkages. Thus the chapters by Shea, Hugo Mercier, and Cecilia Heyes are organized around social learning and intergenerational transmission. Social learning is clearly of special importance in human psychology and social life, but it is not uniquely hominin; indeed, one of the themes of Heyes's chapter is that the cognitive significance of imitation learning has been much overemphasized.

Another cluster of chapters focuses on a distinctive and perhaps unique feature of human cognition: our capacity to think about, speak about, and be motivated by *norms* (moral norms; norms of disgust, of religion, and of custom). Humans have views not just about how the world is, but how it ought to be. There has been an enormous recent interest in the evolution and function of normative cognition (see Hauser, 2006; Joyce, 2006; Kitcher, 2011). All these theories of the evolution of normative cognition connect normative thought to the expansion of cooperation in the hominin lineage. This collection reflects and responds to these ideas. In Part I, the evolution of norms is touched on in the chapters by Ross, Sterelny, Seabright, and Gintis. But it is in Part II that the topic is faced more squarely. Cushman, Tangney (with her coauthors Jeffrey Stuewig, Elizabeth T. Malouf, and Kerstin Youman), Fessler and Quintelier, Henrich (with his coauthors Maciek Chudek and Wanying Zhao), Daniel Kelly, Matteo Mameli, and Richard Joyce all focus to a greater or lesser degree on the role of norms, especially moral norms and their connection with human cooperation and social life.

Henrich et al. argue that cultural group selection added to hominin social psychology a set of cognitive mechanisms for recognizing and internalizing norms that support large-scale cooperation. (Thus they seek to explain the same explosion of Holocene cooperation that is the focus of Sterelny's and Seabright's chapters.) The human capacity to internalize norms, according to this picture, importantly involves emotion. Emotion is also central to Kelly's discussion of norms. His focus is disgust. Like Henrich et al., Kelly advances a gene-culture coevolution hypothesis: Basic disgust responses have been co-opted to serve social functions. Kelly briefly considers the role of emotions like disgust in specifically *moral* norms and deliberation. Moral norms themselves are the focus of the chapters by Joyce and Mameli. Joyce is primarily concerned with clarifying rather than advocating moral nativism. He maps in detail the conceptual landscape covered by the term "moral nativism" and ends with a caution: moral nativism and moral nonnativism may *both* be perfectly defensible positions, even when all data are in. Mameli's chapter on moral motivation both criticizes and complements Joyce's earlier (2006) advocacy of moral nativism. Mameli's goal is to identify the best way to make sense of the claim that moral beliefs are motivationally powerful. His discussion is an important extension of work to date on moral nativism, and it also contributes more generally to an understanding of how morality allows and encourages cooperation among humans.

Future Directions

There is still a great deal about the evolution of cooperation that is not understood. While it is risky in the extreme to predict the development of a living field, these chapters signpost at least four apparent routes that future research might take. One task will be to untangle the connection between complexity and cooperation. In some respects, more complex social worlds are less friendly to cooperation than smaller and simpler ones. For example, the much-celebrated tit-for-tat solution to the free rider problem simply fails to work when there are many players instead of just two. This is a symptom of a more general problem: As the size and differentiation of a group increase, so too do problems of policing and coordination. If one is interacting with ten agents each of whom is supposedly making a different contribution to a collective action problem, rather than two others each of whom is supposed to be doing the same thing, it is much more difficult both to know what others should do and to know whether they have done it. But equally, the combination of size and complexity is a potential lever of cooperation, offering more opportunities for powerful collective action, and for specialization and the division of labor. As Hart reminds us, teams exert more power over their environment than mere aggregates.

Second, we noted earlier in this introduction that there has been a bias in the literature in its focus on defection and free riding: on explaining why cooperation is not destabilized by conflicts over the division of the profits. There is a much less systematic attempt to identify the conditions in which cooperation is profitable in the first place (a notable exception is Queller, 2000). This collection does something to redress this balance, but there is still a long way to go. For example, there are prima facie gaps in cooperation in the natural world, where it appears that cooperation would provide an important advantage, yet it hasn't evolved. For example, it would seem that most social animals—even minimally social animals such as those that live in a herd—would profit from collective defense. Wildebeest would have nothing to fear from African wild dogs if only they acted cooperatively. Yet collective defense is relatively rare. Why? Is the apparent benefit an illusion or are these cases where an otherwise profitable form of cooperation cannot become established because of free riding?

There is another possible obstacle blocking the evolution of the cooperative wildebeest, mention of which takes us to our third line of development. Perhaps wildebeest lack some critical prerequisite, making it impossible for them to reach that adaptive peak: Their evolution is constrained by their extant proximate biology. In their paper exploring the strange lack of clear examples of reciprocal cooperation in the wild, Stevens and Hauser (2004) focus on the cognitive preconditions of policing and of deferring gratification. In most environments, reciprocation requires cooperating agents to recognize others, to recall and evaluate their behavior, and to resist temptation. But of course it also requires the capacity to recognize and evaluate cooperative versus solitary courses of action, and to signal and coordinate behavior. Perhaps many collectively living agents lack even rudimentary versions of these capacities. We still have a lot of to learn about the interaction between proximate mechanism and environmental trajectory.

Finally, it is all too obvious that we know very little about the evolutionary history of some central cooperative transitions in the history of life. Most critically of all, the evolution of the first cells, with their sets of cooperating genes, is still very poorly understood. The evolution of multicellularity is almost as important, and a general understanding of its origins is perhaps more easily attained, for multicellular organisms have evolved many times in the history of life. The exact number is debatable and depends in part on choice of criteria, but it is certainly over ten (Bonner, 1998). Despite this multiplicity of origins, multicellular life that is complex and differentiated has evolved only three times. This combination is promising; there are enough comparative data to (potentially) reveal the conditions in which simple multicelled cooperation evolves, and perhaps also the added conditions that make more complex multicellularity possible (Calcott & Sterelny, 2011).

In summary, the five themes we discuss above demonstrate how puzzles surrounding cooperation are ubiquitous in evolutionary biology. The contributions in this collection do not fit neatly under these themes; the research represented here is too rich and diverse to allow for that. What the collection exemplifies is how tremendously fertile the topic of the evolution of cooperation is, how surprisingly varied are the interdisciplinary approaches it encourages, and how enormously much we still have to learn about it.

References

Arthur, W. (2004). *Biased embryos and evolution*. Cambridge: Cambridge University Press.

Bonner, J. T. (1998). The origins of multicellularity. *Integrative Biology, 1*, 27–36.

Bowles, S., & Gintis, H. (2011). *A cooperative species: Human reciprocity and its evolution*. Princeton: Princeton University Press.

Burt, A., & Trivers, R. (2006). *Genes in conflict*. Cambridge, MA: Harvard University Press.

Buss, L. W. (1987). *The evolution of individuality*. Princeton: Princeton University Press.

Calcott, B. (2008). The other cooperation problem: Generating benefit. *Biology and Philosophy, 23*, 179–203.

Calcott, B., & Sterelny, K. (Eds.). (2011). *The major transitions in evolution revisited*. Cambridge, MA: MIT Press.

Fehr, E., & Fischbacher, U. (2003). The nature of human altruism. *Nature, 425*, 785–791.

Gardner, A. (2009). Adaptation as organism design. *Biology Letters, 5*, 861–864.

Godfrey-Smith, P. (1996). *Complexity and the function of mind in nature*. Cambridge: Cambridge University Press.

Godfrey-Smith, P. (2009). *Darwinian populations and natural selection*. Oxford: Oxford University Press.

Hamilton, W. D. (1996). *Narrow roads of gene land: The collected papers of W. D. Hamilton* (Vol. 1): *Evolution of social behaviour*. Oxford: Oxford University Press.

Hammerstein, P. (Ed.). (2003). *Genetic and cultural evolution of cooperation*. Cambridge, MA: MIT Press.

Hauser, M. (2006). *Moral minds: How nature designed our universal sense of right and wrong*. New York: HarperCollins.

Holldobler, B., & Wilson, E. O. (2007). *The superorganism: The beauty, elegance, and strangeness of insect societies*. Cambridge, MA: Harvard University Press.

Hrdy, S. B. (2009). *Mothers and others: The evolutionary origins of mutual understanding*. Cambridge, MA: Harvard University Press.

Joyce, R. (2006). *The evolution of morality*. Cambridge, MA: MIT Press.

Kitcher, P. (2011). *The ethical project*. Cambridge, MA: Harvard University Press.

Maynard Smith, J., & Szathmáry, E. (1995). *The major transitions in evolution*. Oxford: W. H. Freeman Spektrum.

Mayr, E. (1961). Cause and effect in biology. *Science*, 134(3489), 1501–1506.

Michod, R. E. (2007). Evolution of individuality during the transition from unicellular to multicellular life. *Proceedings of the National Academy of Sciences of the United States of America, 104*, 8613–8618.

Nowak, M. A., Tarnita, C. E., & Wilson, E. O. (2010). The evolution of eusociality. *Nature, 466*, 1057–1062.

Odling-Smee, J., Laland, K., & Feldman, M. W. (2003). *Niche construction: The neglected process in evolution*. Princeton: Princeton University Press.

Okasha, S. (2006). *Evolution and the units of selection*. Oxford: Oxford University Press.

Orr, H. A. (2005). The genetic theory of adaptation: A brief history. *Nature Reviews: Genetics, 6*, 119–127.

Queller, D. (2000). Relatedness and the fraternal major transitions. *Philosophical Transactions of the Royal Society, Series B, 355*, 1647–1656.

Scott-Phillips, T., Dickins, T., & West, S. A. (2011). Evolutionary theory and the ultimate–proximate distinction in the human behavioral sciences. *Perspectives on Psychological Science, 6*, 38–47.

Skyrms, B. (2010). *Signals: Evolution, learning, and information*. Oxford: Oxford University Press.

Sober, E., & Wilson, D. S. (1998). *Unto others: The evolution and psychology of unselfish behavior*. Cambridge, MA: Harvard University Press.

Sterelny, K. (Forthcoming). Co-operation in a complex world. *Biological Theory*.

Stevens, J. R., & Hauser, M. D. (2004). Why be nice? Psychological constraints on the evolution of cooperation. *Trends in Cognitive Sciences, 8*, 60–65.

Tomasello, M. (2009). *Why we cooperate*. Cambridge, MA: MIT Press.

Trivers, R. (1985). *Social evolution*. Menlo Park, CA: Benjamin/Cummings.

Trivers, R. (2002). *Natural selection and social theory: Selected papers of Robert Trivers*. Oxford: Oxford University Press.

West, S., El Mouden, C., & Gardner, A. (2011). Sixteen common misconceptions about the evolution of cooperation in humans. *Evolution and Human Behavior, 32*, 231–262.

West-Eberhard, M. J. (2003). *Developmental plasticity and evolution*. Oxford: Oxford University Press.

Williams, G. C. (1966). *Adaptation and natural selection*. Princeton: Princeton University Press.

Williams, G. C. (1975). *Sex and evolution*. Princeton: Princeton University Press.

Zahavi, A., & Zahavi, A. (1997). *The handicap principle: A missing piece of Darwin's puzzle*. Oxford: Oxford University Press.

I Agents and Environments

1 The Evolution of Individualistic Norms

Don Ross

1 Descriptive and Normative Individualism

The venerable doctrine of "individualism" comes in two tropes: descriptive and normative. Often they have been subsumed—and not infrequently confused—under the label of "methodological" individualism. This chapter is about the relationships among these ideas in light of the genetic and cultural evolutionary history of humans. It argues that we best appreciate the persuasiveness of normative individualism to most modern people by understanding why their evolutionary history has made working, everyday descriptive individualism hard to achieve and maintain.

A normative individualist is someone who maintains that the justification of all values ultimately lies in the normative judgments of individual people, and in assessments about the effects of change on the welfare of individuals. This view has polemical bite against one version or another of normative collectivism, according to which groups—clans, nations, ethnic communities, classes—may have and promote valuable objectives that transcend the preferences of their individual members. In nonsecular traditions it has sometimes been maintained that religious communities are obligated by value considerations that might, at least in dark periods, animate *none* of their members. But this is an extreme case. Politically and philosophically relevant versions of normative individualism allow that there must be some relationship between individual and collective valuations. But normative collectivists typically suppose that groups can have goals that are conceived only vaguely by most of their members. At least as importantly, collectivists typically endorse some second-order normative beliefs to the effect that members of groups *should* at least sometimes avoid pursuing their individual objectives when these conflict with the good of their teams.

Someone is persuaded by descriptive individualism to the extent that she thinks that, as a matter of fact, people are generally motivated by considerations that refer to their private welfare, or to the welfare of a restricted set of specific individuals that they value *as* individuals. Descriptive collectivists, by contrast, emphasize the capacity they attribute to at least as many people of putting aside personal interests when these conflict with what they consider best for collectives that mainly include other members with whom they are specifically unacquainted.

Framed in these terms, the conflict between the descriptive individualist and the descriptive collectivist is likely to seem to rest on a simple false dichotomy. It is widely supposed that most people are relatively self-interested across a wide range of common decision settings and problems but also recognize various duties to collectives with which they are affiliated, and often subordinate their private goals to satisfaction of these duties. Descriptive individualism and collectivism come into genuine conflict only insofar as someone seeks to prioritize one class of motives over the other in the context of promoting a general view about the best way to explain and predict broad tendencies in human action. This is why most scholarly discussions of descriptive individualism focus on it as a *methodological* thesis.

The social science tradition that has been most closely associated with methodological individualism is economics. Some important economists—though not as large a proportion of them as popular imagination, and stereotypes prevalent in other disciplines, routinely suppose—have favorably contrasted their profession with neighboring social sciences such as sociology by insisting that economists keep a clearer and more rigorous eye on the principle that actions of collectives must ultimately decompose into, and thus be explained by reference to, actions of their individual members. This basis for descriptive individualism has often been expressed as an application to human action of a more general principle of atomism, the idea that the causal capacities of composite structures should be explained by reference to interactions among the causal capacities of their parts. This is frequently given what philosophers call an ontological interpretation, according to which composites have only derivative reality as constructs out of what more basically or "really" exists, namely, the articulated constituents. Margaret Thatcher famously said that there is no such thing as society, but only individuals. The motivation for this assertion was likely her normative conviction that the *welfare* of society is nothing over and above the *welfare* of individuals added together; but like a great many normative individualists over the years, she reached straight for an ontological trope in order to seem to justify it.

More sophisticated thinkers than Thatcher have joined her in blurring the line between normative and descriptive individualism. I will concentrate on the basis of this synthesis in economics. According to many economists, the ultimate subject of their discipline is the comparative efficiency of alternative allocations of scarce resources. Unlike in thermodynamics, efficiency in economics refers necessarily to the relationship between energy expenditure and *value*; and this in turn tends naturally to prompt the question, at least in reflective moments: value to *whom*? The majority of economists have been at least vaguely utilitarian in their convictions, and this is reflected in the existence of the subdiscipline of welfare economics that studies conditions under which a society as a whole can achieve the highest value of output from different combinations of inputs. No welfare economist would regard a society as producing efficiently if individuals didn't want to consume the products it churned out; the legendary Soviet factory that made hundreds of thousands of unmated boots for right feet is never taken to be a model of happy industrial organization. What seems essential to saying that one economic process is more efficient than another is that it does a better job of satisfying consumers' wants. And consumers are typically supposed to be individuals in the end, even if in practice the consumption behavior that is actually measured is more often that of households.

Noting that a commodity is consumed by somebody, or by lots of somebodies, hardly puts an end to our inquiries about its value, or about the relative value of having devoted resources to producing it. Welfare economists pressed to further justify their emphasis on satisfaction of consumer demand often invoke a doctrine known as *consumer sovereignty*. It is at this point that we find the fusion of normative and descriptive individualism: according to the consumer sovereigntist, the individual person is the most *accurate* possible evaluator of what is valuable to *her*; and the ultimate source of *all* value is value to individuals.

Referring to this as a fusion will be seen to be an overstatement. The first, descriptive, clause is a logically independent proposition from the second, normative, one. Lukes (1968) is one among many philosophers who have pointed out that this applies to all conjunctions of normative and descriptive individualism that have thus far been articulated. However, this somewhat technical philosophers' point should not stop us from recognizing that descriptive individualism would likely not have attracted the interest or support that it has were it not for the fact that most normative individualists have appealed to it in formulating their arguments.

A main linking idea that has been emphasized by more thoughtful individualists, such as von Mises (1949), is the *uniqueness* of individuals. On the normative side, it is clear enough how uniqueness relates to value; one might simply ask oneself whether most people who lost spouses would feel that their loss could be made whole by a replacement with similar preferences and looks. The relationship between descriptive individualism and uniqueness requires more development. As Miller (1978) discusses, thinkers who put descriptive individualism to work in support of normative individualism have tended to emphasize a specific constellation of properties that individuals, but not groups, are taken to enjoy. Specifically, von Mises and his followers present individual *minds* as the essential sites of rational, self-conscious, explicit, effortful deliberation among possible actions and projects. They acknowledge, of course, that groups also work at explicit rational deliberation, but insist that this is parasitic on the occurrence of that kind of mental processing in individuals. In folk psychology, choices arrived at by such processing are generally regarded as the clearest basis for full normative responsibility—"I wasn't thinking" is regarded as a kind of excuse, even if not a fully adequate one. It seems evident that part of the basis for this cultural norm is an assumption that explicit individual ratiocination is under closer personal control than other varieties of action selection. Among its typically ascribed functions, by philosophers going at least as far back as Plato, is management of morally obtuse subconscious or "automatic" will. This most morally serious kind of thought is then connected to uniqueness by way of the assumption that it is essentially private and only imperfectly communicable. Thus another leading apologist for methodological individualism, Hayek (1949), argues that individuals should be sovereign with respect to their welfare because only they have full information about the subtle priorities and detailed relationships of mutual justification among their preferences.

Even if we grant that rational deliberation contributes to the moral and other normative weight of a choice, we may still diagnose an element of circularity in the synthesis of descriptive and normative individualism just sketched. Everyone acknowledges that although individual deliberation is a typical input to group deliberation, influence also strongly prevails in the opposite direction. There is arguably no consensus among scientific psychologists as to which direction of influence is in general more powerful, but several disciplines include rich literatures that promote the predominant influence of public reason (along, of course, with public *un*reason). Still, I would maintain that we might take the political philosopher Philip Pettit as representative of the dominant view on this question. Few writers have accorded greater moral importance than has Pettit (1997) to public deliberation. Nevertheless, in his 1993 explicit book-length treatment of the relationship between individual and collective intentionality, he argues for the traditional ontological *and* moral priority of the individual level.

There is an element of irony in the prevailing association of methodological individualism with the discipline of economics. Though both von Mises and Hayek were economists, they are associated with one specific, minority tradition in the field, the so-called Austrian school. The dominant tradition in microeconomic theory, with its most important historical highlights in the works of Walras (1874), Pareto (1927), and Samuelson (1947/1983), has been carefully agnostic on the question of whether rational, self-conscious deliberative processes have *any* causal relevance to economic behavior or are, often or even always, mere epiphenomena or post facto rationalizations. This fact is obscured by economists' overwhelming rhetorical emphasis on "choice," and often "rational choice," as the central subject matter of their discipline. However, as I argue elsewhere (Ross 2011b), "choice" in mainstream microeconomics does not mean what it does for the folk, psychologists, and most philosophers, namely, a process of comparison of alternatives that unfolds, consciously or unconsciously, over time in a mind or brain. In mainstream economics, a behavioral pattern is regarded as *chosen* just in case it is influenced, through any kind of channel, by incentives. It is important to emphasize the use of the word "pattern" here. Most economic choices are identified only statistically, as tendencies observed over runs of instances, usually in pooled sets of agent responses across a population, when incentivizing environmental influences change exogenously. An alternative description would be that choices for mainstream economists are equivalence classes of behaviors selected by common cost-benefit ratios. This may or may not involve any explicit representation in any consumer's computational or neural processes of the alternatives over which valuations are assigned by the economic modeler.[1] Thus, for example, household consumption patterns may be consistent with downward-sloping demand curves not because any consumers explicitly weigh opportunity costs in marginal terms, but simply because households with smaller budgets tend to buy less of everything in their consumption baskets.

Recently, implicit individualist themes in economics have been revived from an unexpected source: heterodox economists who reject mainstream theory and method and refer to themselves as "behavioral economists." According to promoters of this research program, such as Thaler (1992), Camerer et al. (2005), Ariely (2008), and Akerlof and Shiller (2009), explanations and predictions in economics should advert directly to the beliefs and desires of actual individuals, which generally do not correspond to the hypothetical knowledge or preferences of the "representative agents" in which orthodox economics traffics. Because behavioral economists use game theory to model the interactions of individuals, outcomes often differ from states sought by any of the individuals in question; but individual-scale properties, as captured in the utility functions attributed to people on the basis of experiments or observation, are the foundation stones for modeling, predicting, and explaining social-scale effects.

Nonindividualism also finds its methodological proponents in contemporary experimental economics. A leading example of empirically driven anti-individualist research in contemporary microeconomics is the "ecological rationality" program championed by Vernon Smith (2008), which emphasizes influences on choice (in the economist's sense described above) that are stored in the social and institutional environment, and that may have no representational analogues in the idiosyncratic representational spaces of typical individuals. This research program comports naturally with the "distributed social cognition" and "extended mind" perspectives that have recently been defended by some methodologists and philosophers of cognitive science (Hutchins, 1995; Clark, 1997; Ross et al., 2007).

A view that perches explicitly between methodological individualism and anti-individualism is defended by Hollis (1998) and Bacharach (2006). They join behavioral economists in appealing to individual representational states in explaining choice. However, they argue on theoretical and empirical grounds that people are strongly disposed to frame many of the alternatives they face in terms of the welfare of collective entities with which they identify. Such "team reasoning," beginning from questions about "What is best for us?" rather than "What is best for me?," reframes people's strategic situations and typically changes the equilibria of the formal games by which analysts model the situations in question. A favorite example is of players in a team sport, such as soccer; on a cohesive and effective team, players choose actions directly by reference to maximization of the prospects of collective victory. A game-theoretic representation that ignored this, perhaps by including players' interests in personally scoring goals in their decision sets, would be empirically incorrect as a model of such a team.[2] The rhetoric of team-reasoning theorists reflects the legacy of methodological individualism insofar as they generally refer to team representations as "reframings" of representations couched in terms of individual utility functions, implying that the latter are ontologically or psychologically basic. However, this rhetoric is, from the formal point of view, strictly incidental; so far as any of the proffered evidence is concerned, we might just as naturally regard individual utility functions as reframings of more basic team utility functions. I will argue that, in light of evolutionary considerations, this inversion is the more natural perspective.

2 Cooperation, Coordination, Imitation, and Human Evolution

Basic elements of Darwinian theory are often thought to be crucial to philosophical debates over individualism. Just as often, however, philosophers working in these precincts tend to derive overly sweeping conclusions from consideration of under-specified models.

Natural selection favors genes that have higher inclusive fitness than competitors (Hamilton 1964). Individualists often try to score debating points by correctly insisting that it can favor no others. As West, Mouden, and Gardner (2010) emphasize, there is no group selection in the sense of an evolutionary pressure that is *opposed to* the statistical maximization of inclusive fitness by individuals. However, genes in all multicellular organisms maximize the inclusive fitness of their bearers by entangling their fortunes with those of other genes. Nothing restricts such gene associations to the boundaries of individual organisms, and there is no limit in principle to the complexity of networks of effects through which genes indirectly promote the inclusive fitness of organisms. In *this* sense, "group selection" should not be controversial, but it also does not have the exciting philosophical consequences often attributed to it.

Where social science is concerned, basic Darwinian theory supplies a constraint on modeling: No model should be accepted that requires genes to systematically *dispromote* the inclusive fitness of their bearers. West et al. show that this constraint gains a surprising amount of traction against some currently popular but underformalized models of specific, speculative, dynamic effects in the evolution of human sociality, such as models that invoke "strong reciprocity" (Gintis, 2000; Gintis et al., 2003, 2008) and models that appeal to "greenbeard" genes (Frank, 1987; Bergstrom, 2002; Bowles & Gintis, 2004). However, basic Darwinian theory is perfectly compatible with the idea that individuals often promote their own fitness by promoting the prospects of groups to which they belong. It therefore erects no barriers against stories of human evolution that emphasize competition between groups. Such accounts motivate taking seriously the possibility that team reasoning was the historical default frame, or even the only available frame, for early human ancestors in strategic situations. Of course, basic Darwinian theory is equally compatible with accounts based on the opposite proposal.

West et al.'s emphasis on Hamilton's model as the complete generalization of Darwinian theory allows us to identify a feature that all of the models undermined by the inclusive fitness maximization constraint have in common: They are led to hypothesize novel evolutionary mechanisms by supposing that cooperative behavior is harder for natural selection to support than is actually the case. This is closely related to the widespread view that humans are uniquely cooperative as a species, at least among noneusocial animals. West et al. challenge this second supposition directly. Humans, they observe, are less altruistic than a number of species scattered liberally around the tree of life, are by no means special in establishing cooperative relationships with nonrelatives, and are not unique in incentivizing cooperation by punishment of noncooperators.[3]

This is immediately relevant to the individualism debate. Stories of human evolution that rest primary weight on overcoming obstacles to cooperation effectively presuppose individualism.[4] They take the problem of the origin of human sociality to be: How do basically selfish individuals evolve commitment devices against their default Darwinian dispositions to defect against one another in prisoner's dilemmas, public goods games, and similar strategic settings that preoccupy behavioral economists? This inspires an explicit search for an evolutionary discontinuity that allowed individual human utility functions to be composed into group dynamics of a more interesting and complex character than pure competition. Seabright (2010) explicitly elaborates the entire history of humanity around this trope.

This is precisely the individualist style of explanation identified in the previous section. It is the basis for Gintis's (2006) tireless campaign for the hypothesis that modern humans arose through a genetic adaptation that produced a disposition for strong reciprocity. Burnham and Johnson (2005), Ross (2006a), and Guala (2012) provide direct counterarguments against this hypothesis. When West et al. (2010) identify sixteen "common misconceptions" about the evolution of human cooperation, they charge that Gintis falls into all of them; however, they do not offer a general underlying conceptual diagnosis for the attribution of so many alleged confusions. The individualist style of explanation I earlier associated with revisionist behavioral economics is a good candidate for this diagnosis. Individualism, I suggest, inflects many theorists' entire views of human evolutionary history.

The demand for a specific explanation of how selfish, cognitively sophisticated individual hominids achieved cooperative dispositions is misplaced. All apes live in family groups.[5] In such groups, inclusive fitness of individuals is typically best promoted by at least some level of resource-sharing and communal protection of young. Many different degrees of cooperativeness with respect to different behavioral modalities are equilibria, depending on the subtle interplay of a limitless range of environmental factors. The most basic mechanism maintaining cooperativeness is a simple feedback loop. Animals that forage and nest in groups are likely to be more closely related to nearby conspecifics than they are to geographically distant ones. By a direct implication of Hamilton's rule, maximization of individual fitness will then typically indicate cooperative dispositions (alongside competitive ones, of course). These dispositions in turn contribute to the maintenance of group-living patterns. Seabright (2010) is therefore right to emphasize that a major transition among some humans has been the development of institutions that promote cooperation among strangers.

It is important to distinguish between dispositions to cooperate *in general* and capacities to process information that facilitate *specific forms* of cooperation. That is to say, we must keep an eye on the difference between *cooperation* and *coordination*. If the perceptual–cognitive apparatus of a species is not designed to track and respond to the cues from which possibilities for cooperation can be identified, then we can infer nothing about the extent of dispositions to team framing among the organisms in question, or any other aspect of their preference structures, if they miss opportunities for group projects. We do not conclude from the absence of symphony orchestras among gorillas that they are individualists when it comes to musical expression and prefer singer-songwriters.

A major theme in the literature on cooperation in intelligent social animals is untangling questions about motivations from questions about cognitive capacities. The key source of leverage has mainly been carefully sequenced experiments in which scientists first establish that members of a species understand and can be motivated to respond to an action-goal contingency, and also understand that a conspecific is in an analogous situation to themselves, with respect to this same contingency. One can then put two animals in a situation where they can only realize their goal by acting together. Only if the animals behaviorally manifest this further understanding can one infer that the capacity for the specific form of coordination being tested is present. One can then launch experimental manipulations, such as varying relative costs and benefits, intended to uncover dispositions to cooperate. Studies of this kind that have been conducted with nonhuman primates, particularly chimpanzees, have generally supported the conclusion that although chimpanzees can understand what behavior would serve the interests of a conspecific—knowing, for example, how to respond positively to directly incentivized assistance—only rare individuals show any disposition to take costless actions that would improve social welfare (Silk et al., 2005; Jensen, Call & Tomasello, 2006; Vonk et al., 2008; Silk, 2009). Other experiments have compared human and chimpanzee infants at similar stages of cognitive development and concluded that the former, but not the latter, focus on and take up opportunities to complete others' goals when they are unable to, and are disposed to supply information that would assist others in completing goals (Warneken & Tomasello, 2006; Warneken et al., 2007). On these bases it is widely inferred that chimpanzees are less disposed to cooperative behavior, once capacities for coordination are controlled for, than humans.

One might press questions over whether these experiments truly separate coordination capacities from cooperative dispositions. The fact that chimpanzees will often respond helpfully to explicit requests for assistance might be taken to suggest that opportunities to promote the social good are simply not salient to them, but that their prosocial preferences can be induced to shine through their obtuseness. However, it is important to note in this context that, in the mammalian brain areas associated with reward learning and control of reward-directed action, cognitive salience and motivation are almost inextricably linked (McClure, Daw & Montague, 2003); so hypothesizing morally communitarian but socially dense chimpanzees might rest on a folk psychological distinction that neuroscience undermines.

The chimpanzee results are often cited in support of the idea that modern humans' ancestors were cognitively sophisticated but selfish, like chimpanzees, and that at some point in the hominid line dispositions evolved that enhanced our socialization. Such inferences are hasty. It seems clear enough on the basis of the evidence to date that modern humans cooperate more extensively than our nearest *living* relatives. It would, however, be rash to infer from this that modern humans are therefore more disposed to team framing than our *ancestral* relatives. Chimpanzees may now coexist with *Homo sapiens*, whereas more closely related hominid species no longer do, precisely because chimpanzees' distinctive ecology has confined them to a niche in which they have not strongly competed with our line of descent, at least until recently. It is every bit as compatible with the evidence to postulate that chimpanzee sociality has atrophied in their stable and food-rich forest environment as to speculate that human cooperative dispositions are exaggerated relative to such dispositions in early hominids. Of course, it is not in doubt that the technology available to contemporary humans has vastly enhanced their coordinative capacities, which in turn allows them to achieve uniquely extended cooperative projects, in time, space, and number of cooperators. This in itself speaks neither for nor against individualism.

The underlying coordination failures that impede chimpanzee opportunities for cooperation indicate one central behavioral dispositional property that distinguishes them not only from humans, but from such other socially intelligent animals as dolphins, parrots, and corvids (Hurley & Chater, 2005): Chimpanzees do not spontaneously imitate one another. This is linked to evidence about perceptual salience; even when motivated, chimpanzees cannot pay sufficiently careful anticipatory attention to one another to perform such cognitively nondemanding tasks as jointly carrying a bucket of water without spilling it (Tomasello et al., 2004).

Merlin Donald (1991) proposes a comprehensive theory of the evolution of the modern human mind, based on interpretation of physical and cultural anthropological evidence, that promotes the central historical importance of a specific form of elaborate imitation. Donald argues that the career of *Homo erectus* involved a major transition from the *episodic*[6] representations of social situations on which contemporary great apes rely to *mimetically* structured representations. The latter are presented as a necessary platform for the later transition to the fully semiotic, abstract representations characteristic of the modern human mind and expressed in modern humans' distinctive use of languages with structured grammars. The core difference between an episodic and a mimetic representation is that the latter, but not the former, involves perceiving and storing memories of specific behaviors of others by reference to general stylistic features that allow them to be subsequently reenacted. Such mimesis is held to be the basis for human artistic, ritual, and religious expression; and it is by reference to these that one best inductively grasps what mimesis is in the first place. According to Donald's hypothesis, mimetic performances in turn provided the first basis for limited cross-generational learning and cultural accumulation among hominids. Donald argues that this significantly increased the selection advantage of greater memory capacity, and that the flowering of mimesis in *Homo erectus* thus predicts and partially explains the major advance in encephalization that make this species the pivotal anatomical transition figure between apelike hominids and modern humans.

Donald's theory is an exemplary instance of inference to the best unifying explanation, since all of the evidence for it is indirect and suggestive, but there is a lot of it, carefully assembled from strong sources. If the thesis is correct, it supplies a cognitive disposition by which the ecological *effectiveness* of human coordination was amplified, through exercise of new cognitive capacities that allowed humans to fuse their agency to a greater extent than is possible for chimpanzees. A disposition to imitate potentially undermines descriptive individualism, since in the limit a group of organisms that slavishly imitated everything they remembered having seen conspecifics do would more closely resemble a single distributed organism than a collection of individuals. Of course, the forms of individualistic expression that most people most admire normatively are those that find surprising scope for novel variation on the platform of substantive imitation (Elster, 2000). For example, jazz group improvisations require maintenance of some shared structures as constraints in order to be interesting.[7] I will argue in section 3 that this pattern offers an important clue to the true relationship between descriptive and normative individualism.

It is an advantage of Donald's hypothesis that, unlike Gintis's, it does not posit an evolutionary development that promotes or requires a novel *kind* of behavior. Capacities for imitation are widespread in nature (Hurley & Chater, 2005), and their basis in neural computation is becoming steadily clearer. The well-known literature on mirror neurons is part of this developing set of mechanistic foundations (Gallese, 2003, 2007), though its significance is dogged by philosophical uncertainties over the question of whether identifying some neurons as functional mirrors amounts to anything more than suggesting neural correlates in advance of explanation. Understanding imitation as a form of coordination offers a clearer avenue to explanation, both mechanistic and adaptive, than conceiving imitation simply as basing bodily movements on another's perceived template. Neural-computational mechanisms for coordination are more powerful than mechanisms for mirroring, because the former might explain how groups of organisms find equilibria in games where there is no single dominant strategy shared by all.

The computational basis for such coordination is emerging from a research program initiated by Paul Glimcher and collaborators (summarized in Glimcher, 2003), based on single-cell recordings in monkeys while they play games against computers that implement strategies the experimenters systematically vary. This work strongly suggests that, at least in primate brains, individual neurons in the circuit that estimates comparative reward values directly compute statistical variations in choice that track mixed-strategy Nash equilibria (NE), that is, optimal responses in situations in which different agents do best by doing different things, and indeed where each agent should vary their behavior even across occurrences of strategically identical situations. Subsequent work by Lee et al. (2004) and Lee, McGreevy, and Barraclough (2005) has extended the implications of this result. In general, individual groups of monkey neurons quickly learn new NE responses, and overall monkey behavior adjusts accordingly. Unlike much or all mirror neuron work, the implications of these studies go beyond identification of neural correlates. Revealingly, monkey neurons don't quite learn NE strategies when the unique best reply to the computer is pure randomization. Instead, they come as close to randomization as implementation of a classical Rescorla-Wagner conditioning rule can get. This makes evolutionary sense, because outside of some rigorously monitored asset markets and game theorists' experimental labs, it is unlikely that humans, let alone monkeys, often face opponents that can detect and exploit the difference between true NE play and NE approximation. However, the key implication here is that neurons can implement "good enough" coordination, in complex strategic settings, through long-understood processes of conditioned learning if these are supplemented by drift diffusion processes that exploit statistical relationships between stochastic behavior control mechanisms and variations in reward frequencies and rates (Lee & Wang, 2009).

Coordination around NE strategy mixes based on neural conditioning and drift diffusion explains the superficially paradoxical dynamic by which learning based essentially on copying leads to distributions of variable behavior that are stochastically stable. On the basis of these considerations, we should regard human achievement of the capacity for mimesis as surprising—since nature did not replicate these capacities in nonhumans—but as not *mysterious*, since all that was needed was extra memory to amplify the power of basic mechanisms found in other primates and probably, given the ubiquity of classical conditioning responses, in nervous systems generally. This buttresses Donald's identification of the development of mimesis with the first phase of rapid enchephalization in the hominid line; adding neurons and synaptic connections is the basic way, in a neural network, to increase memory capacity.

As Donald stresses, this takes us only to a first plateau on the road to modern human ecology, culture, and cognition. The game structures tracked by conditioned learning in the experiments just reviewed are themselves stable and exogenous to the estimation task presented to the neural mechanisms. However, as economists implicitly recognize in regarding Adam Smith as their discipline's founder, the key to expanding the productivity of resources, and hence, in evolutionary terms, the capacity of the global environment to support ever larger numbers and communities of people, is specialization of labor accompanied by exchange. The economist Haim Ofek (2001) argues persuasively that specialization and exchange were a precondition rather than a consequence of the evolutionary trajectory from *Homo erectus* to modern *Homo sapiens*. Like Donald, Ofek assembles physical evidence from paleontology to build his case, but surveys it with an economist's attention to opportunity costs of alternative behavioral strategies available to our ancestors.

Opportunity costs are based on scarcity, measured as a function of budget constraints given fixed technology for resource exploitation. A constraint that faced the species as a whole concerned the metabolic demands of the larger brain. The only comparably expensive organ that could be traded off to support such increasing metabolic pressure—which selection would then tend to reduce *if* circumstances rendered such reduction compatible with Hamilton's rule—was the complex gut needed to digest raw plant food. Thus Ofek argues, in company with Wrangham et al. (1999) and Wrangham (2009), that mastery of fire was a specific precondition for at least the later and most rapid stage of human encephalization. He then marshals reasons to believe that fire-keeping was the first specialized occupation in the hominid social ecology. This involves interpretation of paleontological evidence in light of an economic analysis according to which, for *Homo erectus* and his immediate successors, it was much more efficient for specialists to maintain fires from which bands of local hunter-gatherers could draw in exchange for food and pelts than for each small band of hunter-gatherers to search for suitable kindling each day—which would have severely restricted their foraging ranges—and then endure the high-risk, failure-prone ordeal of starting a nightly fire without modern ignition technology. Caves, Ofek argues, were not primarily used as homes by early humans, as popular imagination supposes, but as fire service stations. This naturally leads one to speculate, though he does not, that cave art, exploiting early humans' mimetic dispositions, might have had the intended function of distinguishing places of business from competing shops.[8] Since fire maintenance requires steady presence but not steady labor, once fire service centers were established, it would be natural for their operators to diversify into making hand axes, body ornaments, and other products that would be of value to hunter-gatherers but are not most efficiently manufactured *while* moving around to follow prey and locate fruit and vegetable patches. The pattern of human settlement of challenging environments such as ice-age Europe, Ofek argues, was constrained by considerations of economies of scale: hunter-gatherers could not begin to productively work a new territory until there were enough of them to support a local fire station. The expansion of frontiers of settlement in modern times according to this pattern—on much faster timescales, of course—is a familiar one to economic historians.

Ofek's project is not merely to explain the origins of markets. Rather, his thesis is that market exchange was the basic behavioral adaptation that allowed humans to construct a distinctive ecological niche, and the only such niche that tends by its own endogenous dynamic to expand indefinitely. Of central importance to the present argument, this adaptation is primarily one of social organization, and only secondarily one of individual cognitive and preference dispositions. Like Seabright (2010), but without any need for the hypothesis of a genetic discontinuity to support "strong reciprocity," Ofek observes that the progress of cross-band exchange in turn required the partial displacement of natural xenophobic violence by diplomacy, thus promoting the enhanced strategic competence in which social intelligence partly consists.

As noted, by their nature markets *grow* and *change*. This makes coping with their dynamics a more complex problem than that faced by groups of monkey neurons in the experiments surveyed by Lee and Wang (2009). On an evolutionary scale, the power of market participants to change the outcome spaces of games destabilizes agency itself, by making utility functions dynamic, and by embedding games within metagames.[9] The evolution of modern societies is characterized by dizzying acceleration in the special human capacity for niche construction; by their behavior, traders don't merely adapt to markets, but change their structures. If, as we should expect to be typical, people approach their strategic interactions with both asymmetric information and the ability to exploit this information to dynamically influence outcome spaces, why and how should we imagine that agents converge on a shared model of outcomes? It is no gain in explanation to suggest that early people coordinated through constrained variations on imitative patterns if the patterns to be imitated were themselves unstable and relative to interpretations of available payoffs on which we cannot understand how they might have jointly converged.

Game theorists have formally studied this problem using a family of models referred to as "global games" (Carlsson & van Damme, 1993). In a global game, players receive slightly noisy, nonpublic signals about uncertain states of the environment. If players have correct beliefs about the sources of noise, when each one observes her own signal she can estimate the distributions of signal values received by other players. Not knowing their background beliefs, she assumes that these are randomly distributed about the unit interval, because in her ignorance this is the least arbitrary prior. On this basis, the player estimates the probable distribution of actions by others and chooses her best reply. Carlsson and van Damme show that given some plausible technical restrictions, this setup mimics the solution space of standard classical game theory while nevertheless taking into account that players choose actions in light of uncertain, conjectural beliefs about the beliefs of others.

A leading domain of application for global game theory has been to speculative crises in financial markets (e.g., Morris & Shin, 1998). Global game theory is, among other things, a tool for formally exploring circumstances under which coordinators can converge on inferior equilibria, or traps—for example, in the Morris and Shin model, bank runs that can only be stopped by exogenous interventions. It is interesting to reflect on this against the background of the history of the individualism debates in economic methodology. The neoclassical tradition began with Walras's problem of trying to determine the circumstances under which atomic individuals with uncorrelated utility functions could efficiently coordinate on prices and trades in a market. The global game theorist turns this venerable question on its head: Given coordination that is as efficient as possible in light of background uncertainties among players, under what circumstances might players "overcoordinate"—that is, converge on inefficient game structures and eliminate strategic variance within the population that might otherwise have provided the basis for discovery of paths to sets of equilibria containing superior outcomes? Note the close relationship between these contrasting modeling approaches and our main topic: The Walrasian tradition assumes distinct individuals and must explain how they converge on a shared information structure, whereas the global game theorist makes players' identities strategically endogenous to the structures of their games but then faces the problem of market instability due to insufficient interindividual strategic variation.

A specific version of overcoordination has been studied experimentally, under the label of the "herding" problem. Suppose that an agent, i, who is unsure about the distribution of private information in a market observes a number of other agents all choosing strategies that would be explained by the hypothesis that those agents share belief β. Suppose that i has private information, ι, that contradicts β. Under certain plausible conditions, it can be expected-utility-maximizing for i to choose to imitate the other participants' β-based strategies and ignore her own private information. But in that case ι is lost to the market. This must, in general, decrease the expected efficiency of the market. Furthermore, initial observers in the chain might happen to be unlucky, and falsely attribute β. The result can be a "reverse cascade" in which everyone converges on an incorrect model that throws away *all* private information.

Experimental tests of herding have generated equivocal but interesting results. Anderson and Holt (1997) found significant cascading and reverse cascading even when participants knew that their own ability to estimate the market, based on their private information, was at least as good as anyone else's. Sgroi (2003) replicated this result in situations where subjects could decide to wait to choose until they had observed the choices of others. Sgroi also tested the effect of correcting errors incorporated in reverse cascades. In these instances, participants tended to move further away from Bayesian rationality than recognition of the identified errors warranted, suggesting failure by participants to fully recognize that rational choice can produce suboptimal outcomes. On the other hand, Huck and Oechssler (2000), Nöth and Weber (2003), and Spiwoks, Bizer, and Hein (2008) found general failures of Bayesian rationality and overweighting of private signals, and therefore fewer cascades. That is, in these experiments people departed from individual rationality by taking their own private information too seriously—and thereby behaved in a way that, given some plausible informational structures, could *increase* expected market efficiency.

Two experimental reports are especially interesting in the context of our present main topic. Hung and Plott (2001) found prevailing near-rational behavior (from the individual's point of view), and broad confirmation of Anderson and Holt's findings, when subjects were encouraged to frame their decisions as individuals. ("Near" rationality refers to the fact that subjects produced somewhat fewer cascades than fully rational agents would be predicted to do.) However, when subjects understood that the majority decision would bind all participants, and were thus given incentive to reframe the choice problem as one confronting a team, subjects paid more attention to their private signals. This "contrarian behavior" will tend to improve social efficiency in a very noisy environment, while lowering it in a highly transparent one. However, as a further complication, when Corazzini and Greiner (2007) encouraged subjects to frame their choices in the familiar context of independent choices over lotteries, herding collapsed and individually irrational but socially efficient contrarian behavior abounded.

The Hung and Plott finding carries a nice warning about reliance on intuitions in thinking about the relationship between descriptive and normative individualism. We might describe their setting in philosophical terms as follows: subjects were incentivized to adopt nonindividualistic norms, in a context where these were best served by cultivating idiosyncratic individual strategies, such that an individualistic frame is essential for adequately describing behavior. Page (2007) provides evidence for the efficiency of this pattern of framing, and its tendency to self-stabilizing dynamics, in contemporary firms, schools, and other complex organizations.

Let us summarize. In an early human environment where most groups of relatives hunted and gathered, but some formed households that maintained fire services and general merchandise shops in caves, simple imitation could not tell a family what to do. If it sought to optimize, the family should in the first place have focused not on the special properties of its individual members, but on a social property: What were the local marginal costs and benefits of being, respectively, the next foraging group in one's area, the next foraging group in a new area, and going into retail? If the family opted for business, it *then* needed a basis for stable specialization among its members; who will cultivate the craft of hand-axe manufacture, who will concentrate on cave art, who will gather kindling for the fire? Basic principles of organizational psychology tell us that stability within the production unit is best served if people imaginatively identify themselves with their assigned roles. This gives all household members incentive to collaborate in reinforcing one another's professional identifications. We might not be surprised to discover, though of course we never can, that cave artists, as possessors of the most rare skill among those of value, and who required emotional creativity to anticipate the tastes of hypothetical customers from among nonkin, were encouraged to think it natural for artists to be relatively narcissistic and temperamental, while the personality of a fire-maintenance officer should be the opposite.

Specialization of labor thus promotes shared normative framing of individual differences. Such differences may sometimes have their basis in genetically produced variations in talent or temperament, but where they do, the members of a corporate entity have incentive to exaggerate these, and where they do not, to create them. We now turn to consider processes by which these incentives are channeled into stable patterns of behavior.

3 The Social Sculpting of Individuals and Norms of Individuality

There is at least one body of scholarly literature that seems clearly committed to the conjunction of descriptive anti-individualism and normative individualism. This is work by historians of social and cultural organization. It is a familiar theme in such history that "the individual" emerged gradually, and very recently, as an idea that governed people's normative expectations about one another and about themselves. Of course, historians have many different opinions as to when and how quickly this happened in various parts of the world. It is also widely disputed as to whether "the individual" was a bit of social technology invented in Europe, which then spread from there to other continents. Although some versions of such stories connote ethnocentric triumphalism that could be thought to imply racism, plausible mechanisms for them are available. If capitalism is promoted by individualistic norms, as stressed in a venerable tradition going back to Weber ([1905] 2002), then one might expect that capitalism and individualism would generally coevolve, and that individualism would have spread, geographically and temporally, along with the other member of the dyad. (For a recent account that is explicitly coevolutionary in just this way, see Clark, 2007.)

A representative historian's text that takes the gradual emergence of the individual for granted as a phenomenon that needs explanation and contextualizing is that by Morris (1972). Joining a controversy already in progress over whether individualism arose with the Renaissance or during the medieval period, Morris mounts a monograph-length defense of a critical 150-year episode between 1050 and 1200. He opens by sketching his target explanandum:

> We think of ourselves as people with frontiers, our personalities divided from each other as our bodies visibly are. Whatever ties of love or loyalty may bind us to other people, we are aware that there is an inner being of our own; that we are individuals. To the Western reader it may come as a surprise that there is anything unusual in this experience. It is to us a matter of common sense that we stand apart from the natural order in which we are set, subjects over against its objectivity, and that we have our own distinct personality, beliefs, and attitude to life. . . . [I]t is true that Western culture, and the Western type of education, has developed this sense of individuality to an extent exceptional among the civilizations of the world. . . .
>
> [The] relative weakness of the sense of individuality is not confined to those societies which we normally call primitive. The student of the Greek Fathers or of Hellenistic philosophy is likely to be made painfully aware of the difference between their starting-point and ours. Our difficulty in understanding them is largely due to the fact that they have no equivalent to our concept "person," while their vocabulary was rich in words which express community of being. (Morris, 1972, pp. 1, 2)

Of particular interest in the context of the present essay is Morris's effort to distinguish individualism as a contested political ideology from a deeper sense of individualism that he takes to be a normative construct, though one that now constitutes an objective description of modern "Western" people:

> This book will [be concerned with] . . . that respect for individual human beings, their character and opinion, which has been instilled into us by our cultural tradition, and with its implications for personal relationships and beliefs. The hard core of this individualism lies in the psychological experience with which we began; the sense of a clear distinction between my being and that of other people. The significance of this experience is greatly increased by our belief in the *value* of human beings in themselves. (Ibid., p. 3)

Very interestingly, Morris takes as a key indicator of the presence of the "new" sense of individuality in the cognitive formation of a person the ability to write biography and—especially—inwardly reflective autobiography:

> What cannot be verbalized can scarcely be thought, and before 1050 the capacity of most writers to express themselves lucidly was poor. When, in the ninth century, Einhardt attempted to describe Charlemagne's personal appearance—a bold undertaking, for there were few recent precedents to guide him—he built up a pastiche of quotations from Suetonius, to such an extent that some commentators have suspected that the passage is not a description of Charlemagne

from the life, but a merely literary construction. . . . If we seek for genuinely individual description from the life, we must look to men who were able to write down fluently and naturally what they saw. . . . The same is true of the art of self-expression. The meditations of Anselm or Aelred of Rievaulx, who were able to express their affections and longings in a practiced way, moving easily from one idea to another, would have been literally unthinkable a century before. (Ibid., pp. 7–8)

I have no side to take in the historians' controversy over when individualistic norms and literary expressions arose in Europe. The interest of Morris's discussion for present purposes lies in two aspects: first, the confidence on both sides of the argument that individualistic *norms* were culturally constructed at an identifiable, recent time and place; and, second, the explicit link that Morris makes, on which the whole method of argument in his book depends, between individualism and autobiographical narrative.

The significance of autobiographical narrative in both historical-cultural and individual development has been extensively discussed by psychologists and cognitive scientists. Donald (1991) follows Bruner (1986) in defending the view that narrative structure as the basis for logical organization and explanation of events arose with human language and "develops early and naturally in children" (Donald 1991, 256), whereas "analytic" or "theoretical" reasoning, based on postulated timeless and abstract principles, arose with the ancient Greeks, sometime between Homer and Herodotus. However, the disposition to produce narratives can in principle predate a disposition to produce narratives centered on the narrator and taken to express a partial and distinctive subjective point of view. Jaynes (1976) proposed the radical thesis that Greeks in the time of Homer did not engage in autobiographical narrative and in consequence lacked self-awareness. As we saw above, this seems to be Morris's assumption, expressed a few years before Jaynes's book appeared.[10] Dennett (1991), though he is agnostic with respect to Jaynes's dating of the development of self-awareness, argues on the basis of considerations from the neuroscience of consciousness that Jaynes is correct to tie self-awareness to the production of reflective autobiographical narrative and to see this capacity as essentially relying on cultural scaffolding. Dennett doubts that narrative arises "naturally" in children, if "natural" is taken to mean "without cultural exemplification and reinforcement"; and his account of consciousness depends on the idea that, at least, autobiographical narrative does not so arise. Similarly, Hutto (2008) argues that folk psychology is essentially a learned facility with the narrative construction of others, which adepts reflexively apply to themselves.

Based on game-theoretic logic, Ross (2004, 2005, 2006b, 2007, 2008a,b) has extensively characterized the processes by which people learn, over the course of childhood and adolescent development, to construct narrative selves that have the following properties:

(1) They are adapted to local cultural expectations, so that they facilitate location of equilibria in global games with others who share a similar cultural background.
(2) The dimensions along which their variance is culturally salient form the basis for a prevailing typology of personalities and linked aptitude sets that are normatively and statistically associated with types of economic occupations and social roles.
(3) They are attractive to others, and so encourage cooperative activities that exploit specialized, complementary roles, to the extent that they display creative uniqueness within the boundaries of local normative conventions.
(4) They develop inconsistency, which tends in the limit to incoherence, if they are not reinforced by a person's recurrent interactants; and inconsistent or incoherent narrative selves are regarded by others as diagnostic of unreliability at best and insanity at worst.
(5) Their relative inconsistency or incomprehensibility to others will be associated with ostracism and exclusion from collaborative projects, including opportunities for mutually advantageous exchange.
(6) Their general comprehensibility to at least a subset of the community sufficient for maintenance of the person's economic niche is a precondition for material flourishing in a society based on division of labor.
(7) They are more closely controlled and influenced, at least from adolescence, by cohort peers than by living ancestors (Harris, 2006).
(8) They may be drastically revised in the course of a lifetime by appeal to the occurrence of milestone events recognized as such by culturally stable metanarratives. Examples of such milestones in contemporary Western societies are college graduation, marriage, first parenthood, religious conversion, and acknowledged recovery from addiction.

Properties (1) through (3) explain the economic function of narrative selves as structures that facilitate the organization of specialized perspectives and capacities so as to avoid overcoordination and the resulting inefficient loss of information. Of course, the economic efficiency of a structure does not predict its existence unless it is supported by equilibrium dynamics. Properties (4) through (6) indicate the incentive structures that lead all cognitively and conatively competent people to devote significant resources to narrative self construction and maintenance, and indeed to defend these constructions tenaciously, not infrequently choosing biological death as preferable to self-undermining actions such as shaming one's family, treason to country, or abandonment of religious commitments. Properties (7) and (8) indicate how cultural conventions on allowed self-narratives may avoid locking into conservative traps that cannot keep pace with environmental or technological change. These properties do not *guarantee* avoidance of conservative traps, and of course many communities and subcommunities fall into such traps to varying degrees.

One of the core capacities that human parents must nurture in potentially successful offspring is that of self-narration. This skill mainly consists in the ability to engage in recurrent generate-and-test cycles in social interactions, and to track shifting local norms that define the range within which distinctive styles of behavior pass from being celebrated, to barely tolerated, to resented. As described by McGeer (2001) and others, parents reward their children's' adoption of consistent focused interests and forms of expertise, and encourage them to explicitly identify with "signature" clusters of activities and domains of knowledge. However, as property (7) indicates, and as recounted in detail by Harris (2006), while parents nurture the capacity for self-narration, peers play a larger role in suggesting and constraining content, at least from early adolescence. This is popularly appreciated as the child's drive to establish "independence," and social institutions express strong normative interest in the extent to which it is legitimate for parents to resist it.

5 Conclusion

At three scales—that of the evolution of the modern human species, that of the cultural emergence of values adapted to giant industrial communities, and that of the etiology of distinctive personal characters—I have identified arcs of development from behavioral spaces with little individual variation to spaces characterized by emphasis on special capacities and characteristics of individuals. All of these developmental arcs are both driven and constrained by largely implicit and nondeliberative normative considerations. Specialization of labor was culturally promoted because it made people richer, and the promotion of such specialization in turn made people smarter. The tendency of team reasoners to inefficiently overcoordinate was resisted by the cultural evolution of pressures to use the new resource of language for individual differentiation. But since the point of this differentiation was ultimately high-order coordination, we should expect to find, as we do, that it is controlled by cultural norms about which people care a lot. In most historical human societies, people who are less than ideally unique are merely regarded as boring and shuffled a few steps backward in the mate selection sweepstakes, whereas people who carry their self-making art to avant garde lengths, where general comprehensibility and predictability to others break down, are often savagely persecuted.[11]

In light of this developmental vector, normatively sculpted human individuals tend to celebrate individuality as a principle, up to a point. At the scale of cultural development, this is manifest as cultural pride that often tips into aggressive ethnocentrism, but is generally regarded as a good thing when it is channeled into peoples' fascination with their own history and forms of art. At the scale of individual development, people regard their narrated selves as among their most precious assets; in general, only the persons of very close kin, especially offspring, are assigned competing levels of value. Contrary to parochial but widespread perspectives such as that of Morris reviewed above, these dynamics are universal among human communities and not restricted to Western cultures. At the same time, since pressure for specialization of labor, in communities extending beyond family groups, is the primary exogenous pressure that drives the evolutionary dynamics, the *extent* to which individuals are encouraged to differentiate themselves is indeed correlated with the growth of complexity in economic production.

Thus, from the perspective of evolution and development, Bacharach's rhetorical presentation of individual utility maximization as default strategic framing, and team utility maximization as "reframing," reverses figure and ground. Atomism is an upside-down explanatory stance in all social sciences, including historical anthropology and economics. The history of attempts to base normative individualism on descriptive individualism, such as can be attributed to the Lockean tradition in political philosophy, appear profoundly confused from Darwinian and historicist perspectives.

However, normative individualism is not entirely independent of the relation of explanatory priority as between individual and collective scales of description. Strong normative individualism makes sense precisely *because* maintenance of distinctive individuality is a kind of *achievement*. If strong individual distinctiveness were the human biological default, it would be gratuitous to normatively celebrate and defend it. Protective and promotional norms do not generally arise around assets that require no effort to acquire or maintain.

These considerations do not license a metaphysical argument in favor of caring about the welfare of collective entities only as derivative of the welfare of individual people, of the kind sought by the Austrian economists. I am persuaded by such naturalistic metaethicists as Hume ([1748] 1977) and Joyce (2001) that no justified metaphysical arguments are in principle available for either side of this enervating dispute. The best we can do is explain why most people find normative individualism attractive and indeed emotionally irresistible. If nothing is made "ultimately good" by the nature of the universe, then it is sound procedure to appeal, in policy disagreements, to values that shape the majority of human judgments, as contingent consequences of history. It is reasonable to expect proposers of policies to indicate the individuals whose welfare the policies in question will promote, and to tell us how those who suffer welfare losses will be compensated.

Notes

1. See Ross (2011a). Macroeconomists confuse the relevant associations in a special way by assigning high methodological importance to "microfoundations." The general validity of this concern is controversial; see Hoover (1988) for the clearest treatment, and Hartley (1997) for a sustained criticism. For present purposes it suffices to say that what is "foundational" about the sought-after microfoundations of macroeconomics is not ontological *or* normative; the demand for them arises from macroeconomists' understandable unease with any model that posits statistically extractable data that markets fail to extract. So the concern is ultimately about

social structures and processes, not individuals. See Janssen (1993) for a detailed workout of this argument, by an author who seems to hope, in exact opposition to my own attitude, that economists *will* eventually succeed in taking ontological individualism more seriously.

2. The model of more selfish players will of course sometimes be correct—team cohesion not infrequently unravels, as in the case of the French national side in the 2010 World Cup. But the game theorist would not be able to correctly model the difference between the French team and, for example, the superb Spanish winning side, without resort to Bacharach's innovation.

3. West, Mouden, and Gardner (2010) acknowledge that humans use special proximate mechanisms to coordinate their cooperation—particularly language. This is important to issues raised later in the chapter.

4. This point is forcefully made by Thalos and Andreou (2009).

5. Orangutans were once thought to be solitary. This has turned out to be inaccurate as a generalization, and such solitude as is observed in some orangutans now appears to be a recent adaptation to habitat changes. See Dunbar (1988).

6. Psychologists sometimes understand episodic memory as necessarily involving narrative memory, which would confine it to humans. Donald's use of the term is more general, but still in the standard conceptual ballpark; he presents evidence that modern apes remember particular social situations involving specific individuals.

7. This is even true of so-called free jazz. Furthermore, it is not evident that the freest jazz would have value to any listeners except by way of contrasts with less free jazz.

8. Given the inaccessibility of much cave art, we can rule out a billboard function. But many modern businesses regard stylish and expensive customer service areas as essential.

9. This complexity explains why economists did not know how to model markets with imperfect competition—as opposed to markets in which all agents are price takers—until a few decades ago, and why economists have become increasingly interested in evolutionary game theory as an important part of the analytical toolkit in addition to classical game theory.

10. Jaynes's thesis had been suggested by him in articles that predate Jaynes (1976) by decades. However, Morris does not cite Jaynes, or indeed anyone, as authorities for his remarks about the ancient Greeks.

11. Happy are societies that, like England, evolved second-order norms favoring amused appreciation of eccentricity.

References

Akerlof, G., & Shiller, R. (2009). *Animal spirits*. Princeton: Princeton University Press.

Anderson, L., & Holt, C. (1997). Information cascades in the laboratory. *American Economic Review, 87*, 847–862.

Ariely, D. (2008). *Predictably irrational*. New York: Harper.

Bacharach, M. (2006). *Beyond individual choice*. Princeton: Princeton University Press.

Bergstrom, T. (2002). Evolution of social behavior: Individual and group selection. *Journal of Economic Perspectives, 16*, 67–88.

Bowles, S., & Gintis, H. (2004). The evolution of strong reciprocity: Cooperation in heterogeneous populations. *Theoretical Population Biology, 65*, 17–28.

Bruner, J. (1986). *Actual minds, possible worlds*. Cambridge, MA: Harvard University Press.

Burnham, T., & Johnson, D. (2005). The biological and evolutionary logic of human cooperation. *Analyse & Kritik, 27*, 113–135.

Camerer, C., Loewenstein, G., & Prelec, D. (2005). Neuroeconomics: How neuroscience can inform economics. *Journal of Economic Literature, 43*, 9–64.

Carlsson, H., & van Damme, E. (1993). Global games and equilibrium selection. *Econometrica: Journal of the Econometric Society, 61*, 989–1018.

Clark, A. (1997). *Being there: Putting brain, body, and world together again*. Cambridge, MA: MIT Press.

Clark, G. (2007). *A farewell to alms*. Princeton: Princeton University Press.

Corazzini, L., & Greiner, B. (2007). Herding, social preferences, and (non-)conformity. *Economics Letters, 97*, 74–80.

Dennett, D. (1991). *Consciousness explained*. Boston: Little, Brown.

Donald, M. (1991). *Origins of the modern mind*. Cambridge, MA: Harvard University Press.

Dunbar, R. (1988). *Primate social systems*. London: Croom Helm.

Elster, J. (2000). *Ulysses unbound*. Cambridge: Cambridge University Press.

Frank, R. (1987). If *Homo economicus* could choose his own utility function, would he want one with a conscience? *American Economic Review, 77*, 593–604.

Gallese, V. (2003). The roots of empathy: The shared manifold hypothesis and the neural basis of intersubjectivity. *Psychopathology, 36*, 171–180.

Gallese, V. (2007). Before and below "theory of mind": Simulation and the neural correlates of social cognition. In N. Emery, N. Clayton, & C. Frith (Eds.), *Social intelligence* (pp. 179–196). Oxford: Oxford University Press.

Gintis, H. (2000). Strong reciprocity and human sociality. *Journal of Theoretical Biology, 206*, 169–179.

Gintis, H. (2006). Behavioral ethics meets natural justice. *Politics, Philosophy & Economics, 5*, 5–32.

Gintis, H., Bowles, S., Boyd, R., & Fehr, E. (2003). Explaining altruistic behavior in humans. *Evolution and Human Behavior, 24*, 153–172.

Gintis, H., Henrich, J., Bowles, S., Boyd, R., & Fehr, E. (2008). Strong reciprocity and the roots of human morality. *Social Justice Research, 21*, 241–253.

Glimcher, P. (2003). *Decisions, uncertainty, and the brain*. Cambridge, MA: MIT Press.

Guala, F. (2012). Reciprocity: weak or strong? What punishment experiments do (and do not) demonstrate. *Behavioral and Brain Sciences 35*, 1-59.

Hamilton, W. (1964). The genetical evolution of social behavior I & II. *Journal of Theoretical Biology, 7*, 1–52.

Harris, J. (2006). *No two alike*. New York: Norton.

Hartley, J. (1997). *The representative agent in macroeconomics*. London: Routledge.

Hayek, F. (1949). *Individualism and economic order*. Chicago: University of Chicago Press.

Hollis, M. (1998). *Trust within reason*. Cambridge: Cambridge University Press.

Hoover, K. (1988). *The new classical macroeconomics*. Oxford: Blackwell.

Huck, S., & Oechssler, J. (2000). Informational cascades in the laboratory: Do they occur for the right reasons? *Journal of Economic Psychology, 21*, 661–671.

Hume, D. [1748] (1977). *An enquiry concerning human understanding*. Indianapolis: Hackett.

Hung, A., & Plott, C. (2001). Information cascades: Replication and an extension to majority rule and conformity-rewarding institutions. *American Economic Review, 91*, 1508–1520.

Hurley, S., & Chater, N. (Eds.). (2005). *Perspectives on imitation*. Cambridge, MA: MIT Press.

Hutchins, E. (1995). *Cognition in the wild*. Cambridge, MA: MIT Press.

Hutto, D. (2008). *Folk psychological narratives*. Cambridge, MA: MIT Press.

Janssen, M. (1993). *Microfoundations: A critical inquiry*. London: Routledge.

Jaynes, J. (1976). *The origin of consciousness in the breakdown of the bicameral mind*. Boston: Houghton Mifflin.

Jensen, K., Call, J., & Tomasello, M. (2006). What's in it for me? Self-regard precludes altruism and spite in chimpanzees. *Proceedings: Biological Sciences, 273*, 1013–1021.

Joyce, R. (2001). *The myth of morality*. Cambridge: Cambridge University Press.

Lee, D., Conroy, M., McGreevy, B., & Barraclough, D. (2004). Reinforcement learning and decision making in monkeys during a competitive game. *Brain Research: Cognitive Brain Research, 22*, 45–48.

Lee, D., McGreevy, B., & Barraclough, D. (2005). Learning and decision-making in monkeys during a rock-paper-scissors game. *Brain Research: Cognitive Brain Research, 25*, 416–430.

Lee, D., & Wang, X.-J. (2009). Mechanisms for stochastic decision making in the primate frontal cortex: Single-neuron recording and circuit modeling. In P. Glimcher, C. Camerer, E. Fehr, &

R. Poldrack (Eds.), *Neuroeconomics: Decision making and the brain* (pp. 481–501). London: Elsevier.

Lukes, S. (1968). Methodological individualism reconsidered. *British Journal of Sociology, 19,* 119–129.

McClure, S., Daw, N., & Montague, R. (2003). A computational substrate for incentive salience. *Trends in Neurosciences, 26,* 423–428.

McGeer, V. (2001). Psycho-practice, psycho-theory, and the contrastive case of autism. *Journal of Consciousness Studies, 8,* 109–132.

Miller, R. (1978). Methodological individualism and social explanation. *Philosophy of Science, 45,* 387–414.

Morris, C. 1972. *The discovery of the individual: 1050–1200.* New York: Harper & Row.

Morris, S., & Shin, H.-S. 1998. A theory of the onset of currency attacks. Economics Group, Nuffield College, University of Oxford *Economics Papers* Working Paper Series no. 149.

Nöth, M., & Weber, M. (2003). Informational aggregation with random ordering: Cascades and overconfidence. *Economic Journal, 113,* 166–189.

Ofek, H. (2001). *Second nature.* Cambridge: Cambridge University Press.

Page, S. (2007). *The difference.* Princeton: Princeton University Press.

Pareto, V. (1927). *Mannuel d'économie politique.* Paris: Marcel Giard.

Pettit, P. (1993). *The common mind.* Oxford: Oxford University Press.

Pettit, P. (1997). *Republicanism.* Oxford: Oxford University Press.

Ross, D. (2004). Meta-linguistic signalling for coordination amongst social agents. *Language Sciences, 26,* 621–642.

Ross, D. (2005). *Economic theory and cognitive science* (Vol. 1): *Microexplanation.* Cambridge, MA: MIT Press.

Ross, D. (2006a). Evolutionary game theory and the normative theory of institutional design: Binmore and behavioral economics. *Philosophy, Politics & Economics, 5,* 51–79.

Ross, D. (2006b). The economics and evolution of selves. *Cognitive Systems Research, 7,* 246–258.

Ross, D. (2007). *H. sapiens* as ecologically special: What does language contribute? *Language Sciences, 29,* 710–731.

Ross, D. (2008a). Economics, cognitive science and social cognition. *Cognitive Systems Research, 9,* 125–135.

Ross, D. (2008b). Classical game theory, socialization, and the rationalization of conventions. *Topoi, 27,* 57–72.

Ross, D. (2011a). The economic agent: Not human, but important. In U. Maki (Ed.), *Handbook of the philosophy of science* (Vol. 13): *Economics*. London: Elsevier.

Ross, D. (2011b). Estranged parents and a schizophrenic child: Choice in economics, psychology, and neuroeconomics. *Journal of Economic Methodology, 18*, 217–231.

Ross, D., Spurrett, D., Kincaid, H., & Stephens, G. L. (Eds.). (2007). *Distributed cognition and the will*. Cambridge, MA: MIT Press.

Samuelson, P. [1947] (1983). *Foundations of economic analysis* (enlarged ed.). Cambridge, MA: Harvard University Press.

Seabright, P. (2010). *The company of strangers* (rev. ed.). Princeton: Princeton University Press.

Sgroi, D. (2003). The right choice at the right time: A herding experiment in endogenous time. *Experimental Economics, 6*, 159–180.

Silk, J. (2009). Social preferences in primates. In P. Glimcher, C. Camerer, E. Fehr, & R. Poldrack (Eds.), *Neuroeconomics: Decision making and the brain* (pp. 269–284). London: Elsevier.

Silk, J., Brosnan, S., Vonk, J., Henrich, J., Povinelli, D., Richardson, A., et al. (2005). Chimpanzees are indifferent to the welfare of other group members. *Nature, 435*, 1357–1359.

Smith, V. (2008). *Rationality in economics*. Cambridge: Cambridge University Press.

Spiwoks, M., Bizer, K., & Hein, O. (2008). Informational cascades: A mirage? *Journal of Economic Behavior & Organization, 67*, 193–199.

Thaler, R. (1992). *The winner's curse*. New York: Free Press.

Thalos, M., & Andreou, C. (2009). Of human bonding: An essay on the natural history of agency. *Public Reason, 1*, 46–73.

Tomasello, M., Carpenter, M., Call, J., Behne, T., & Moll, H. (2004). Understanding and sharing intentions: The origins of cultural cognition. *Behavioral and Brain Sciences, 28*, 691–735.

Vonk, J., Brosnan, S., Silk, J., Henrich, J., Richardson, A., Lambeth, S., et al. (2008). Chimpanzees do not take advantage of very low cost opportunities to deliver food to unrelated group members. *Animal Behaviour, 75*, 1757–1770.

von Mises, L. (1949). *Human action*. New Haven: Yale University Press.

Walras, L. (1874). *Élements d'économie politique pure ou théorie de la richesse sociale*. Lausanne: L. Corbaz.

Warneken, F., & Tomasello, M. (2006). Altruistic helping in human infants and young chimpanzees. *Science, 311*, 1301–1303.

Warneken, F., Hare, B., Melis, A., Hanus, D., & Tomasello, M. (2007). Spontaneous altruism by chimpanzees and young children. *PLoS Biology, 5*, e184. doi:10.1371/journal.pbio.0050184.

Weber, M. [1905] (2002). *The Protestant ethic and the spirit of capitalism* (Baehr, P., & Wells, G., Trans.). Harmondsworth: Penguin.

West, S., El Mouden, C., & Gardner, A. (2011). Sixteen common misconceptions about the evolution of cooperation in humans. *Evolution and Human Behavior, 32*, 231–262.

Wrangham, R. (2009). *Catching fire*. New York: Basic Books.

Wrangham, R., Jones, J., Laden, G., Pilbeam, D., & Conklin-Brittain, N. L. (1999). The raw and the stolen: Cooking and the ecology of human origins. *Current Anthropology, 40*, 567–594.

2 Timescales, Symmetry, and Uncertainty Reduction in the Origins of Hierarchy in Biological Systems

Jessica C. Flack, Doug Erwin, Tanya Elliot, and David C. Krakauer

Introduction

An outstanding question in biology is why life has evolved to be hierarchically organized. From genomes, to cells, tissues, individuals, societies, and eco-systems, evolution generates structures with nested spatial and temporal levels (e.g., Feldman & Eshel, 1982; Buss, 1987; Campbell, 1990; Maynard Smith & Szathmary, 1995, Valentine & May, 1996, Jablonski, 2000; Michod, 2000, Gould, 2002; Frank, 2003; Jablonka & Lamb, 2005, Frank, 2009). Typically, with each new structural level comes new functionality—a new feature with positive payoff consequences. This new functionality can be in the form of a new behavioral output such as a feeding response to a previously inaccessible resource. Or, it can arise when coarse-grained information encapsulated at the new level through some mechanism feeds back to lower levels, changing the accessibility of strategies for components and allowing the space of functions that components can perform to increase. An example is the protein kinases, which have been repeatedly recruited into many different functional pathways over the course of their evolutionary history (e.g., Manning et al., 2002).

In this chapter we present a novel approach to the origin of levels problem. We suggest that a primary driver of evolutionary change is the reduction of environmental uncertainty through the construction of dynamical processes with a range of characteristic time constants, or nested slow variables. Slow variables arise from mechanisms that naturally integrate over fast, microscopic dynamics. Proteins, for example, have a long half-life relative to RNA transcripts, and can be thought of as the summed output of translation. Cells have a long half-life relative to proteins, and are a function of the summed output of arrays of spatially structured proteins. Both proteins and cells represent some average measure of the noisier activity of their constituents and processes of formation. Slow variables, then, can be thought of as coarse-grained variables encoding statistics that are informative about the state of the system. Hence a median protein density is often thought to be informative about the rate of gene expression and RNA translation. Further examples include the steady state density of cells in multicellular organisms, and, as we will discuss in this chapter, power structures and related stable interaction networks in societies of individuals, and properties of environmental architecture built by organisms such as leafcutter ants and corals.

As a consequence of integrating over abundant microscopic processes, features that can serve as slow variables provide better predictors of the local future configuration of a system than the states of the fluctuating microscopic components. We propose that *when detectable by the system or its components*, slow variables can reduce environmental uncertainty and, by increasing predictability, promote accelerated rates of microscopic adaptation. The reduced uncertainty facilitates adaptation in two ways: It allows components to fine-tune their behavior and it frees components to search at low cost a larger space of strategies for extracting resources. This phenomenon has been studied extensively in relation to neutral networks in RNA folding. Many different sequences can fold into the same secondary structure. This implies that over time, structure changes more slowly than sequence, thereby freeing sequences to explore many configurations under normalizing selection (Schuster & Fontana, 1998).

The preliminary data analysis and theory that we discuss in this chapter suggest that slow variables *arise* through the accumulation of physically instantiated (e.g., through cementation processes or nest building, which result in physical structures, or through changes to neural circuits) memory[1] of asymmetric, and typically competitive, outcomes.[2] At some threshold, this asymmetry is reinforced though feedback to the lower levels from the integrated output of higher levels amplifying the asymmetry. We use the term "consolidation of slow variables" to capture the insight that as this history of competitive outcomes builds up, the coarse-grained representations of these dynamics become more robust predictors of the system's future state.

As we shall discuss, one obvious danger of slow variables is lock-in. If a process changes very slowly relative to some underlying microscopic dynamic, it can cease to be good predictor over short time frames, and will lag behind critical shifts at the microscopic level. We propose that hierarchy is a solution to the dual problem of informational noise and informational inertia. Our thesis is that *evolution has led to systems that operate over multiple timescales to balance the trade-off between robustness and adaptability*. Slow variables serve as reliable referents for decision making. Faster variables track changes at the microscopic level, providing a mechanism through which information can percolate up through levels to facilitate trajectory correction when environments change. If this thesis is correct, evolution will have had to fine-tune the timescales of adaptive dynamics (for examples with respect to mutation, see Ishii et al., 1989; Baer, Miyamoto & Denver, 2007).

The origins of the theory of slow variables can be found in a variety of fields within the evolutionary literature. We briefly review these fields to show how the theory builds on existing ideas. We then ground the thesis of slow variables in two case studies. Providing a formal, mathematical framework for the theory of slow variables is beyond the scope of this chapter; but, in the section entitled "Slow Variable Features, Function, and Detection," we discuss two mathematical concepts—the concept of the macrostate and the concept of symmetry breaking—that we believe are required to formalize the theory.

Brief Review of Related Ideas in Evolutionary Theory

Since the late 1980s, interest in the origins of hierarchy has grown within the evolutionary biology community. This can be attributed to developments in four areas of research: the mechanisms of inheritance; the levels of selection; the evolution of development; and the theory of niche construction.

From the empirical perspective, interest in the origins of hierarchy has grown out of the observation that several times in the history of life, highly integrated and coordinated aggregations arose out of collections of self-replicating, autonomous components with only partially aligned interests (e.g., Feldman & Eshel, 1982; Buss, 1987; Maynard Smith & Szathmary, 1995; Frank, 2003; Jablonka & Lamb, 2005; Frank, 2009).

These transitions have been called the major transitions in evolution (Maynard Smith & Szathmary, 1995) and in some cases are thought to reflect transitions to new levels of individuality (Buss, 1987; Michod, 2000, 2007). Much of the work in this area has focused on the role of novel mechanisms of inheritance and transmission mechanisms, including the evolution of epigenetic, learning, and symbolic transmission mechanisms (see also Feldman & Eshel, 1982; Jablonka & Lamb, 2005). The transition from unicellular to multicellular organisms, for example, has been hypothesized to have been made possible by the evolution of a germ line, whereas the transition from loose, homogeneous aggregates of organisms to integrated, differentiated societies is thought to have been facilitated by the evolution of an elaborate combinatorial symbol system—natural language.

Unfortunately, principled, information-theoretic definitions of individuality remain elusive,[3] and so there is substantial controversy and misunderstanding with respect to what constitutes evidence for a new type of individual (see Santelices, 2008; Krakauer & Zanotto, 2009), as well as controversy over what constitutes a major transition. In addition, there have been very few studies of the dynamical process (e.g., models of phase transitions or symmetry breaking) underlying the origins of new levels outside the pattern formation literature with obvious bearing on the evolution of development (e.g., Palmer, 2004).[4]

Within the multilevel selection community, interest in the origins of hierarchy is readily appreciated in relation to a decades-long debate on the preferred level at which selection operates (as illustrated by the reception of the recent paper by Nowak, Tarnita & Wilson, 2010). Traditional multilevel selection theory is based largely on equilibrium solutions operating on aggregate variables. Often the variables are assumed to be fundamental when in fact they are only nominal (Krakauer & Flack, 2010). As an example, consider the Hamilton kin-selection framework. The aggregate variables in kin-selection models correspond to benefit, cost, and relatedness. Because these variables are not typically derived through modeling of microscopic interactions, it can be unclear what in nature they correspond to, or how they should best be measured (see supplement in Nowak, Tarnita & Wilson, 2010).

Consensus is slowly building that to justify the sound choice of macroscopic or aggregate variables as a principled representation of some coarse-grained microscopic processes, we are better off starting with observations at the microscopic level. Hence, within the community of researchers working on the multilevel selection theory, there is a push to model mechanisms that give rise to different degrees of assortative mixing[5]—correlation among individuals in space or time (e.g., Roussett, 2004; supplement in Nowak, Tarnita & Wilson, 2010), and restricting kin selection to those mechanisms that can be grounded only in genetic assortativity.

Another area of evolutionary theory in which hierarchy is a theme might broadly be called "construction dynamics." Areas of research that fall under construction dynamics are those that are explicitly concerned with understanding processes giving rise to ordered phenotypic states. The most rigorous empirical work in biology of this kind includes studies of developmental biology and, in particular, studies connecting cellular differentiation to the gene regulatory architecture underlying aspects of development (e.g., Davidson et al., 2002; Davidson & Levine, 2008). One goal of this work to give a principled computational account—describe the logic of gene activation in terms of Boolean operations—of the emergence of morphological features from gene-gene and gene-protein interactions. And an ultimate goal is to determine whether aspects (e.g., subcircuits) of this process have been conserved over evolutionary time (Erwin & Davidson, 2009; Davidson, 2009).

A fourth area—which perhaps also falls under the rubric of construction dynamics—of evolutionary theory that is concerned with understanding how increasingly inclusive states arise is niche construction (Lewontin, 1982; Dawkins, 1982; Odling-Smee, Laland & Feldman, 2003, Laland & Sterelny, 2006). Niche construction, which finds its origins at the interface of ecology and evolution, posits that organisms, by modifying variables in their ecological and social[6] environments (Laland, Odling-Smee & Feldman, 2003; Borenstein, Kendal & Feldman, 2006; Flack et al., 2006; Flack & Krakauer, 2009, Boehm & Flack, 2010), can partially control the selection pressures to which they are subject. Another way to put this is that by controlling the rate of change or trajectory of environmental variables, organisms are better able to predict their environments (Boehm & Flack, 2010).

The observation that organisms modify environmental processes suggests that the environmental and organismal timescales cannot easily be separated, as is assumed to be true in a typical adiabatic treatment of evolutionary ecology (for a discussion of the adiabatic assumption in evolutionary dynamics, see Krakauer, Page & Erwin, 2009). The majority of niche construction studies have focused on the consequences of niche construction on processes of adaptation (e.g., Odling-Smee, Laland & Feldmann, 2003; Kylafis & Loreau, 2008). Many basic questions remain. Among them are how coupling comes about and what factors promote persistence of coupling.

Answering these questions requires an understanding of the precise procedures organisms use to adaptively modify the environment, and how these procedures and their associated outputs or behaviors are encoded. When modification is a collective process with multiple individuals and species contributing, the problem becomes one of collective social computation. When posed this way, this problem has much in common with central issues in the evolution of development, pattern formation, and collective animal behavior. Within the niche construction literature, such issues have been somewhat marginal, showing up largely in work on social niche construction (Flack, Krakuaer & de Waal, 2005; Flack et al. 2006; Flack & Krakauer, 2006; Flack & de Waal, 2007; Flack & Krakauer, 2009; Boehm & Flack, 2010) where the integration of the population cannot be taken for granted.

Two Case Studies of Distributed Slow Variables

As the examples at the start of this chapter were designed to allow, slow variables can include environmental features that can be described in terms of mass, volume, chemical, and energetic observables, as well as social features that are better described in terms of information or relational coordinates from node degree in a network, to interaction frequency, through more sophisticated measures of conflict, coordination, or cooperation.

We review in detail two case studies for which the ideas of multiple timescales, and adaptive slow variables, are informative. In the first example, from social evolution, the slow variable—statistical features of a power distribution—reduces uncertainty about the behavioral strategies of group members. This is achieved by increasing the predictability of the cost of social interactions. Power structure arises from social interactions, and, hence, can be thought of as a *fully constructed* slow variable. In the second example, from paleoecology, the slow variable—the mass and volume of a reef—reduces organismal uncertainty about ecological features including ocean currents and resource availability. The reef arises from the interaction of organisms with one another and with properties of the physical environment. Hence we call it a *partially constructed* or *modified* slow variable. Each of these studies is associated with a biological time series:[7] one on the order of years and the other on the order of millennia. We have chosen these two examples because they evolve along fundamentally different time and space scales and represent aggregations of fundamentally different kinds of components. This allows us to gauge the generality of the framework.

Readers not interested in details of the case studies can skip to the section entitled "General Features of the Case Studies," which provides a summary of the important points.

A Case Study from Social Evolution: The Consolidation of Power Structures

Within the social evolution and animal behavior communities, studies of the emergence[8] of structure have centered on spatial pattern formation in flocking, schooling, and swarming species (e.g., Couzin, 2009). Equally impressive are collective social phenomena—the arrangement of individuals in relation to one another in a social coordinate space, typically defined in terms of networks of conflict, cooperation, and coordination. Observation and experiment in primates suggest that slowly changing social networks arise from and feed down to influence individual behavior and interaction patterns by changing the cost of interaction (Flack et al., 2006; Flack & Krakauer, 2006; Flack, Krakauer & de Waal, 2005, Boehm & Flack, 2010). We review how these kinds of slowly changing social structures arise and introduce predictability into individual interactions. We explore these issues in the specific context of the consolidation of power structures in macaque societies, providing enough empirical detail to make the case study comprehensible to a nonspecialist.[9] We begin with some background on the concept of power.

We operationally define an individual's power as the degree of consensus among group members that it can use force successfully during agonistic interactions (Bierstedt, 1950; Flack & Krakauer, 2006). We have suggested in previous work (Flack & Krakauer, 2006; Boehm & Flack, 2010; Brush, Krakauer & Flack, in prep.) that there are four critical properties of power: perception—individuals in the given system are said to have power if others in that system perceive them as capable of using force; consensus—if power is to be useful in the social domain, it is the collective perception of group members that matters; temporal stability—the perception that an individual is powerful needs to be stable over time if power is to be predictive and therefore useful for decision making; and computability—individuals must be able to assess their relative power, and this estimate must be correlated in the face of error with their inherent fighting ability, if the resulting power structure is to be stable and predictive.

An individual's estimate of its power can predict—when the estimate is a good one—the cost it will pay if a fight erupts, thereby changing the probability and predictability of social strategy use (Flack, Krakauer & de Waal, 2005; Flack, de Waal & Krakauer, 2005; Flack et al., 2006, Flack & Krakauer, 2006). A positively skewed distribution of power with a long tail (e.g., a power law tail) describes a society in which members of a nonvanishing minority of individuals are collectively perceived as disproportionately powerful. The power structure in our study group is best described by this kind of heavy-tailed distribution (Flack & Krakauer, 2006). Our data suggest that these power structures can support the implementation of novel, beneficial conflict regulatory mechanisms, such as policing (Flack, de Waal & Krakauer, 2005), in which individuals intervene in and impartially break up fights among group members. For any individual to adopt this policing role, it must be able to estimate with relatively low error the cost it will pay for intervening, and that cost must be low. We have found (Flack, de Waal & Krakauer 2005) that in our study group this cost is negligible for the individuals in the tail of the power distribution.

A behavioral knockout experiment (Flack, Krakauer & de Waal, 2005; Flack et al., 2006), in which the policing mechanism was temporarily disabled, showed that regulatory mechanisms like policing are critical contributors to social robustness. Knockout resulted in a destabilization of the groups' social networks: The cost of social interaction increased; investment in social capital acquisition, like alliance partners, decreased; and the cliquishness and assortative structure of the group's social networks increased. Data from this experiment and related studies suggest that the predictive utility of power for estimating the cost of social interaction depends on the extent of the coupling between the power distribution and the underlying conflict and signaling networks from which it arises. Although the power structure must change slowly to be useful as a predictor of interaction cost, it cannot change too slowly because it needs to approximately represent the underlying distribution of asymmetries in fighting ability, and this changes over time (Boehm & Flack, 2010).

Information about power in pigtailed macaque societies is encoded in a status signaling network. Power structure arises from a collective process in which each individual integrates over the status signals it has received to estimate how it is perceived by the group (Flack & Krakauer, 2006; Brush, Krakauer & Flack, in prep.). This signaling network arises in turn from an underlying aggression network, which in turn arises out of a social interaction network (see fig. 2.1). We sketch below the process generating each network and, ultimately, the power structure.

Individuals interact. An interaction is any event in which there is an opportunity for immediate contact or in which a signal has been exchanged. Interaction patterns vary in time, such that at any given moment some fraction of group members is in contact, proximity, or signaling from afar. Sampling these interactions at a regular interval gives a time series in which the successive "events" can be represented as interaction networks. The nodes in these networks correspond to individuals, and the presence of an (undirected) edge between two nodes indicates an interaction was observed. These time-sampled networks are sparsely connected. If we collapse these temporally resolved network data into a single static network, we can calculate the overall probability of individual i interacting with individual j. A simple measure, for example, of i's probability of interacting with j is the number of times i interacts (the weight of the ij edge) with j divided by the total number of times i interacts with all of its partners (i's degree).[10]

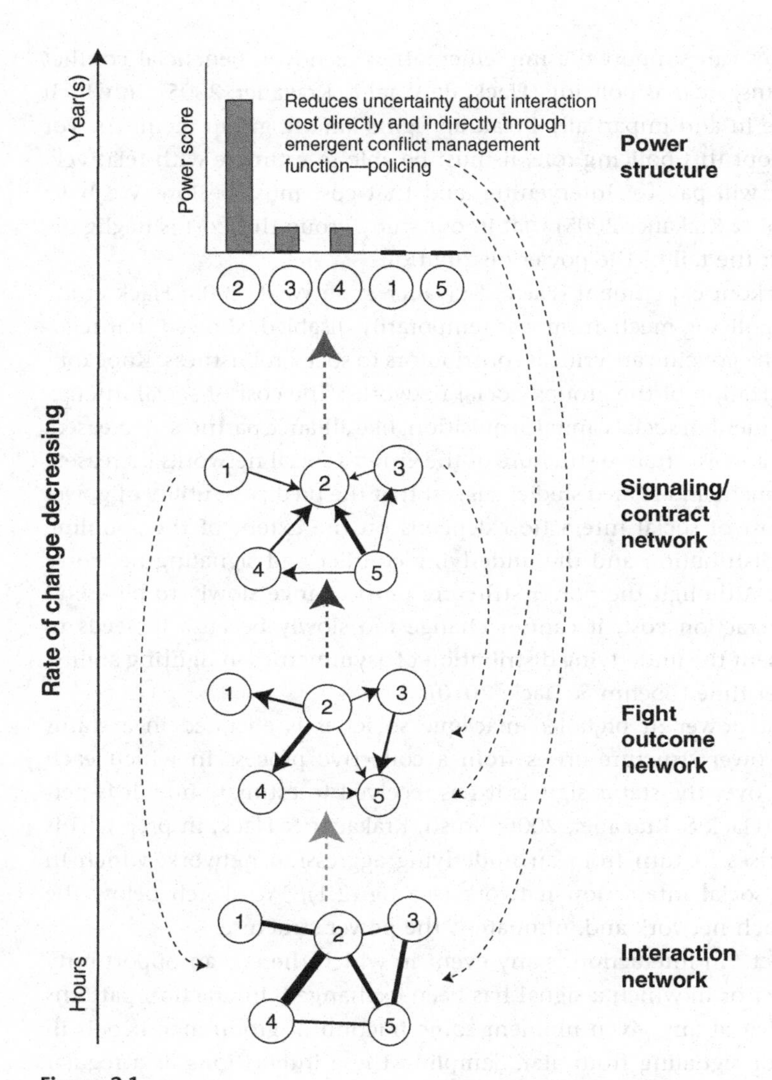

Figure 2.1
Schematic illustrating the dynamics and proliferation of temporal scales underlying the consolidation of power structure and the emergence of a new conflict management function through the buildup and amplification of asymmetries resulting from competitive interactions among individuals. See text for details.

Macaques fight for dominance and other resources. Whether individuals win or lose fights depends on temporally stable factors, including body size, fighting experience, and size of alliance networks. Contextual factors, including fatigue, variation in priorities, leverage, the presence of coalition partners, and immediate past successes or failures in fights (reviewed in Boehm & Flack, 2010), are also important insofar as they generate stochasticity in fight outcomes. The average probability of winning can be defined as some function of the ability of the opponents to perceive asymmetries in fighting ability due to body size, experience, size of alliance network, and so on, and to correctly estimate the implications of these asymmetries weighted by information obtained from a past history of fighting (see also Preuschoft & van Schaik, 2000).

In the absence of any prior information, each individual starts with a 50–50 chance of winning. This symmetry will be broken once a fight history has accumulated through direct experience or observations confirming any perceived differences in body size or fighting ability have been informative.[11] Presumably symmetry will be broken faster (e.g., it will take fewer fights to verify) the larger the inherent asymmetry.[12]

This process results in a fight outcome network. The nodes in this network correspond to individuals. The presence of an edge means that a fight occurred between node i and node j. The edge flows toward the individual who lost the fight. If the fight was a draw, the edge is undirected. The edge can be annotated with a vector containing elements that are updated with respect to the sampling interval, with the first element giving the number of fights, and the second element giving the number of fights won by individual i. Because we are interested in the rate at which fight outcomes reverse, the time evolution of the network must be tracked. Hence the aggregated network is computed from many network snapshots collected (for example, on an hour timescale) over the course of the study.

Once a pattern of losing exceeds a certain threshold, an expressed (as opposed to an inherent) asymmetry is established. In our study group, the individual perceiving itself as likely to lose signals this recognition using what is called a subordination signal (de Waal, 1982, 1986; Preuschoft & van Schaik, 2000; Flack & de Waal, 2007). In pigtailed macaques the signal is a peaceful variant of the silent bared-teeth display (Flack & de Waal, 2007). Subordination signals are unidirectional, meaning only the animal perceiving itself as the subordinate emits them (de Waal, 1986; de Waal & Luttrell, 1985; Flack & de Waal, 2007). Our analyses suggest that subordination signals communicate agreement to a primitive social contract in which the signaler agrees for some time period to the subordinate role, and thus to yield when a conflict arises in the future (Flack & de Waal, 2007). The contract, which is cost-free (Bergstrom & Lachmann, 1998), is upheld as long as two conditions are satisfied: The signaler must yield during periods of scarce resources or when the receiver expresses an interest in a resource, and the underlying asymmetry must continue to be perceived by the subordinate as large. The advantage of the contract is that it establishes a new conditional symmetry, in which the sender and receiver are free to interact with a reduced concern that a fight will erupt—a form of context-specific equality (Flack & de Waal, 2007).

These signaling interactions generate the slowly changing status-signaling network, by which we mean that edges in the signaling network are deleted, added, and reversed relatively slowly compared to the rate at which fights erupt or individuals interact. As with the fight outcome network one level down, the nodes in this network correspond to individuals. The presence of an edge indicates that a signaling event occurred between node *i* and node *j*. The edge flows to the receiver of the signal and is weighted by the number of signals exchanged.[13,14] Because we are interested in the rate at which contracts change and the number of signals emitted by the subordinate under a given contract, the time evolution of the network must be tracked. Hence the aggregated network is computed from many network snapshots collected (on, for example, the hour timescale) over the course of the study.

Our data suggest that it takes several reversals of edges in the fight network before the signal is withheld, and many more before signal sender and receiver reverse their contract (de Waal, 1982; de Waal & Luttrell, 1985; de Waal, 1986). This means that statistical features—for example, the rate of edge flipping and network level statistics, such as mean weighted in-degree—of the signaling network should be relatively impervious to fluctuations in fight outcomes (Boehm & Flack, 2010).

Encoded in the signaling network is information about the degree of consensus among group members that any individual is capable of successfully using force (Flack & Krakauer, 2006; Brush, Krakauer & Flack, in prep.). We have shown that individuals, by taking into account how many signals they receive in total from their population of signalers weighted by an estimate of the diversity of the signaling population, can estimate (using an appropriate heuristic) how much power others collectively perceive them to have. They also seem to be capable of roughly estimating the population distribution of power in the group (Flack, de Waal & Krakauer, 2005; Flack & Krakauer, 2006).

Three factors appear to conspire to generate power structures with slowly changing statistical features (e.g., mean, variance, kurtosis, rank order of individuals in the distribution). First, many fight reversals have to "build up" to reverse each subordination contract. Second, multiple pairs of individuals need to change whether they have a contract or change the "direction" of that contract. And, finally, the probability of a reversal occurring is reduced by virtue of the fact that the contracts reduce the number of agonistic interactions, which slows the rate at which a history can accumulate that supports the inverse pattern of asymmetry (Flack & de Waal, 2007).

Hence we have three, hierarchically organized networks—an interaction network, a fight outcome network, and a status signaling, or social contract, network—and a power distribution that is read off the signaling network.

Our data suggest that the advantage to building multiple networks is that their variable rates of change can be used to maximize objectives that would be at odds if there were only a single timescale (Flack & de Waal, 2007; Boehm & Flack, 2010). Individuals, by temporarily agreeing to a contract, can reference the contract for strategic decisions concerning the receiver (or dominant partner) rather than the rapidly changing and potentially misleading fluctuations in fight outcomes with that individual. Individuals can use information about their relative power encoded in the signaling network to make decisions about how to behave during polyadic conflicts, as described at the beginning of this case study. Allowing a low level of fighting to continue even after contracts and a stable power structure have been established is advantageous because it allows learning to continue and so provides a mechanism by which the social contract can be reversed (and relative power can change) when new asymmetries have been established.

In summary, the value of a social slow variable such as power lies in the fact it reduces uncertainty about the cost of social interactions. In doing so, it allows for more efficient social, and presumably ecological, resource extraction, thereby facilitating the implementation of conflict management mechanisms that amplify these effects.

A Case Study from Paleoecology: Reefs as Ecological Aggregates

Ecological assemblages in the fossil record display greater long-term persistence than would be expected from the apparently transitory nature of many ecological communities. Such patterns of persistence can range from thousands to even several million years, despite ongoing environmental disturbances and species invasions (DiMichele et al., 2004; Brett & Baird, 1995; Miller, 1997). The interpretation of the data supporting this argument has been controversial, with some proponents favoring a high degree of species coevolution so that the assemblage acts as a superorganism (a Clemensian view) while others view the pattern as reflecting overlapping environmental requirements (a Gleasonian view) or an epiphenomenon of poor data analysis. The relationship between stability and diversity has long been controversial within ecology (Ives & Carpenter, 2007), and our intent here is not to review this debate. The paleoecological record, however, supports four factors that may explain the prolonged persistence and biodiversity of ecological communities (DiMichele et al., 2004): (1) coadapted ecological relationships and mutualisms that constrain the number of interacting species; (2) coexistence of species with strongly overlapping environmental requirement networks, so that species co-occur even in the absence of coevolved relationships; (3) biogeographic control of the regional species pool, so that abundant species are likely, simply on a statistical basis, to co-occur in ecological assemblages; and (4) the law of large numbers (Hubbell, 2001) that may produce patterns of apparent stability through the neutral model of biodiversity.

The hypothesis advanced in this chapter provides a different perspective on long-term ecological persistence, which we illustrate with a discussion of reefs. The concept of slow variables introduces another explanation for the persistence of ecological communities over evolutionary timescales: it may reflect the generation of ecological structures that respond to environmental changes over much longer timescales than individual organisms, and thus provide a degree of ecological stability. In other words, ecological structures such as reefs and forests, while they persist, influence the environment sufficiently to reduce environmental uncertainty and allow a variety of species to coexist. This results in an increase in biodiversity (fig. 2.2).

Although today reefs are commonly viewed as wave-resistant structures dominated by scleractinian corals and found in sunny, tropical climes, reefs have been produced for at least 2 billion years, far longer than scleractinian corals have been around. Throughout the history of life, reefs have been built by a variety of organisms, including microbial assemblages, sponges, a variety of corals, bryozoans, and bivalves, among other groups. In almost all cases reefs are spatially confined structures built by sessile organisms (and thus composed of carbonate) and are wave-resistant structures that stand above the surrounding sediment and consequently influence current patterns. Depending on the organisms that construct the reef, they often produce a complex three-dimensional structure with a host of diverse environments for other organisms (Wood, 1999). Thus reefs act as ecosystem engineers (Jones, Lawton & Shachak, 1997; Wright & Jones, 2006) to modify the environment of other species and enhance overall biodiversity. Both modern (e.g., the Great Barrier Reef, Australia; Belize) and fossil reefs are some of the largest biogenic structures on Earth, and can have an influence over thousands of square kilometers. Moreover, these structures can persist for thousands to tens of thousands of years. Remarkably, Pleistocene reef communities have remained stable across substantial glacially induced drops in sea level (Pandolfi, 1996). In addition, throughout the Phanerozoic (542 million years ago to today), reef communities have displayed ecological persistence on evolutionary timescales (Kiessling, 2005, 2009).

Initiation of a reef involves interaction between reef-constructing organisms and the substrate. For microbes, this may involve trapping and binding sediment or secreting carbonate to form a microbially bound structure. For many reef-forming animals such as corals, however, the critical interaction involves larval settling, often on hard substrate. The dominant ecological force thought to be shaping this process and coral recruitment more generally is competition among larva and juvenile coral on one hand, and macroalgae on the other (Dubinksy & Stambler, 2011). Many different mechanisms of competition have been reported, including preemption of space, basal encroachment, and shading (see papers in Dubinksy & Stambler, 2011).

As the reef develops, it attracts a variety of other organisms, many of which are preferentially associated with reef ecosystems and may be virtually absent in nonreef environments. This produces a trophic network. As reefs grow in size, they increasingly modify the surrounding environment, including wave and current activity, sedimentation

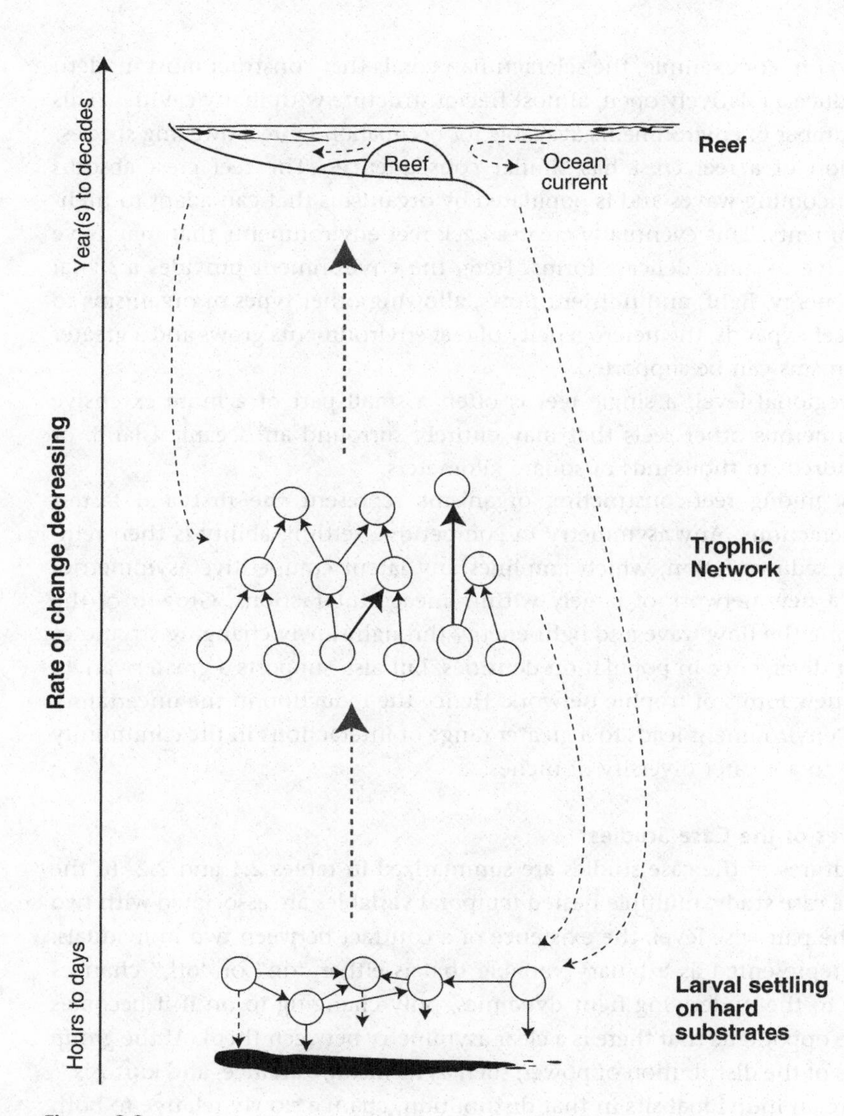

Figure 2.2
A schematic illustrating the dynamics by which reefs arise from density-dependent interactions among reef organisms. See text for details.

patterns, and so on. For example, the scleractinian corals that construct most modern reefs often produce a relatively open, almost fractal structure with many cavities. This increases the number of environments available for occupation by reef-dwelling species.

The formation of a reef crest has similar consequences. The reef crest absorbs the energy of incoming waves and is populated by organisms that can adapt to high-energy environments. This eventually creates back reef environments that may have cavities conducive to more delicate forms. Here, the environment provides a signal through wave energy, light, and nutrient flows, allowing other types of organisms to thrive. As the reef expands, the heterogeneity of reef environments grows and a greater variety of organisms can be supported.

At a more regional level, a single reef is often a small part of a more extensive platform of numerous other reefs that may entirely surround an oceanic island, or may cover hundreds to thousands of square kilometers.

Interactions among reef-constructing organisms represent the first and fastest network of interactions. Any asymmetry in competitive settling ability is then reinforced through sedimentation, which amplifies any latent competitive asymmetries and generates a new network of largely within-lineage interactions. Growth of the reef by modifying the flow wave and light energy through slowly changing structures leads to further divergence in population densities, but also supports a greater variety of species and new forms of trophic network. Hence the reduction in the uncertainty of the physical environment leads to a greater range of interactions in the community and eventually to a greater diversity of niches.

General Features of the Case Studies

The general features of the case studies are summarized in tables 2.1 and 2.2. In the power structure case study, multiple nested temporal variables are associated with two processes. At the pair-wise level, the existence of a contract between two individuals, which can be represented as a binary variable that is either "on" or "off," changes slowly relative to the underlying fight dynamics, only changing to on if it becomes apparent to the opponents that there is a clear asymmetry between them. At the group level, moments of the distribution of power, such as its mean, variance, and kurtosis,[15] as well as where an individual sits in that distribution, change slowly relative to both the timescale on which fight outcomes fluctuate and the timescale on which the contracts reverse.

In the reef case study, the hierarchical levels of reef ecosystems represent different networks of interaction: between substrate and larvae; among the species within a food web; between an individual reef and the surrounding environment, and beyond that, among individual reefs across a larger reef tract as they exchange larvae and collectively affect currents and other environmental variables (see fig. 2.2). As with the social networks, the reef interaction networks change over progressively slower

Table 2.1
Summary of the slow and fast variables for the power and reef case studies.

	Social Dynamics	Ecological Dynamics
Fast Variable	Fight outcomes	Ecological interactions (e.g. competition and cementation)
Slow Variable(s)	Contract and power distribution	Reef architecture and mass
Uncertainty Reduction	Interaction cost	Hydrodynamic energy

Table 2.2
Summary of the process by which symmetry is broken and the slow variable consolidated, shown for the power and reef case studies.

Symmetry Breaking	Social Dynamics	Ecological Dynamics	Summary
Stage 1:	Fights/time	Species abundance/time	Rapid rate of individual interactions, with randomly distributed competitive abilities
Stage 2:	Signal/time	Cementation/time	Individuals signal or secrete into environment as a function of outcome of stage 1.
Stage 3:	Power/time	Volume/time	Stochastic fluctuations fixed in stage 2 accumulate slowly in the form of social power or material mass, consolidating initial variation and amplifying competitive asymmetry through feedback to stages 1 and 2.

timescales. Larval settling happens on timescales of hours to days, albeit generally at certain times of the year. Individual trophic interactions occur on a timescale of days to weeks, although the resulting food webs may persist for years, depending on the changing composition of species and their relative abundances.

The macroscopic, physical structure of the reef can endure on a timescale of years or decades. Indeed, since many of the effects of the reef are a consequence of its size and mass, these effects would persist even if all organisms disappeared, until the reef slowly eroded. The volume or mass of a reef is thus a candidate slow variable, and the broader reef tract is an even slower variable on a larger spatial and temporal scale.

In both case studies, the construction of the slow variables reduces uncertainty about environmental states: in the power case, by increasing the predictability of social interaction cost, and in the reef case, by increasing predictability of wave energy and the hydrodynamic environment.

In both cases, the slow variables arise as asymmetries are established as competitive interactions grow. In the power example, the asymmetry builds up through memory of past outcomes of competitive interactions. Once it becomes clear that one individual in a pair is more likely to lose a fight (e.g., a threshold is crossed), the individual that perceives itself to be the likely loser signals. The signal can then be said to feed down to consolidate the history and reify the asymmetry. In the reef example, the asymmetry builds up through larval settling and cementation resulting from a complex competition dynamic between macroalgae and larvae and juvenile coral.[16]

Slow Variable Features, Functions, and Detection

Slow Variables as Macroscopic Variables

The simplest examples of macroscopic properties illustrative of the kinds of dynamical process generating slow variables come from the study of physical macroscopic variables like temperature, pressure, and entropy. These variables are averages over abundant fast variables, which correspond to many microstates and constitute macroscopic properties of a system (Callen, 1985). Temperature, for example, is an average over the microscopic motions of single particles. Technically it is related to an average over the energy in each degree of freedom of the particles (say, translational and rotational motion). The value of this average is that it allows us to think of temperature as an effective variable, which describes how heat is transferred via conduction, convection, and radiation. These are all statistical mechanisms constructed at the scale of the "effective" variable. For example, Fourier's law of heat conduction predicts how heat is transferred from high-temperature to low-temperature material without describing the individual particles. This represents a statistical law expressed through an equality that predicts the average behavior of a system—an effective degree of freedom—without attending to the microscopic degrees of freedom of the system. When we consider a number of macroscopic properties together, sufficient to explain the behavior of our system, they constitute a macrostate of our system.

Just as with physical systems, we can identify macroscopic variables in biological systems. Consider a social or ecological system with a given number of individuals or species. We can coarse-grain the variables (e.g., individuals can be grouped according to age, sex, and fighting ability, and species can be grouped into genera). We can then calculate suitable averages that form candidate macroscopic properties of our adaptive system. Collections of these properties that are predictive of future states of our system constitute candidate macrostates. For example, it might be that knowledge of the sex ratio, the population size, and the average age constitutes a very useful macrostate for predicting resistance to disease.

One problem is that in systems with numerous parts and properties, many different macroscopic functions can be calculated. As observers of the system, we may find many or all of these macroscopic functions to be of interest. However, only some of these properties will be useful coarse-grainings from the system's (or its components') perspective. In the case of biological and social systems, the macroscopic variables of interest are those against which system components can tune their strategies, either in evolutionary or ontogenetic time.

Slow variables are macroscopic states that (1) change slowly with respect to the underlying dynamics generating the state, (2) are associated with specific biological or behavioral updating mechanisms,[17] and (3) have features that change on a scale that can be *detected*, directly or through a proxy, by components in the system. For example, variability of glacial cycles over 10,000+ years is too long to be useful for adaptation for even the longest-lived organisms. Hence *slow variables can be understood as macroscopic variables with properties that allow for the evolution of parameters governing microstate behavior.*[18] In the power case, moments of the distribution of power, including its mean, variance, and kurtosis, change more slowly than changes in fight outcomes and the formation and dissolution of subordination contracts. In the case of reefs, reef properties such as mass change more slowly than the cementation processes that produce these features.

Uncertainty Reduction

As averages, macroscopic variables provide information about the future state of the system. Detection of these variables, or their proxies, by an adaptive system produces a reduction in uncertainty.[19] By detection, we simply mean that the variables can affect components in the system. Detection is required for adaptation, as illustrated with the above example on variability of glacial cycles, whether in evolutionary or ontogenetic time.

When the components in a system produce slowly changing features like reefs or power structures with detectable features, they essentially produce a regular, predictable environment. Hence the reef reduces variability in ocean currents, making them more predictable. It also ensures a diversity of niches and high biodiversity, compared with, say, the range of organisms that might be found in the open ocean. Power structure in monkey societies predicts the cost of social interaction and consequently can modulate the frequency and diversity of affiliative interactions.

Slow variables can serve as *reference states* (Flack & de Waal, 2007; Boehm & Flack, 2010): states that are insensitive to fluctuations at lower levels or in the environment and so serve as a good "data points" to use in decision making and adaptation. Ants navigate according to pheromone trails cumulatively deposited by other ants, and not by following single ants or invariant features of the environment. Monkeys use social contracts consolidated through status signal exchange, rather than fight outcomes, to decide how to interact. And corals and other reef organisms, by contributing to reef mass through cementation processes, create a predictable, resource-rich environment against which they can adapt.

It is our hypothesis that evolution produces temporal hierarchies (adaptive macrostates) because these provide economical, sufficient predictors of environmental states for organisms, thereby reducing uncertainty and facilitating adaptation. Hence collections of cells or individuals emerge as more efficient means of responding to environmental fluctuations.[20]

Whereas temperature and pressure are our way of efficiently describing and responding to changing environments, aggregates in biology are nature's way of doing so. This stands in contrast to the conventional view in evolutionary theory that—because aggregates typically reduce the reproductive output of lower levels—the origin of aggregation is a problem to be explained. If the above hypothesis is correct, aggregation under certain conditions is a *solution* to uncertainty.

Detecting Slow Variables

Can slow variables be optimized by the components building them? And, can the components evolve or learn better means for detecting properties of slow variables with the greatest predictive utility?

Consider the case of power. Studies of the consolidation of power structure, as well as studies of timescales in other systems (e.g., Fairhall et al., 2001; Honey et al., 2007; Kim et al., 2010; Shirvalkar, Rapp & Shapiro, 2010), suggest that an advantage to building multiple networks with multiple timescales is that their variable rates of change can be used to maximize objectives that would be at odds if there were only a single timescale (Flack & de Waal, 2007; Boehm & Flack, 2010). In the case of power, individuals, by temporarily agreeing to a contract, can reference the contract for strategic decisions concerning the receiver (or dominant partner) rather than the rapidly changing and potentially misleading fluctuations in fight outcomes with that individual. Individuals can use information about their relative power encoded in the signaling network to make decisions about how to behave during polyadic conflicts (see below). Of course maintaining a low level of fighting even after the contracts and power structure are established is useful because it allows learning to continue and so provides a mechanism by which the coupling between the power structure and fight distribution can be tuned.

We can summarize this process in more general terms. The process by which slow variables are constructed is collective: the slow variables arise over time through the joint behavior of multiple individuals or components acting strategically. The consolidated variable reduces social uncertainty by feeding down to direct component behavior along certain trajectories. This facilitates adaptation through learning dynamics or genetic inheritance. Ongoing variation in fast competitive processes percolates up to the slow processes, allowing for error-correction through either analog or digital mechanisms, so that slow variables continue to be useful predictors of the underlying distribution of traits or abilities, or environmental features as these underlying distributions shift in time or with changes to context.[21] In a concrete sense, the system is computing with multiple timescales.

The extent to which a slow variable is a good predictor of the average fast dynamics depends on how quickly changes at the microscopic level can percolate up. Hence we do not want too much timescale separation. On the other hand, reduced coupling between the aggregate properties of the system and the microscopic dynamics—which occurs as the aggregate variables change more and more slowly—can result in lock-in to a constructed environment that is suboptimal. The problem for the system then is to find the adaptive degree of coupling between the higher and lower levels (see also Fairhall et al., 2001).

This brings us to our second question: Can the coupling can be modulated through collective dynamics? To answer this question, we need to understand how components can evolve or learn better means for detecting slow variables. *The only slow variables that matter are those that the system can use*. Since the system is both using and constructing slow variables, we might call the process of finding the right slow variables a process of *endogenous coarse-graining*.[22]

In the case of power, individuals, after learning they are likely to lose, emit a signal that communicates agreement to a subordination contract. They then use the contract—through memory of the signaling event—to make decisions about how to interact, rather than using the most recent fight outcome, or recalling their perception of the entire history of wins and losses each time a decision needs to be made. In this case, it appears that making decisions using knowledge of the slow variable that summarizes the history of fights—the contract—is cognitively less demanding than remembering the entire history of events. However, the data suggest that stabilizing these contracts required the evolution of a signal that could effectively stand in for the fight history—more generally, the evolution of a mechanism by which the average could be updated once a critical threshold was passed. We do not yet understand how difficult it is to evolve such signals and mechanisms. The fact that in the existing data[23] these signals are not widespread among social primates with nonegalitarian structures suggests that this problem is nontrivial.

In addition to evolving mechanisms for communicating agreement to social contracts, individuals need means for estimating their power. Without some capacity to estimate power, information about power cannot be used to make social decisions. Before we can determine how individuals might make such an estimate, we need a principled statistical description of the distribution of power so that we know what it is they are trying to estimate. Recall that in systems with many parts and properties, many different macroscopic functions can be calculated. To test the utility of a coarse-graining, or in some cases, an average, we rely on the property of *statistical sufficiency* (Fisher, 1922).

A garden-variety statistic is simply a function of a sample where the function itself is independent of the sample's distribution. A sufficient statistic takes this idea further and is a statistic that is just as informative about some statistical parameter as any other statistic, and as informative as the complete sample from which the statistic was calculated. In the case of power we would like to determine whether some property of its distribution is a sufficient statistic for other variables of social importance. Once we identify a potential sufficient statistic, we can ask if the individuals in the system, given their information-processing capabilities, can estimate the statistic using a heuristic (e.g., Kahneman, Slovic & Tversky, 1982; Pearl, 1984) and local information, which serves as input to the computation.

Although many technical and conceptual issues are raised by framing detection of slow variables as a computational problem, the basic idea that individuals can use heuristics to estimate summary statistics is not problematic in the case of power or in social systems in which learning plays an important role.[24] It is less clear how computational ideas might help us understand adaptation to slow variables in systems such as reefs, in which strategy updating occurs in evolutionary time rather than ontogenetic time.

Reef organisms are unlikely to be making estimates of reef mass to determine the likely impact of ocean currents when making decisions about where to forage, take shelter, or mate. However, the genotype encodes the history of environmental states the organism experienced in evolutionary time. It can consequently be thought of as a hypothesis about the likely environmental state the organism will encounter (Krakauer & Zanotto, 2009), with the phenotype serving as the experiment to test a selective hypothesis. In this sense, information about slow variables such as reef mass can be said to be encoded in the genotype insofar as adaptations that arose in the context of large reefs are more useful on large reefs than small ones or in the open ocean. The extent to which this process is formally similar to the slow variable feature estimation in systems with learning is one of the many open questions that remain for future work.

The Consolidation of Slow Variables and Emergence of New Functionality through Symmetry-Breaking

For our purposes, symmetry is an invariance under a specified group of changes, which can be orientations, actions, or other properties. For example, competitive ability might be invariant on average assuming a change in identity.

The breaking of symmetry is often associated with the occurrence of some new phenomenon and an increase in effective complexity (Nicolis & Progogine, 1989). In physics, for example, superconductivity is thought to be critical to the emergence of new physical properties of a system (Tinkham, 2004). In the competitive example, if a small fluctuation in outcome meant that resources could be acquired by excluding others, thereby promoting an increase in competitive ability, this would represent a broken symmetry. It would also represent an increase in effective complexity and a decrease in Shannon entropy, as we will have transitioned from a state of maximum ignorance (a uniform distribution over abilities) to a state of minimum ignorance, where one competitor wins.

Spontaneous symmetry-breaking occurs when the laws or equations of a system are symmetrical but specific solutions do not respect the same symmetry. For example, in a superconductor, the symmetry possessed by free electrons in a material is broken at low temperatures because the lowest energy solution is one in which the electrons pair up. *Spontaneous* here simply means endogenous to the dynamics of the system and not catalyzed by some exogenous input. Explicit symmetry-breaking occurs when the rules governing a system are not manifestly invariant under the symmetry group considered. Over a century ago the geneticist Bateson suggested that as symmetry is broken, more information is expressed by an organism, and greater structure is observed. This is consistent with the idea from physics that symmetry-breaking is accompanied by a transition to a less probable but more structured state. In this process, fluctuations from mean behavior are important and can give rise to unexpected behavior that biases the equilibrium state of a system. In other words, asymmetry in a macrostate can emerge from the collective actions of local or transient symmetry-breaking in the microscopic degrees of freedom. Noisy outcomes of competition can lead to long-lasting inequalities. Perhaps the best-known example of this process in biology comes from the study of spontaneous symmetry-breaking during developmental pattern formation, introduced by Turing (1951).

We are currently exploring whether we can model the emergence of power structure as a symmetry-breaking process. The intuition is as follows. The initial condition is one of equivalence: two individuals who are unfamiliar with each other's fighting abilities think they are equally likely to lose a fight if a fight were to erupt between them. As a history of fighting and a corresponding memory for the outcomes build up, the individuals learn who in the pair is likely to lose. It is an open question whether the learning is the result of small fluctuations in fight outcomes and hence can be characterized as a process of spontaneous symmetry-breaking, or whether individuals are learning about inherent underlying asymmetries they did not initially perceive. In that case, the symmetry-breaking would be explicit.

Once an individual learns that it is likely to lose, symmetry is broken, and the relationship shows increased structure. Now there are two possible states (dominant and subordinate) instead of just one (adversarial). This new state is consolidated through the exchange of the signal, which changes slowly and so serves as a good reference state for decision making. Without it, there is only the memory for outcomes distributed over multiple brains, and so any state change that results from the buildup of the memory is not verifiable. As the network of subordination signals builds up, encoding a power structure, asymmetry is consolidated through feedback to the two lower levels. The power structure amplifies the asymmetries by changing the cost of social interactions. Although this increases the mean rate of interaction, the effect is more pronounced for higher-power animals, allowing them to pick up more signals.

Thus the statistics of the power structure, derived from the signals, stabilize around values that change on a much slower timescale than the underlying fight dynamics and rate at which contracts are reversed.

In the case study of the reef, the intuition goes as follows. In an unbroken soft substrate, reef initiation is dependent on a piece of shell, or a similar hard substrate on which coral larvae can settle. That hard substrate and the resulting larval settlement breaks the environmental symmetry, inducing a positive feedback process that may result in the formation of a larger structure and which can be amplified by the buildup in competitive asymmetries between larvae, juvenile coral and macroalgae. As the reef-builders construct a larger structure, they generate a variety of subenvironments, in effect progressively breaking the environmental symmetries. These new subenvironments contain information about environmental conditions, which influence either the morphology of the organisms that grow there (if they are phenotypically plastic, as many corals are) or their probability of persistence. This has the effect of consolidating the environmental signals through feedback to the trophic interactions and larval settling patterns. As in the social example (albeit on far longer timescales), this has the effect of stabilizing the reef architecture around structures that change on much slower timescales relative to the underlying ecological dynamics. In table 2.1 we summarize the slow and fast variables in the case studies. In table 2.2 we compare the symmetry-breaking processes in the two case studies.

Conclusions

We have hypothesized that the construction of slow variables, which increase the predictability of environmental variables, represents the first step in the evolution of hierarchical biological organization. Slow variables are macroscopic products of a system capable of feeding back to constrain the behavior of individual components. Slow variables arise through a process of symmetry-breaking, whereby either small fluctuations in competitive ability at the individual level are amplified through memory, generating long-lived asymmetries in competitive ability, or individuals come to learn underlying differences in fighting ability. The slow variables serve to increase the predictive ability of the system components by serving as aggregate reference states for behavior or strategy choice.

Identifying reliable macroscopic properties in most living systems is a nontrivial problem for both observers and the system itself. The system might not have the search capacity or computational power to identify predictive features. In the case of power, if individuals do not interact frequently enough to build up a long enough history of fights, symmetry-breaking might not occur and the slow variable social contract never arises. These kinds of problems raise various empirical and theoretical questions. For example, a sizable data set is required to validate that a posited slow variable is actually a good predictor. These data are almost never available to observers and are almost certainly not available to components of the system itself. Hence we must develop concepts that reflect both mathematical rigor and biological parsimony: We need to take into account the largest data set that can be recorded and processed by any component of the system given its cognitive capacity.

The formal challenge presented by the theory of slow variables is therefore fourfold. First, the time constants of posited slow variables must be measured quantitatively and their associated statistics must be shown empirically to constitute informative macroscopic variables. By this we mean that they are computable or estimable by individuals (or perhaps species) so that they can be used for decision making or to inform strategy choice in either evolutionary or ontogenetic time. Second, formal models for competitive symmetry-breaking in biological and social systems are required in order to develop a mechanistically grounded theory for slow variable evolution. Finally, the mechanistic theory should predict time constants for the slow variables given factors like the degree of environmental uncertainty, life span of individuals in a population, and the complexity of their perceptual systems. Hence the kinds of issues that need to be addressed in future work include (i) development of good measures for decomposing networks and time formally describing the role of spontaneous symmetry-breaking in hierarchy origination, (ii) evaluating the formal relation between slow variables and adaptive order parameters describing adaptive macrostates, (iii) exploring the adaptive benefits of reducing environmental uncertainty through suitable averaging processes, and (iv) given trade-offs between robustness and evolvability, predicting how much timescale separation across the emerging hierarchy of levels is adaptive in the long term.

Acknowledgments

We thank Dan Rockmore, Karen Page, Simon DeDeo, Fred Cooper, Walter Fontana, Olaf Sporns, and Eric Smith for helpful discussion, and Dan Rockmore, Brett Calcott, Kim Sterelny, Richard Joyce, and Ben Fraser for their comments on an early draft of this chapter. Support for this project was provided by NIH Grant 1 R24 RR024396.

Notes

1. An adaptive system can be said to have memory any time there is a mechanism by which pattern can be extracted from a time series or network, encoded and stored, such that the representation of the pattern changes slowly compared to the time series or network itself.

2. We consider a behavior or process to be competitive if it reduces the density of another type of individual, organism, or species.

3. For a discussion of individuality as it relates to autonomy, see Bertschinger, Olbrich, and Ay (2008).

4. The pattern formation work in development is not conventionally considered to address the levels question, but to us it seems quite relevant.

5. Family, group, species, and clade reflect different degrees of assortative mixing in space and time.

6. Social niche construction describes the process whereby individuals through their interactions with one another coconstruct their social environments, building social structures and changing the social selection pressures to which they are subject.

7. A time series is a sequence of data points measured at successive times, ideally at uniform intervals if the data are sampled. If the data comprise a complete or nearly complete representation of the events that occurred over a given observation period, the spacing between data points reflects the spacing between the events that the time series describes, as perceived by the data collector. A high-dimensional time series is one in which each time step is associated with a vector, which captures multiple features of the event, rather than a single data point.

8. An emergent property has the following characteristics: (1) the process generating it involves compression or coarse-graining of lower-level dynamics and (2) the coarse-graining is associated with some surprise value (e.g., a new function). It remains under debate whether a phase transition is required or consolidation of a slow variable is sufficient.

9. This sketch of what is presently understood about how power structures emerge in macaque societies is based on a series of studies by Flack, Krakauer, and collaborators on agonistic interactions, status signaling, social networks, and conflict management in pigtailed macaques and what is generally understood about macaque dominance relations (see Thierry, Singh & Kaumanns, 2004).

10. These probabilities can be corrected for variability across individuals in interaction rate.

11. We discuss symmetry breaking in the section entitled "The Consolidation of Slow Variables and the Emergence of New Functionality through Symmetry Breaking."

12. Inherent asymmetry means the actual difference in size, fighting ability (assessed using independent data), or alliance network, regardless of how this is perceived by either opponent.

13. The sampling procedure for collecting signals should be the same for all signalers.

14. Although the signals are given infrequently compared to the interaction rate, they are repeated. The number of signals given by the subordinate to the dominant appears to contain information about the perceived magnitude of the expressed asymmetry (Flack & Krakauer, 2006).

15. Kurtosis is the skewedness of a distribution.

16. It should be noted that the details of the role of this process in reef formation are less well understood than they are in the case of pigtailed macaque power structures. Hence the longer case study for power.

17. The subordination contracts have an additional feature worth noting. The contracts are not simply averages over the aggression dynamics that change continuously. They are more like digital

place-holders, as they require a long history of fighting to be built up and a threshold-level of asymmetry to be established before a contract can be established or one can be substituted for another.

18. We discuss the issue of detection in greater detail in the section entitled "Detecting Slow Variables."

19. For a similar view on the role of uncertainty reduction in the evolution of collective behavior, see Bettencourt (2009).

20. Biological systems also have spatial scales. Whether a similar logic can account for spatial hierarchies is an open question.

21. For a similar explanation for the range of timescales associated with neural codes, see Fairhall et al. (2001).

22. "Endogenous coarsegraining" is a term also used by Fontana and colleagues in their search for a natural representation of molecular dynamics (Feret et al., 2009). Crutchfield (1994) has referred to this process as "intrinsic emergence."

23. A caveat: Researchers look for submission/subordination signals almost exclusively in the context of aggression. Our studies (Flack & de Waal, 2007) suggest that submission signals take on the meaning of subordination when used outside the context of aggression and that researchers should also be looking for such signals in peaceful contexts.

24. We are currently exploring how well individuals can estimate their power using heuristics, as well as how robust the predictive utility of the power structure is to estimation errors. In other words, how "off" can an individual's estimate be before the estimate is no longer informative about its power or predictive of its cost during social interactions?

References

Aureli, F., & de Waal, F. B. M. (Eds.). (2000). *Natural conflict resolution*. Berkeley: University of California Press.

Baer, C. F., Miyamoto, M. M., & Denver, D. R. (2007). Mutation rate variation in multicellular eukaryotes: Causes and consequences. *Nature Reviews: Genetics*, *8*, 619–631.

Bergstrom, C. T., & Lachmann, M. (1998). Signaling among nonrelatives. II. Talk is cheap. *Proceedings of the National Academy of Sciences of the United States of America*, *95*, 5100–5105.

Bertschinger, N., Olbrich, E., Ay, N., & Jost, J. (2008). Autonomy: An information-theoretic perspective. *Bio Systems*, *91*, 331–345.

Bettencourt, L. M. A. (2009). The rules of information aggregation of collective intelligent behavior. *Topics in Cognitive Science*, *1*, 598–620.

Bierstedt, R. (1950). An analysis of social power. *American Sociological Review*, *15*, 730–738.

Boehm, C., & Flack, J. C. (2010). The emergence of simple and complex power structures through social niche construction. In A. Guinote & T. Vescio (Eds.), *The social psychology of power* (pp. 46–86). New York: Guilford Press.

Borenstein, E., Kendal, J., & Feldman, M. (2006). Cultural niche construction in a meta-population. *Theoretical Population Biology, 70*, 92–104.

Brett, C. E., & Baird, G. C. (1995). Coordinated stasis and evolutionary ecology of silurian to middle devonian faunas in the appalachian basin. In D. H. Erwin & A. L. Anstey (Eds.), *New approaches to speciation in the fossil record* (pp. 285–315). New York: Columbia University Press.

Brush, E., Krakauer, D.C., and Flack, J.C. (In preparation). Social computation: Consensus processes generating power structure under imperfect information.

Buss, L. W. (1987). *The evolution of individuality*. Princeton, NJ: Princeton University Press.

Callen, H. B. (1985). *Thermodynamics and an introduction to thermostatistics* (2nd Ed.). New York: Wiley.

Campbell, D. T. (1990). Levels of organization, downward causation, and the selection-theory approach to evolutionary epistemology. In G. Greenberg & E. Tobach (Eds.), *Theories of the evolution of knowing* (pp. 1–15). T. C. Schneirla Conference Series. Hillsdale, NJ: Erlbaum.

Couzin, I. (2009). Collective cognition in animal groups. *Trends in Cognitive Sciences, 13*, 36–43.

Crutchfield, J. (1994). The calculi of emergence: Computation, dynamics, and induction. *Physica D: Nonlinear Phenomena, 75*, 11–54.

Davidson, E. H. (2009). Network design principles from sea urchin embryo. *Current Opinion in Genetics & Development, 19*, 535–540.

Davidson, E. H., & Levine, M. (2008). Properties of developmental gene regulatory networks. *Proceedings of the National Academy of Sciences of the United States of America, 325*, 20063–20066.

Davidson, E. H., Rast, J. P., Oliveri, P. Ransick, A., et al. (2002). A genomic regulatory network for development. *Science, 295*, 1669–1678.

Dawkins, R. (1982). *The extended phenotype*. San Francisco: W. H. Freeman.

de Waal, F. B. M. (1982). *Chimpanzee politics: Power and sex among the apes*. London: Jonathan Cape.

de Waal, F. B. M. (1986). The integration of dominance and social bonding in primates. *Quarterly Review of Biology, 61*, 459–479.

de Waal, F. B. M., & Luttrell, L. M. (1985). The formal hierarchy of rhesus monkeys: An investigation of the bared-teeth display. *American Journal of Primatology, 9*, 73–85.

DiMichele, W. A., Behrensmeyer, A. K., Olszewski, T. D., Labandeira, C. C., Pandolfi, J. M., Wing, S. L., & Bobe, R. (2004). Long-term stasis in ecological assemblages: evidence from the fossil record. *Annual Review of Ecology Evolution and Systematics, 35*, 285–322.

Dubinksy, Z., & Stambler, N. (Eds.). (2011). *Coral reefs: An ecosystem in transition*. Dordrecht: Springer.

Erwin, D., & Davidson, E. H. (2009). The evolution of hierarchical gene regulatory networks. *Nature Reviews: Genetics, 10*, 141–148.

Fairhall, A. L., Lewen, G. D., Bialek, W., & de Ruyter van Steveninck, R. R. (2001). Efficiency and ambiguity in an adaptive neural code. *Nature, 412*, 787–792.

Feldman, M., & Eshel, I. (1982). On the theory of parent-offspring conflict: A two-locus genetic model. *American Naturalist, 119*, 285–292.

Feret, J., Danos, V., Krivine, J., Harmer, R., & Fontana, W. (2009). Internal coarse-graining of molecular systems. *Proceedings of the National Academy of Sciences of the United States of America, 106*, 6453–6458.

Fisher, R. A. (1922). On the mathematical foundations of theoretical statistics. *Philosophical Transactions of the Royal Society of London, Series A, 222*, 309–368.

Flack, J. C., & de Waal, F. B. M. (2007). Context modulates signal meaning in primate communication. *Proceedings of the National Academy of Sciences of the United States of America, 104*, 1581–1586.

Flack, J. C., de Waal, F. B. M., & Krakauer, D. C. (2005). Social structure, robustness, and policing cost in a cognitively sophisticated species. *American Naturalist, 165*, E126–E139.

Flack, J. C., Girvan, M., de Waal, F. B. M., & Krakauer, D. C. (2006). Policing stabilizes construction of social niches in primates. *Nature, 439*, 426–429.

Flack, J. C., & Krakauer, D. C. (2006). Encoding power in communication networks. *American Naturalist, 168*, 97–102.

Flack, J. C., & Krakauer, D. C. (2009). The evolution and construction of moral systems. In S. Levin (Ed.), *Games, groups, and the global good*. Princeton, NJ: Princeton University Press.

Flack, J. C., Krakauer, D. C., & de Waal, F. B. M. (2005). Robustness mechanisms in primate societies: A perturbation study. *Proceedings of the Royal Society of London, Series B: Biological Sciences, 272*, 1091–1099.

Frank, S. (2003). Repression of competition and the evolution of cooperation. *Evolution: International Journal of Organic Evolution, 57*, 693–705.

Frank, S. (2009). Evolutionary foundations of cooperation and group cohesion. In *Games, groups, and the global good*. Berlin: Springer.

Gould, S. J. (2002). *The structure of evolutionary theory*. Cambridge, MA: Harvard University Press.

Gould, S. J., & Eldredge, N. (1993). Punctuated equilibrium comes of age. *Annual Review of Ecology Evolution and Systematics, 366*, 223–227.

Hamilton, W. D. (1975). Innate social aptitudes of man: An approach from evolutionary genetics. In R. Fox (Ed.), *Biosocial anthropology* (pp. 133–153). London: Malaby Press.

Fox, R. (Ed.). (1975). *Bioosocial anthropology*. London: Malaby Press.

Honey, C. J., Kotter, R., Breakspear, M., & Sporns, O. (2007). Network structure of cerebral cortex shapes functional connectivity on multiple timescales. *Proceedings of the National Academy of Sciences of the United States of America, 24,* 10240–10245.

Hubbell, S. P. (2001). *The united neutral theory of biodiversity and biogeography*. Princeton, NJ: Princeton University Press.

Ishii, K., Matsuda, H., Iwasa, Y., & Sasaki, A. (1989). Evolutionary stable mutation rate in a periodically changing environment. *Genetics Society of America, 121,* 163–174.

Ives, A. R., & Carpenter, S. R. (2007). Stability and diversity of ecosystems. *Science, 317,* 58–62.

Jablonka, E., & Lamb, M. (2005). *Evolution in four dimensions: Genetic, epigenetic, behavioral, and symbolic variation in the history of life*. Cambridge, MA: MIT Press.

Jablonksi, D. (2000). Micro-and macroevolution: scale and hierarchy in evolutionary biology and paleobiology. *Paleobiology, 26* (Supplement to Issue 4):15–52.

Jablonksi, D. (2007). Scale and hierarchy in macroevolution. *Paleontology, 50,* 87–109.

Jackson, J. B. C. (1994). Constancy and change of life in the sea. *Philosophical Transactions of the Royal Society of London, Series B: Biological Sciences, 344,* 55–60.

Jones, C. G., Lawton, J. H., & Shachak, M. (1997). Positive and negative effects of organisms as physical ecosystem engineers. *Ecology, 78,* 1946–1957.

Kahneman, D., Slovic, P., & Tversky, A. (Eds.). (1982). *Judgment under uncertainty: Heuristics and biases*. Cambridge: Cambridge University Press.

Kiessling, W. (2005). Long-term relationships between ecological stability and biodiversity in phanerozoic reefs. *Nature, 433,* 410–413.

Kiessling, W. (2009). Geologic and biological controls on the evolution of reefs. *Annual Review of Ecology Evolution and Systematics, 40,* 173–192.

Kim, J. R., Shin, D., Jung, S. H., Heslop-Harrison, P., & Cho, K. H. (2010). A design principle underlying the synchronization of oscillations in cellular systems. *Journal of Cell Science, 123,* 537–543.

Krakauer, D. C., & Flack, J. C. (2010). Better living through physics. *Nature, 467.*

Krakauer, D. C., Page, K., & Erwin, D. H. (2009). Diversity, dilemmas, and monopolies of niche construction. *American Naturalist, 2009,* 26–40.

Krakauer, D. C., & Zanotto, A. P. (2009). Virality and the limitations of the life concept. In S. Rassmussen, M. A. Bedau, X. Chen, D. Deamer, D. C. Krakauer, N. H. Packard, et al. (Eds.), *Protocells: Bridging non-living and living matter* (pp. 513–536). Cambridge, MA: MIT Press.

Kylafis, G., & Loreau, M. (2008). Ecological and evolutionary consequences of niche construction for its agent. *Ecology Letters, 11,* 1072–1081.

Laland, K., & Sterelny, K. (2006). Seven reasons not to neglect niche construction. *Evolution: International Journal of Organic Evolution, 60*, 1751–1762.

Lewontin, R. C. (1982). Organism and environment. In E. C. Plotkin (Ed.), *Learning, development, and culture*. London: Wiley.

Manning, G., Plowman, G. D., Hunter, T., & Sudarsanam, S. (2002). Evolution of protein kinase signaling from yeast to man. *Trends in Biochemical Sciences, 10*, 514–520.

Maynard Smith, J., & Harper, D. G. C. (1995). Animal signals: Models and terminology. *Journal of Theoretical Biology, 177*, 305–311.

Maynard Smith, J., & Szathmary, E. (1995). *The major transitions in evolution*. Oxford: W. H. Freeman Spektrum.

Michod, R. E. (2000). *Darwinian dynamics: Evolutionary transitions in fitness and individuality*. Princeton, NJ: Princeton University Press.

Michod, R. E. (2007). Evolution of individuality during the transition from unicellular to multicellular life. *Proceedings of the National Academy of Sciences of the United States of America, 104*, 8613–8618.

Miller, A. I. (1997). Coordinated stasis or coincident relative stability. *Paleobiology, 23*, 155–164.

Nicolis, G., & Prigogine, I. (1989). *Exploring complexity: An introduction*. New York: W. H. Freeman.

Nowak, M., Tarnita, C. E., & Wilson, E. O. (2010). The evolution of eusociality. *Nature, 466*, 1057–1062.

Odling-Smee, J., Laland, K., & Feldman, M. (Eds.). (2003). *Niche construction: The neglected process in evolution*. Princeton, NJ: Princeton University Press.

Okasha, S. (2006). *Evolution and the levels of selection*. Oxford: Oxford University Press.

Palmer, A. R. (2004). Symmetry breaking and the evolution of development. *Science, 306*, 828–833.

Pandolfi, J. M. (1996). Limited membership in Pleistocene reef coral assemblages from the Huon Peninsula, Papua New Guinea: Constancy during global change. *Paleobiology, 22*, 152–196.

Pearl, J. (1984). *Heuristics*. Boston: Addison-Wesley.

Preuschoft, S., & van Schaik, C. (2000). Dominance and communication. In F. Aureli & F. B. M. de Waal (Eds.), *Natural conflict resolution* (pp. 77–105). Berkeley: University of California Press.

Roussett, F. (2004). *Genetic structure and selection in subdivided populations*. Princeton, NJ: Princeton University Press.

Santelices, B. (2008). How many kinds of individual are there? *Trends in Ecology & Evolution, 14*, 152–155.

Schuster, P., & Fontana, W. (1998). Continuity in evolution: On the nature of transitions. *Science, 280*, 1451–1455.

Shirvalkar, P. R., Rapp, P. R., & Shapiro, M. L. (2010). Bidirectional changes to hippocampal theta-gamma comodulation predict memory for recent spatial episodes. *Proceedings of the National Academy of Sciences of the United States of America, 107*, 7054–7059.

Smith, D. E. (2010). The interaction of diversity and transience with universality in the complexity of the biosphere. SFI Working Paper Series.

Tinkham, M. (2004). *Introduction to superconductivity*. Mineola, NY: Dover.

Thierry, B. (2000). Conflict management patterns across macaque species. In F. Aureli & F. B. M. de Waal (Eds.), *Natural conflict resolution* (pp. 106–128). Berkeley: University of California Press.

Thierry, B., Singh, M., & Kaumanns, W. (Eds.). (2004). *Macaque societies: A model for the study of social organization*. Cambridge: Cambridge University Press.

Turing, A. (1951). The chemical basis of morphogenesis. *Philosophical Transactions of the Royal Society of London, Series B: Biological Sciences, 237*, 37–72.

Valentine, J. W., & May, C. L. (1996). Hierarchies in biology and paleontology. *Paleobiology, 22*, 23–33.

Wood, R. (1999). *Reef evolution*. Oxford: Oxford University Press.

Wright, J. P., & Jones, C. G. (2006). The concept of organisms as ecosystem engineers ten years on: Progress, limitations, and challenges. *Bioscience, 56*, 203–209.

3 On Depending on Fish for a Living, and Other Difficulties of Living Sustainably

Hanna Kokko and Katja Heubel

Mark Kurlansky's *Cod: Biography of the Fish That Changed the World* (Kurlansky, 1997) is one of those books that you wish every decision maker and policy planner would read. On its pages we read a description of how for centuries, Basque seafarers brought in salted and dried cod from nobody knew where (except for the Basques themselves). By the time explorers like John Cabot described their "New Found Land" for the king, their findings were certainly not news for the Basques who had found that land ages ago (and had been using it for drying cod) but preferred keeping their fishing secrets to themselves.

The subsequent colonization of the east coast of Canada ("New France" as it was called then) was massively influenced by the abundance of cod in the Great Banks region in particular. A legend of the time stated that a person could walk across the water on their backs, and Cabot's statements were similar, up to the point of complaining that the schools of cod were so thick they slowed the speed of the ship. Cod in this part of the world may indeed have been superabundant, but similar statements about a virtually unlimited resource were made centuries later, even when the improved efficiency of vessels led to discussions that fisheries might have to be regulated.

In 1882 Great Britain, Germany, France, Denmark, and Belgium signed the North Sea Fisheries Convention, but eminent scientists of the time expressed views that such developments were unnecessary in principle. Thomas Henry Huxley, the man known as "Darwin's Bulldog" as he never hesitated to defend the idea of natural selection in public, also served as the Inspector of Fisheries and on several Royal Commissions dealing with fisheries. In his inaugural address to the Fisheries Exhibition in London 1883, he stated that there certainly are conditions where sustainability requires laws and agreements:

> all river fisheries . . . can be exhausted by man because man is, under ordinary circumstances, one of the chief agents of destruction; and, for the same reason, its exhaustion can usually be prevented, because man's operations may be controlled and reduced to any extent that may be desired by force of law.

But then he carried on to state that seas are simply too large for such an argument to apply:

> I believe, then, that the cod fishery, the herring fishery, the pilchard fishery, the mackerel fishery, and probably all the great sea fisheries, are inexhaustible; that is to say, that nothing we do seriously affects the number of the fish. And any attempt to regulate these fisheries seems consequently, from the nature of the case, to be useless.

Huxley may have been one of the quickest to accurately assess the value of Darwinian thinking, but his estimate of the ability of humans to exploit their natural resources was too low. Fast-forwarding to the 1990s, Huxley's position on river fisheries now also applies to marine fisheries. By now the Canadian cod stock had collapsed, forcing a 1992 decision by the Canadian government to impose a moratorium on all northern cod fisheries. Approximately 30,000 fishermen became unemployed overnight. To this date, the cod has not recovered sufficiently for commercial fishing to resume. (The Fisheries Museum of the Atlantic, based in Lunenburg, Nova Scotia, intriguingly does not appear to want to convey this sobering message; the closest it gets is to allude to bigger and fancier vessels and equipment being needed in the ever-toughening conditions that fishermen face.)

The collapse of the northern cod is a sobering tale of what happens to an overexploited public good. The conceptual framework for understanding what happens dates back to Hardin's (1968) famous essay on the tragedy of the commons. An abundance of fish, clean air, or an uncongested road are examples of *public goods*: something that is essentially freely available to anyone, but requiring some contributions to maintain (where a "contribution" may simply mean restraint from overexploiting). Public goods are not easy to maintain when it is more profitable for each user to exploit the goods a little more—or much more—than what the system can maintain in the long term. Hardin (1968) saw no other solution to this problem than "mutual coercion, mutually agreed upon." He also specifically stated that he did not mean "coercion" to be a dirty word; he was simply pointing out that we need to be cooperative, and that this may be best achieved by agreeing on suitable sanctions.

A century after Huxley spoke about regulation being unnecessary and irrelevant, it has certainly become accepted that the potentially dirty word (though revamped as "management" and "negotiation" rather than "coercion") has to be employed if fish stocks are to be maintained for future generations (Worm et al., 2009). Humans possess capabilities of intellect and cooperation like no other species on this planet, and indeed much of this volume is devoted to understanding how we accomplish such feats. Intellect and cooperation are precisely what is needed when managing a public good such as cod: It is obvious that avoiding overfishing requires restraint, which in turn relies on being able to learn from the past and/or extrapolate to predict the consequences of our future actions. It also relies on the ability to negotiate mutually binding agreements. Regardless of *why* we evolved large brains, it appears that we are uniquely equipped to deal with the sort of problems that the ever-increasing efficiency of fishing vessels (in itself a product of human intellect) throws at us.

But how much have we learned from the past? And how much are we able to cooperate? It is instructive to look at what has happened in European waters since the Canadian moratorium. The International Council for the Exploration of the Sea (ICES) has, since 2002, recommended that the EU ban cod fishing in the North Sea. In 2004, the repetition of this recommendation led to an EU decision that quotas will be diminished by 15 percent *if* North Sea stocks fall below 70,000 tons. In 2006, politicians faced a new estimate of 50,000 tons. A new meeting in December 2006 led to the decision that the quota should indeed be reduced, not quite as much as promised in 2004, but by 14 percent. Obviously, much negotiation effort goes into achieving even modest conservation measures, but this complexity should not cloud the fact that goalposts are shifting in a way that is clearly detrimental for long-term sustainability (and the consequent livelihood of fishermen, too). Reports from other fisheries do not show unanimous gloom and doom, but true recovery always appears rather hard to reach (Worm et al., 2009). But then, Hardin (1968) did not predict that this would be easy.

Evolutionary Arguments

There is something curious about our inability to see the big picture—or rather, our failure to act even in those cases where the big picture is quite easy for us to see. It emerges from the efforts of fisheries, scientists, and from experience. In some ways, however, it is not surprising that humans act in such uncooperative ways favoring the individual profit over the collective long-term optimum. Humans, like any animal, are products of evolution, and evolution is known to be a process that typically fails to optimize system performance.

In some respects, such a statement can appear odd. In 1930, R. A. Fisher published a book containing his famous *fundamental theorem of natural selection*, according to which natural selection will lead to increasing fitness in a population at a rate equal to genetic variance in fitness. Sewall Wright's walk on the adaptive landscape, where populations reach local peaks, is a similar concept (Wright, 1932). Of course, caveats apply before one can trust that evolution produces organisms that are more fit: The fundamental theorem has the immediate corollary that genetic variation has to be present on which natural selection can act. If there isn't, then the fundamental theorem correctly predicts that evolutionary change will be zero, no matter how much better (nonexistent) alternative genotypes would perform. Another caveat is that many genes have pleiotropic effects, which means that a single gene can have effects on multiple traits of an organism. As a consequence, it is difficult to select for, say, large wings without changing the overall body size. If the best evolutionary outcome happens to be a creature with large wings attached to a small body, achieving this may take quite a bit of time.

Although such constraints on evolutionary speed are much discussed in the contemporary evolutionary literature (e.g., Kirkpatrick, 2010), they actually do not form the main reason why one expects tragedies of the commons to play out in nature as well as in human societies (Rankin, Bargum & Kokko, 2007). The main reason is that evolution utterly lacks foresight. It proceeds in ridiculously small steps. Evolutionary progress is hampered by constraints, but its direction is additionally determined by fitness changes over a very short timescale. And crucially, each step—every generation—can change the landscape on which the adaptive walk happens. Hardin's metaphor of the tragedy can be visualized as a population attempting to run up a hill (e.g., become ever better at capturing ever-rarer fish), but the running effort itself makes the entire hill slump downward.

What happens to such a population? Genes that make an individual better at running upward become more common, but if individuals trample the landscape (capture all the fish), the absolute height of the top that the entire population reaches can be lower than it was before the competitive run began. Better vessels, but no fish for them to catch. Importantly, this prediction can follow from a process during which, at every single point in time, individuals who were better at striving upward reaped higher benefits from their efforts than their inferior competitors. Every time step reveals optimization going on within the population (as any snapshot will reveal), yet the bigger picture leads to deterioration of the environment.

Again, a human analogy is clear. Want to travel fast to work? Choose a car instead of public transport. But if everyone switches to doing this, the congestion will mean that drivers of private cars may spend more time on the road than they would have done initially if all had stuck to the idea of the bus. Still, at every level of congestion the drivers still get to work faster than the users of public transport (for the mathematical derivation of this example, see Kokko, 2007).

Indeed, the tragedy of the commons has a forerunner, the *tyranny of small decisions* (Kahn, 1966). It describes the same outcome—an undesirable change in the big picture—but specifically focuses on the insignificance of each individual decision. Kahn's favorite example is the Ithaca railroad, which could not be sustained because individuals too often chose to drive instead of maintaining the profitability of the railroad. If asked, they might very well have voted to live in a world where the railroad still existed. Likewise, the more individuals that invest in having nice houses in beautiful outdoorsy settings, the higher the potential that these environments disappear entirely. The impact of one house is minimal—the problem is that there will be more than one (Haraldsson et al., 2008).

Intriguingly, Fisher (1930) wrote about a similar process in his evolutionary writings, although the message is somewhat hidden. He included an explicit term in his equations that refers to environmental deterioration throughout the adaptive process. Population fitness improves, but the environment deteriorates; thus the net effect can mean no change in observable population growth. This is to prevent the absurdity that populations of, say, rodents now would have exponentially increased their reproductive rate each generation ever since the end of the Paleocene when rodents first appeared on this planet—imagine rats reproducing at a rate that expands the surface of this planet ever-faster into space.

Although Fisher's deterioration argument precedes Kahn (1966) and Hardin (1968) by decades, he perhaps never emphasized enough that the environmental deterioration of his equations is a direct consequence of the adaptive process, not a property of the external environment. This issue has rarely been fully appreciated in discussions of Fisher's fundamental theorem. To this day, models of evolutionary processes that take environmental feedback into account tend to be offshoots of game theory where the "what others do" aspect of fitness has always been in the limelight (Maynard Smith, 1982; Kokko, 2007; McGill & Brown, 2007). Such work only rarely refers to Fisher's way of modeling the adaptive process. Recently, however, in quantitative genetics too—a more direct offspring of Fisher's way of thinking about evolution—the study of feedback due to actions of conspecifics has finally gained momentum (e.g., Bijma & Wade, 2008).

What If You're a Fish Yourself?

Natural populations combine three characteristics that make tragedies likely. First, evolution proceeds along the adaptive landscape in a very local fashion. Second, the traits (or actions) of one individual rarely have a big impact on the shape of the entire adaptive landscape. (A fisherman "gets away" with catching too many fish; his own efforts do not shoot him in his leg. A cooperative and conscious fisherman does not get the future profit of a better fish supply during his individual lifetime.) Third, when the traits (or actions) of the entire population change, even minutely, the adaptive landscape does react and change. The prediction follows that evolution will often favor traits that lead to solutions that appear grossly shortsighted to any outsider who is able to perceive the entire adaptive landscape and its changing peaks.

Here, we have not been too explicit at defining what the "height of a peak" means. It should refer to population fitness, but this is a somewhat slippery concept. It relates to growth potential, but we already know that no population can grow indefinitely since food or other resources become limiting before the entire universe consists of rats. And as we have just seen, better individuals simply deplete their resources faster; thus, as they take over, they build populations that are not necessarily bigger or "better surviving" than their ancestors. In this process, *net* population fitness might not change, yet its intrinsic component (Fisherian fitness *sans* its environmental deterioration component) can nevertheless have changed in a very real sense.

For example, consider a hypothetical organism that initially is not very mobile. The ancestral sluggish forms find enough food to reproduce, however, because none of them is very good at depleting the local food supply. On average at a population dynamic equilibrium, each female produces one female offspring who survives to maturity. Later versions of this organism are faster at finding food (since within each generation the faster ones were a bit more successful than the average individual at producing surviving young), but these fast foragers now deplete the food resources to a much greater degree. Their fast foraging may increase reproductive rates, but producing more mouths to feed means that food gets depleted even faster. As a result, the population will be food limited again, with each female only producing one mature female offspring on average. This kind of net fitness has not changed at all, but if it was possible to stage a competition experiment across generations, the sluggish ancestor would not be able to reach the reproductive rate that it did back when fast foragers had not yet evolved to grab food quickly. This is how net fitness can remain constant while a component of fitness, the ability to find food and convert it into offspring, has greatly increased. Intriguingly, this kind of competition experiment is precisely what researchers have achieved with experiments involving microbial populations—these truly can be put into the freezer and later fully recovered (e.g., Visser & Lenski, 2002).

But what if we are interested in overall population fitness that *includes* the environmental deterioration caused by evolution, instead of the optimist's approach that corrects for deterioration to reveal the intrinsically improving component? Overall population fitness is the relevant quantity for asking questions about the tragedy of the commons, as it takes "landscape trampling" into account. And a story of another fish shows that tragedies can become surprisingly dramatic when this is done.

The Amazing Amazon Molly

The Amazon molly, *Poecilia formosa*, is a species that should provide a bit of a challenge to anyone who likes to promote vague ideas of "intelligent design." First, its way of reproducing is so convoluted that one can only shake one's head at the intelligence of its designer. Second, the very same system also makes the species prone to rapidly depleting a very central resource that it requires for reproduction—in this case, another closely related fish species. Like a fisherman, this fish needs fish to survive, but not because of any piscivorous habits. Instead, the Amazon molly needs *males* of other fish species to make its eggs develop. The Amazon molly never produces any males of its own.

A few clarifying remarks are immediately needed here. The Amazon molly does not use any genes from the sperm to form new individuals: All such genes are discarded. Thus this is parthenogenetic reproduction with just a little bit of superficial contact with sperm (Schlupp, 2010). The use of female-only genetic material to form new females is not unheard of in the animal kingdom. It occurs in a lot of species, from aphids to certain lizards (Avise, 2008; Schwander & Crespi, 2009). Thus it is not obvious why the Amazons have not evolved "pure" parthenogenesis (all-female reproduction) and instead still rely on old-fashioned contact with males: If they are not after the genes, why bother?

The answer seems to lie in the genetic evidence that Amazons are a product of a chance hybridization event. The "parental" species, a different type of molly, is sexual in the ordinary way. The first ever Amazon molly thus inherited a mixed bag of genetic assumptions about how reproduction should occur: For instance, for millennia, eggs have been programmed to get a kick-start when sperm touches them. Before this kick-start signal, the embryo will not start developing.

Nothing in the evolutionary process says that this kick-start has to be associated with the *genes* in sperm (rather than some surface protein or any other physical property of the contact). The Amazon molly has invented only one evolutionary novelty at a time: the discarding of the genes carried by sperm. There is no immediate rush for it to get rid of other aspects of sexual reproduction, such as the assumption that males will be nearby—and willing to mate—to get reproduction started. It lives, after all, in the same streams and ponds as its parental species, and how would the males of those species spot the difference between an Amazon female and any other female molly?

Thus, the beginning looks benign enough. While stopping sperm production without checking if sperm might still be needed for something is decidedly daft, it is not immediately penalized as an evolutionary dead end. Over time, of course, one might expect males of the "host" species to evolve discriminatory abilities: They waste time (and sperm) in matings that lead to no offspring for them. If they discriminated strongly enough, then the Amazons could be in trouble. However, once again evolution does not deterministically come up with solutions that appear sensible when the system is viewed as a whole. As long as males of the sexual species do not lose too many mating opportunities with their conspecifics, the selection pressure to avoid matings with Amazons can remain very mild. What matters to a male is how much he reproduces, not how many additional zero results he scored as a by-product of his eagerness to mate. In fact, males mating with Amazons may become more attractive to conspecific females via mate-copying (Heubel et al. 2008), and a male who wastes time scrutinizing each female before mating with her may lose out in competition with his more eager conspecifics; thus the selection pressure to avoid mating with Amazons can be *negative* (though it is hard to know for sure—in reality a little bit of discrimination has been shown to exist, but not nearly enough to make all Amazons die a virgin death; Aspbury & Gabor, 2004; Aspbury, Espinedo & Gabor, 2010).

So, males are not selected to worry too much about which molly species a female belongs to, and female Amazons are not selected to worry about the lack of males of the correct species. Silly, but perhaps it works without problems? Not so fast. This system has a built-in time bomb: the Amazons, with a bit of harmless mating, deplete the stocks of their sperm-donating species at an amazing speed, without much foresight in realizing that their reproduction cannot occur independently of the survival of these other species.

The built-in time bomb operates because all-female reproduction is an incredibly efficient affair. In "normal" species with a 1:1 sex ratio, only half of babies have the potential to develop into adults that can themselves produce babies. Asexual species bypass the production of males and double the production of baby-producers (females). The growth rate of an asexual lineage is therefore double that of a sexual one. Amazon mollies retain the need to have some contact with males, but since males are "borrowed" from other species rather than any effort being spent in producing them, Amazons have a strong tendency to become more common than the other species that they coexist with.

And this "being more common" does not stop at any reasonable level. As long as males are available, and as long as they have enough sperm to mate with Amazons as well as properly sexual females, the next generation has ever more Amazons compared with the sexual species. Since the total number of fish in any local water body is necessarily limited, the increasing ratio of Amazons to other fish can only mean that eventually the local supply of the sexual species runs out. Extinction of the sexuals is then followed by extinction of the Amazons. Effectively, the Amazons have succumbed to a tragedy of the commons through a failure to make sure that their behavior is compatible with the long-term persistence of a public good (males of the sexual species). Intriguingly, the males of the sexual species suffer a tragedy of their own as they fail to evolve to categorically avoid mating with Amazons.

The impossibility of this breeding system was mathematically proven decades ago (Kiester et al., 1981), and extinctions are easily observed in the wild too (Heubel, 2004). Thus, there is a slightly puzzling aspect of the system: the fact that Amazons and their sexual "host" species exist at all for us to study. Given that local extinctions really do occur, it appears that the system persists at a dynamic equilibrium where local extinctions are followed by colonizations from elsewhere (Kokko, Heubel & Rankin, 2008). This resembles a general ecological principle that very local interactions can be difficult to maintain (try to keep a predatory fish together with its prey in a small fish tank) while larger scales offer enough refugia for local populations to recover. In addition to predators and their prey, similar principles play out in the dynamics of hosts and parasites. Say, a virus may run out of locally susceptible host individuals after which it can persist only by moving to wreak havoc elsewhere. The local host population then has time to recover.

The Tyranny of Small Steps Revisited

The Amazon molly story does not make sense at any level, at least if we wish to believe that evolution effortlessly optimizes population fitness. The system of reproducing is so crazy in this fish that it keeps hitting evolutionary dead ends, perhaps only to be rescued by the hide-and-seek game of exploiters and exploitees being played out asynchronously in different subpopulations. It is even possible that the only reason that this system still persists as a whole on this planet is that Amazon mollies deplete populations of other fish so fast that local Amazon molly populations are doomed—which then leaves space and time for the sexual fish to recolonize and form a supply of unsuspecting males for newly invading Amazons. Mathematical modeling of this type indeed shows that a parasitic species can sometimes benefit, at a species level, if its own local extinction rate is high (Kokko, Heubel & Rankin, 2008).

It is not easy for evolution to solve problems that involve very different timescales. Amazons would profit in the long run if they reproduced more prudently and did not outcompete their relatives, the sexual species. But in any one generation a less-fecund Amazon female would lose out to its rampantly reproducing sisters. Conversely, a male of the sexual species would create a wonderful environment for its offspring and grandoffspring if he always refused to mate with any Amazons. However, in a typical generation, one male's effect on whether the Amazon danger intensifies is minor. If one male appears with a mutation that makes him a prudent scrutinizer of females, and others still go for all the females, the latter are likely to fertilize the eggs of most of the conspecific females. The prudent male's genes do not get passed to the next generation with equal ease, and he alone has not managed to save the population from the Amazon molly danger anyway—his less prudent competitors will have mated with them and allowed all Amazon eggs to develop.

The prediction is that the male population will fail to develop a strong discrimination against Amazons, and this laidback mating behavior allows the Amazons to take over. Mating eagerness is typically selected for in males, but in the molly case it tramples the adaptive landscape in the worst imaginable way: It leads to deterministic extinction. Each mating decision is a small step, but the net effect is a disaster for future generations.

Does It Ever Work?

The intriguing aspect of all this is that some degree of inefficiency and waste is clearly permitted by natural selection. The Amazons appear to live fairly precariously, but they have persisted for tens or hundreds of thousands of years (according to estimates based on genetic evidence, the hybridization occurred 280,000 years ago; Schlupp, 2010). One could argue that if their reproduction were any sillier, their very first attempt to establish a population would have failed, and thus evolution weeds out solutions that are too outrageous. Evolutionary theoreticians are busy working out details of such principles (Lehmann, 2010).

There is ongoing debate among evolutionary biologists about the correct way to interpret these principles. Everyone who currently takes part in this so-called group selection controversy appears to agree that selection can act on multiple levels (e.g., Gardner, 2010; Korb, 2010; Leigh, 2010). Also, there is nothing particularly controversial about extinction hitting species in a nonrandom way (Jablonski, 2008). For example, lineages of brachiopods persist for longer in the fossil record if they have lifestyles that permit larvae to disperse widely (Powell, 2007). Easy dispersal permits a large geographic range and—relatively unsurprisingly—this makes a species (or genus) more extinction-proof than one with a more sedentary lifestyle. To what extent species selection and multilevel selection are conceptually similar is debated (Okasha, 2006), but an even more important key question is that of adaptation (Gardner, 2010), that is, what kind of problems entities evolve to solve. Does it make sense to say "larvae disperse widely to avoid extinction"? Most biologists would agree that it does not. Discussing species traits as if they evolved to benefit the species is a schoolboy error because it ignores the fact that short-term selection acting on individuals is free to modify a species' dispersal ability upward or downward, entirely depending on what is adaptive at the very moment. Amazon mollies or brachiopod larvae simply lack the foresight to modify their behavior according to the longer-term needs of their species. Every biologist needs to be clear about this fact, and countless generations of undergraduates still struggle to get rid of the lay person's idea that evolution "aims to" ensure the survival of the species.

At another level, however, it also remains true that the ability of selection to weed out what does not work does not stop at the short term. The very fact that the faster-acting processes can be terribly short sighted is the reason why evolution can lead to outcomes that involve tragedies and turn out to be bad in the long run. These outcomes get removed, albeit only stochastically and not perhaps very efficiently. High-level selection takes place, but whenever it is a slow and stochastically occurring process, current adaptation will to a large extent reflect processes that have occurred over more recent timescales than slowly emerging disasters.

It follows that evolutionary biologists, as well as anyone interested in issues of sustainability, should think explicitly about different timescales. From the above one might predict that as long as individuals of any given species exist, they will remain relatively ignorant of the possibility of their lifestyle being unsustainable. Here "ignorance" can mean a cognitive process in a conscious being, or short-sighted trait evolution in the more normal setting of nonconscious existing. As a male fish would think if it suddenly did develop consciousness: "We seem to be doing just fine despite this slowly increasing Amazon proportion in the population; why believe those who say that the future will be dire when nothing bad has happened?" The problem, of course, is that disasters are not easy to believe in beforehand, and when they are imminent, it may be too late to change one's behavior. In the case of humans, the problem is perhaps exacerbated because we are so often able to delay it. An Amazon molly will not be able to switch to using sperm from too distantly related fish species, but Canadian fishermen were able to find profit in selling prawn. It still would have been nicer to keep the cod there too.

Conclusions

Kahn laments the loss of a railroad, we all will have to lament the loss of cod populations (or a stable climate), and Amazon mollies and their host species simply carry on without having the ability to perceive the ridiculousness of their system to lament anything.

I began to work on this chapter during a time when world leaders gathered in Copenhagen in an attempt to avert an unprecedented disaster of more rapid climate change than this planet has perhaps ever experienced before, while the popular media kept expressing doubt that anything is happening beyond ordinary fluctuations in weather. It appears very hard for humans to believe that the current state of affairs might not be sustainable, even if an incredible amount of evidence points to the contrary. The reason why Amazon mollies and their hosts have to live a very labile existence is that they face very different selective pressures in one generation than, say, ten generations later. What is a small "environmental" problem for male sailfin mollies (a species of molly whose males mate with the Amazons) in one generation (a couple of Amazons coexisting with the "proper" females) can become a life-and-death issue a couple of generations later.

It is almost too easy to find examples in human current affairs where the change of conditions from one generation to the next works in exactly analogous ways. Shifting goalposts and tacit acceptance of the gradual erosion of the commons quality are ubiquitous features of the problems of humankind. To quote Sir Robert May, in the context of combating one aspect of reaching the Millennium Development Goals, "A pessimist could argue that, above all else, we need a better understanding of the evolution of altruistic or cooperative behaviour, which unfortunately remains the biggest unsolved problem in evolutionary biology" (May, 2007, p. 502).

We could add that even though much theoretical and empirical effort has been spent on this issue, we lack a good understanding of what happens under nonequilibrium scenarios where the future truly can be quite different from the past. The big problem, of course, is that natural selection is based on counting on what worked well in the past. How much can humans deviate from the shortsightedness of all our ancestors? That might depend on how much we can learn to like a little bit of mutually agreed coercion. The intriguing finding is that there is some evidence that such a thing can happen without brainwashing or other forms of cruelty. In an experimental setting, at least, people have been shown to prefer joining an institution that uses sanctions over one where bad behavior goes unpunished—at least when they are allowed to switch and cancel their initially expressed aversion for rules (Gürerk, Irlenbusch & Rockenbach, 2006). The negative lesson, of course, is that the study also indicates that we start believing in the reality of the longer-term consequences only once we see with our own eyes that bad things do happen in the nonsanctioning societies. In any environmental setting with a tragedy well on its way, that may often be a little too late.

References

Aspbury, A. S., & Gabor, C. R. (2004). Discriminating males alter sperm production between species. *Proceedings of the National Academy of Sciences of the United States of America, 101*(45), 15970–15973.

Aspbury, A. S., Espinedo, C. M., & Gabor, C. R. (2010). Lack of species discrimination based on chemical cues by male sailfin mollies, *Poecilia latipinna. Evolutionary Ecology, 24*(1), 69–82.

Avise, J. C. (2008). *Clonality: The genetics, ecology, and evolution of sexual abstinence in vertebrate animals*. Oxford: Oxford University Press.

Bijma, P., & Wade, M. J. (2008). The joint effects of kin, multilevel selection, and indirect genetic effects on response to genetic selection. *Journal of Evolutionary Biology, 21*(5), 1175–1188.

Fisher, R. A. (1930). *The genetical theory of natural selection*. Oxford: Oxford University Press.

Gardner, A. (2010). Adaptation as organism design. *Biology Letters, 5*(6), 861–864.

Gürerk, Ö., Irlenbusch, B., & Rockenbach, B. (2006). The competitive advantage of sanctioning institutions. *Science, 312*, 108–111.

Haraldsson, H. V., Sverdrup, H. U., Belyazid, S., Holmqvist, J., & Gramstad, R. C. J. (2008). The tyranny of small steps: A reoccurring behaviour in management. *Systems Research, 25*(1), 25–43.

Hardin, G. (1968). The tragedy of the commons. *Science, 162*, 1243–1248.

Heubel, K. U. (2004). *Population ecology and sexual preferences in the mating complex of the unisexual Amazon molly Poecilia formosa (GIRARD, 1859)*. Dissertation. Hamburg: University of Hamburg.

Heubel, K. U., Hornhardt, K., Ollmann, T., Parzefall, J., Ryan, M. J., & Schlupp, I. (2008). Geographic variation in female mate-copying in the species complex of a unisexual fish, *Poecilia formosa*. *Behaviour, 145*(8), 1041–1064.

Jablonski, D. (2008). Species selection: Theory and data. *Annual Review of Ecology Evolution and Systematics, 39*(1), 501–524.

Kahn, A. E. (1966). The tyranny of small decisions: Market failures, imperfections, and the limits of economics. *Kyklos, 19*(1), 23–47.

Kiester, A. R., Nagylaki, T., & Shaffer, B. (1981). Population dynamics of species with gynogenetic sibling species. *Theoretical Population Biology, 19*, 358–369.

Kirkpatrick, M. (2010). Limits on rates of adaptation: Why is Darwin's machine so slow? In M. A. Bell, D. J. Futuyma, W. F. Eanes, & J. S. Levinton (Eds.), *Evolution since Darwin: The first 150 years* (pp. 177–195). Sunderland, MA: Sinauer.

Kokko, H. (2007). *Modelling for field biologists and other interesting people*. Cambridge: Cambridge University Press.

Kokko, H., Heubel, K. U., & Rankin, D. J. (2008). How populations persist when asexuality requires sex: The spatial dynamics of coping with sperm parasites. *Proceedings. Biological Sciences, 275*, 817–825.

Korb, J. (2010). Social insects, major evolutionary transitions and multilevel selection. In P. Kappeler (Ed.), *Animal behaviour: Evolution and mechanisms* (pp. 179–211). Berlin: Springer.

Kurlansky, M. (1997). *Cod: A biography of the fish that changed the world*. New York: Walker.

Lehmann, L. (2010). Space-time relatedness and Hamilton's rule for long-lasting behaviors in viscous populations. *American Naturalist, 175*, 136–143.

Leigh, E. G., Jr. (2010). The group selection controversy. *Journal of Evolutionary Biology, 23*, 6–19.

May, R. M. (2007). Parasites, people, and policy: Infectious diseases and the Millennium Development Goals. *Trends in Ecology & Evolution, 22*(10), 497–503.

Maynard Smith, John. (1982). *Evolution and the theory of games*. Cambridge: Cambridge University Press.

McGill, B. J., & Brown, J. S. (2007). Evolutionary game theory and adaptive dynamics of continuous traits. *Annual Review of Ecology, Evolution, and Systematics, 38*, 403–435.

Okasha, S. (2006). *Evolution and the levels of selection*. Oxford: Clarendon Press.

Powell, M. G. (2007). Geographic range and genus longevity of late Paleozoic brachiopods. *Paleobiology, 33*(4), 530–546.

Rankin, D. J., Bargum, K., & Kokko, H. (2007). The tragedy of the commons in evolutionary biology. *Trends in Ecology & Evolution, 22*(12), 643–651.

Schlupp, I. (2010). Mate choice and the Amazon molly: How sexuality and unisexuality can coexist. *Journal of Heredity*, *101*(Suppl. 1), S55–S61.

Schwander, T., & Crespi, B. J. (2009). Twigs on the tree of life? Neutral and selective models of integrating macroevolutionary patterns with microevolutionary processes in the analysis of asexuality. *Molecular Ecology*, *18*(1), 28–42.

Visser, J. Arjan G. M. de, & Lenski, R. E. (2002). Long-term experimental evolution in Escherichia coli. XI. Rejection of non-transitive interactions as cause of declining rate of adaptation. *BMC Evolutionary Biology*, *2*, 19.

Worm, B., Hilborn, R., Baum, J. K., Branch, T. A., et al. (2009). Rebuilding global fisheries. *Science*, *325*(5490), 578–585.

Wright, S. G. (1932). Evolution in Mendelian populations. *Genetics*, *16*, 97–159.

4 Life in Interesting Times: Cooperation and Collective Action in the Holocene

Kim Sterelny

1 Prologue: The Pleistocene

When our lineage split from the other great apes, our ancestors' lives were probably rather similar to those of living chimps: social, intimate, small scale, but not very cooperative. Chimps (and bonobos) form coalitions in social competition, in male-male intergroup rivalry, and perhaps in hunting. So there is some cooperation in conspecific interactions, and perhaps in some aspects of ecological interaction. But there is virtually no informational or reproductive cooperation. Since that split there have been two great revolutions in hominin social life. One is the human revolution: the gradual transition from chimplike social lives to culture, cognition, communication, and cooperation in intimate, egalitarian social worlds; the social worlds of Pleistocene foragers. The human revolution took perhaps 3.5 million years, beginning when stone tools became central to hominin life; complete perhaps 50 to 100 thousand years ago, when the human foraging cultures of antiquity began to resemble those known from the ethnographic record (Sterelny, 2011). The second was the Holocene revolution: a rapid transition from the intimate and egalitarian social worlds of foragers to large-scale, hierarchically organized social worlds. Both transitions pose explanatory problems. But the burden of this chapter is to show that the Holocene transition is not at all well understood. In particular, I aim to show that the preservation of the social contract—respect for norms and participation in collective action—is an unresolved paradox. The Pleistocene social contract, I shall argue, was stable because it was mutually profitable. But the social contract of early farming and herding societies is much more puzzling.[1] But first, the Pleistocene. Until about 10,000 years ago, *sapiens* populations lived in forager societies, and those forager groups were cooperative in many mutually reinforcing ways.

(i) There were staghunt returns to cooperative hunting (Skyrms, 2003). Large game was a crucial resource for many forager cultures, and harvesting such game depended on cooperation and coordination among hunting parties, especially before the domestication of dogs and the invention of high-velocity projectile weapons (bows and javelins, plus throwers). Projectile weaponry is quite recent. There is some evidence of bows at about 75,000 years ago (Lombard & Phillipson, 2010); it is perhaps as recent as the last 50,000 years (Shea, 2003, 2009; Marlowe, 2005).

(ii) Cooperation is an effective form of risk reduction. Large game hunting is very rewarding when successful, but hunts often fail. In cultures without effective storage, a boom-and-bust foraging strategy is a recipe for disaster, unless supplemented by insurance against a run of bad luck. Sharing with other hunting parties helps to manage risk, especially as the marginal value of shares of a large kill declines, as the hunters and their dependents become satiated.

(iii) The complementarity of male and female roles also helps to manage risk. In most forager societies, women focus on plant-based resources and on small game. Insofar as we have quantitative data on foraging returns, their average return is less than that from hunting. But their returns are significant, and they are more reliable on a day-to-day basis. This reliability is important in itself, and the adaptive rationale of this low-risk foraging strategy is further underwritten by the difficulties and risks that would come in an attempt to combine pregnancy and infant care with large game hunting. The resources a woman finds in gathering are, thus, important to the woman herself, her children, and her male partner. Likewise, in most forager ecologies, male input is critical to the family economy. Moreover, when resources are found in different places, and when harvesting them demands different skills sets, it can pay to specialize. Thus male and female roles are complementary.[2] Economic cooperation reinforces and supports reproductive cooperation: If a partnership is successful economically, it often is best to avoid the costs and risks of repartnering.

(iv) Human foraging cultures depend on extensive intergenerational family cooperation, not just cooperation between fathers and mothers. Unlike the other great apes, humans take a long time to reach peak efficiency as foragers, often not until their late teens or early twenties. Most foragers live in environments with seasonal shortfalls: There is a hard time of the year in which children and younger adolescents cannot support themselves. In such seasonally challenging environments, the foraging life is viable only because peak efficiency, once reached, produces a significant surplus for many years (especially from men), and that surplus supports the growing generation. Parents and grandparents support children and adolescents through their long period of partial dependence.

(v) There is some controversy about the slow fruition of foraging competence, and, in particular, on the relative importance of size and skill. Its slow decline—foragers are still near-peak in their forties and early fifties—convinces me that skill is the key ingredient. In any case, few would deny that foraging competence depends on skill and on deep knowledge of the local terrain and biota: The open question concerns the time needed for learning these skills. These informational resources are acquired culturally, from the local community. So forager life depends on cultural learning and hence on informational cooperation. The specific pathways are probably highly variable, from skill to skill and culture to culture. Very likely, vertical, oblique, and horizontal information flows are all important in some contexts.[3]

(vi) Finally, Kristin Hawkes, Sarah Blaffer Hrdy, and their colleagues have argued that reproductive cooperation has played a foundational role in human life (Hawkes & Bird, 2002; Hawkes, 2003; Hrdy, 2009). They point to an apparent paradox: Human infants are even more metabolically expensive than great ape infants. They are even more helpless when born: They cannot even cling to their mother. And they remain helpless for a considerable period. Moreover, birth poses a severe, dangerous, and sometimes prolonged challenge. And yet they are weaned more rapidly than great ape infants, and forager women typically proceed to breed again more rapidly than their great ape counterparts. Women clearly are being helped in ways great ape females are not. The sources of that help surely vary over time, place, and social context, but Hrdy argues persuasively that both a woman's female kin and her cohort of peers are important sources of help.

There is much we do not know about ancient foragers. We have only rough estimates of group size and of the origins of art and ideology. We do not know when the division of labor and specialization—itself an important form of cooperation—became important within and between groups (though there is persuasive evidence that some trade is ancient; Ofek, 2001). But ancient forager life depended on a complex web of cooperative relationships, and despite the puzzle cooperation poses to evolutionary biology, we have a persuasive model of the stability of cooperation in these ancient forager worlds—a persuasive model of its resilience in the face of temptations to defect and to expropriate.[4] But the plausibility and power of this picture of Pleistocene cooperation makes what happened next even more puzzling. Between about 10,000 and 6,000 years ago, in many areas of the world, a revolution in human social life took place. That revolution, it seems, undermined the foundations of cooperation. And yet cooperation did not collapse. The aim of this chapter is to expose the paradox of the Holocene in the light of the Pleistocene.

Thus the evolution of human social life saw two profound transformations. The first established the social life of late hominin foragers: cooperative, egalitarian, encultured, knowledge-dependent, and organized around family units. This social life is very different from the great ape social life from which it evolved. The second saw complex and stratified societies develop, and develop very quickly, over a few millennia, from simple, flat ones. Almost certainly, the shift from production systems based on foraging to production systems based on farming cereals (and, a little later, domesticated animals) played a crucial role in this transition from intimate to larger, more complex social worlds. To understand this second revolution, we need to answer three questions: (a) why did farming establish and spread early in the Holocene? (b) Why does farming push social life toward stratification and complexity? And (c) why does the social contract—respect for norms and participation in collective action and the provision of public goods—survive the transition to social complexity? Participation in collective action in unequal societies has an obvious explanation, once these have developed top-down command and control systems backed by coercion. Once influence and position are reinforced by the police, armies, and spies, even for those low in the hierarchy, participation in collective action may well be the least bad option. But such command by coercion systems took many generations to develop.

The first two of these problems are recognized, but by no means solved. Our best guess is that farming emerged through decreasing returns from foraging conjoined with new climatic conditions that made farming a viable alternative. But this is not much more than a guess. There is no direct evidence that forager subsistence strategies were failing where and when agriculture originated; indeed, they originated in highly fertile and productive river flood plans, not in habitats that were at all marginal. Two recent special issues on the origins of agriculture (*Current Anthropology*, 50 [5], 2009; *Current Anthropology*, 52, [S4], 2011) make it clear that we lack an empirically well-supported causal model of the origin and spread of agriculture. We know a good deal about where and when agriculture started, and we can identify early crops. But we have no well-confirmed theory of why it began. As we shall see in section 3, stratification is linked to farming through a change in the relative importance of embodied and material capital. So we know something about the origins of agriculture and its consequences for inequality. In my view, the third problem has hardly been recognized, let alone solved. But to see the scope of the problem, we need to understand the foundations of the Pleistocene social contract, and I turn to that now.

2 The Pleistocene Social Contract

Forager cooperation was intimate world cooperation. Forager communities were (we think) small and hence relatively homogeneous. Reciprocation often involved like-for-like trades. Male hunters provided resources for others broadly similar to those they and their families would need when they were unlucky. Similarly, reproductive cooperation among a female cohort probably involved reciprocated exchange of similar aid: suckling one another's children; protection and babysitting; aid and protection during the birth itself; perhaps helping carry infants when the group itself was in motion. When reciprocation consists in like-for-like returns, fair return is easier to judge. The simpler the social environment, and the smaller the range of commodities in circulation, the easier it is to be confident that you are not being exploited. Moreover, interactions are typically long repeated, and in most cases fairly small costs and benefits are at stake in each interaction. Iterated interactions will not give agents incentive to cooperate, if there are enormous rewards for a single successful defection, or horrific costs if you are the victim of defection. Iterated interactions do give agents incentive to cooperate if the summed profit of cooperation over repeated interaction is large, compared to the resources at stake in any one of the interactions, and that will often be true of forager cooperative partnerships. Except perhaps in a dire and immediate crisis, a woman will not gain much if she refuses to suckle another baby, if she refuses support and protection at another's birth, or if she refuses to keep a protective eye on another's four-year-old. The same is true of sharing among male hunters. Famously, forager societies are also egalitarian, not just in having no positions of formal leadership with coercive power, but also in having relatively flat wealth distributions, and, typically, monogamous mating systems. There are doubtless more or less successful foragers, but the variation in reproductive success in forager cultures is not extreme, and to the extent that it exists, it is not the result of entrenched social advantage. As a consequence, the profits of collective action are not expropriated by elites.[5] All have a stake in the gains of collective action: game drives, collective defense, and collective niche construction. Forager niche construction can be on a surprisingly large scale: For example, aboriginal Australian groups have constructed extensive stone fish traps on suitable sites on the coast and on inland water ways (see, e.g., Stockton, 1982; Bandler, 1995).

Small size has information and coordination advantages, too. Forager groups face whole-group collective action problems: of peace and war; when and where to move; perhaps decisions on when and where to hunt. A Neanderthal large game hunt may well have been a whole-community project, and when *sapiens* tribes were equipped only with stabbing spears and heavy, short-range javelins, the same may have been true of them. It is certainly easier to solve coordination problems bottom up, by persuasion and consensus, in smaller groups. Forager social worlds are intimate: Each agent will be well known to most others in the band. Collective foraging exposes every member of a hunting band under conditions of fatigue, stress, and danger: very revealing conditions. Likewise, if Hrdy is right about the importance of cooperation between women in childbirth and childcare (see Hrdy, 2009), the women in the group will have many opportunities to see one another cope (or fail to cope) with stress, anxiety, exhaustion, and pain. By the time *sapiens* has evolved, these direct information linkages were magnified by gossip.[6] It is notoriously hard to keep secrets in a village; harder still, I suspect, in a forager band. Much forager economic activity is public.

The infamous "folk theorem" of cooperation requires that interactions be iterated, but it also requires that free riders and other cheats be identified, and that they be sanctioned. To my mind, there is no doubt that forager social worlds were awash with information about other agents. But it is not enough to identify lazy free riders and exploitative bullies: They have to be controlled by punishment, and by the credible threat of punishment. Ken Binmore and Don Ross do not see this as a problem: They have argued that much social sanction is both effective and free (Binmore, 2006; Ross,

2006). The targets of sanction are sensitive to social signals of approval and disapproval for both instrumental reasons—our social reputations are precious assets—and through the intrinsic psychology of human minds. With the exception of a few psychopaths, approval and disapproval really matter to people. So signals of social assessment have punch, but they are essentially free.

I think that view is a mistake. It ignores risk costs. Even when the initial norm violation is relatively mild, and the initial signals of disapproval, likewise, are mild, conflicts can and do escalate. Disapproval triggers resentment and verbal retaliation, and conflicts can both intensify and drag in third parties. We all know of small communities (academic departments, for example) that have fractured as a result of initially trivial disagreements becoming cancerous.[7] This may well be an especially serious risk in forager societies, for they do not have institutional mechanisms that limit conflict escalation: Paul Seabright points out that forager societies have very high murder rates, and uncontrolled escalation may in part explain those rates (Seabright, 2010). Moreover, some norm violations—and not just those to do with sex—are serious. Chris Boehm's survey of forager egalitarianism details many examples of free riding being kept in check by just the informal mechanisms that Binmore and Ross have in mind. But it also shows that much more serious sanctions—including assassination—are not rare, especially when the defection in question is expropriative bullying (Boehm, 1999). But although sanctions are not free, often, and for many agents, they are worth paying, both to preserve the cooperative networks on which their life depends and to signal that they themselves are reliable cooperators—that they are staunch when times are tough. Thus, in some circumstances, the cost of punishment can be co-opted as an honest signal of commitment to cooperation within the group.

In summary: Among Pleistocene foragers there was enough overlap of interest at a time and over time for cooperation to be stable, enough mutual knowledge to recognize these commonalities of interest and to solve coordination problems from the bottom up, by persuasion and consensus. The informational transparency of the group made it easy to recognize free riders and exploitative bullies. Although punishment of such individuals was not free, and sometimes not cheap, the value of reputation and the returns on investment in cooperative partnerships were often high, and so, often enough, it was worth paying the costs of punishment.

Fast-forward the tape a few thousand years to the stratified but sill localized prestate societies of Europe, the Middle East, Mesopotamia, China, or America (though American stratified societies emerged somewhat later).[8] There are major differences in wealth and influence. Polygyny has replaced monogamy in many cultures. Social worlds are larger. They are also less transparent: Economic activity is often household based, so the automatic flow of information that working together generates has become a trickle. Emotions of affiliation and cooperation are built through shared work and danger, and that mechanism of social solidarity is weaker too, for the same reason. Exotic, high-prestige luxury goods are in circulation, but they are concentrated in a few hands, and are often interred with their wealthy owners. The conditions that sustained the Pleistocene social contract are no longer in play. Yet the contract seems not to have collapsed.

3 Inequality and Inheritance

Prior to the Holocene, about 10,000 years ago, human social experience was of cooperation in intimate worlds, and in such worlds, cooperation contingent on reciprocation is both profitable and stable. It is very likely that human psychology—especially the moral emotions and their role in our life—shows the marks of a long history of intimate world cooperation. The folk theorem describes the world of Pleistocene folk, and, of course, of many more recent human cultures. But between the beginning of the Holocene and the first establishment of agrarian state societies—perhaps 6,000 years ago—revolutionary changes took place in the social and economic lives of many people. The conditions that made contingent cooperation stable and profitable were eroded, but many stratified but prestate groups left impressive physical traces of collective action. Some of these are utilitarian constructions: irrigation systems (for example, the Hohokam and Anasazi cultures in North America) or defensive fortifications. Many have ceremonial or ideological functions; for example, many cultures cooperated in building large-scale burial mounds (some very large scale indeed). There is no mystery about the persistence of cooperation and collective action in early state societies. States are equipped with effective top-down mechanisms of command and control, backed by the credible threat of lethal violence. Moreover, in such states, the small political elites that wielded power were physically distant from most of those subject to these command-and-control mechanisms, and hence safe from lethal counterstrike. It is worse than useless to kill the king's local lackey; he will just send another, and a punishment party as well. In such circumstances, with resistance crippled by intractable coordination problems, often the least bad option will be to cooperate in collective action, despite the expropriation of its profits.

However, there was a temporal and structural intermediate: larger, stratified inegalitarian social worlds, but without a developed state apparatus. In some places this intervening stage appears to have been brief, perhaps a few hundred years. But in other places, thousands of years go past between the first establishment of sedentary communities with significant wealth differentials (shown, for example, by variations in grave goods and house size) and the first archaeological signatures of states—regional centers, with public buildings, elite housing, and a dominating regional role. How did the social contract survive the erosion of the conditions that made it stable in intimate worlds? Cooperation in these stratified worlds still largely depended on consent and on respect for the norms and customs of the group. Indeed, Seabright emphasizes the fact that even in the world of states, laws, and police, norms and consent still play an essential and ineliminable role in the survival of the social contract. Thus in the interregnum, the conditions that support intimate-world cooperation have eroded. The command-and-control institutions that coerce large world cooperation have yet to become efficient (and will never become sufficient in themselves to enforce cooperation). There are local elites, but these are conveniently close and gratifyingly mortal. So how did stratification without top-down command and control emerge? Once it emerged, what prevented the collapse of cooperation, followed by fragmentation back to intimate worlds?

Importantly, the pathway to inequality combined with continuing collective action was not fragile, nor contingent on a specific constellation of local conditions. For farming, together with a shift to larger and more stratified social worlds, seems to have multiple points of origin, and to be resilient once established. Resilience is not complete, of course, as a recent survey of Polynesia and Melanesia shows (Currie, Greenhill et al., 2010). Stephen Mithen suggests that in some places, the Younger Dryas may have forced a shift from agriculture back to foraging (Mithen, 2003). But if so, that is an externally forced rather than endogenous reversion to a more intimate world. Likewise, Peter Bogucki details a number of examples in which complex, highly stratified societies reverted to simpler ones, though not to foraging (Bogucki, 1999). Nevertheless, while the archaeological record is certainly not one of a unidirectional increase in complexity, once cultures become fully committed to farming as a mode of production, stratification develops, and often increases. Stratification does not seem to be constrained or contained by the collapse of the social contract, or by the threat of collapse.

Two factors seem to be central to the growth of elites. One is a change in the nature of wealth, to a form that can be accumulated at and across generations. The second is the stability of the customs, norms and values that recognize and protect individual control of wealth. We need to explain both factors to understand elite escape from egalitarian control, but I begin with wealth. In an insightful analysis of the evolution of human social structures and of the replacement of egalitarian forager societies by an array of stratified social systems, Kaplan, Hooper, and Gurven (2009) identify the central role of women in risk management in forager economies. This role helps give women leverage in the implicit bargains between the sexes. However, as Kaplan, Hooper, and Gurven see it, women have bargaining power not just because their product is important; their work is also highly skilled. In foraging worlds, the most crucial factor of production is human capital: embodied capital. Skill reduces the leverage of physical power, for skilled labor is difficult to coerce effectively. Foragers are rarely slaves, because efficient foragers must be mobile, independent in their decision making, and often armed. Effective, close supervision is expensive, perhaps impossible. Moreover, embodied capital (skill, strength, health) is inalienable. Injury and illness are always possible; aging is inevitable. But an agent's connection to his or her embodied capital is much less fragile than his or her ownership of lands, herds, crops, or physical plant. Herds can be stolen, crops devastated in raids or storms. A fisherman's boat can be wrecked; his nets can be lost. But in most cases, the skills an agent brings to the table are retained through seasonal fluctuations of fortune. In economies in which embodied capital plays the pivotal role, the distribution of assets does not change markedly as a result of bad luck and trouble. It follows that the distribution of wealth in forager cultures is fairly even, because natural variation in embodied capital is modest, and because the effects of natural variation are not greatly amplified by luck within a generation, or by intergenerational transfer and accumulation of difference across generations.

It follows that farming leads to inequality in part because the most critical resource a farmer needs is good land, and land is alienable: It can be acquired and accumulated. And it leads to inequality because much farming labor is not skilled, and so can be coerced. Like forager societies, slash-and-burn horticultural societies are relatively egalitarian. While land is clearly critical to horticultural economies, it is not yet in short supply, and clearing land for gardens is a skilled activity that must be performed regularly. So foragers and horticulturalists depend primarily on embodied and social capital, that is, their networks of support from family and allies. These forms of wealth cannot be transferred to others by social convention. But with intensive farming, fertile land is the critical factor of production. Likewise, herding cultures absolutely depend on their herds. Herds are transferable: They are rich, dense, but fragile resources.

If Gurven and his colleagues are right, societies are liable to become stratified if life success begins to depend on material resources, especially when these resources can be concentrated, either by natural physical processes or by social convention. The salmon runs of the Pacific Northwest are a paradigm of natural resource concentration: Salmon coming upstream to breed are concentrated at particular, known choke points. These offer very rich resources indeed. And because they are so localized, they can be defended. Once defended, they become heritable sources of wealth. Land and herd ownership makes far greater wealth inequalities possible. Windfall inequalities are entrenched and accumulated by intergenerational transfer, as the lucky of one generation leverage the life prospects of the next. Thus the stratified "complex forager cultures" that are built around the control of salmon runs and similarly dense resource concentrations prefigure the larger and even more stratified societies built around the control of fertile alluvial river basins and around herd ownership.

I am persuaded that inequality in pastoralist and agricultural societies does indeed depend on the fact that material resources are more important in these societies, and that these resource differences intensify over generations, as initial symmetry-breaking perturbations of good and ill fortune are entrenched and amplified by the lucky of the next generation having a head start while others face handicaps. Kaplan, Hooper, and Gurven treat the stratified, polygynous societies that result as examples—often extreme examples—of "resource defense polygyny." This breeding system is famously exemplified by beach master elephant seals: They control a limited resource, namely the females' need for a safe beach, and they use this control to lever exclusive sexual access to the females using their territory. Polygyny derives from control of these rich, alienable, heritable resources.

The growth of inequality is surely linked to the increased importance (in both absolute and relative terms) of heritable forms of wealth, and hence to accumulating differentiation. But this explanation of stratification is seriously incomplete; we need an explanation of the continued respect for ownership. To put the same point in somewhat different terms, we need an explanation of what does not happen, as well as an explanation of what does happen. There is an option that seems available, adaptive, but not seized by the less wealthy members of stratified societies; an option that if taken, would destabilize stratification: the formation of coalitions. The fragile connection between agent and resource—the possibility of alienation—is a double-edged sword. Subordinate male seals do not have the social, cognitive, and communicative capacities to organize a coalition that would dispossess the beachmaster and share the sexual spoils. Most males never mate, so a share of the spoils would much enhance their fitness. Manifestly, humans do have the ability to form such coalitions, and inequality emerges from social worlds in which coalitions formed and acted against embryonic elites. Chris Boehm has shown that the threat of such coalitions (and occasionally, their reality) was important in stabilizing forager egalitarianism. So we should expect rich, dense, predictably located resources to be expropriated and shared by highly motivated coalitions of the less wealthy. It would seem to be a staghunt game: They each would get more acting together than any would get truckling to local elites. The greater the inequalities leveraged by possession of rich resources, the more highly motivated redistributing coalitions should be. Yet the commoners of stratified prestate societies do not only tolerate ownership norms that exclude them from wealth; they are often willing to engage in major projects of collective action, aiding in the construction of massive burial platforms and the like.

Expropriation seems to be a genuine option, because the ownership of land, salmon runs, or herds of cattle is not a natural property of an agent—not a property like skill with weapons or the strength to use them. An agent has such skills and capacities independently of their recognition by other agents. Such skills, together with physical strength and social allies, are the crucial variable resources among foragers and horticulturalists (Bowles, Smith & Borgerhoff Mulder, 2010; Smith et al., 2010).[9] The same studies show that material wealth is pivotal for pastoralists and agriculturalists (Borgerhoff Mulder et al., 2010; Shenk et al., 2010). As foragers and horticulturalists rely on less heritable forms of wealth, wealth differences are relatively modest and do not accumulate over time. Pastoralists and agriculturalists rely on more heritable forms of wealth, so these economic systems tend to be more unequal at a generation, and that inequality tends to ramp up over time. Yet as we have seen, there is a gap in the analysis here. Bradburd notes in his response to the inheritance model of inequality that ownership depends on consent—on its recognition by others: "Ownership is a social institution. People own things because other members of society recognise their rights. If a society has material wealth and transmits it, then that occurs through social institutions" (Bradburd, 2010, p. 100).

Thus resource wealth or, at least, the secure and exclusive enjoyment of wealth, depends on the conventions, norms, and values of others: Land, cattle, and the like do have to be physically defended by their owners in stratified, prestate societies. But if the need to physically defend was incessant rather than occasional, these resources would be of little value to their owners. So though wealth inequality and its inheritance is probably the critical variable in the establishment of societies that are markedly unequal in wealth and power, we need an explanation of the origin and stability of the norms, customs, and values that support the claims of particular agents (or small groups) to exclusive use of resources. Since land and livestock cannot literally be transferred, the higher heritability of material capital over social capital is an explanatory target rather than an explanatory resource. Norms and customs can make social capital highly heritable too: Thus place in a social network often becomes highly heritable, as institutions like hereditary chiefdoms emerge. Yet social capital—place in a network—is even more obviously a social rather than an individual trait.

Inequality, then, is explained by a shift in the productive basis of life from foraging to agriculture or husbandry, together with a cultural system that recognizes individual (or family) control over specific resources, and that continues to recognize that control even as it becomes concentrated in fewer hands. To explain inequality, we need to explain the stability of that cultural system, not just the change in the productive basis of life. It is true that humans internalize and act on norms, but until we have an explanation of the origin and persistence of the norms of ownership, we have no explanation of stratified societies. The problem inherent in the explanatory appeal to norms is nicely made vivid by an example from Dubreuil. He discusses the case of the Baruya, a Papuan and New Guinean culture with profound gender inequalities that seem to be sustained by an ideology that presents semen as the source of life, strength, and health, whereas women and their bodily secretions are seen as a source of pollution and disorder (Dubreuil, 2010, pp. 182–183). Breast milk, for example, on this construct, is transformed semen: Nursing mothers must fuel their motherhood through fellatio. Dubreuil concludes that in this culture, "gender inequality is explained at least in part by the fact that women consider it immoral to refuse the sperm of their husbands (and thus deprive their children of high quality milk) or to pollute their husband during menstruation" (ibid., p. 182). Not so: We have no explanation here at all, without some explanation of the stability and power of this set of norms and practices.

It is possible to see how an explanation of the origins of property norms might be developed. The first element of the explanation is to see farming as akin to gathering rather than hunting, and hence legitimately produced and consumed by households. The returns to large game hunting are highly variable, and often that variation is a result of factors outside individual control. So pooling to control risk is adaptive. Variation in returns from farming may well reflect variation more in effort than luck. And local pooling will not help much: If bad luck in the form of drought or flood strikes, it will strike everyone. So farming is a household responsibility. But farming also depends on discount rates. Winterhalder and Kennett point out that farming poses new risk management problems, both because of farmers' dependence on a few key foods and because of long delays between investment and return (Winterhalder & Kennett, 2009). Hunting parties and foraging trips deliver payoffs over timescales that vary from a few hours to a week or so. Farmers must expect to wait months before they harvest crops. Moreover, farming (and herding, though perhaps to a lesser extent) demands heavy investment in preparing and maintaining soils and defending the growing crops against pests. So the shift to farming seems to depend on certainty—on low discount rates of future profit. How did early farmers have a rational confidence in low discount rates?[10]

A natural suggestion is that farming coevolved with norms and customs that entrenched agents' exclusive access to the product of lands (or herds) that they and their family worked. The social and cultural institution of property rights supported low discount rates, thus making it rational to invest time, resources, and skills in making land productive. If (and it is a big if) such norms emerged at the beginning of transitions to farming—before farming led to highly stratified societies—they may well have become generatively entrenched (Wimsatt, 2007), with norms of property and private ownership becoming foundational to many other customs and practices of the culture. Moreover, at least in some parts of the world, there was a long period—hundreds, perhaps thousands of years—of mixed production systems (if Bogucki [1999] is right, the North American "Hopewell Interaction Sphere" is one example, and temperate Europe provides others). These cultures seem to have been fairly flat, without marked differences of wealth and power. Thus, although the Hopewell culture produced many significant ceremonial structures, there does not seem to be evidence of marked wealth differences—for example, no burials with rich grave goods.

So property norms over land worked by a single household might have evolved and become entrenched before wealth differences became marked. Furthermore, as Stephen Shennan and others have noted, norms governing one aspect of a group's activities often become connected to others and become part of a group's self-identity. Once norms become linked into local ideological systems, the effects of specific norms are less salient to individual learning (or Darwinian filtering). In this respect, they are quite different from technique and technology. Individuals within a group might differ in the crops they plant, their weeding regimes, and the extent to which they invest time and energy improving their soils. And those individuals, and third parties, can observe those differences and learn from them. But property customs are a feature of a group as a whole rather than of individual behavioral dispositions, and they are typically part of a package deal that includes other customs about the organization of social and family life. Once norms become established, there is much less of the variation in practice that gives agents' epistemic leverage.[11] Moreover, it is obviously often expensive for agents (especially lone agents) to violate customary expectations, however maladaptive those customs may be.

Even if communal recognition of individual property rights is essential if farming is to emerge as a production system, this explanation is at best a partial one. It seems irrelevant to those complex foraging cultures where wealth and status depend on naturally occurring, high-value, predictably located resources. The value of a salmon run does not depend on high investments with delayed returns. Moreover, entrenched norms are never completely immune from internal challenge. That is especially true when a culture is under stress, internally or externally. One famous example is the Xhosa cattle-killing of 1856 to 1857. The Xhosa were a culture under stress through protracted and debilitating war with colonial settlers, and in that context a millenarian movement developed insisting that the Xhosa appease their ancestors by killing their own cattle (and by not sowing crops). The gods, thus appeased, would raise the dead from their ashes and drive the invaders into the sea. The result was catastrophe: Perhaps by 1857, 400,000 cattle had been slaughtered and 40,000 Xhosa had died of starvation (Peires, 1989). When cultures are in painful transition, even entrenched norms and customs are vulnerable to challenge that can lead to utter social collapse. We should not succumb to the fallacy of thinking that stratified societies are stabilized by the catastrophic costs of cultural meltdown, should cooperation and collective action collapse. Meltdown can happen.

The take-home message of this section, then, is that the growth of highly stratified societies depends on a shift to economic systems in which material capital is more important than embodied and social capital, and on cultural institutions that allow material capital to be accumulated and inherited. But the stability of these cultural institutions is itself an explanatory problem. As societies become more stratified, they serve the interests of fewer and fewer agents.

4 Cooperation in Stratified Worlds

I have argued that in the development of complex, stratified societies, there was a long period of the coexistence of (a) collective action involving many or most members of local groups; (b) significant inequality, and hence significantly unequal returns from collective action; and (c) minimal and inefficient mechanisms of coercive control. The survival—indeed, the elaboration—of cooperation and collective action through this period is puzzling. Prima facie, the gradual evolution of elites once farming and husbandry spreads is a triumph of free riding, eventually entrenched by statelike entities. Given the long history of the control of free riding, we need to explain why this trajectory was not derailed by the punishment of incipient elites, or by withdrawing cooperation from them, or both. There seem to be four mechanisms that might explain the survival of the social contract in these stratified societies. One possibility is that low-ranked agents are still making the optimal choice in cooperating; other options are even worse. The second is that low-ranked agents suffer from adaptive lag. Cooperation is not in their interests, but they lack the cognitive resources to respond appropriately in a new world of winners and losers. The third is that low-rank agent cooperation is adaptive, but for the group, not the individual agent: It is the result of cultural group selection. The fourth is a devil's bargain between the top and the middle against the landless. Early farming, as noted, is a high-risk activity, and subsistence farmers need to be risk averse. One way those with some land might express risk aversion is through endorsing and defending social norms that give them security over their minimally sufficient holdings. They might have just enough to lose to ally themselves with the influential but less numerous wealthy, rather than the more numerous but less influential land-poor (I owe this suggestion to Campbell McLachlan).

I mentioned above the possibility that participation in collective action by commoners is making the best of a bad hand. Perhaps even in this transition period, the less wealthy, the less influential, the less rich were better off cooperating, albeit with a reduced share of cooperation's profits. As noted above, that might especially be true of those in an intermediate state, with something to lose if the social world were turned upside down. It is easy to see how this might be true even for the land-poor, if collective response is off the table. A land-poor farming family might well be best off accepting a subordinate status to a rich neighbor, perhaps marrying their daughter off as a second or third wife. In general, it rarely pays a lone individual to defect from the customs of his or her local group. But why would collective response be off the table? Commoners had the cognitive and cultural machinery for collective action, and elites lacked efficient means of coercion. So what prevented the less wealthy majority from ganging up on or excluding the wealthy minority? One possibility is that farming cultures expanded in size before becoming markedly socially differentiated. The larger and less intimate the social world, the more difficult bottom-up coordination becomes. Another possibility is that the costs of social disruption—of the fissures in the local community—would impose very high costs on the members of the coalition; I shall return to this idea in considering the potential role of group selection.

There has been very little explicit discussion of the robustness of the social contract in stratified societies, but those who have considered the issue have appealed to adaptive lag. Peter Richerson (with Robert Boyd) and Paul Seabright develop ideas along these lines. They suggest that features of mind and culture that evolved in Pleistocene-scale worlds were co-opted to support collective action in large-scale worlds, independently of whether such action was adaptive for the agent. Thus Richerson and Boyd argue that some large-scale social institutions support a cognitive and affective illusion: the illusion of still being part of a tribal world (Richerson & Boyd, 2001). Paul Seabright develops a different version of the same idea, one that accords a greater weight to the role of conventions, norms, and fictive kinship systems in motivating prosocial action, as the size of human social worlds expands (Seabright, 2010). As I argued in section 3, although norms and customs no doubt play central roles in stabilizing social contracts, their power and stability is part of our explanatory target.

Perhaps intergroup selection has a role to play in explaining the prevalence of such norms and customs. Only those social entities whose norms, conventions, religions, and kinship systems promoted collective action survived in the dog-eat-dog world of the Holocene. Sam Bowles has recently defended the idea that selection for human cooperation in the Pleistocene depended on intense and often violent intergroup competition. In my view, the analysis is unconvincing: The archaeological evidence of intergroup violence is late Pleistocene and Holocene, and the cost-benefit analysis overstates the benefits of aggression (mobile, armed foragers are difficult targets with not much to steal) and understates the risks. But if we think of this as a picture of the early to mid-Holocene, it is much more plausible. The evidence of conflict is not universal even there: Holocene farming cultures are not always organized around fortified villages. But there are many early Holocene cultures that have clearly invested heavily in fortification. Herding cultures are even more vulnerable: Livestock is mobile and vulnerable to theft; hence the extreme "culture of honor" norms of revenge associated with pastoralism. So there is some plausibility in seeing the prestate Holocene as an arena of intense cultural group selection.

If cultural groups are strongly filtered for norms promoting cooperation and collective action, we might have an explanation of the psychological power and stability of norms of collective action in these stratified societies—norms that induce agents to engage in collective action even when their own interests are not promoted by adhering to the social contract. But if the environment was one of tense, often violent intergroup relations, it might also be true that even for the poor adhering to the social contract is the best of a grim set of options. At the beginning of this section, I suggested that elites are free riders who have escaped sanction and are successfully expropriating the profits of collective action. But there is considerable debate about the social role of elites in stratified, prestate societies.[12] On one view, they are indeed social parasites running a protection racket, taking advantage of coordination failure among the poorer majority. On the other, elites play a crucial managerial role, organizing essential, large-scale collective action projects. Of course, both views could be right. In particular, in environments of severe competition between local communities, elites may well exploit their critical managerial function to extract a disproportionate share of the group's resources.

Forager social worlds probably do not face urgent, large scale coordination and management problems. Their foraging practices do not offer important economies of scale. There may be a few circumstances in which very large forager groups can assemble and hunt: for example, when hunting large migratory herds. But in general, the profit of hunting does not rise with group size. Likewise, much collective decision making can be leisurely. It will not usually be critical that a decision about camp location is made today rather than tomorrow. Bottom-up, consensus decision making usually suffices. In a world of serious intergroup strife, none of this is true. Famously, God is on the side of the big battalions, and so there are very great economies of scale in military interactions. And military decisions are often urgent decisions. Intergroup hostility may well select for top-down mechanisms of coordination and decision. We thus get a somewhat heterodox view of the interactions between within-group and across-group conflict. David Wilson has often pictured group-group competition as selecting for harmonious cooperation within a group (Wilson, 2002). We play nicely with one another, for otherwise we all lose. But on the picture imagined here, group-group competition sets up the conditions that permit elite takeovers. The cost of competitive failure at the group level is so high that low-ranking agents are induced to accept exploitative outcomes of collective action. In effect, elites are playing and winning a game of brinkmanship: Empower and enrich us, or we all face disaster. The commoners have blinked.

Clearly, practices of collective action did survive the transitions from foraging, to small-scale farming, and thence to larger, more vertically complex, inegalitarian, command-and-control societies. Presumably, making the best of a poor set of options, devils' bargains, adaptive lag, and intergroup competition have all played some role in the survival of the social contract in stratified worlds. Group selection in favor of cultural traits that sustain collective action, the co-option of cultural mechanisms that once supported adaptive choices, and making the best of a poor hand have all been important. Even so, there is an unsolved chicken-and-egg problem in the transition from the bottom-up organization of collective action in small, relatively egalitarian worlds to the top-down, coercion-backed organization of collective action in larger, inegalitarian social worlds. Egalitarian, intimate world cooperation is stable: It pays most agents to cooperate most of the time. State society cooperation is stable, because elites can coerce commoners. Paradoxically, though, the historical record is one of incremental, hill-climbing transitions from egalitarian to stabilized stratified societies, rather than rapid flips from one stable system to the other.

Notes

1. For the cooperation problem in general, see Okasha (2006); for an insightful overview of cooperation and increasing social complexity in the hominin lineage, see Foley and Gamble (2009) and Tomasello, Melis, et al. (forthcoming).

2. This view of hunting remains controversial. It has most recently been defended in detail by Gurven and Hill (2009). Kristin Hawkes and her coworkers continue to defend a signaling view of the role of hunting in forager society: see Hawkes and Bird (2002); they respond directly to Gurven and Hill in Hawkes, O'Connell, and Coxworth (2010). I evaluate this debate in my *The Evolved Apprentice* (Sterelny 2012, ch. 4.5), conceding something to Hawkes but defending a version of the provisioning view of hunting.

3. Vertical flow is parent–to-offspring transmission; oblique flow is from adults to other children in the next generation; horizontal flow is learning from others of the same generation.

4. I think we also have a quite persuasive picture of the origins of hominin cooperation, but that would require a separate argument.

5. These ethnographic impressions have recently been confirmed in quantitative detail, though with small sample sizes, in a special issue of *Current Anthropology* on wealth differentials in prestate societies. See Bowles, Smith & Borgerhoff Mulder (2010) and the following papers (*Current Anthropology* 51, [1], February 2010).

6. As Dunbar (1996), in particular, has emphasized.

7. Benoit Dubreuil argues that punishing intimates is cheaper than punishing strangers, because they are so well known to the punishing agent: He or she will know what punishment will balance the transgression. With strangers, it is easy to over- or underpunish through ignorance. Indeed, this claim is fundamental to his picture of the development of hierarchy (Dubreuil, 2010, pp. 157–161). I think this is quite wrong: It again ignores the risk costs. The cost of fracturing social bonds with intimates is very high, especially as such fractures have third-party effects: Alienating one friend will often alienate others.

8. For a good survey of these changes, with many examples, see Bogucki (1999).

9. This case is made in empirical detail in the case studies on wealth inequality in a special issue of *Current Anthropology*: vol. 51, no. 1, February 2010.

10. See Bogucki (1999, pp. 193–195) for an emphatic defense of the importance of family unit control over production and its profits to the origins of agriculture. He takes this to require a major shift away from forager sharing norms. But those norms, as I read the literature, apply much more to large game hunting than to gathered resources (see, e.g., Gurven, 2004). So I think he overstates the ideological break from forager life. But he is certainly right that no one will rationally invest in the labor and experimentation required for farming unless he or she is fairly sure that he or she will get the rewards.

11. I discuss the relative insensitivity of norms to trial and error filtering in some detail in Sterelny (2007).

12. Briefly and clearly reviewed by Bogucki (1999, pp. 263–267), with Sahlins defending the more benign view of elites, and Gilman skeptical.

References

Bandler, H. (1995). Water resources exploitation in Australian prehistory environment. *Environmentalist, 15*, 97–107.

Binmore, K. (2006). Why do people cooperate? *Politics, Philosophy & Economics, 5*, 81–96.

Boehm, C. (1999). *Hierarchy in the forest*. Cambridge, MA: Harvard University Press.

Bogucki, P. (1999). *The origins of human society*. Oxford: Blackwell.

Borgerhoff Mulder, M., Fazzio, I., Irons, W., McElreath, R. L., Bowles, S., Bell, A., Hertz, T., & Hazzah, L. (2010). Pastoralism and wealth inequality: Revisiting an old question. *Current Anthropology, 51*(1), 35–49.

Bowles, S., Smith, E. A., & Borgerhoff Mulder, M. (2010). The emergence and persistence of inequality in premodern societies: Introduction to the special section. *Current Anthropology, 51*(1), 7–17.

Bradburd, D. (2010). Studying wealth transmission and inequality in premodern societies: Some caveats. *Current Anthropology, 51*(1), 99–100.

Currie, T., Greenhill, S., et al. (2010). Rise and fall of political complexity in island South-East Asia and the Pacific. *Nature, 467*, 801–804.

Dubreuil, B. (2010). *Human evolution and the origin of hierarchies*. Cambridge: Cambridge University Press.

Dunbar, R. (1996). *Grooming, gossip, and the evolution of language*. London: Faber & Faber.

Foley, R., & Gamble, C. (2009). The ecology of social transitions in human evolution. *Philosophical Transactions of the Royal Society of London, Series B: Biological Sciences, 364*, 3267–3279.

Gurven, M. (2004). To give and to give not: The behavioral ecology of human food transfers. *Behavioral and Brain Sciences, 27*, 543–583.

Gurven, M., & Hill, K. (2009). Why do men hunt? A reevaluation of "man the hunter" and the sexual division of labor. *Current Anthropology 50*(1): 51–74.

Hawkes, K. (2003). Grandmothers and the evolution of human longevity. *American Journal of Human Biology, 15*(3), 380–400.

Hawkes, K., & Bird, R. (2002). Showing off, handicap signaling, and the evolution of men's work. *Evolutionary Anthropology, 11*(1), 58–67.

Hawkes, K., O'Connell, J. F., & Coxworth, J. E. (2010). Family provisioning is not the only reason men hunt. *Current Anthropology, 51*(2), 259–264.

Hrdy, S. B. (2009). *Mothers and others: The evolutionary origins of mutual understanding*. Cambridge, MA: Harvard University Press.

Kaplan, H., Hooper, P., & Gurven, M. (2009). The evolutionary and ecological roots of human social organization. *Philosophical Transactions of the Royal Society of London, Series B: Biological Sciences, 364*, 3289–3299.

Lombard, M., & Phillipson, L. (2010). Indications of bow and stone-tipped arrow use 64 000 years ago in KwaZulu-Natal, South Africa. *Antiquity, 84*, 635–648.

Marlowe, F. W. (2005). Hunter-gatherers and human evolution. *Evolutionary Anthropology, 14*, 54–67.

Mithen, S. (2003). *After the ice: A global human history 20000–5000 BC*. London: Weidenfeld & Nicolson.

Ofek, H. (2001). *Second nature: Economic origins of human evolution*. Cambridge: Cambridge University Press.

Okasha, S. (2006). *Evolution and the units of selection*. Oxford: Oxford University Press.

Peires, J. P. (1989). *The dead will arise: Nongqawuse and the great Xhosa cattle-killing movement of 1856–7*. Bloomington: Indiana University Press.

Richerson, P., & Boyd, R. (2001). The evolution of subjective commitment to groups: A tribal instincts hypothesis. In R. Nesse (Ed.), *Evolution and the capacity for commitment* (pp. 186–220). New York: Russell Sage Foundation.

Ross, D. (2006). Evolutionary game theory and the normative theory of institutional design: Binmore and behavioral economics. *Politics, Philosophy & Economics, 5*(1), 51–80.

Seabright, P. (2010). *The company of strangers: A natural history of economic life*. Princeton: Princeton University Press.

Shea, J. (2003). Neanderthals, competition, and the origins of modern human behavior in the Levant. *Evolutionary Anthropology, 12*, 173–187.

Shea, J. (2009). The impact of projectile weaponry on Late Pleistocene hominin evolution. In J. J. Hublin and M. P. Richards (Eds.), *The evolution of hominid diets* (pp. 187–198). Berlin: Springer Science.

Shenk, M. K., Borgerhoff Mulder, M., Beise, J., Clark, G., Irons, W., Leonetti, D., Low, B. S., Bowles, S., Hertz, T., Bell, A., and Piraino, P. (2010). Intergenerational wealth transmission among agriculturalists: Foundations of agrarian inequality. *Current Anthropology, 51*(1), 65–83.

Skyrms, B. (2003). *The stag hunt and the evolution of social structure*. Cambridge: Cambridge University Press.

Smith, E. A., Hill, K., Marlowe, F., Nolin, D., et al. (2010). Wealth transmission and inequality among hunter-gatherers. *Current Anthropology, 51*(1), 19–34.

Sterelny, K. (2007). SNAFUS: An evolutionary perspective. *Biological Theory, 2*(2), 317–328.

Sterelny, K. (2011). From hominins to humans: How sapiens became behaviourally modern. *Philosophical Transactions of the Royal Society, London, Series B, 366*(1566), 809–822.

Sterelny, K. (2012). *The evolved apprentice: How evolution made humans unique*. Cambridge, MA: MIT Press.

Stockton, J. (1982). Stone-wall fish traps in Tasmania. *Australian Archaeology, 14*, 107–114.

Tomasello, M., Melis, A., et al. (Forthcoming). Two key steps in the evolution of human cooperation: The interdependence hypothesis. *Current Anthropology*.

Wilson, D. S. (2002). *Darwin's cathedral: Evolution, religion, and the nature of society*. Chicago: University of Chicago Press.

Wimsatt, W. C. (2007). *Re-engineering philosophy for limited beings: Piecewise approximations to reality*. Cambridge, MA: Harvard University Press.

Winterhalder, B., & Kennett, D. (2009). Four neglected concepts with a role to play in explaining the origins of agriculture. *Current Anthropology, 50*(5), 645–648.

5 The Birth of Hierarchy

Paul Seabright

Archaeologists and anthropologists are now in broad agreement that forager societies were substantially more egalitarian than virtually all the societies that succeeded them after the widespread adoption of agriculture. It is no longer tenable to claim that this is because the naturally egalitarian instincts of humankind have been corrupted by modern society. Christopher Boehm (1999), who surveys the evidence for the egalitarian nature of forager societies, argues that in fact their members seem to have been at least as status-conscious and competitive as their modern descendants. However, human beings also share both a taste and a talent for collaborating to restrain the behavior of individuals who press their competitive instincts too far, who claim too large a share of the benefits of social life. What made forager societies different from those that succeeded them was that "coalitions of losers" were able to keep the self-aggrandizement of winners very substantially in check. The most notable exceptions were complex hunter-gatherer societies like those of the Pacific Northwest, which through control over salmon runs enjoyed some of the benefits of the sedentary life without engaging in cultivation (see Hayden, 1995). Let's call the more typical forager outcome "the egalitarianism of countervailing power," to distinguish it from other kinds of egalitarianism, such as the kind Jean-Jacques Rousseau believed had once existed in the absence of the competitive instincts allegedly instilled by modern social living. What were the conditions that made possible the egalitarianism of countervailing power, and why were they undermined by the arrival of agriculture?

It is also widely agreed among scholars that the conditions that made forager egalitarianism possible included the need for cooperation in foraging (especially but not only in hunting) and the infeasibility of coercion in bringing that cooperation about. Hunting and gathering cannot be effectively performed at the point of a spear or while wearing a ball and chain. Agricultural societies, by contrast, can and to a massive extent did put slaves or simply very poor laborers to work tilling fields or building monuments. And the earliest states had substantial organized coercion: in Bruce Trigger's words, "in all the early civilizations for which we have adequate documentation, the privileges of the upper classes were protected by armed forces and the legal system" (Trigger, 2003, p. 265).

It seems reasonably clear why these two sets of technological and social conditions (foraging-plus-low-inequality and agriculture-plus-high-inequality) could each have represented an equilibrium of social behavior. The equilibrium would have been maintained in the former case by rewards for cooperation and social, rather than physically coercive, penalties for ordinary selfish behavior. Interesting questions would remain about why human beings had evolved to be so responsive to social penalties, but it is hard to doubt that they have.

In the latter case, the equilibrium would have been maintained by coercion, with those responsible for enforcing the coercive penalties motivated in turn by fear of coercion. It also seems reasonably clear why high inequality could not have been a characteristic of a forager economy—those at the bottom of the hierarchy would simply not have had enough stake in the outcome to be willing to cooperate. Thus in a forager economy *only* the egalitarianism of countervailing power was a feasible equilibrium.

However, Kim Sterelny argues in his chapter in this volume that it does not follow that we can understand how hierarchy arose, nor *a fortiori* why it became so widespread in such a comparatively short time after the spread of agriculture. Though others have seen the need for specific theories of the transition (see, e.g., Kennett et al., 2009), this is the first time I have see this argument so clearly and explicitly made. Hierarchy was *one* equilibrium of Holocene life, but this does not make it the only equilibrium. Nothing in the logic of hierarchy rules out the possibility that the egalitarianism of countervailing power remained an alternative equilibrium of Holocene life, because unless established institutions of coercion *already* existed, potential hierarchs might have had no way to impose their will. If in fact the egalitarianism of countervailing power was no longer an equilibrium once agriculture became widespread (and the evidence of the near-universal movement of human societies toward substantially greater hierarchy strongly suggests this), we need to understand what it was about agriculture that made this happen. Appealing to conditions after the establishment of hierarchy (when coercion was readily available to reinforce the position of those who benefited from it) will not tell us why social entrepreneurs could get away with appropriating a far larger share of the rewards of cooperation for themselves before the conditions for coercion of the losers were present. What made such entrepreneurs able to succeed in the Holocene (after agricultural technology but before the institutions of coercion) with initiatives that could never have succeeded in the Pleistocene?

Figures 5.1 and 5.2 illustrate this difficulty. Figure 5.1 shows a relation between the development of agricultural technology and the average degree of inequality in society, according to the model that most historians and anthropologists appear implicitly to

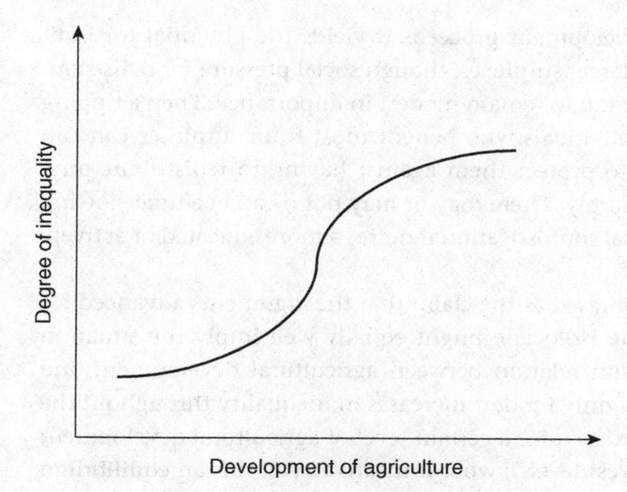

Figure 5.1
A single social equilibrium.

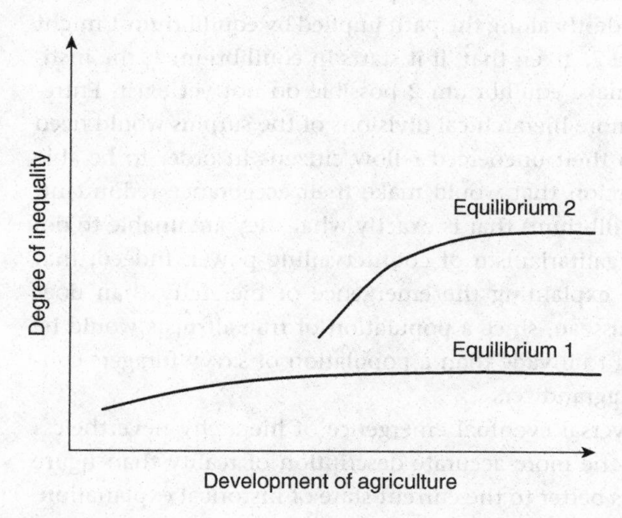

Figure 5.2
Two social equilibria.

have adopted. As agricultural development proceeds it yields the potential for individuals to accumulate somewhat larger surpluses, though social pressure for redistribution initially causes this phenomenon to remain modest in importance. Then a tipping point is reached at which the individuals who benefit most from surpluses can call on the instruments of coercion to protect them against having to redistribute anything, and inequality increases sharply. There may or may not be an eventual slowing of the process as greater agricultural sophistication requires a more educated or actively participative workforce.

Sterelny's objection can be rephrased as the claim that the arguments advanced for the possibility of hierarchy in the Holocene might equally well imply the situation in figure 5.2. Here the equilibrium relation between agricultural development and inequality continues to hold with only modest increases in inequality throughout the Holocene. Another equilibrium exists once a certain level of agricultural development is reached (though not at the lowest levels); what it means to call this an equilibrium is that *if* the institutions of coercion exist then members of society will comply with the hierarchy, and the existence of those institutions of coercion is one of the features of the hierarchy that is in turn assured by their compliance. However, it is not at all clear how a society advancing modestly along the path implied by equilibrium 1 might find itself jumping to equilibrium 2, given that, if it starts in equilibrium 1, the institutions of coercion that would make equilibrium 2 possible do not yet exist. Entrepreneurs who seek to introduce more hierarchical divisions of the surplus would need first to make these acceptable to their uncoerced fellow citizens in order to be able subsequently to pay for the coercion that would make their acceptance redundant. And (by the definition of an equilibrium) that is exactly what they are unable to do, according to the model of the egalitarianism of countervailing power. Indeed, that model has a much harder time explaining the emergence of hierarchy than does egalitarianism in the style of Rousseau, since a population of true altruists would be much easier for a selfish hierarch to invade than a population of savvy foragers continually on the lookout for self-aggrandizers.

The evidence of the near-universal eventual emergence of hierarchy nevertheless suggests that figure 5.1 must be the more accurate description of reality than figure 5.2, even if figure 5.2 corresponds better to the current state of historical explanation. In addition, figure 5.2 implies that when hierarchy emerged it did so in a relatively rapid and discontinuous jump. This is something that Currie et al. (2010) find to be a very implausible model of the evolution of 84 Austronesian-speaking societies, analyzed using phylogenetic methods. They estimate closeness of these societies in time via closeness in certain basic vocabulary terms, estimated probabilistically over a large number of samples of vocabulary terms, and note that societies close in time are also only incrementally different in degree of political complexity. The evidence would not look like this if, for example, societies had passed from relatively egalitarian to very hierarchical in a short space of time.

So how might it have happened that entrepreneurs could introduce hierarchy even before a technology of coercion existed? Four possible explanations, not mutually exclusive, strike me as worth exploring.

First, coercion might have been easier even without the existence of armies and police forces, simply through the threat of exclusion from the community for those who did not accept the new hierarchical division. A forager community cannot really afford to exclude more than one or two of its members: all hands, or nearly all hands, are needed for the collective tasks of hunting and gathering. A community of farmers has fewer tasks that are strictly collective (Shennan, 2011, makes this point), though it should not be thought that it has none at all: Labor sharing between households is often important to smooth out seasonal fluctuations as well as yielding important risk-sharing benefits. If, in addition, it is hard for individuals or small groups who are expelled from a farmer community to survive on their own, the mere threat of expulsion might have had a more coercive influence than it had previously done for foragers. Kennett et al. (2009) report evidence from California's Northern Channel Islands suggesting that social stratification was more likely where there was rapidly declining marginal productivity of land. Evidence from other species supports this: Tim Clutton-Brock has argued that the extreme reproductive skew of females in meerkat communities (subordinate females have no offspring of their own but care for the offspring of the dominant female) is made possible not just by the kinship relation of the subordinates to the dominant, who is usually their own mother, but also by the fact that expulsion of subordinates in groups smaller than three or four leads to their almost certain death (Clutton-Brock, 2011).

Second, it may be that the egalitarianism of countervailing power was an equilibrium only in a statistical sense. Perhaps not all attempts at establishing hierarchy produced a successful backlash against the aggrandizers. Indeed, Brian Hayden (1995) has argued that it was the emergence of a degree of hierarchy in resource-rich environments that itself provided the spur for domestication through intensified status competition. The arguments for the egalitarianism of countervailing power imply only that most such attempts failed, not that they all did. In a forager economy, the exceptions remained unusual because the exceptions had no greater ability to reproduce themselves than the others. But in an agricultural economy, a successful hierarchical exception might then have been able to use its surpluses to build armies that could conquer and subjugate other, more egalitarian groups. Egalitarian solidarity might have been very effective militarily against hierarchical foragers but very ineffective against hierarchical agriculturists. This effectiveness could only have come from the higher populations characteristic of agricultural communities, leading armies of farmers to outnumber armies of foragers. It would have been unlikely to come from superior technologies or better motivation. Indeed, Hanson (1989) has emphasized the greater military effectiveness of the relatively egalitarian Greek armies against the Persians—evidence that, though coming from a much later period, illustrates the general difficulty that armies of unwilling conscripts have in clashes with armies of comparative volunteers. Likewise, the greater mobility and more practiced tracking skills of foragers would have given them an advantage that only weight of numbers could offset. But weight of numbers would have been a possible source of advantage for hierarchical agricultural societies in a way that it was not for foragers.

Third, would-be hierarchs might have been able to exploit for their own purposes characteristics in their fellow citizens that had evolved for other purposes, much as parasites can exploit features of their host's metabolism that were originally adaptive for the host even though in their capacity as recruited by the parasite they reduce the host's fitness. So, for example, individual members of forager groups may have developed a willingness to sacrifice their own interests in warfare against rival groups, an argument developed in detail by Bowles and Gintis (2011) in their recent book. Then would-be hierarchs who could most effectively appeal to sentiments of out-group hostility might have been able to use these to justify "necessary" sacrifices of many kinds. Unfortunately, though, this suggestion does not explain why such a parasitic exploitation of in-group altruism and out-group hostility should have become so much easier with the development of agriculture. Furthermore, it ignores the fact that sacrifices are easier to justify if they are shared among all group members, which is difficult to square with the presence of marked hierarchy. Other features of ideological justification of sacrifice (see Yoffee, 2005, esp. pp. 38–41) may have been important later on, but we still need to explain how the previously egalitarian foragers came to be susceptible to their appeal.

A fourth possible explanation builds on this notion that in-group altruism may have coevolved with out-group hostility, by suggesting a way in which a social innovation could have introduced hierarchy without triggering the usual countervailing reactions against aggrandizers, namely, via the institution of slavery. The ethnography of modern-day foragers suggests, and the archaeology of the late Pleistocene confirms, that warfare among forager groups was probably frequent, and frequently lethal (Bowles, 2009). When a group was defeated by rivals, its women would sometimes have been incorporated into the victor population, but the males would usually have been killed, since they would have constituted a negative resource given the impossibility of ensuring their uncoerced acquiescence and the high costs of watching and guarding them. Agriculture changed all this, by creating work that could be productively performed by coerced individuals—by slaves, in short. Plowing, weeding, and harvesting can all be undertaken by workers whose legs are restrained sufficiently to prevent them from running away, and the work can be supervised by many fewer people than are actually working—none of this is true of most foraging activity. The first hierarchies may well have been those in which indigenous farmers lorded it over slaves abducted from rival groups. It would then have been relatively easy to construct institutions of coercion aimed at creating and reinforcing hierarchy among the remaining farmers. Slaves could be offered as an inducement to those individuals who helped the first hierarchs to establish their dominance. This could have tipped the balance for many subordinate males between choosing to join a coalition to restrain an aggrandizer and choosing to join the aggrandizer.

What kind of evidence might help corroborate or refute these four hypotheses? We cannot replay the tape to watch the rise of hierarchy, but each hypothesis has testable implications. The first hypothesis (the effectiveness of the threat of exclusion) implies that, where we do see exclusion practiced by existing forager communities, mortality among the excluded individuals should be high, and we should see exclusion practiced as a disciplinary mechanism among existing agricultural communities that lack the formal institutions of coercion. The second implies that, where we find ethnographic or archaeological evidence of conflict between forager and farmer communities, mortality among foragers should be substantially higher, on average, than among farmers. The third implies, at a minimum, that we should find evidence of cults of worship or otherwise honorific treatment of high-status individuals even in societies where no institutions of coercion exist. Finally, the fourth implies that we should find evidence of slavery very early in the Neolithic, well before the rise of those states in which the existence of slavery is undisputed. Slavery should precede institutionalized coercion and not vice versa. I am not aware of systematic surveys of the evidence on any of these predictions, but the very existence of testable predictions means these hypotheses need not rest forever conjectural.

To summarize, it is not enough to show that hierarchy would have been self-sustaining once agricultural development permitted large enough surpluses to pay for the institutions of coercion. We need to show how hierarchy could initially have become established before such institutions existed. Various explanations are possible; choosing between them provides an intriguing challenge for future research.

References

Boehm, C. (1999). *Hierarchy in the forest: The evolution of egalitarian behavior*. Cambridge, MA: Harvard University Press.

Bowles, S. (2009). Did warfare among ancestral hunter-gatherer groups affect the evolution of human social behaviors? *Science, 324*, 1293–1298.

Bowles, S., & Gintis, H. (2011). *A cooperative species: Human reciprocity and its evolution*. Princeton: Princeton University Press.

Clutton-Brock, T. (2011). Animal societies. Plenary lecture to Human Behavior and Evolution Society annual congress, Montpellier, Vermont.

Currie, T. E., Greenhill, S. J., Gray, R. D., Hasegawa, T., & Mace, R. (2010). Rise and fall of political complexity in island South-East Asia and the Pacific. *Nature*, *467*, 801–804.

Hanson, V. D. (1989). *The Western way of war: Infantry battle in classical Greece*. Berkeley: University of California Press.

Hayden, B. (1995). A new overview of domestication. Chapter 10 of T. Douglas Price & A. Gebauer (Eds.): *Last hunters—first farmers: New perspectives on the prehistoric transition to agriculture*. Santa Fe, NM: School of American Research Press.

Kennett, D., Winterhalder, B., Bartruff, J., & Erlandson, J. M. (2009). An ecological model for the emergence of institutionalized social hierarchies on California's Northern Channel Islands. Chapter 20 of Stephen Shennan (Ed.): *Pattern and process in social evolution*. Berkeley: University of California Press.

Shennan, S. (2011). Property and wealth inequality as cultural niche construction. *Philosophical Transactions of the Royal Society, Series B: Biological Sciences, 366*, 918–926.

Trigger, B. (2003). *Understanding early civilizations*. Cambridge: Cambridge University Press.

Yoffee, N. (2005). *Myths of the archaic state: Evolution of the earliest cities, states, and civilizations*. Cambridge: Cambridge University Press.

6 Territoriality and Loss Aversion: The Evolutionary Roots of Property Rights

Herbert Gintis

The *endowment effect* is the notion that people value a good that they possess more highly than the same good when they do not possess it. Experimental studies (see sec. 1) have shown that subjects exhibit a systematic endowment effect. The endowment effect gives rise to *loss aversion*, according to which agents are more sensitive to losses than to gains. The leading analytical model of loss aversion is *prospect theory* (Kahneman & Tversky, 1979). This chapter suggests a plausible argument for the existence and importance of the endowment effect and loss aversion in humans and links this with territoriality in nonhuman animals.

By *natural property* I mean respect for property in the absence of legal institutions ensuring third-party contract enforcement. The endowment effect can be modeled as natural property. Natural property has been observed in many species, in the form of the recognition of territorial possession. I will develop a model of natural property loosely based on the Hawk-Dove game (Maynard Smith & Parker, 1976) and the War of Attrition (Maynard Smith, Parker & Price, 1973) to explain the natural evolution of property. Jones (2001) and Stake (2004) have developed analyses of the evolution of property stressing similar themes.

I show that if agents in a group exhibit the endowment effect, then property rights in that resource can be established on the basis of incumbency. The model assumes the agents know the present value π_g of incumbency, as well as the present value π_b of nonincumbency, measured in units of biological fitness. We assume utility and fitness coincide, except for one situation, described below: this situation explicitly involves loss aversion, where the disutility of loss exceeds the fitness cost of loss. When an incumbent faces an intruder, the intruder determines the expected value of attempting to seize the resource, and the incumbent determines the expected value of contesting versus ceding incumbency when challenged. These conditions will not be the same, and in plausible cases there is a range of values of π_g/π_b for which the intruder decides not to fight, and the incumbent decides to fight if challenged.

I call this a (natural) *property equilibrium*. We will assume all agents have equal fighting size and power, and hence equal fighting ability. In a property equilibrium, since the potential contestants are of equal power, it must be the case that individuals are *loss averse*, the incumbent being willing to expend more resources to hold the resource than the intruder is to seize it. Note that in a property equilibrium, potential usurpers respect property rights not based on moral obligation, even in part, but rather based on a purely self-regarding cost-benefit analysis. Similarly, incumbents defend property rights because they are genetically predisposed to do so, and not at all because they want to punish norm-violators or protect valuable social institutions. Of course, the property equilibrium would persist if we relaxed the equal power assumption, provided that power differences are appropriately small or are not systematic (e.g., older adults are not systematically larger and stronger than younger adults). Indeed, some of our examples illustrate the (limited) respect for property in hierarchically ordered groups where unequal power is prominent.

The present values of incumbency and its absence, π_g and π_b, will generally be endogenous in a fully specified model. Their values will depend on the supply of the resource relative to the number of agents, the intrinsic value of the resource, the ease in finding an unowned unit of the resource, and other issues extensively explored by Gintis (2007).

Our model of decentralized property is like the "Bourgeois" equilibrium in the Hawk-Dove game, in that agents contest for a unit of an indivisible resource, contests may be very costly, and in equilibrium, incumbency determines who holds the resource without costly contests. Our "property equilibrium," like the Bourgeois strategy in the Hawk-Dove game, is both evolutionarily stable and can invade the Hawk-Dove equilibrium. Our model also fills critical gaps in the Hawk-Dove game. The central ambiguity of the Hawk-Dove game is that it treats the cost of contesting as exogenously given and taking on exactly two values, high for Hawk and low for Dove. Clearly, however, these costs are in large part under the control of the agents themselves and should not be considered exogenous. In our model, the level of resources devoted to a contest is endogenously determined, and the contest itself is modeled explicitly as a modified War of Attrition, the probability of winning being a function of the level of resources committed to combat. One critical feature of the War of Attrition is that the initial commitment of a level of resources to a contest must be *behaviorally ensured by the agent*, so that the agent will continue to contest even when the costs of doing so exceed the fitness benefits. Without this precommitment, the incumbent's threat of "fighting to the death" would not be credible (i.e., the chosen best response of the agent would not be subgame perfect). From a behavioral point of view, this precommitment can be summarized as the incumbent having a degree of *loss aversion* leading his utility to differ from his fitness.

Our fuller specification of the behavioral underpinnings of the Hawk-Dove game allows us to determine the conditions under which a property equilibrium will exist while its corresponding antiproperty equilibrium (in which a new arrival rather than the first entrant always assumes incumbency) does not exist.

1 The Endowment Effect and Territoriality

The endowment effect, according to which a good is more highly prized by an agent who is in possession of the good than one who is not, was first documented by the psychologist Daniel Kahneman and his coworkers (Tversky & Kahneman, 1981; Kahneman et al., 1991; Thaler, 1992). Thaler describes a typical experimental verification of the phenomenon as follows. Seventy-seven students at Simon Fraser University were randomly assigned to one of three conditions: Seller, Buyer, or Chooser. Sellers were given a mug with the University logo (selling for $6.00 at local stores) and asked whether they would be willing to sell at a series of prices ranging from $0.25 to $9.25. Buyers were asked whether they would be willing to purchase a mug at the same series of prices. Choosers were asked to choose for each price between receiving a mug or that amount of money. The students were informed that a fraction of their choices, randomly chosen by the experimenter, would be carried out, thus giving the students a material incentive to reveal their true preferences. The average Buyer price was $2.87, while the average Seller price was $7.12. Choosers behaved like Buyers, being on average indifferent between the mug and $3.12. The conclusion is that owners of the mug valued the object more than twice as highly as nonowners.

The aspect of the endowment effect that promotes natural property is known as *loss aversion*: the disutility of giving up something one owns is greater than the utility associated with acquiring it. Indeed, losses are commonly valued at about twice that of gains, so that to induce an individual to accept a lottery that costs $10 when one loses, it must offer a $20 payoff when one wins (Camerer, 2003). Assuming that an agent's willingness to engage in combat over possession of an object is increasing in the subjective value of the object, owners will be prepared to fight harder to *retain* possession than nonowners are to *gain* possession. Hence there will be a bias in favor of recognizing property by virtue of incumbency, even where third-party enforcement institutions are absent.

We say an agent *owns* something (is incumbent with respect to that thing) if the agent has exclusive access to it and the benefits that flow from this privileged access. We stress that respect of ownership as modeled in this chapter does not depend on moral reasoning. Ownership equals incumbency, and it is *respected* if it is rarely contested and, when contested, generally results in ownership remaining with the incumbent.

The dominant view in Western thought, from Hobbes, Locke, Rousseau, and Marx to the present, is that property is a human social construction that emerged with the rise of modern civilization (Schlatter, 1973). However, evidence from studies of animal behavior, gathered mostly in the past quarter-century, has shown this view to be incorrect. Various territorial claims are recognized in nonhuman species, including butterflies (Davies, 1978), spiders (Riechert, 1978), wild horses (Stevens, 1988), finches (Senar, Camerino & Metcalfe, 1989), wasps (Eason et al., 1999), nonhuman primates (Ellis, 1985), lizards (Rand, 1967), and many others (Mesterton-Gibbons & Adams,

2003). There are, of course, some obvious forms of incumbent advantage that partially explain this phenomenon: The incumbent's investment in the territory may be idiosyncratically more valuable to the incumbent than to a contestant, or the incumbent's familiarity with the territory may enhance its ability to fight. However, in the above-cited cases, these forms of incumbent advantage are unlikely to be important. Thus, a more general explanation of territoriality is needed.

In nonhuman species, that an animal owns a territory is generally established by the fact that the animal has occupied and altered the territory (e.g., by constructing a nest, burrow, hive, dam, or web, or by marking its limits with urine or feces). In humans there are other criteria of ownership, but physical possession and first to occupy remain of great importance. According to John Locke, for example,

> every man has a property in his own person. The labour of his body, and the work of his hands, we may say, are properly his. Whatsoever then he removes out of the state that nature hath provided, and left it in, he hath mixed his labour with, and joined to it something that is his own, and thereby makes it his property. (Locke, *Second Treatise on Government*, 1690)

Since property in human society is generally protected by law and enforced by complex institutions (judiciary and police), it is natural to view property in animals as a categorically distinct phenomenon. In fact, however, decentralized, self-enforcing types of property, based on behavioral propensities akin to those found in nonhuman species (e.g., the endowment effect), are important for humans and arguably lay the basis for more institutional forms of property rights. For instance, many developmental studies indicate that toddlers and small children use behavioral rules similar to those of animals in recognizing and defending property rights (Furby, 1980).

How respect for ownership has evolved and how it is maintained in an evolutionary context is a challenging puzzle. Why do loss aversion and the endowment effect exist? Why do humans fail to conform to the smoothly differentiable utility function assumed in most versions of the rational actor model? The question is equally challenging for nonhumans, although we are so used to the phenomenon that we rarely give it a second thought.

Consider, for instance, the sparrows that built a nest in a vine in my garden. The location is choice, and the couple spent days preparing the structure. The nest is quite as valuable to another sparrow couple. Why does another couple not try to evict the first? If they are equally strong, and both value the territory equally, each has a 50 percent chance of winning the territorial battle. Why bother investing if one can simply steal (Hirshleifer, 1988)? Of course, if stealing were profitable, then there would be no nest building, and hence no sparrows, but that heightens rather than resolves the puzzle.

One common argument, borrowed from Trivers (1972), is that the original couple has more to lose since it has put a good deal of effort already into the improvement of the property. This, however, is a logical error that has come to be known as the *Concorde* or the *sunk cost* fallacy (Dawkins & Brockmann, 1980): To maximize future returns, an agent ought consider only the future payoffs of an entity, not how much the agent has expended on the entity in the past. Of course, if the incumbent has specifically tailored the property to its needs, and these needs are not shared by the intruder, the loss aversion asymmetry can be explained without recourse to a theory of property rights. But in none of the cases of which I am aware is this alternative at all plausible, because the property is equally valuable to all owners.

The Hawk-Dove game was offered by Maynard Smith and Parker (1976) as a logically sound alternative to the sunk cost argument. In this game, Hawks and Doves are phenotypically indistinguishable members of the same species, but they act differently in contesting ownership rights to a territory. When two Doves contest, they posture for a bit, and then each assumes the territory with equal probability. When a Dove and a Hawk contest, however, the Hawk takes the whole territory. Finally, when two Hawks contest, a terrible battle ensues, and the value of the territory is less than the cost of fighting for the contestants. Maynard Smith and Parker showed that, assuming that there is an unambiguous way to determine who first found the territory, there is an evolutionarily stable strategy in which all agents behave like Hawks when they are *first* to find the territory, and like Doves otherwise.

The Hawk-Dove game is an elegant contribution to explaining the endowment effect, but the cost of contesting for Hawks and the cost of display for Doves cannot plausibly be taken as fixed and exogenously determined. Indeed, it is clear that Doves contest in the same manner as Hawks, except that they devote fewer resources to combat. Similarly, the value of the ownership is taken as exogenous, when in fact it depends on the frequency with which ownership is contested, as well as other factors. As Grafen (1987) stresses, the costs and benefits of possession depend on the state of the population, the density of high-quality territories, the cost of search, and other variables that might well depend on the distribution of strategies in the population.

First, however, it is instructive to consider the evidence for a close association, as Locke suggested in his theory of property rights, between ownership and incumbency (physical contiguity and control) in children and nonhuman animals.

2 Property Rights in Young Children

Long before they become acquainted with money, markets, bargaining, and trade, children exhibit possessive behavior and recognize the property rights of others on the basis of incumbency. In one study (Bakeman & Brownlee, 1982), participant observers studied a group of 11 toddlers (12 to 24 months old) and a group of 13 preschoolers (40 to 48 months old) at a day care center. The observers found that each group was organized into a fairly consistent linear dominance hierarchy. They then cataloged *possession episodes*, defined as a situation in which a *holder* touched or held an object and a *taker* touched the object and attempted to remove it from the holder's possession. Possession episodes averaged 11.7 per hour in the toddler group, and 5.4 per hour in the preschool group.

For each possession episode, the observers noted (a) whether the taker had been playing with the object within the previous sixty seconds (prior possession), (b) whether the holder resisted the take attempt (resistance), and (c) whether the take was successful (success). They found that success was strongly and about equally associated with both dominance and prior possession. They also found that resistance was associated mainly with dominance in the toddlers, and with prior possession in the preschoolers. They suggest that toddlers recognize possession as a basis for asserting control rights, but do not respect the same rights in others. The preschoolers, more than twice the age of the toddlers, use physical proximity both to justify their own claims and to respect the claims of others. This study was replicated and extended by Weigel (1984).

3 Respect for Possession in Nonhuman Animals

In a famous paper, Maynard Smith and Parker (1976) noted that if two animals are competing for some resource (e.g., a territory), and if there is some discernible asymmetry (e.g., between an "owner" and a later animal), then it is evolutionarily stable for the asymmetry to settle the contest conventionally, without fighting. Among the many animal behaviorists who put this theory to the test, perhaps no research is more elegant and unambiguous than that of Davies (1978), who studied the speckled wood butterfly (*Pararge aegeria*), found in the Wytham Woods, near Oxford, England. Territories for this butterfly are shafts of sunlight breaking through the tree canopy. Males occupying these spots enjoyed heightened mating success, and on average only 60 percent of males occupied the sunlit spots at any one time. A vacant spot was generally occupied within seconds, but an intruder on an already occupied spot was invariably driven away, even if the incumbent had occupied the spot only for a few seconds. When Davies "tricked" two butterflies into thinking each had occupied the sunny patch first, the contest between the two lasted, on average, ten times as long as the brief flurry that occurs when an incumbent chases off an intruder.

Stevens (1988) found a similar pattern of behavior for the feral horses occupying the sandy islands of the Rachel Carson Estuarine Sanctuary near Beaufort, North Carolina. In this case, it is fresh water that is scarce. After heavy rains, fresh water accumulates in many small pools in low-lying wooded areas, and bands of horses frequently stop to drink. Stevens found that there were frequent encounters between bands of horses competing for water at these temporary pools. If a band approached a water hole occupied by another band, a conflict ensued. During 76 hours of observation,

Stevens observed 233 contests, of which the resident band won 178 (80%). In nearly all cases of usurpation, the intruding band was larger than the resident band. These examples, and many others like them, support the presence of an endowment effect and suggest that incumbents are willing to fight harder to maintain their position than intruders are to usurp the owner.

Examples from nonhuman primates exhibit behavioral patterns in the respect for property rights much closer to that of humans. In general, the taking of an object held by another individual is a rare event in primate societies (Torii, 1974). A reasonable test of the respect for property in primates with a strong dominance hierarchy is the likelihood of a dominant individual refraining from taking an attractive object from a lower-ranking individual. In a study of hamadryas baboons (*Papio hamadryas*), for instance, Sigg and Falett (1985) hand a food-can to a subordinate who was allowed to manipulate and eat from it for five minutes before a dominant individual who had been watching from an adjacent cage was allowed to enter the subordinate's cage. A "takeover" was defined as the rival taking possession of the can before thirty minutes had elapsed. They found that (a) males never took the food-can from other males; (b) dominant males took the can from subordinate females two-thirds of the time; and (c) dominant females took the can from subordinate females half of the time. With females, closer inspection showed that when the difference in rank was one or two, females showed respect for the property of other females, but when the rank difference was three or greater, takeovers tended to occur.

Kummer and Cords (1991) studied the role of proximity in respect for property in long-tailed macaques (*Macaca fascicularis*). As in the Sigg and Falett study, they assigned ownership to a subordinate and recorded the behavior of a dominant individual. The valuable object in all cases was a plastic tube stuffed with raisins. In one experiment, the tube was fixed to an object in half the trials and completely mobile in the other half. They found that with the fixed object, the dominant rival took possession in all cases and very quickly (median one minute), whereas in the mobile condition, the dominant took possession in only 10 percent of cases, and then only after a median delay of 18 minutes. The experiment took place in an enclosed area, so the relative success of the incumbent was not likely due to the ability to flee or hide. In a second experiment, the object was either mobile or attached to a fixed object by a stout two-meter or four-meter rope. The results were similar. A third case, in which the nonmobile object was attached to a long dragline that permitted free movement by the owner, produced the following results. Pairs of subjects were studied under two conditions, one where the rope attached to the dragline was 2 meters in length, and a second where the rope was 4 meters in length. In 23 of 40 trials, the subordinate maintained ownership with both rope lengths, and in 6 trials the dominant rival took possession with both rope lengths. In the remaining 11 trials, the rival respected the subordinate's property in the short-rope case, but took possession in the long-rope case. The experimenters observed that when a dominant attempts to usurp a subordinate when other group members are around, the subordinate will scream, drawing the attention of third parties, who frequently force the dominant individual to desist.

Hauser (2000) relates an experiment run by Kummer and his colleagues concerning mate property, using four hamadryas baboons, Joe, Betty, Sam, and Sue. Sam was let into Betty's cage, while Joe looked on from an adjacent cage. Sam immediately began following Betty around and grooming her. When Joe was allowed entrance to the cage, Joe kept his distance, leaving Sam uncontested. The same experiment was repeated with Joe allowed into Sue's cage. Joe behaved as Sam had in the previous experiment, and when Sam was let into the cage, he failed to challenge Joe's proprietary rights with respect to Sue.

No primate experiment, to my knowledge, has attempted to determine the probability that an incumbent will be contested for ownership by a rival who is, or could easily become, closely proximate to the desired object. This probability is likely very low in most natural settings, so the contests described in the papers cited in this section are probably rather rare in practice. At any rate, in the model of respect for property developed in the next section, we will make informational assumptions that render the probability of contestation equal to zero in equilibrium.

4 Conditions for Property Equilibrium

Suppose that two agents, prior to fighting over possession, simultaneously precommit to expending a certain level of resources to the contest. As in the War of Attrition (Bishop & Cannings, 1978), a higher level of resource commitment entails a higher fitness cost but increases the probability of winning the contest. We assume throughout this chapter that the two contestants, an incumbent and an intruder, are *ex ante* equally capable contestants in that the costs and benefits of battle are symmetric in the resource commitments s_o and s_u of the incumbent and intruder, respectively, and $s_o, s_u \in [0,1]$. To satisfy this requirement, we let $p_u = s_u^n / (s_u^n + s_o^n)$ be the probability that the intruder wins, where $n > 1$. Note that larger n implies resource commitments are more decisive in determining victory. We assume that combat leads to injury $\beta \in (0,1]$ to the losing party with probability $p_d = (s_o+s_u)/2$, so $s = \beta\, p_d$ is the expected cost of combat for both parties.

This chapter uses a territorial analogy throughout, some agents being incumbents and others being migrants in search of either empty territories or occupied territories that they may be able to occupy by displacing current incumbents. Let π_g be the present value of being a currently uncontested incumbent, and let π_b be the present value of being a migrant searching for a territory. We assume throughout that $\pi_g > \pi_b > 0$. Suppose a migrant comes upon an occupied territory. We show in Gintis (2007) that a property equilibrium occurs precisely when

$$\frac{s_o^n}{s_u^n}s > \frac{\pi_g}{\pi_b(1-c)} - 1 > \frac{s_u^n}{s_o^n}s, \tag{1}$$

and s_o=1 maximizes π_c. The condition for a potential usurper not contesting the incumbent is then given by

$$\frac{\pi_g}{\pi_b(1-c)} - 1 < \beta \tag{2}$$

Finally, by an *antiproperty* equilibrium we mean a situation where intruders always contest and incumbents always relinquish their possessions without a fight. We then have:

Theorem 1 If $\pi_g > (1+\beta)\pi_b(1-c)$, there is a unique equilibrium in which a migrant always fights for possession and an incumbent always contests. When the reverse inequality holds, there exists both a property equilibrium and an antiproperty equilibrium.

Theorem 1 implies that property rights are more likely to be recognized when combatants are capable of inflicting great harm on one another, so β is close to its maximum of unity, or when migration costs are very high, so c is close to unity.

The result that there exists an antiproperty equilibrium exactly when there is a property equilibrium is quite unrealistic since few, if any, antiproperty equilibria have been observed. Our model, of course, shares this anomaly with the Hawk-Dove game, for which this weakness has never been analytically resolved. In our case, however, when we expand our model to determine π_g and π_g, the conditions for the existence of the different antiproperty equilibria generally are not satisfied. Moreover, we find that if investment by an incumbent can improve the productivity of the property, and if two groups differ only in that one plays the property equilibrium and the other plays the antiproperty equilibrium, the former will grow faster and hence displace the latter, provided that there is some scarcity of resources leading to a limitation on the combined size of the two the groups. Finally, if there is no profitable investment by the incumbent, the property and antiproperty equilibria differ in only one way: The identity of patch owner changes in the latter more rapidly than in the former. It is quite reasonable to add to the model a small cost d of ownership change, for instance, because the intruder must physically approach the patch and engage in some sort of display before the change in incumbency can be effected. With this assumption, the antiproperty equilibrium again has a lower average payoff than the property equilibrium, so it will be disadvantaged in a competitive struggle for existence.

5 The Nature of a Property Equilibrium

To determine π_g and π_b, we must flesh out the above model of incumbents and migrants. Consider a field with many patches, each of which is indivisible and hence can have only one owner. In each time period, a fertile patch yields a benefit $b > 0$ to the owner, and dies with probability $p > 0$, forcing its owner (should it have one) to migrate elsewhere in search of a fertile patch. Dead patches regain their fertility after a period of time, leaving the fraction of patches that are fertile constant from period to period. An agent who encounters an empty fertile patch invests an amount $\upsilon \geq 0$ in preparing the patch for use and occupies the patch. An agent suffers a fitness cost $c > 0$ each period he is in the state of searching for a fertile patch. An agent who encounters an occupied patch may contest for ownership of the patch, according to the War of Attrition structure analyzed in the previous section.

Suppose there are n_p patches and n_a agents. Let r be the probability of finding a fertile patch, and let w be the probability of finding a fertile unoccupied patch. If the rate at which dead patches become fertile is q, which we assume for simplicity does not depend on how long a patch has been dead, then the equilibrium fraction f of patches that are fertile must satisfy $n_p fp = n_p(1 - f)q$, so $f = q/(p + q)$. Assuming that a migrant finds a new patch with probability ρ, we then have $r = f\rho$. If φ is the fraction of agents who are incumbents, then writing $\alpha = n_a/n_p$, we have

$$\omega = r(1-\alpha\varphi). \tag{3}$$

Assuming the system is in equilibrium, the number of incumbents whose patch dies must be equal to the number of migrants who find empty patches, or $n_a\varphi(1 - p) = n_a(1 - \varphi)\omega$. Solving this equation gives φ, which is given by

$$\alpha r\varphi^2 - (1 - p + r(1 + \alpha))\varphi + r = 0. \tag{4}$$

It is easy to show that this equation has two positive roots, exactly one lying in the interval (0,1).

In a property equilibrium, we have

$$\pi_g = b + (1 - p)\pi_g + p\pi_b(1 - c), \tag{5}$$

and

$$\pi_b = \omega\pi_g(1 - \upsilon) + (1 - \omega)\pi_b(1 - c). \tag{6}$$

Note that the cost υ of investing and c of migrating are interpreted as fitness costs, and hence as probabilities of death. Thus, the probability of a migrant becoming an incumbent in the next period is $\omega(1 - \upsilon)$, and the probability of remaining a migrant is $(1 - \omega)$. This explains (6). Solving these two equations simultaneously gives equilibrium values of incumbency and nonincumbency:

$$\pi_g^* = \frac{b(c(1-\omega)+\omega)}{p(c(1-\upsilon\omega)+\upsilon\omega)} \text{ and} \tag{7}$$

$$\pi_b^* = \frac{b(1-\upsilon)\omega}{p(c(1-\upsilon\omega)+\upsilon\omega)}. \tag{8}$$

Note that $\pi_g, \pi_b > 0$, and

$$\frac{\pi_g^*}{\pi_b^*} - 1 = \frac{c(1-\omega)+\omega\upsilon}{\omega(1-\upsilon)}. \tag{9}$$

By Theorem 1, the assumption that this is a property equilibrium is satisfied if and only if this expression is less than β, or

$$\frac{c(1-\omega)+\omega\upsilon}{\omega(1-\upsilon)} < \beta. \tag{10}$$

This inequality shows that, in addition to our previous result, low fighting cost and high migration cost undermine the property equilibrium, a high probability ω that a migrant encounters an incumbent undermines the property equilibrium, and a high investment υ has the same effect.

6 Conclusion

Humans share with many other species a predisposition to recognize property. This takes the form of *loss aversion*: an incumbent is prepared to commit more vital resources to defending his property, *ceteris paribus*, than an intruder is willing to commit to taking the property. The major proviso is that if the property is sufficiently valuable, a property equilibrium will not exist (Theorem 1).

History is written as though property is a product of modern civilization, a construction that exists only to the extent that it is defined and protected by judicial institutions operating according to legal notions of ownership. However, it is likely that property in the fruits of one's labor existed for as long as humans lived in small hunter-gatherer clans, unless the equality in Theorem 1 holds, as might plausibly be the case for big game. The true value of modern property, if the argument in this chapter is valid, is fostering the accumulation property even when $\pi_g > (1 + \beta)\pi_b(1 - c)$. It is in this sense only that Thomas Hobbes may have been correct in asserting that life in an unregulated state of nature is "solitary, poor, nasty, brutish, and short." But even so, it must be recognized that modern notions of property are built on human behavioral propensities that we share with many species of nonhuman animals. Doubtless, an alien species with a genetic organization akin to our ants or termites would find our notions of individuality and privacy curious at best, and probably incomprehensible.

References

Bakeman, R., & Brownlee, J. R. (1982). Social rules governing object conflicts in toddlers and preschoolers. In K. H. Rubin & H. S. Ross (Eds.), *Peer relationships and social skills in childhood* (pp. 99–112). New York: Springer-Verlag.

Bishop, D. T., & Cannings, C. (1978). The generalised war of attrition. *Advances in Applied Probability, 10*(1) (March), 6–7.

Camerer, C. (2003). *Behavioral game theory: Experiments in strategic interaction.* Princeton: Princeton University Press.

Davies, N. B. (1978). Territorial defence in the speckled wood butterfly (*Pararge aegeria*): The resident always wins. *Animal Behaviour, 26,* 138–147.

Dawkins, R., & Brockmann, H. J. (1980). Do digger wasps commit the *Concorde* fallacy? *Animal Behaviour, 28,* 892–896.

Eason, P. K., Cobbs, G. A., & Trinca, K. G. (1999). The use of landmarks to define territorial boundaries. *Animal Behaviour, 58,* 85–91.

Ellis, L. (1985). On the rudiments of possessions and property. *Social Sciences Information. Information Sur les Sciences Sociales, 24*(1), 113–143.

Furby, L. (1980). The origins and early development of possessive behavior. *Political Psychology, 2*(1), 30–42.

Gintis, H. (2007). The evolution of private property. *Journal of Economic Behavior & Organization, 64*(1), 1–16.

Grafen, A. (1987). The logic of divisively asymmetric contests: Respect for ownership and the desperado effect. *Animal Behaviour, 35,* 462–467.

Hauser, M. (2000). *Wild minds.* New York: Henry Holt.

Hirshleifer, J. (1988). The analytics of continuing conflict. *Synthese 76,* 201–233.

Jones, Owen D. (2001). Time-shifted rationality and the law of law's leverage: Behavioral economics meets behavioral biology. *Northwestern University Law Review, 95,* 1141–1206.

Kahneman, D. & Tversky, A. (1979). Prospect theory: An analysis of decision under risk. *Econometrica, 47,* 263–291.

Kahneman, D., Tversky, A., Knetsch, J. L. & Thaler, R. H. (1991). The endowment effect, loss aversion, and status quo bias. *Journal of Economic Perspectives, 5*(1), 193–206.

Kummer, H. & Cords, M. (1991). Cues of ownership in long-tailed macaques, *Macaca fascicularis. Animal Behavior, 42,* 529–549.

Maynard Smith, J. & Parker, G. A. (1976). The logic of asymmetric contests. *Animal Behaviour, 24,* 159–175.

Maynard Smith, J., Parker, G. A. & Price, G. R. (1973). The logic of animal conflict. *Nature, 246,* 15–18.

Mesterton-Gibbons, M. & Adams, E. S. (2003). Landmarks in territory partitioning. *American Naturalist, 161*(5), 685–697.

Rand, A. S. (1967). Ecology and social organization in the iguanid lizard *Anolis lineatopus*. *Proceedings of the US National Museum, 122*, 1–79.

Riechert, S. E. (1978). Games spiders play: Behavioural variability in territorial disputes. *Journal of Theoretical Biology, 84*, 93–101.

Schlatter, R. B. (1973). *Private property: History of an idea*. New York: Russell.

Senar, J. C., Camerino, M. & Metcalfe, N. B. (1989). Agonistic interactions in siskin flocks: Why are dominants sometimes subordinate? *Behavioral Ecology and Sociobiology, 25*, 141–145.

Sigg, H. & Falett, J. (1985). Experiments on respect of possession and property in hamadryas baboons (*Papio hamadryas*). *Animal Behaviour, 33*, 978–984.

Stake, J. E. (2004). The property instinct. *Philosophical Transactions of the Royal Society of London, Series B: Biological Sciences, 359*, 1763–1774.

Stevens, E. F. (1988). Contests between bands of feral horses for access to fresh water: The resident wins. *Animal Behaviour, 36*(6), 1851–1853.

Thaler, R. H. (1992). *The winner's curse*. Princeton: Princeton University Press.

Torii, M. (1974). Possession by non-human primates. *Contemporary Primatology*, 310–314.

Trivers, R. L. (1972). Parental investment and sexual selection, 1871–1971. In B. Campbell (Ed.), *Sexual selection and the descent of man* (pp. 136–179). Chicago: Aldine.

Tversky, A. & Kahneman, D. (1981). Loss aversion in riskless choice: A reference-dependent model. *Quarterly Journal of Economics, 106*(4), 1039–1061.

Weigel, R. M. (1984). The application of evolutionary models to the study of decisions made by children during object possession conflicts. *Ethnology and Sociobiology, 5*, 229–238.

7 Cooperation and Biological Markets: The Power of Partner Choice

Ronald Noë and Bernhard Voelkl

The Problem of Cooperation in Nature

Cooperation is a phenomenon that has attracted interest from scholars of many different disciplines, some of which concentrate on human behavior (anthropology, economics, sociology, and psychology), while others focus on interactions among nonhuman organisms (behavioral ecology, evolutionary ecology). In the biological sciences it attracts so much attention because its emergence and maintenance, both on an evolutionary timescale and on the timescale of individual relationships, form an enigma that is difficult to explain. Cooperation seems to contravene the basic notion of evolutionary biology that natural selection favors selfish entities that promote only their own wellbeing. Darwin (1859) already noted that several forms of cooperation as they can be observed in nature—especially in the social insects—represent a potential stumbling block for his theory of evolution by means of natural selection.

Early attempts to explain the existence of altruistic behavior, defined in biological terms as an act that results in a net loss to the actor in terms of fitness and a net gain to the recipient, were usually based on the idea that such a behavior would be "for the good of the species" or advantageous for the group (Wynne-Edwards, 1962). However, this naive group-selectionists' view of cooperation was faulted since any group of cooperative individuals would be vulnerable to being exploited by defectors, who profit from the cooperative tendencies of others without contributing to cooperative actions themselves (Williams, 1966). In two hallmark papers, Hamilton (1964) and Maynard Smith (1964) demonstrated convincingly that cooperation can be evolutionary stable when the cooperative individuals are genetically related. While many forms of cooperation in animals—including the sterile worker casts in social insects—could be largely explained by kin selection, the puzzle of cooperation between unrelated individuals belonging to the same species or between individuals of different species remained unsolved.

Trivers (1971) was one of the first to propose the use of the two-player Iterated Prisoners' Dilemma (IPD) as a paradigm for what he called "reciprocal altruism." (Although oddly enough, reciprocal altruism stresses the alternation of choices made by the repeatedly cooperating partners and the problems caused to them by the time lag between their choices, whereas simultaneous choice and thus a lack of information on the last move of the partner at the moment the choice is made are typical for a classic IPD.) The basic idea of this paradigm is that two individuals can maximize their benefit in terms of fitness gains by taking turns in altruistically supporting their partner. The idea seemed elegant and worked out well on paper. Moreover, it got convincing support from a computer tournament carried out by Axelrod (Axelrod & Hamilton, 1981). In this tournament 16 strategies played repeatedly against each other in IPDs. The winning strategy was "Tit-for-Tat" (TFT) submitted by Anatol Rapaport, which cooperates in the first move and thereafter always reciprocates the behavior of its partner.

The elegance and general applicability of this idea fueled the production of a rapidly increasing number of further theoretical papers on reciprocity based cooperation. These, in turn, fostered the interest of empirical researchers to explain their observations in terms of reciprocity or to find new evidence for reciprocal help or exchange of commodities. Yet, even after three decades of active search for fitting examples, unequivocal evidence for reciprocal altruism in nature remains very thin (Packer, 1986; Noë, 1992; Enquist & Leimar, 1993; Noë & Hammerstein, 1994; Clements & Stephens, 1995; Roberts, 1998; Hammerstein, 2003; Bshary & Bronstein, 2004; Sachs et al., 2004; Leimar & Hammerstein, 2006; Bergmüller et al., 2007; Silk, 2007; West et al., 2007a,b; Clutton-Brock, 2009). Besides the obvious lack of evidence, the idea of reciprocal altruism as an explanation for cooperation between unrelated animals was also criticized on theoretical grounds (Noë, 2006a). Nevertheless we observe an overwhelming amount of cooperative interactions and relationships in nature, notably in the form of mutualistic interactions between members of different species, but also among members of the same species. In this chapter we try to identify which mechanisms, other than TFT-like partner control strategies, ensure the stability of these naturally occurring forms of cooperation.

Partner Control versus Partner Choice

Models based on iterated two-player games like the Prisoners' Dilemma, the Game of Chicken (which is equivalent to the Snow Drift game and the Hawk-Dove game with certain parameter values) or the Stag Hunt game can be characterized as "partner control" models. They take the formation of cooperating partnerships for granted and concentrate on the mechanisms that each participant uses to prevent being cheated by its partner. They aim at a continuation of net gain, or an improvement in net gain, in a series of interactions with one specific partner. Each individual can have such series of interactions going with multiple partners, but the silent assumption of these models is that these different partnerships do not influence each other. The participants must somehow adjust their behavior conditionally upon fluctuations of their returns that coincide with changes in the partner's behavior. A player reacts to a failure of its partner to provide return benefits by aborting the relationship, which is usually labeled "defecting," thereby losing the advantages of cooperation in the process. Yet, the potential benefits lost by ending a relationship can be at least partially be compensated if one switches to another partner, even if the potential maximal profit of the latter is lower.

We refer to models that put the emphasis on the option of choosing and switching partners as "partner choice models" (Bshary & Noë, 2003). These include several extensions of the IPD-model (Dugatkin & Wilson, 1991; Batali & Kitcher, 1995; Ashlock et al., 1996; Roberts, 1998), a partner choice model based on selective abortion proposed by Bull & Rice (1991); now better known under the label "sanctioning" coined by Denison (2000) and the biological markets paradigm (Noë et al., 1991; Noë & Hammerstein, 1994, 1995). These models point out that the presence of alternative partners can provide leverage over the current partner, forcing them into outbidding competition, even in the absence of an actual switch. Partner choice and partner switching can be either based on a comparison of offers available at the moment a choice can be made, or on a comparison of the net gain from multiple partners over past interactions. The latter option implies a form of bookkeeping, but of a different nature than the tallying bookkeeping that forms the basis of what de Waal (2000) called "calculated reciprocity." The mechanisms used in comparative bookkeeping will be very different for different species. For cognitively operating species, such as non-human primates, one can expect a form of "attitudinal partner choice," a term inspired by de Waal's (2000) attitudinal bookkeeping, based on fluctuating titers of neurohormones and neurotransmitters (Fruteau et al., 2009).

In nature few animals have only a single potential partner or pair up randomly to cooperate. Exceptions are forms of cooperation in dyads that form for reasons external to the cooperation, e.g., couples that form in order to reproduce together, such as pairs of cleaner fish (Bshary et al., 2008). The preference for specific partners as observed in some predator inspection studies, be it on the basis of previous experience with that partner (Milinski et al., 1990; Dugatkin & Alfieri, 1991) or on the basis of size (Walling et al., 2004), adds an argument for the use of partner choice models rather than partner control models. The opportunity for switching partners can have two opposite consequences: on the one hand being exploited can be prevented by exchanging notorious freeloaders for other partners, but on the other hand free-riding can be facilitated if exploiters can switch to new partners before the old ones can react effectively (Enquist & Leimar, 1993).

Biological Markets

Noë and Hammerstein (1994) coined the term "biological markets" to stimulate the development of new models for cooperation based on knowledge accumulated in two rather distant fields—sexual selection theory and economics. From an economic point of view, most forms of cooperation can be represented as trading situations where (1) two or more individuals exchange goods and/or services in such a way that the participants involved are usually better off after the interaction than before it, and (2) the participants have to invest something in the interaction without a full guarantee of net gain. The biological market paradigm stresses some aspects of cooperation that were ignored in earlier models, notably partner choice and partner switching, competition in the form of outbidding among potential partners, and the division of benefits and the exchange rates of commodities. In an analogy to sexual selection theory, biological market theory predicts that partner choice can lead to the selection for specific traits. The importance of partner choice as a selective force has long been recognized in the context of sexual selection, but its importance has not yet been generally acknowledged in the case of cooperation. This form of selection through partner choice in the context of cooperation was labeled 'market selection' by Bshary and Noë (2003).

One imminent consequence of partner choice is that it generates competition for the best partners. This competition is a ubiquitous theme in both animal and human mating markets. When a competitor cannot be excluded by brute force, the favors of an attractive partner must be won by offering a higher potential benefit than all other competitors do. A skilled competitor should outbid the competition without making concessions that are unnecessarily high. Thus, not only the choosers but also the bidders have to be able to estimate the relative market value of themselves and their potential partners (cf. Pawlowski & Dunbar, 1999; Penke et al., 2007).

Partner choice implies the assessment of different partners, which will usually be associated with certain search costs. Both finding potential partners and assessing their cooperative value needs time, energy, and—at times—taking risks. The search costs will usually depend on several factors: the density of the population, the speed with which an individual can move in the environment, the decision rule used, the proportion of potential partners that are not already fully engaged in a cooperative relationship with another partner, and eventually signaling costs needed to establish contact with a potential partner. The variation in benefit between partners must be large enough to make an investment in sampling worthwhile; if differences are very small, the cost of sampling does not outweigh the benefits of making a better choice. Sampling problems have been modeled extensively in the sexual selection literature (Gibson & Langen, 1996) and in the vast optimal foraging literature (Stephens & Krebs, 1986). For cooperation problems other than reproduction, Enquist and Leimar (1993) have shown that higher search costs can render the temptation to defect as a free-rider an unattractive option, thus fostering cooperation.

Finally, the choice of a cooperation partner can be based on its "reputation" (also referred to "image" or "standing"). When formulating his idea of reciprocal altruism, Trivers (1971) suggested that individuals need not base their decision about whether to cooperate or to defect solely on their own past interactions with a specific partner. Instead, they can also use any information that they acquire from observing the behavior of this partner in interactions with others. This idea has been dubbed "indirect reciprocity" and can explain why it can be adaptive to cooperate even when the likelihood of repeatedly interacting with the same individual is low (Sachs et al., 2004; Nowak, 2007). If the chosen strategy depends not only on the last observed interaction with the specific partner in question but on an overall image that builds up over successive observation, this mechanism is then referred to as "image scoring" (Nowak & Sigmund, 1998). Evidence for such image scoring comes both from studies with human (Wedekind & Milinski, 2000) and animal (Bshary, 2002; Bshary & Grutter, 2006; Russell et al., 2008) subjects. In partner choice models, reputation can have an important role if partners interact only once or repeated interactions with the same partner are scarce. However, the emphasis will not be on the question of whether to defect or to cooperate, but on whether to choose this particular individual as a trading partner over others.

Power Asymmetry

In the simplest form of a market, traders play symmetrical roles: any trader can choose any other as a partner and initiate a transaction. Interactions take place in the larger context of similar interactions by all traders. Yet, many markets will be characterized by interactions between traders belonging to two (or sometimes more) distinct classes. These markets will often be asymmetrical, because one class of traders is (much) smaller than the other, more mobile than the other, or both. Well-studied markets with trader classes differing considerably in size include: primate mothers trading grooming for permission to touch infants (Henzi & Barrett, 2002; Barrett & Henzi, 2006; Gumert, 2007b; Slater et al., 2007; Fruteau et al., 2011), obligate "nursery" pollination mutualisms (James et al., 1994; Pellmyr & Huth, 1994; Fleming & Holland, 1998; Kato et al., 2003; Marr & Pellmyr, 2003; Kawakita & Kato, 2009), ant-plant (Bronstein, 1998; Edwards et al., 2006) or ant-insect (Fischer et al., 2001, 2005) protection mutualisms, and nutrient exchange mutualisms between plants and either mycorrhizal fungi (Wilkinson, 2001; Kiers & van der Heijden, 2006; Kummel & Salant, 2006; Cowden & Peterson, 2009; Peay et al., 2010; Bever et al., 2009; Kiers et al., 2011) or rhizobia (Simms & Taylor, 2002; West et al., 2002; Simms, 2006; Heath & Tiffin, 2007, 2009; Ryoko et al., 2009; Gubry-Rangin et al., 2010; Friesen, 2012). Trader classes of comparable, typically large, size but of different mobility are found in non-specific pollination and seed dispersal interactions, where food is exchanged for transport services. An interesting system in which both mobility and class size are asymmetric is the cleaner fish—client mutualisms in which small and stationary cleaner fish feed on ecto-parasites found on larger "client fish." The latter can either also be stationary ("resident clients"), or highly mobile ("roaming clients"). Both the cleaners and the roaming clients can exert choice, but resident clients can do little more than visit the cleaner at the only cleaning station in their territory (Bshary & Noë, 2003).

Even in asymmetric markets, traders should exchange goods and services at an exchange rate that is determined by the total supply and demand for both commodities, because the individuals of the more powerful class are still in competition with each other. One important consequence of power asymmetries is, however, that the powerful partner can demand a certain commitment from its partners in advance of their own investment. That is, investments are not necessarily made simultaneously, but can also be made sequentially. Sequential partner choice changes the structure of the cooperation game, creating different dilemmas for those who have to offer first and those who can respond to the initial offer.

Gradually Adjustable Investment

Controlling the partner by defecting is the only sanction available in classical IPD-based models. However, several variations on the theme have been proposed, such as investing slightly more or less depending on the partner's behavior (Roberts & Sherratt, 1998) or splitting up the investments of each round in small portions (Connor 1992, 1995b; Leimar & Connor, 2003). According to Roberts (1998), it is only the combination of variation in the investment—what he calls variable "generosity"—and partner choice that leads to competition for partners. Buyers can only challenge sellers to outbid each other when they can choose freely between their offers (Noë, 1990, 1992). This pressure can be an incentive to give as much as possible while still making a small net gain. Without competition the sellers' offers are likely to converge to zero.

In the simplest trading scenario, two traders have to invest in the cooperative action in the form of the commodity they offer and both benefit from the received commodities. While the possibility of continuously adjusting investments or the partitioning of benefits facilitates bargaining, it is, however, not necessary that both commodities are divisible. As long as at least one commodity is divisible, bargaining can lead to the stabilization of exchanges at an equilibrium price, provided that one partner can adjust the amount to invest and the other the threshold value at which to accept the offer. Even if the investment in, or the returns of, a cooperative interaction are of a dichotomous all-or-nothing type, this does not necessarily mean that it cannot be subject to a bargaining process.

The most obvious consequence of divisible investments is that for each amount of supply of a specific commodity and the demand for this commodity, a specific equilibrium price exists (Smith, 1776; Samuelson, 1983). Honest signaling or book-keeping over past interactions can assure that commodities are exchanged at this equilibrium price. The effect of changes in supply and demand is the subject of many studies that investigate proximate mechanisms of biological markets. In nonhuman primates, grooming another individual is a commodity that can be exchanged for different services like tolerance at food sites, access to newborns, compliance of females, and support in conflicts. Several observational studies have suggested that the time a specific monkey was groomed increased when this individual could offer a commodity that was in short supply or decreased when the commodity was at plenty (Barrett et al., 2002; Henzi & Barrett, 2002; Henzi et al., 2003; Gumert 2007a,b; Chancellor & Isbell, 2008; Fruteau et al., 2011; Balasubramaniam et al., 2011; Barelli et al., 2011). In a field experiment, Fruteau and colleagues (2009) could show that the reward that a single provider received was only half as high when the number of suppliers was doubled.

Communication and Honest Signaling

One of the basic features of the original Prisoners' Dilemma (PD) is that both partners have to make their decision simultaneously without communicating with their partner (Luce & Raiffa, 1957). While most models of cooperation that are based on the PD adopt this as an assumption, it is in fact in stark contrast to what can be observed in nature. Empirical researchers studying cooperative behavior find indications for communication everywhere—from the quorum-sensing molecules of bacteria (Diggle et al., 2007; West et al., 2007b) to the sophisticated signals and gestures in birds and mammals (Dawkins & Krebs, 1978; Maynard Smith, 1982; Harper, 1991). On the one hand communication can clearly facilitate cooperation, but on the other hand it creates a new dilemma—the problem of honest signaling. This problem has been studied extensively in the context of sexual selection (Zahavi, 1975, 1977; Grafen, 1990; Johnstone, 1995), but hardly at all in the context of biological markets. Two factors might facilitate the emergence of honest signaling in the context of cooperative exchange of commodities: sequential choice as a consequence of a power asymmetry and the divisibility of a commodity.

Noë (2006a) used the cooperative hunt of lionesses to illustrate how the divisibility of the investment or the returns can facilitate bartering—that is, communication about the distribution of returns. After the hunt, the spoils have to be split, which is a problem of conflict resolution rather than cooperation, but the problem is very similar. Communication plays a major role: one takes a bite too much, the other growls in threat, and so on. Connor (1995a) called this "parcelling" and proposed it as a way of reducing the costs of being cheated in cooperative interactions. Because individual bites are very small relative to the whole zebra, they function not only as simple elements in the mechanism of reaping benefit, but also as signals during the bargaining over splitting the spoils, signaling the intention to claim more of the prey. Furthermore the "parcelling" of the zebra is a lengthy process embedded in a wealth of further signals such as growls, movements, facial expressions and whipping tails. Likewise the small jerky moves that are typical for pairs of small fish approaching a predator they want to inspect are both part of the approach and a signal of the intention to move on and invest further in the cooperative inspection (Noë, 2006a).

In many cases, the benefits obtained in interactions with different partners in the past will provide the best possible indication for future benefits. The sexually selected parallel of this is repeated courtship feeding before mating, which can be used by females to assess the ability of a male to feed their future offspring (Wiggins, 1986). The potential for deceit is obvious, but a mate that is not even able to bring in food before mating will almost certainly be a lousy caretaker after mating. The direct benefits that females receive before the actual mating takes place make this a low-cost form of mate sampling. An individual choosing between several cooperation partners faces a similar problem: information about a past difference between the partners may not be worth much, but is better than no information at all and provides at least information about potential contributions. During periods in which competitors try to outbid each other, they may be forced to produce their commodity at the maximum possible level. Thus, while their output from this competition cannot be taken as a proxy for how much they will invest later on, it provides reliable information about their potential.

Repeated versus Single Interactions and Sequential Choice

While IPD models emphasize the dynamics of repeated interactions between pairs of individuals, the biological market paradigm emphasizes the context in which cooperative interactions take place. That means that repeated interactions are a possible option but not strictly necessary to establish cooperative exchange of commodities. In extreme cases it should be possible that an individual of one trader class interacts only once with a specific individual of the other class. Bull and Rice (1991) suggested such a model. Of course, systems without repeated interactions are more vulnerable to defectors, as partners cannot retaliate against being short-changed in a future interaction. In one-shot interaction models one needs, therefore, other mechanisms than retaliation to enable continued trading relationships. In the model proposed by Bull and Rice, cooperation can be established because of an asymmetry between the trading partners: one partner can assess the cooperative value of the other before cooperating and this discriminating individual has many partners to choose from. This means that the two trading partners make their choice not at the same time but sequentially. Simultaneous decision making without communication is a fundamental paradigm of most game theoretic analyses of cooperative scenarios. However, as we have argued earlier, it is questionable whether such a rule reflects the reality of naturally occurring cooperation. If it does not, it would be advisable to abandon this assumption.

In situations with sequential choice we can distinguish "auctioneers" (sometimes also referred to as "sellers"), i.e., the traders making the initial offer, and "buyers," the discriminating partners who can chose among the auctioneers' offers. If the auctioneers differ in the value they offer, the buyers might then adopt the strategy of always choosing the most cooperative auctioneer, i.e., the one with the highest offer. The same argument has been employed by Dawkins (1976) to describe the evolution of male investment under female choice in the mating market. As soon as a female copulates with a male (and the egg is internally fertilized), the female is committed to the well-being of the egg much more so than the male, who might just wander off and try to fertilize more eggs somewhere else. However, if males differ in the amount of investment they are willing to make *before* copulation, then females preferring males that have invested most will have the highest fitness and their strategy will thus emerge under selection.

An alternative to the strategy "select the one with the best offer" is the strategy "reject the one with the worst offer." The latter strategy, originally proposed by Bull and Rice (1991), is now usually referred to as "sanctioning" (Denison, 2000; Kiers et al., 2003). In sanctioning, it is typically the case that large and long-lived individuals of one species dispose of the least profitable partners among many small partners belonging to a species with a short generation time. Examples are interactions between yuccas and yucca-moths (James et al., 1994; Pellmyr & Huth, 1994) and between various plants and mycorrhizal fungi (Kummel & Salant, 2006) or rhizobia (Simms, 2006). Sanctioning can, therefore, be seen as an extreme and rather drastic form of "partner choice."

At a first glance, sanctioning is reminiscent of punishing, which is a central topic in the partner control literature. Yet, there is an essential difference between punishment and sanctioning: punishing pays because the punished individual is likely to interact with the punisher again. A plant sanctioning unprofitable partners will typically not interact with the rejected individuals again. But by cutting its losses, the plant obtains immediate benefits, which makes sanctioning akin to getting rid of parasites or infected tissue. Plants and trees usually have mechanisms to abort rotten fruits and unprofitable parts of their root system independent of a need to eliminate unprofitable partners. Strong selection on the sanctioned partners is thus a by-product of a purely self-regarding strategy. While the punishment concept stresses effects on individual behavior, the sanctioning literature emphasizes the selective effect on sanctioned species at evolutionary time scales. Of course, punishment can have a selective effect, as well, albeit probably weak, through its effect on the fitness of punished individuals.

Market Selection

Like sexual selection theory, the biological market paradigm predicts that partner choice can lead to selection for specific traits. Yet, the exact mechanisms involved are likely to differ. In sexual selection theory, genetic linkage can have powerful effects on the selection of specific traits (Fisher, 1930; Lande, 1981; Andersson, 1982). The offspring of a female who has a preference to mate with males with a certain characteristic will contain both their mother's gene for the preference and their father's gene for the characteristic. Because these two genes become non-randomly associated, this can then lead to a runaway process (Fisher, 1930) where costly secondary sexual characters, such as the peacock's tail, can be selected for. Clearly, this cannot happen in mutualistic systems where the trading partners belong to different species. Some sort of linkage might perhaps occur in mutualisms with vertical transmission of endosymbionts, where the parents of one species (e.g., polyps) inherit their partners (e.g., algae) from their parents (Herre et al., 1999). The inheritance mechanism in such cases is, however, a different one and needs separate treatment (Sachs & Wilcox, 2006; Vautrin et al., 2008; Aanen et al., 2009).

In the purple martin, dominant territory holders can monopolize human-built martin houses with several nest-holes, but cannot exploit them fully by installing several females in them because of the obligate biparental care in this species (Morton, 1987; Morton & Derrickson, 1990; Morton et al., 1990). The solution is to tolerate subordinate satellite males with their mates and to perform extra-pair copulations with the latter. The territory holder thus profits from additional reproductive opportunities, the satellite from obtaining a place to breed. For this form of cooperation, territory holders should preferentially accept yearling males, because it is highly unlikely that they will challenge the territory holder or successfully court his mate. Noë and Hammerstein (1994) identified such a constellation as a system where market selection, the result of the choice of the house owner for subtenants with the dullest plumage, should exert a selection pressure toward delayed plumage in yearlings. Interestingly, exactly such an effect, albeit in a somewhat different and more natural ecological context, was found in a different species—the lazuli bunting (Greene et al., 2000). Lazuli buntings are songbirds with obligate biparental care too that breed in early successional habitats throughout western North America. Males defend territories that include shrubs in which the nests are built. Shrub cover is extremely patchy, ranging from single isolated shrubs to larger areas of dense bushes. Older males return to the breeding grounds in the spring before yearling males, and competition among males for the limited areas with dense shrub cover is intense (Muehter et al., 1997; Greene et al., 2000). The opportunity to settle into territories for the later arriving yearlings depends heavily on their interactions with the already settled older males. Adult males are aggressive to colorful yearlings but tolerate dull ones, thus "choosing" dull males as their territory neighbors. The result is that dull colored males occupy territories of better quality than most of their more colorful peers. As in most territorial songbirds females are attracted in the first place by the quality of the territory and only secondarily by the male's appearance, although this may play a large role when she exerts female choice during extra-pair copulations (cf. Kempenaers et al., 1992, for blue tits). Pairing success of yearling lazuli buntings shows a bimodal relation with plumage brightness: the brightest yearlings who can manage to establish their own territory and the most inconspicuous colored yearlings that are accepted in the neighborhood of older adults are likely to get a female partner while intermediate males have low reproductive success (Greene et al., 2000). This is a very neat example of a system where two opposing selection pressures are at work at the same time: sexual selection favors brightly colored males, while market selection favors dull looking ones. The outcome is disruptive selection resulting in high variability among yearlings.

Biological versus Mating Markets

In principle, mating markets can be considered as a special form of biological markets and certain general rules for biological markets will apply equally to mating markets. Reproductive cooperation involves, by definition, partners belonging to the opposite sex of the same species. The operational sex ratio (OSR), i.e., the relative numbers of males and females available for mating, is the basic supply-demand parameter of the mating markets (Emlen & Oring, 1977; Noë & Hammerstein, 1995; Noë, 2006b). The OSR is largely determined by the amount of time each sex is bound by the production of a batch of offspring. In species with lek-mating systems, the relationships do not last much longer than the copulation itself and males return to the market immediately, whereas females may return only in the next season. In such skewed markets, sexual selection results in competitiveness in one sex (usually males) and choosiness in the other (usually females). In species with obligate biparental care, on the other hand, the operational sex ratio is much more balanced and in general both sexes are choosy. Of course, as both partners might be tempted to induce the partner to invest more than they do themselves, partner control mechanisms should evolve that ensure a continuation of the cooperative relationship. We would expect that the same general principle applies not only to mating markets but also to biological markets in general: while in "one-shot" or short-term relationships partner choice is essential; the longer a relationship lasts the more the emphasis should shift from partner choice to partner control. The higher the threshold for partner switching, the longer lasting relationships will tend to be.

The distinction between "mating markets" and "economic markets" becomes blurred where differences in market value between the sexes are compensated for by goods and services offered by the members of the competing sex, for example, in the form of protection, nuptial gifts, safe nesting locations, or territories containing resources (e.g., Gumert, 2007a; Metz et al., 2007). Human mating markets have been explained with the help of biological market theory by several authors (Pawlowski & Dunbar, 1999; Voland, 2000; Roberts et al., 2004; Woodward & Richards, 2004; Penke et al., 2007; Pollet & Nettle, 2008, 2009).

Conclusion

Models of cooperation can be divided into those focusing on partner control and those focusing on partner choice. The biological market paradigm strongly emphasizes partner choice as the pivotal force in the selection for traits that foster cooperation. By pointing out the analogies between cooperation among non-human organisms and trading among humans, Noë and Hammerstein (1994, 1995) encouraged the introduction of theoretical insights derived from economics into the field of behavioral ecology.

These early papers, together with some precursors (Eshel & Cavalli-Sforza, 1982; Bull et al., 1991; Noë et al., 1991), have set the stage. Since then our understanding of biological markets has increased due to theoretical contributions by several authors (Schwartz & Hoeksema, 1998; Barrett et al., 1999; Hoeksema & Bruna, 2000; Bowles & Hammerstein, 2003; Hoeksema & Kummel 2003; Hoeksema & Schwartz 2003; Kiers & van der Heijden 2006; Kummel & Salant, 2006; Cowden & Peterson, 2009; Golubski & Klausmeier, 2010; Mazancourt & Schwartz, 2010; Archetti, 2011; Aktipis, 2011; André & Baumard, 2011).

Overall, the number of papers focusing on theoretical aspects of biological markets is low compared with large and rapidly growing body of empirical evidence for market effects in biological systems (Johnstone & Bshary, 2008). This situation contrasts with the one for reciprocal altruism, where research papers reporting empirical evidence are scarce while theoretical papers are still produced in large numbers (Leimar & Hammerstein, 2006).

The biological market paradigm has motivated two distinct lines of investigation. On the proximate level, researchers have investigated how individuals gain information about the value of their commodities, how supply and demand of commodities influence exchange rates, and which mechanisms can help to stabilize exchange markets and protect against free riders. Research paradigms based on repeated partner control games have been useful in the study of certain aspects of cooperation, but they have several severe limitations. We, therefore, believe that it will be more fruitful to focus more on paradigms that allow for: (1) the option for partner choice and partner switching, (2) communication and/or sequential decisions, and (3) the quantification of investments and returns.

On the ultimate level, partner choice should lead to the operation of market selection just like mate choice drives sexual selection. We have argued above that market selection might often be weaker or more subtle than sexual selection, but it should nevertheless exist. In obligatory exchange markets—as are found in some inter-specific mutualisms—market selection might equal sexual selection in strength. Few studies thus far have addressed the question of market selection in vertebrate species. The best example might be the selection for delayed plumage in lazuli buntings (Greene et al., 2000). Apart from this example, market selection is a clearly underexplored area, and we hope that we will hear more about it in the future.

References

Aanen, D. K., de Fine Licht, H. H., Debets, A. J. M., Kerstes, N. A. G., Hoekstra, R. F., & Boomsma, J. J. (2009). High symbiont relatedness stabilizes mutualistic cooperation in fungus-growing termites. *Science, 326*, 1103–1106.

Aktipis, C. A. (2011). Is cooperation viable in mobile organisms? Simple Walk Away rule favors the evolution of cooperation in groups. *Evolution and Human Behavior, 32*(4), 263–276.

Andersson, M. (1982). Female choice selects for extreme tail length in a widowbird. *Nature, 299*, 818–820.

André, J.-B., & Baumard, N. (2011). The evolution of fairness in a biological market. *Evolution, 65*(5), 1447–1456.

Archetti, M. (2011). Contract theory for the evolution of cooperation: The right incentives attract the right partners. *Journal of Theoretical Biology, 269*, 201–207.

Ashlock, D., Smucker, M. D., Stanley, E. A., & Tesfatsion, L. (1996). Preferential partner selection in an evolutionary study of prisoner's dilemma. *Bio Systems, 37*, 99–125.

Axelrod, R. A., & Hamilton, W. D. (1981). The evolution of cooperation. *Science, 211*, 1390–1396.

Balasubramaniam, K. N., Berman, C. M., et al. (2011). Using biological markets principles to examine patterns of grooming exchange in Macaca thibetana. *American Journal of Primatology, 73*(12), 1269–1279.

Barelli, C., Reichard, U. H., et al. (2011). Is grooming used as a commodity in wild white-handed gibbons, Hylobates lar? *Animal Behaviour, 82*(4), 801–809.

Barrett, L., Henzi, S. P., Weingrill, T., & Hill, R. A. (1999). Market forces predict grooming reciprocity in female baboons. *Proceedings of the Royal Society, Series B: Biological Sciences, 266*, 665–670.

Barrett, L., Gaynor, D., & Henzi, S. P. (2002). A dynamic interaction between aggression and grooming reciprocity among female chacma baboons. *Animal Behaviour, 63*, 1047–1053.

Barrett, L., & Henzi, S. P. (2006). Monkeys, markets and minds: Biological markets and primate sociality. In P. M. Kappeler & C. P. van Schaik (Eds.), *Cooperation in primates and humans* (pp. 209–232). Berlin: Springer.

Batali, J., & Kitcher, P. (1995). Evolution of altruism in optional and compulsory games. *Journal of Theoretical Biology, 175*, 161–171.

Bergmüller, R., Johnstone, R. A., Russell, A. F., & Bshary, R. (2007). Integrating cooperative breeding into theoretical concepts of cooperation. *Behavioural Processes, 76*, 61–72.

Bever, J. D., Richardson, S. C., Lawrence, B. M., Holmes, J., & Watson, M. (2009). Preferential allocation to beneficial symbiont with spatial structure maintains mycorrhizal mutualism. *Ecology Letters, 12*, 13–21.

Bowles, S., & Hammerstein, P. (2003). Does market theory apply to biology? In P. Hammerstein (Ed.), *Genetic and cultural evolution of cooperation* (pp. 153–165). Cambridge, MA: MIT Press.

Bronstein, J. L. (1998). The contribution of ant plant-protection studies to our understanding of mutualism. *Biotropica, 30*, 150–161.

Bshary, R. (2002). Biting cleaner fish use altruism to deceive image-scoring client reef fish. *Proceedings of the Royal Society, Series B: Biological Sciences, 269*, 2087–2093.

Bshary, R., & Noë, R. (2003). Biological markets: The ubiquitous influence of partner choice on the dynamics of cleaner fish-client reef fish interactions. In P. Hammerstein (Ed.), *Genetic and cultural evolution of cooperation* (pp. 167–184). Cambridge, MA: MIT Press.

Bshary, R., & Bronstein, J. L. (2004). Game structures in mutualistic interactions: What can the evidence tell us about the kind of models we need? *Advances in the Study of Behavior, 34*, 59–101.

Bshary, R., & Grutter, A. S. (2006). Image scoring and cooperation in a cleaner fish mutualism. *Nature, 441*, 975–978.

Bshary, R., Grutter, A. S., Willener, A. S. T., & Leimar, O. (2008). Pairs of cooperating cleaner fish provide better service quality than singletons. *Nature, 455*, 964–966.

Bull, J. J., Molineux, I. J., & Rice, W. R. (1991). Selection of benevolence in a host-parasite system. *Evolution: International Journal of Organic Evolution, 45*, 875–882.

Bull, J. J., & Rice, W. R. (1991). Distinguishing mechanisms for the evolution of co-operation. *Journal of Theoretical Biology, 149*, 63–74.

Chancellor, R. L., & Isbell, L. A. (2008). Female grooming markets in a population of graycheeked mangabeys (*Lophocebus albigena*). *Behavioral Ecology, 20*, 79–86.

Clements, K. C., & Stephens, D. W. (1995). Testing models of non-kin cooperation: Mutualism and the prisoner's dilemma. *Animal Behaviour, 50*, 527–535.

Clutton-Brock, T. (2009). Cooperation between non-kin in animal societies. *Nature, 462*, 51–57.

Connor, R. C. (1992). Egg-trading in simultaneous hermaphrodites: An alternative to Tit for Tat. *Journal of Evolutionary Biology, 5*, 523–528.

Connor, R. C. (1995a). Altruism among non-relatives: Alternatives to the "prisoner's dilemma." *Trends in Ecology & Evolution, 10*, 84–86.

Connor, R. C. (1995b). Impala allogrooming and the parcelling model of reciprocity. *Animal Behaviour, 49*, 528–530.

Cowden, C. C., & Peterson, C. J. (2009). A multi-mutualist simulation: Applying biological market models to diverse mycorrhizal communities. *Ecological Modelling, 220*, 1522–1533.

Darwin, C. (1859). *On the origin of species by means of natural selection, or the Preservation of favoured races in the struggle for life*. London: Murray.

Dawkins, R. (1976). *The selfish gene*. Oxford: Oxford University Press.

Dawkins, R., & Krebs, J. R. 1978. Animal signals: information or manipulation. In J. R. Krebs & N. B. Davies (Eds.), *Behavioural ecology: An evolutionary approach* (3rd Ed.). Oxford: Blackwell.

de Waal, F. B. M. (2000). Attitudinal reciprocity in food sharing among brown capuchin monkeys. *Animal Behaviour, 60*, 253–261.

Denison, R. F. (2000). Legume sanctions and the evolution of symbiotic cooperation by rhizobia. *American Naturalist, 156*, 567–576.

Diggle, S. P., Griffin, A. S., Campbell, G. S., & West, S. A. (2007). Cooperation and conflict in quorum-sensing bacterial populations. *Nature, 450*, 411–415.

Dugatkin, L. A., & Alfieri, M. (1991). Guppies and the TIT FOR TAT strategy: Preference based on past interaction. *Behavioral Ecology and Sociobiology, 28*, 243–246.

Dugatkin, L. A., & Wilson, D. S. (1991). ROVER: A strategy for exploiting cooperators in a patchy environment. *American Naturalist, 138*, 687–701.

Edwards, D. P., Hassall, M., Sutherland, W. J., & Yu, D. W. (2006). Selection for protection in an ant–plant mutualism: Host sanctions, host modularity, and the principal–agent game. *Proceedings of the Royal Society of London, Series B: Biological Sciences, 273*, 595–602.

Emlen, S. T., & Oring, L. W. (1977). Ecology, sexual selection, and the evolution of mating systems. *Science, 197*, 215–223.

Enquist, M., & Leimar, O. (1993). The evolution of cooperation in mobile organisms. *Animal Behaviour, 45*, 747–757.

Eshel, I., & Cavalli-Sforza, L. L. (1982). Assortment of encounters and evolution of cooperativeness. *Proceedings of the National Academy of Sciences of the United States of America, 79*, 1331–1335.

Fischer, M. K., Hoffmann, K. H., & Völkl, W. (2001). Competition for mutualists in an ant-homopteran interaction mediated by hierarchies of ant attendance. *Oikos, 92*, 531–541.

Fischer, M. K., Völkl, W., & Hoffmann, K. H. (2005). Honeydew production and honeydew sugar composition of polyphagous black bean aphid, *Aphis fabae (Hemiptera: Aphididae)* on various host plants and implications for ant-attendance. *European Journal of Entomology, 102*, 155–160.

Fisher, R. A. (1930). *The genetical theory of natural selection*. Oxford: Clarendon Press.

Fleming, T. H., & Holland, J. N. (1998). The evolution of obligate pollination mutualisms: Senita cactus and senita moth. *Oecologia, 114*, 368–375.

Friesen, M. L. (2012). Widespread fitness alignment in the legume–rhizobium symbiosis. *New Phytologist, 194*, 1096–1111.

Fruteau, C., Voelkl, B., van Damme, E., & Noë, R. (2009). Supply and demand determine the market value of food providers in wild vervet monkeys. *Proceedings of the National Academy of Sciences of the United States of America, 106*, 12007–12012.

Fruteau, C., van de Waal, E., van Damme, E., & Noë, R. (2011). Infant access and handling in sooty mangabeys and vervet monkeys. *Animal Behaviour, 81* (1), 153–161.

Gibson, R. M., & Langen, T. A. (1996). How do animals choose their mates? *Trends in Ecology & Evolution, 11*, 468–470.

Golubski, A. J., & Klausmeier, C. A. (2010). Control in mutualisms: Combined implications of partner choice and bargaining roles. *Journal of Theoretical Biology, 267*, 535–545.

Grafen, A. (1990). Sexual selection unhandicapped by the Fisher Process. *Journal of Theoretical Biology, 144*, 473–516.

Greene, E., Lyon, B. E., Muehter, V. R., Ratcliffe, L., Oliver, S. J., & Boag, P. T. (2000). Disruptive sexual selection for plumage coloration in a passerine bird. *Nature, 407*, 1000–1003.

Gubry-Rangin, C., Garcia, M., & Béna, G. (2010). Partner choice in Medicago Truncatula–Sinorhizobium symbiosis. *Proceedings of the Royal Society, Series B: Biological Sciences, 277*, 1947–1951.

Gumert, M. D. (2007a). Payment for sex in a macaque mating market. *Animal Behaviour, 74*, 1655–1667.

Gumert, M. D. (2007b). Grooming and infant handling interchange in Macaca fascicularis: The relationship between infant supply and grooming payment. *International Journal of Primatology, 28*, 1059–1074.

Hamilton, W. D. (1964). The genetical evolution of social behaviour. *Journal of Theoretical Biology, 7*, 1–52.

Hammerstein, P. (2003). Why is reciprocity so rare in social animals? A protestant appeal. In P. Hammerstein (Ed.), *Genetic and cultural evolution of cooperation* (pp. 83–94). Cambridge, MA: MIT Press.

Harper, D. G. C. 1991. Communication. In J. R. Krebs & N. B. Davies (Eds.), *Behavioural ecology: An evolutionary approach* (3rd Ed.) (pp. 374–397). Oxford: Blackwell.

Heath, K. D., & Tiffin, P. (2007). Context dependence in the coevolution of plant and rhizobial mutualists. *Proceedings of the Royal Society, Series B: Biological Sciences, 274*, 1905–1912.

Heath, K. D., & Tiffin, P. (2009). Stabilizing mechanisms in a legume–rhizobium mutualism. *Evolution: International Journal of Organic Evolution, 63*, 652–662.

Henzi, S. P., & Barrett, L. (2002). Infants as a commodity in a baboon market. *Animal Behaviour, 63*, 915–921.

Henzi, S. P., Barrett, L., Gaynor, D., Greeff, J., Weingrill, T., & Hill, R. A. (2003). Effect of resource competition on the long-term allocation of grooming by female baboons: Evaluating Seyfarth's model. *Animal Behaviour, 66*, 931–938.

Herre, E. A., Knowlton, N., Mueller, U. G., & Rehner, S. A. (1999). The evolution of mutualisms: Exploring the paths between conflict and cooperation. *Trends in Ecology & Evolution, 14*, 49–53.

Hoeksema, J. D., & Bruna, E. M. (2000). Pursuing the big questions about interspecific mutualism: A review of theoretical approaches. *Oecologia, 125*, 321–330.

Hoeksema, J. D., & Kummel, M. (2003). Ecological persistence of the plant-mycorrhizal mutualism: A hypothesis from species coexistence theory. *American Naturalist* (Suppl.), *162*, S40–S50.

Hoeksema, J. D., & Schwartz, M. W. (2003). Expanding comparative-advantage biological market models: Contingency of mutualism on partners' resource requirements and acquisition trade-offs. *Proceedings of the Royal Society of London, Series B: Biological Sciences, 270*, 913–919.

James, C. D., Hoffman, M. T., Lightfoot, D. C., Forbes, G. S., & Whitford, W. G. (1994). Fruit abortion in Yucca elata and its implications for the mutualistic association with yucca moths. *Oikos, 69*, 207–216.

Johnstone, R. A. (1995). Sexual selection, honest advertisement and the handicap principle: Reviewing the evidence. *Biological Reviews of the Cambridge Philosophical Society, 70*, 1–65.

Johnstone, R. A., & Bshary, R. (2008). Mutualism, market effects, and partner control. *Journal of Evolutionary Biology, 21*, 879–888.

Kato, M., Takimura, A., & Kawakita, A. (2003). An obligate pollination mutualism and reciprocal diversification in the tree genus lochidion (*Euphorbiaceae*). *Proceedings of the National Academy of Sciences of the United States of America, 100*(9), 5264–5267.

Kawakita, A., & Kato, M. (2009). Repeated independent evolution of obligate pollination mutualism in the Phyllantheae–Epicephala association. *Proceedings of the Royal Society of London, Series B: Biological Sciences, 276*, 417–426.

Kempenaers, B., Verheyen, G. R., den Broeck, M. V., Burke, T., Broeckhoven, C. V., & Dhondt, A. (1992). Extra-pair paternity results from female preference for high-quality males in the blue tit. *Nature, 357*, 494–496.

Kiers, E. T., Duhamel, M., Beesetty, Y., Mensah, J. A., Franken, O., Verbruggen, E., Fellbaum, C. R., Kowalchuk, G. A., Hart, M. M., Bago, A., Palmer, T. M., West, S. A., Vandenkoornhuyse, P.,

Jansa, J., & Bücking, H. (2011). Reciprocal rewards stabilize cooperation in the mycorrhizal symbiosis. *Science, 333,* 880–882.

Kiers, E. T., Rousseau, R. A., West, S. A., & Denison, R. F. (2003). Host sanctions and the legume–rhizobium mutualism. *Nature, 425,* 78–81.

Kiers, E. T., & van der Heijden, M. G. A. (2006). Mutualistic stability in the arbuscular mycorrhizal symbiosis: Exploring hypotheses of evolutionary cooperation. *Ecology, 87,* 1627–1636.

Kummel, M., & Salant, S. W. (2006). The economics of mutualisms: optimal utilization of mycorrhizal mutualistic partners by plants. *Ecology, 87,* 892–902.

Lande, R. (1981). Models of speciation by sexual selection on polygenic characters. *Proceedings of the National Academy of Sciences of the United States of America, 78,* 3721–3725.

Leimar, O., & Connor, R. C. (2003). Byproduct benefits, reciprocity and pseudoreciprocity in mutualism. In P. Hammerstein (Ed.), *Genetic and cultural evolution of cooperation* (pp. 203–222). Cambridge, MA: MIT Press.

Leimar, O., & Hammerstein, P. (2006). Facing the facts. *Journal of Evolutionary Biology, 19,* 1403–1405.

Luce, R. D., & Raiffa, H. (1957). *Games and decisions: Introduction and critical survey.* New York: Wiley.

Marr, D. L., & Pellmyr, O. (2003). Effect of pollinator-inflicted ovule damage on floral abscission in the yucca-yucca moth mutualism: The role of mechanical and chemical factors. *Oecologia, 136,* 236–243.

Maynard Smith, J. (1964). Group selection and kin selection. *Nature, 201,* 1145–1147.

Maynard Smith, J. (1982). Do animals convey information about their intentions? *Journal of Theoretical Biology, 97,* 1–5.

Mazancourt, C. d., & Schwartz, M. W. (2010). A resource ratio theory of cooperation. *Ecology Letters, 13,* 349–359.

Metz, M., Klump, G., & Friedl, T. (2007). Temporal changes in demand for and supply of nests in red bishops (*Euplectes orix*): Dynamics of a biological market. *Behavioral Ecology and Sociobiology, 61,* 1369–1381.

Milinski, M., Külling, D., & Kettler, R. (1990). Tit for Tat: Sticklebacks Gasterosteus aculeatus trusting a cooperating partner. *Behavioral Ecology, 1,* 7–11.

Morton, E. S. (1987). Variation in mate guarding intensity by male purple martins. *Behaviour, 101,* 211–224.

Morton, E. S., & Derrickson, K. C. (1990). The biological significance of age-specific return schedules in breeding purple martins. *Condor, 92,* 1040–1050.

Morton, E. S., Forman, L., & Braun, M. (1990). Extrapair fertilizations and the evolution of colonial breeding in purple martins. *Auk, 107,* 275–283.

Muehter, V. M., Greene, E., & Ratcliffe, L. (1997). Delayed plumage maturation in Lazuli buntings: Tests of the female mimicry and status signalling hypotheses. *Behavioral Ecology and Sociobiology, 41*, 281–290.

Noë, R. (1990). A Veto game played by baboons: a challenge to the use of the prisoner's dilemma as a paradigm for reciprocity and cooperation. *Animal Behaviour, 39*, 78–90.

Noë, R., van Schaik, C. P., & van Hooff, J. A. R. A. M. (1991). The market effect: An explanation for pay-off asymmetries among collaborating animals. *Ethology, 87*, 97–118.

Noë, R. (1992). Alliance formation among male baboons: shopping for profitable partners. In A. H. Harcourt & F. B. M. d. Waal (Eds.), *Coalitions and alliances in humans and other animals* (pp. 285–321). Oxford: Oxford University Press.

Noë, R., & Hammerstein, P. (1994). Biological markets: Supply and demand determine the effect of partner choice in cooperation, mutualism and mating. *Behavioral Ecology and Sociobiology, 35*, 1–11.

Noë, R., & Hammerstein, P. (1995). Biological markets. *Trends in Ecology & Evolution, 10*, 336–339.

Noë, R. (2006a). Cooperation experiments: Coordination through communication versus acting apart together. *Animal Behaviour, 71*, 1–18.

Noë, R. (2006b). Digging for the roots of trading. In P. M. Kappeler & C. P. van Schaik (Eds.), *Cooperation in primates and humans: Mechanisms and evolution* (pp. 233–261). Heidelberg: Springer.

Nowak, M. A., & Sigmund, K. (1998). Evolution of indirect reciprocity by image scoring. *Nature, 393*, 573–577.

Nowak, M. A. (2007). Five rules for the evolution of cooperation. *Science, 314*, 1560–1563.

Packer, C. (1986). Whatever happened to reciprocal altruism? *Trends in Ecology & Evolution, 1*, 142–143.

Pawlowski, B., & Dunbar, R. I. M. (1999). Impact of market value on human mate choice decisions. *Proceedings of the Royal Society, Series B: Biological Sciences, 266*, 281–285.

Peay, K., Bruns, T., & Garbelotto, M. (2010). Testing the ecological stability of ectomycorrhizal symbiosis: Effects of heat, ash and mycorrhizal colonization on Pinus muricata seedling performance. *Plant and Soil, 330*, 291–302.

Pellmyr, O., & Huth, C. J. (1994). Evolutionary stability of mutualism between yuccas and yucca moths. *Nature, 372*, 257–260.

Penke, L., Todd, P., Lenton, A. P., & Fasolo, B. (2007). How self-assessments can guide human mating decisions. In G. Geher & G. F. Miller (Eds.), *Mating intelligence: New insights into intimate relationships, human sexuality, and the mind's reproductive system* (pp. 37–75). Mahwah: Erlbaum.

Pollet, T. V., & Nettle, D. (2008). Driving a hard bargain: Sex ratio and male marriage success in a historical US population. *Biology Letters, 4*, 31–33.

Pollet, T. V., & Nettle, D. (2009). Market forces affect patterns of polygyny in Uganda. *Proceedings of the National Academy of Sciences of the United States of America, 106*, 2114–2117.

Roberts, G. (1998). Competitive altruism: From reciprocity to the handicap principle. *Proceedings of the Royal Society of London, Series B: Biological Sciences, 265*, 427–431.

Roberts, G., & Sherratt, T. N. (1998). Development of cooperative relationships through increasing investment. *Nature, 394*, 175–179.

Roberts, S. C., Havlicek, J., Flegr, J., Hruskova, M., Little, A. C., Jones, B. C., et al. 2004. Female facial attractiveness increases during the fertile phase of the menstrual cycle. *Proceedings of the Royal Society of London, Series B: Biological Sciences*, 271 [Suppl. 5], S270–S272.

Russell, Y. I., Call, J., & Dunbar, R. I. M. (2008). Image scoring in great apes: Behavioural processes. *Behavioural Processes, 78*, 108–111.

Ryoko, O., Denison, R. F., & Kiers, E. T. (2009). Controlling the reproductive fate of rhizobia: How universal are legume sanctions? *New Phytologist, 183*, 967–979.

Sachs, J., & Wilcox, T. (2006). A shift to parasitism in the jellyfish symbiont *Symbiodinium microadriaticum*. *Proceedings of the Royal Society of London, Series B: Biological Sciences, 273*, 425–429.

Sachs, J. L., Mueller, U. G., Wilcox, T. P., & Bull, J. J. (2004). The evolution of cooperation. *Quarterly Review of Biology, 79*, 135–159.

Samuelson, P. (1983). *Foundations of economic analysis*. Cambridge, MA: Harvard University Press.

Schwartz, M. W., & Hoeksema, J. D. (1998). Specialization and resource trade: Biological markets as a model of mutualisms. *Ecology, 79*, 1029–1038.

Silk, J. (2007). The adaptive value of sociality in mammalian groups. *Philosophical Transactions of the Royal Society, Series B: Biological Sciences, 362*, 539–559.

Simms, E. L., & Taylor, D. L. (2002). Partner choice in nitrogen-fixation mutualisms of legumes and rhizobia. *Integrative and Comparative Biology, 42*, 369–380.

Simms, E. L., Taylor, D. L., Povich, J., Shefferson, R. P., Sachs, J. L., Urbina, M., & Tausczik, Y. (2006). An empirical test of partner choice mechanisms in a wild legume–rhizobium interaction. *Proceedings of the Royal Society, Series B: Biological Sciences, 273*, 77–81.

Slater, K. Y., Schaffner, C. M., & Aureli, F. (2007). Embraces for infant handling in spider monkeys: Evidence for a biological market? *Animal Behaviour, 74*, 455–461.

Smith, A. (1776). *An inquiry into the nature and causes of the wealth of nations*. London: Methuen.

Stephens, D. W., & Krebs, J. R. (1986). *Foraging theory*. Princeton, NJ: Princeton University Press.

Trivers, R. L. (1971). The evolution of reciprocal altruism. *Quarterly Review of Biology, 46*, 35–57.

Vautrin, E., Genieys, S., Charles, S., & Vavre, F. (2008). Do vertically transmitted symbionts co-existing in a single host compete or cooperate? A modelling approach. *Journal of Evolutionary Biology, 21*, 145–161.

Voland, E. (2000). Contributions of family reconstruction studies to evolutionary reproductive ecology. *Evolutionary Anthropology, 9*, 134–146.

Walling, C. A., Dawnay, N., Kazem, Anahita J. N., & Wright, J. (2004). Predator inspection behaviour in three-spined sticklebacks (*Gasterosteus aculeatus*): Body size, local predation pressure and cooperation. *Behavioral Ecology and Sociobiology, 56*(2), 164–170.

Wedekind, C., & Milinski, M. (2000). Cooperation through image scoring in humans. *Science, 288*, 850–852.

West, S. A., Griffin, A. S., & Gardner, A. (2007a). Social semantics: Altruism, cooperation, mutualism, strong reciprocity and group selection. *Journal of Evolutionary Biology, 20*, 415–432.

West, S. A., Griffin, A. S., & Gardner, A. (2007b). Evolutionary explanations for cooperation. *Current Biology, 17*, R661–R672.

West, S. A., Kiers, E. T., Simms, E. L., & Denison, R. F. (2002). Sanctions and mutualism stability: Why do rhizobia fix nitrogen? *Proceedings of the Royal Society, Series B: Biological Sciences, 269*, 685–694.

Wiggins, D. A., & Morris, R. D. (1986). Criteria for female choice of mates: Courtship feeding and parental care in the common tern. *American Naturalist, 128*, 126–129.

Wilkinson, D. M. (2001). Mycorrhizal evolution. *Trends in Ecology & Evolution, 16*, 64–65.

Williams, G. C. (1966). *Adaptation and natural selection*. Princeton: Princeton University Press.

Woodward, K., & Richards, M. H. (2004). The parental investment model and minimum mate choice criteria in humans. *Behavioral Ecology, 16*, 57–61.

Wynne-Edwards, V. C. (1962). *Animal dispersion in relation to social behaviour*. Edinburgh: Oliver & Boyd.

Zahavi, A. (1975). Mate selection—a selection for a handicap. *Journal of Theoretical Biology, 53*, 205–214.

Zahavi, A. (1977). The cost of honesty (further remarks on the handicap principle). *Journal of Theoretical Biology, 67*, 603–605.

8 False Advertising in Biological Markets: Partner Choice and the Problem of Reliability

Ben Fraser

The *partner choice* approach to understanding the evolution of cooperation builds on approaches that focus on *partner control* by considering processes that occur prior to pair or group formation. Proponents of the partner choice approach rightly note that competition to be chosen as a partner can help solve the puzzle of cooperation (Noe, 2006; Miller, 2007; Nesse, 2007). I aim to build on the partner choice approach by considering the role of signaling in partner choice. Partnership formation often requires reliable information. Signaling is thus important in the context of partner choice. However, the issue of signal reliability has been understudied in the partner choice literature. The issue deserves attention because—despite what proponents of the partner choice approach sometimes claim—that approach does face a cheater problem, which we might call the problem of *false advertising* in biological markets. Both theoretical and empirical work is needed to address this problem. I will draw on signaling theory (Maynard Smith & Harper, 2003; Searcy & Nowicki, 2005) to provide a theoretical framework within which to organize the scattered discussions of the false advertising problem extant in the partner choice literature. I will end by discussing some empirical work on cooperation, partner choice, and punishment among humans (Barclay, 2006; Nelissen, 2008; Horita, 2010).

The Problem of Cooperation

Numerous definitions of "cooperation" have been offered in the biological literature. Some researchers use "cooperation" very generally to cover all acts by one individual that benefit one or more other individuals (Sachs et al., 2004, p. 137). Others consider such usage too liberal, since it counts as cooperative behaviors that benefit others only incidentally, and restrict "cooperation" to behaviors that have been selected because they benefit others (West, Griffin & Gardner, 2007, p. 416). Thus, while some researchers would count an elephant that defecates and thereby feeds a dung beetle as cooperating with the beetle, others would not. Minimally, though, the various definitions on offer agree that cooperation involves one organism *A* benefiting another organism

B, where that means *A*'s behavior increases *B*'s fitness. There is also general agreement about when such behavior is, from an evolutionary perspective, *prima facie* puzzling, namely, when the increase in *B*'s fitness comes at an apparent cost to the fitness of *A*. This is the problem of cooperation on which I wish to focus.

One way to solve the puzzle of cooperation is to show how, appearances aside, *A* is not actually sacrificing its own fitness to benefit *B*.[1] Hamilton (1964) put together a good deal of the puzzle when he conceived of the pieces not as individual organisms but as far smaller units: genes. Hamilton's theory of kin selection showed how *A* helping *B* to survive and/or reproduce could increase *A*'s "inclusive" fitness (even if doing so decreased *A*'s "personal" fitness) so long as *A* and *B* were sufficiently close relatives. Even after Hamilton's elegant insight, however, much of the puzzle of cooperation remained fragmentary.

Partner Control Models of Cooperation

Cooperation among unrelated individuals has predominantly been viewed through the lens of reciprocity (Trivers, 1971). The key question here is how an organism *A* should behave toward a given partner *B* over a series of encounters in order to maximize its total payoff over that series. Should *A* cooperate with *B*, or refrain from cooperating, or sometimes do the one and sometimes the other? The important insight grounding the reciprocity-based approach is that, under certain conditions, the immediate cost *A* pays to benefit *B* can be recouped (and more) over time if *B* repays *A*'s help (whether the repayment is in kind, or in a different currency). For example, the Tit-for-Tat strategy in Axelrod's (1984) iterated prisoner's dilemma tournament enjoyed success because it conditionalized its own cooperative behavior on the cooperativeness of the strategies with which it found itself paired.

Studying cooperation from the perspective of reciprocity focuses attention on what has been called "partner fidelity" (Bull & Rice, 1991), "partner verification" (Noe & Hammerstein, 1994), and "partner control" (Noe, 2001). I will use the term "partner control approach" here to mean approaches to understanding the evolution of cooperation that focus on how an individual should manage its interactions with a given partner, and in particular on how individuals engineer the incentive structure of a given interaction partner for their own benefit.

One reason for dissatisfaction with the partner control approach is that it is surprisingly undersupported by empirical data. Uncontroversial examples of reciprocity-based cooperation among nonhuman organisms have proven elusive.[2] A more important reason—and the focus here—is that the partner control approach neglects important aspects of cooperation. It models only one aspect of a complex process. As noted above, the partner control approach focuses on how an individual manages its interactions with a given partner. However, interactions between organisms can be imagined as consisting of three stages: a pair/group *formation* stage; a *decision* stage (e.g., cooperate or not?); and, finally, a *division* stage, in which the yield of the interaction is apportioned among those involved (Noe, 1990, p. 79; Dugatkin, 1995, p. 4).[3] The partner control approach considers only the second of these stages. Importantly for current purposes, that approach typically assumes that individual agents have no control over partnership or group formation, whereas in nature it is likely to often be the case that organisms exert some degree of partner choice (Dugatkin, 1995). The partner control approach is thus limited in scope and makes unrealistic assumptions.

Partner Choice Models of Cooperation

The shortcomings of the partner control approach to the problem of cooperation have left some researchers inclined to explore an alternative: the partner choice approach. The emphasis here is not on how best to deal with a given partner but on "the option of choosing and switching partners" (Noe, 2006, p. 5). It is important to note that partner control and partner choice models are not competing, mutually exclusive alternatives. Rather, the partner choice approach is meant to complement the partner control approach. I will discuss in a later section precisely how the two approaches relate to each other, For now, I am concerned to define some key terms, before critically discussing the partner choice approach. (The following material is drawn from a pair of influential early papers by Noe & Hammerstein, 1994, 1995.)

A "biological market" exists whenever organisms engage in mutually beneficial exchanges of resources and, crucially, when at least one of the trading individuals can exercise choice in selecting a trading partner. Put negatively, this second requirement says that the market metaphor does not apply when desired resources can simply be taken (i.e., theft) or when individuals can force others to partner with them (i.e., coercion).

In many biological markets, organisms can be divided into "trader classes," according to the kind of resource they offer. For example, numerous ant species can be classed together as "protection" traders, and this class forms a market with various species of aphids, butterflies, and plants, which together comprise a (rather heterogeneous) trader class dealing in nutrients. Notice, in some biological markets, there may be only one commodity on offer. For example, in some forms of hunting there may be a marketplace for hunting skill, in which would-be hunters compete to be chosen as participants by hunt-leaders. Even if a hunting group comprises several specialized roles, each requiring a distinct skill-set, there may nevertheless be competition to fill each role. For example, consider turtle hunting among the Meriam Islanders of the Torres Strait (see Smith & Bliege-Bird, 2000): Hunt leaders assemble groups comprising a boat driver, a harpooner, and several "jumpers," and competition to fill each role may result in "submarkets" for specific hunting skills. So, to set up a biological market, it is not necessary that traders fall into two classes, although this is certainly where the emphasis falls in the partner choice literature.

Given the existence of a biological market, competition among individuals to be chosen as a trading partner and pressure on individuals to make good partner choices are both to be expected. There will accordingly be "market selection," defined as "selection of traits that maximize fitness in biological markets, such as the ability to compete with members of the same trading class, the ability to attract trading partners, and the ability to sample alternative partners efficiently" (Noe & Hammerstein, 1995, p. 336).

The partner choice approach to understanding cooperation, then, attempts to explain cooperation as the result of competition to be picked as a partner in profitable exchanges. An early statement of this approach was given by Bull and Rice (1991, p. 68), who defined partner choice models as those in which cooperation is evolutionarily stable because "an individual of species *A* is paired with several members of species *B* for a single interaction, but *A* chooses to reward only the most co-operative members of *B*." Recently, Nesse (2007, p. 151) has discussed the approach under the heading of "the social selection perspective," which "shifts the focus of attention away from decisions to cooperate or defect and abilities to detect cheating, and toward the quite different tasks of selecting carefully among a variety of potential partners, trying to discern what they want, and trying to provide it, so one is more likely to be chosen and kept as a partner." According to the partner choice approach, puzzling cooperation—*A* benefiting *B* at some cost to itself—becomes intelligible when it becomes clear that benefiting *B* allows *A* to either establish or maintain a mutually beneficial interaction with *B*: It is an entry fee.

Although the partner choice approach is presented as complementary to, rather than in competition with, the partner control approach, proponents of the former do claim that it enjoys several advantages over the latter. For one thing, the partner choice approach does not rely on unrealistic assumptions about the nature of cooperative interactions (such as that the interacting individuals have no control over partnership formation). For another, empirical support for partner choice models is comparatively abundant (e.g., see Bshary, 2002; Bshary & Grutter, 2003; Bshary & Noë, 2003; Bshary & Wurth, 2001; and a review in Sachs et al., 2004). Most importantly for current purposes, it is claimed that the partner choice approach, unlike the partner control approach, does not face the so-called cheater problem (or at least faces that problem to a far lesser extent).

The Relationship between Partner Choice and Partner Control Models

As noted above, the partner choice approach is typically presented as complementary to the partner control approach, rather than in competition with it. Both are supposed to be important to fully understanding cooperation. However, the precise way in which the approaches complement each other requires clarification. There are two possibilities here. The two approaches may (1) model different stages of one complex interaction, or (2) model different kinds of interaction.[4] I discuss these possibilities in turn.

One way in which the partner choice approach may complement the partner control approach is by modeling different stages of the complex process that is cooperation. Partner choice focuses on the "formation" stage of the cooperative process, at which partnerships or groups come together. Partner control approaches assume that partnership or group formation is, if not random, at least not under the control of the interacting individuals. Partner choice models supplement partner control models by considering the formation stage of the cooperative process in more realistic detail.

Noë (1990, p. 79) sees the relationship between the partner control and partner choice approaches as complementary in the sense above. There is, though, another way to understand "complementary." Partner choice and partner control models may be relevant—not to different stages of the one extended cooperative interaction—but to importantly different kinds of cooperative interactions.

In some cooperative interactions, the benefits of cooperation are distributed to the partners in sequence, serially rather than simultaneously. This is the cases in so-called reciprocal altruism, such as blood sharing among vampire bats (the classic study being Wilkinson, 1984). In such cases, control mechanisms for preventing or punishing defection are important. (Interestingly, one such mechanism may be the threat of partner switching in response to defection, and here, the distinction between mechanisms of partner control and those of partner choice begins to blur.) So, we may see partner control models as particularly relevant when cooperative interactions involve the sequential distribution of benefits.

In other cooperative interactions, by contrast, the benefits of cooperation are *not* generated and distributed sequentially. Rather, partners reap the rewards of cooperation with each other simultaneously. Crucially, in some (perhaps many) such cooperative interactions, the magnitude of the benefit generated by cooperation may depend on the nature of the partner(s) involved in the interaction. When this is so—when the size of one's reward depends on the quality of one's partners—then partner choice mechanisms will be especially vital, provided of course that there is scope for choice in the first place.

To sum up, one way to understand the claim that partner choice and partner control models of cooperation are complementary is to see each as modeling a different stage of the same extended interaction. In some cases, this may well be the correct way to understand the claim. But, there is another way in which the two approaches can complement each other, namely, by each illuminating a qualitatively different kind of cooperative interaction. In this case, whether a partner choice or a partner control model is likely to be most illuminating will depend (among other things, of course) on the way in which benefits are produced and distributed. We should expect partner choice, rather than partner control, models to be particularly helpful in understanding cooperative interactions that generate benefits simultaneously for the cooperators, where the extent of those benefits depends on the quality of the interacting agents.

Cooperation and the Cheater Problem

In the context of the partner control approach, the problem of cheating is as follows. If *B* receives a benefit from a cooperative individual *A* but does not repay that benefit, or repays less than was received, then *B*-type individuals may eventually replace cooperators like *A* in the population of interacting individuals, since *B*-type individuals enjoy the benefits of cooperation without paying any of the associated costs. Showing how such cheating (defection, exploitation, free riding) can be prevented from undermining cooperation is a central concern for the partner control approach.

Cheating is supposedly not a problem—or is at least less of a problem—for the partner choice approach (see, e.g., Noë & Hammerstein, 1994, p. 2; Nesse, 2007, p. 145). To assess this claim, it is necessary to specify just what the advocates of partner choice mean by "cheating."

Noë and Hammerstein identify one kind of cheating in the context of partner choice as reneging on a proposed trade. They write: "cheating . . . is *changing the value of the commodity offered after the pair has been formed*" (1994, p. 6; emphasis added). The claim is then that such post-choice changes of offer are often impossible, and hence that cheating is often not a problem for the partner choice approach to cooperation. Noë and Hammerstein write:

> To our minds the cheating option can safely be ignored in the large number of cases in which either the commodity cannot be withdrawn or changed in quality or quantity once it is offered on the market. (Ibid., p. 2)

For example, Noë and Hammerstein observe that the "food bodies of myrmecophilous plants are examples of such irretrievable offers [since] once the plant has 'decided' to provide a quantity *x* of food bodies, these remain available to the ants" (ibid., p. 3). In this case, the plant traders *have* a certain amount of the commodity of interest to the ant traders, but cannot withdraw that commodity once it is offered (or at least cannot easily do so).

One problem here is that cheating (in this sense) does not seem precluded for the *ants*. Moreover, there is some evidence that plants in such partnerships can and in fact do withdraw the commodities on offer. For example, Edwards et al. (2006) studied a particular ant-plant mutualism and found that "ant shelters" (domatia) on stems that lose leaves—a sign that protector ants may not be patrolling enough—tend to wither away. So, it may be that the particular example used by Noë and Hammerstein to make the case that cheating is precluded in biological market interactions was poorly chosen.

In any case, even if post-choice changes of offer are impossible in many biological marketplaces, it has become apparent that the term "cheating" is used to pick out different things in the contexts of partner control and partner choice. The partner control approach faces something aptly called a "cheater problem"—free riding—while there is another thing—reneging—which is also aptly described as a cheater problem and which the partner choice approach avoids. Cheater problems come in many varieties.

Even if reneging is impossible in many biological marketplaces, there is a different kind of cheater problem that can arise in the context of partner choice. To appreciate this problem, it will help to first note the importance that proponents of the partner choice approach assign to signaling in the context of partner choice.

Signaling and the Problem of Reliability

Noë and Hammerstein write that "trading may take place on the basis of an honest signal that is correlated with access to a commodity, instead of being based on the commodity itself" (1995, p. 336). Noë also notes that "choosing partners implies a number of mechanisms [including] judging the partner's quality, a memory for the partner's quality and location, searching strategies, *judging the honesty of signals* and so on" (Noë, 2006, p. 5; emphasis added). Indeed, Noë thinks it is important to distinguish between markets "in which commodity values are measured directly and those in which signals play an intermediating role" (Noë, 2001, p. 108). Bull and Rice (1991, p. 69) and Sachs et al. (2004, p. 141) both point out that explanations for cooperation in terms of partner choice depend on there being some way in which individuals can assess and discriminate between potential partners. Partner assessment need not involve signals (as will be discussed below), but signaling is one way that it can be done.

Once it has been acknowledged that signals play an important role in at least some partner choice scenarios, the problem of reliability becomes unavoidable. In brief, the problem of reliability is as follows. Organisms are often interested in unobservable qualities of other organisms: a sexual rival's fighting ability, an offspring's hunger level, the evasive ability of potential prey, or the desirability of a potential mate. Making adaptive decisions depends crucially on estimating these unobvious qualities. As it turns out, potential partners, predators, and prey often provide the relevant information: they roar, beg, stot, sing, or dance, for example, or signal in some other way. The problem of reliability arises when we ask why, given the often strong incentive to mislead signal receivers, signal senders do not do so more often. Why do signalers not exaggerate the relevant quality to their own advantage? Why do signaling systems not collapse as a result of receivers eventually ignoring a cacophony of dishonest signals?

"False Advertising" in Biological Markets

It is now possible to state a cheater problem that faces the partner choice approach, one that is (I claim) currently underappreciated and insufficiently addressed. The problem is that of *false advertising*: A trader of one class may present itself as a better partner than it actually is, in order to increase its chances of being chosen as a partner by members of the other trading class. The problem of false advertising in biological markets is a specific case of the more general problem of reliability in biological signaling systems. The problem of false advertising differs from reneging as described above. False advertising is not a matter of a trader in a biological market genuinely *having* the relevant commodity or property but not delivering it, but rather of the trader *lacking* that commodity or property while convincing others otherwise.

There is some recognition of the false advertising problem in the extant literature on partner choice. For example, Noë (2001, p. 94) noted that in biological markets the "commodities on offer can be advertised [and] as in commercial advertisements there is a potential for false information." Noë also mentions the possibility in some markets for "subtle cheating" where "signals associated with the future transfer of resources are occasionally dishonest" (Noë & Hammerstein, 1995, p. 338; see also Noë, 2001, p. 94). Thus, I do not take myself to be pointing out something overlooked by proponents of the partner choice approach. Rather, I am suggesting that this kind of cheater problem is more pressing than has yet been acknowledged and, further, that the treatments of the problem offered to date are unsatisfactory.

Current Treatments of "False Advertising"

Attempts to address the false advertising problem for the partner choice approach have been too sanguine in dismissing the problem, or have made misdirected efforts to address it, or have been disunified and in need of clarification.

Dismissing the Problem

Nesse (2007, p. 145) in his discussion of partner choice and cooperation claims that social selection "will select for displays of resources and selective altruism that reflect an individual's potential value as a partner." Nesse is too quick to assume selection will favor honest signaling of partner quality. One should wonder why displays that *reflect* partner value would be selectively favored, rather than those that flatteringly *exaggerate* it to the displayer's advantage. Speaking specifically about cooperation among humans, Nesse says that "deception and cheating have been major themes in reciprocity research, and they apply in social selection models, but their effects are limited by inexpensive gossip about reputations and by the difficulty of faking expensive resource displays" (ibid.).

Nesse's claims about the limited scope for deception in human partner choice are questionable. For one thing, not all signals of partner value require the expenditure of large quantities of resources. Displaying qualities like kindness, honesty, and patience—all plausibly valuable qualities in many kinds of cooperative partnerships—may be quite cheap in terms of energy, risk, and material resources. In addition, gossip may not be so cheap. The risk of making enemies is a hard-to-quantify but nevertheless real cost of gossiping, which ought not to be ignored. These brief remarks do not settle matters any more than do Nesse's own, but they serve to show that Nesse is too sanguine regarding the problem of false advertising in biological markets.

Missing the Point

Sachs et al. (2004) note that partner choice models of cooperation must incorporate effective partner assessment systems. They write: "the [partner] assessment system is the biological arena in which one or more potential partners are observed for their cooperative tendencies, such that their level of cooperation in further interactions can be predicted. . . . [It] allows an individual to gain information about which partners are cooperative and how cooperative they are" (2004, pp. 141–142). Sachs et al. identify "parcelling" and "distribution" as two ways in which partner assessment may be conducted. Parcelling and distribution both involve splitting up a resource to be invested. In the former case, the resource is divided temporally, whereas in the latter case, the resource is divided spatially.

There are two problems with taking parceling and distribution to be partner assessment systems that allow effective partner choice. One problem is conceptual, the other empirical. I will discuss them in turn.

The conceptual problem is that parceling and distribution occur *after* members of two trading classes have partnered up. Two impala grooming each other in brief bursts have already formed a grooming pair. A yucca plant selectively aborting those of its many flowers that have been overexploited by selfish yucca moths is already interacting with its many partners. This is not to say that parceling and distribution are unimportant in the context of the partner choice approach. On the contrary, they are good ways of deciding when to do some partner "switching" (Noë & Hammerstein, 1995, p. 337). There is a difference, though, between partner choice, which occurs prior to the formation of a trading pair or group, and partner switching, which is a matter of strategically leaving one's current partner for greener fields. The difference is one that proponents of the partner choice approach are themselves at pains to mark.

Treating parceling and distribution as ways of making effective partner choices is thus conceptually confused and potentially misleading. These are clearly ways of engineering the incentives of partners—and are important as such—but they are not mechanisms of partner choice.

The empirical problem with parceling and distribution as ways of making good partner choices is that neither will be an option in biological markets where indivisible resources are at stake. For example, in mating markets where one trader class consists of monandrous females (those who mate with only one male), the commodity on offer is exclusive reproductive access, which cannot be parceled or distributed. In such cases, the timing of partner assessment matters crucially. Partner switching after dabbling a toe, so to speak, will not be an option. Traders offering indivisible resources must identify who is genuinely a high-value partner and who is not prior to committing to a trade.

Moreover, this problem is not limited to cases in which the resource being traded is indivisible. The problem may also arise in cases where the costs of partner switching are high. For instance, if searching for a new partner is very costly in terms of time, energy, and/or risk, then a choice once made may be effectively fixed. Here again, traders must be able to identify who is genuinely a high-value partner and who is not prior to committing to a trade.

A third means of partner assessment mentioned by Sachs et al. is "image scoring" (Sachs et al., 2004, p. 142; see also Nowak & Sigmund, 1998). For example, potential clients of cleaner fish, while waiting for service, observe the cleaner's interaction with its current client and are much more likely to interact with a cleaner if its current interaction ends peacefully instead of in conflict. This way of making partner assessments can be used prior to pair formation, and is hence potentially a mechanism for genuine partner choice. Waiting clients that see the current cleaning interaction end in conflict can simply swim away.

It is worth drawing a distinction at this point between cues and signals. Signals are behavioral or morphological traits that alter the behavior of other organisms, have evolved because of that effect, and are effective because the response of receivers has also evolved (Maynard Smith & Harper, 2003, p. 15). A cue, by contrast, is any "feature of the world, animate or inanimate, that can be used by an animal as a guide to future action" (ibid.). Showing that a behavioral or morphological trait is a signal is far more demanding than showing that trait to be a cue. In the former case, much must be established about the evolutionary history of the trait. To show something is a cue, though, we need only show that other organisms attend to it when deciding how to act.

Returning now to the case of the cleaner–client fish interaction, it seems that image scoring is better described in terms of cues than of signals. The relevant observation made by potential clients is not of any specific behavioral display by the cleaner that is designed to entice clients. It is rather the observation of a certain state of affairs: an amicable end to the cleaner's current interaction. This seems more akin to predators choosing prey via cues than it is to, say, peahens choosing mates based on signals like the peacock's extravagant tail. A predator choosing which of a herd of prey animals to chase wants to pick one it is likely to catch, and watching to see which ones limp is a good way to find out which ones will be most easily caught: Limping here is a cue. A client fish wants to interact with cooperative cleaners, and watching to see whether a cleaner's current interaction ends peacefully—rather than in a cheating-precipitated chase—seems like a good way to obtain at least some information about the cooperativeness of the cleaner.

The fact that partner assessment is sometimes done using cues is not in itself any kind of problem for the partner choice approach. Indeed, one might think that assessing partners using cues is less problematic than relying on signals from potential partners. Cues can be more or less accurate predictors, but at least they don't provide scope for false advertising. However, such deception in the context of partner assessment is possible even when assessments are cue based, as becomes evident when we pay closer attention to the cleaner–client fish case.

Cleaner fish prefer to eat clients' mucus rather than parasites (Bshary & Grutter, 2003). Large fish have more defection-tempting mucus than do smaller ones. Nonpredatory clients cannot eat a cheating cleaner. Mobile clients—those whose home range encompasses more than one cleaning station—tend not to bother with punitive chases, instead simply swimming away from cheating cleaners. These facts together make large, nonpredatory, mobile clients the perfect "marks" for cheating cleaners. It turns out that the image scoring system in the cleaner-client market is exploited by a certain class of cleaners, dubbed "biting cleaners" (Bshary, 2002, p. 2088).

A biting cleaner servicing a small client while being observed by a large, nonpredatory, mobile client will often rub its pelvic fins around the small client's dorsal fin area (Bshary & Wurth, 2001, p. 1495). This behavior has been termed "host stabilization" and apparently renders the current client quiescent, ensuring that the cleaning interaction ends peacefully (Potts, 1973, p. 274).[5] The biting cleaner thus sets up the score by ensuring that its mark observes the reassuring cue and approaches for service. The cleaner then defects, ignoring the large client's parasites and plundering its abundant mucus. It is unclear whether a dorsal rub provides any benefit to the small client. If it does not—if it merely wastes time and thus inflicts a net loss—then biting cleaners manage an impressive deception indeed, subtly cheating one client while appearing cooperative to another. Biting cleaners should perhaps be dubbed "Machiavellian masseurs."

It is important to note that biting cleaners do not somehow fake the relevant cue: Their interaction with the small client really does end peacefully. Their deception consists in exploiting the cue-based image scoring system of partner assessment operative in this particular biological market. They make sure they are perceived as cooperative under precisely those conditions when being so perceived will open up the most profitable defection opportunities. There is an interesting question here regarding the classification of the Machiavellian masseur's behavior: Should such strategic massaging be counted as a cue, or does it instead qualify as a signal insofar as it has evolved in part in order to influence others' behavior? Even if the behavior is counted as a signal rather than a cue, it should be stressed that the signal is parasitic on the cue-based system of partner assessment. (There is perhaps a parallel of sorts to be drawn here with cases of Batesian mimicry.) Thus, false advertising is an issue even in biological markets where partner assessment and choice is conducted on the basis of cues.

Piecemeal Solutions

Noë's discussions of partner choice emphasize the importance of signaling in biological markets and acknowledge the problem of reliability (what I am calling "false advertising"). The problem with Noë's treatment of the issue is certainly not a lack of ideas. It is rather a lack of unity and detail. Image scoring (as just discussed) is one of many proposals Noë offers about how the problem might be solved (Bshary & Noë, 2003). There are several others.

Noë has sometimes appealed to costly signaling theory in his discussions of partner choice. He claims that

> the handicap principle predicts that in the context of mate choice, agonistic competition or predation, receivers of signals only pay attention to those signals that are costly to produce . . . because only individuals that are fit enough to back-up the signal will produce it at high intensity. (Noë, 2001, p. 109)

Noë is here claiming that advertisements in biological marketplaces where the interests of different trader classes conflict *must* be costly if they are to be believable. Elsewhere, though, he has said things in conflict with this. For example, Noë has suggested that partner choice is facilitated when traders of one class can signal their inability to pursue courses of action detrimental to the interests of traders of the other class. As an example, Noë describes cooperative breeding among purple martins (Noe & Hammerstein, 1994, p. 7). A dominant male will allow "tenant" couples to breed on his territory, in exchange for sexual access to the female tenants. The deal goes sour for the "landlord" if his male tenants sneakily mate with many females on his territory. Noë writes:

> [W]e expect the "choosing" class, i.e. the dominants, to prefer partners with an "honest signal" of inferiority: an easily perceptible character that cannot change overnight, and that constrains its bearer to keep to its role. (Noë & Hammerstein, 1994, p. 6)

As it turns out, landlords prefer male tenants with juvenile plumage. Males bearing juvenile plumage are "sexually handicapped" when it comes to attracting females. By foreclosing his option of mating with numerous tenant females, then, a male with juvenile plumage makes himself a nonthreatening and thus appealing tenant. It is worth pointing out, though, that displaying juvenile plumage is not costly in the way Noë envisions when talking about the handicap principle.

In yet other places, Noë appeals to yet another kind of barrier to false advertising (Noë & Voelkl, this vol.). He notes that during "outbidding competition"—in which members of one trader class vie to be chosen as a partner by a member of the other class—the competitors

> may be forced to produce their commodity at the maximum possible level. Thus, while their output from this competition cannot be taken as a proxy for how much they will invest later on, it provides—at least—reliable information about their potential.

The field of signaling theory has identified several mechanisms that can ensure the honesty of signals (even when sender and receiver interests conflict). Noë's discussions of signaling in partner choice mention many of these, but in a haphazard way, often in the context of specific empirical examples, and without an eye to the bigger picture. The partner choice approach would benefit from having in place a unified theoretical framework for thinking about kinds of solutions to the problem of false advertising in biological markets. Such a framework would help guide investigation of specific market interactions.

A Theoretical Framework for Addressing the False Advertising Problem

In this section, I draw on work in signaling theory to provide a framework within which to organize and clarify Noë's various discussions of the issue of signaling and reliability in biological markets. The mechanisms that can underwrite the reliability of a signal can be divided into three broad classes: costs, constraints, and commitments. Below, I discuss each class of mechanism, and show how the scattered discussions of signal reliability in the partner choice literature can be fitted into the framework this three-way distinction provides. In each case, I suggest research questions that can usefully inform future work on partner choice and cooperation.

Cost and Honest Advertisement

Handicaps are signals kept honest by costs. Amotz Zahavi's (1975, 1977) solution to the problem of reliability in signaling was to point out that signals can be relied upon to be honest if it is prohibitively expensive to send a dishonest signal, that is, if the costs of sending such a signal outweigh whatever benefits might be gained by doing so. Zahavi called his solution to the problem of reliability the "handicap principle."

Noë rightly latches on to the handicap principle as a means of solving the false advertising problem. Many cases of signaling to potential partners in mating markets are amenable to this kind of explanation; for example, the peacock's tail (see, e.g.,

Petrie & Halliday, 1994). However, Noë gives a rather simplistic presentation of the costly signaling idea that is based on Zahavi's initial formulation of the handicap principle.

Zahavi's idea has gone through several incarnations since it was first suggested. He initially claimed that high signal costs impose a test on signalers—a test that only high-quality individuals can pass—and that signaling *and surviving* is thus an effective way to advertise one's quality (Zahavi, 1975). Zahavi later suggested that "the phenotypic manifestation of the handicap is adjusted to correlate to the phenotypic quality of the individual" (1977, p. 603). In the initial formulation of the handicap principle, both high-quality and low-quality individuals were assumed to pay the costs of signaling. In this later version, high signaling costs are paid only by those who can afford those costs (i.e., the genuinely high-quality individuals), while those who cannot afford high signaling costs either do not signal at all or signal at a lower intensity that is affordable given their quality. In light of these three variations on the costly signaling idea, Searcy and Nowicki (2005, p. 10) distinguish between "Zahavi" handicaps, "conditional" handicaps, and "revealing" handicaps.

Future work on signaling to potential partners in biological markets should take into account the advances in discussions of costly signaling theory. In particular, the issue of signal costs should not be treated in too cavalier a fashion. Careful accounting of the costs involved in a behavioral or morphological display is needed to substantiate the claim that the display is a costly signal. A detailed discussion of the challenges posed by such accounting is given by Kotiaho (2001), and a schema for classifying signal costs is given by Searcy and Nowicki (2005). Too narrow a focus on Zahavi's initial—and relatively primitive—statement of the costly signaling idea can only handicap attempts to use this idea to understand signaling in the context of partner choice and cooperation.

Constraints and Honest Advertisement

An index is a signal that is kept honest by constraints against faking. Whereas faking a handicap is possible but unprofitable, faking an index is simply not possible. Indices are signals "whose intensity is causally related to the quality being signalled, and which cannot be faked" (Maynard Smith & Harper, 2003, p. 15).

This is the most likely place to fit Noë's example of members of a trading class engaging in outbidding competition by signaling at maximum output. Actual empirical work investigating this possibility is rather thin, though. For a specific example, consider the production of nectar by caterpillars in order to attract protector ants. Noë mentions studies of this interaction that report that a lone caterpillar's nectar production initially increases with increasing numbers of attending ants, but soon hits a "ceiling" (additional ants don't prompt greater nectar production). This ceiling effect may well show that it is not possible for a caterpillar to produce nectar above some particular level, that is, nectar production is constrained. But this is not yet to show that nectar production is an index used in outbidding competition.

The crucial experiment for determining whether nectar-producing caterpillars really are engaging in outbidding competitions for the services of protector ants has not yet been done. That experiment would keep the number of *ants* fixed, but vary the number of *caterpillars*. If nectar production is a form of outbidding competition, then an increase in caterpillar numbers (i.e., more competing bidders) should generate an increase in each individual caterpillar's level of nectar production (up to some ceiling level that will doubtless vary across individuals).

Outbidding competition conducted via indices is an intriguing theoretical possibility, but it is currently empirically undersupported. Future empirical work investigating this possibility must take into account at least two issues. Obviously, one is the relation between the signal and the quality signaled: Establishing that the possession of the quality constrains the production of the signal is needed in order to show that the signal is an index. The other is the relation between the signal and the context in which it is sent. In cases of outbidding competition, signal intensity should rise as marketplaces become more crowded, that is, as more bidders enter the competition.

Commitment Devices and Cooperation

A commitment device provides reliable information about one's likely future actions in virtue of restricting the space of actions one is able to take, or strongly biasing one toward certain of the available options. The work of economist Robert Frank (1988; cf. Fessler & Quintelier, this vol.) provides a good example of this kind of approach.

Frank's starting point is the idea of a "commitment problem" (Frank, 1988, p. 4). A commitment problem arises whenever an agent can best serve his own interests only by credibly committing himself to act in a way that may later be contrary to his self-interest. An agent might need to make a credible promise of honesty in order to reassure and secure would-be partners in cooperative endeavors where cheating would be profitable and undetectable. An agent might need to make credible threats of revenge in order to deter would-be exploiters in situations where avenging a wrong would be more costly than not doing so. Commitment problems are common and solving them is important.

Frank suggests that evolution has endowed humans with the means to solve commitment problems, namely, "moral sentiments": anger, contempt, disgust, envy, greed, shame, and guilt (Frank, 1988, pp. 46, 53). Moral sentiments help us solve commitment problems because "being known to experience certain emotions enables us to make commitments that would otherwise not be credible" (ibid., p. 5). The promises of an agent known to be prone to guilt will be for that reason more trusted, Frank suggests, and threats from agents known to be prone to anger will be for that reason taken more seriously.

Moral sentiments alone may suffice for solving personal commitment problems, where the goal is for an agent to act in his or her own longer-term interest despite shorter-term temptation. For interpersonal commitment problems, though, more is needed (as Frank recognized). If the commitment device is internal to the agent ("subjective," as it is put by Fessler & Quintelier, this vol.), then there must be some way for other agents to tell—and tell reliably—that one is committed to being honest, or punitive, or cooperative, as the case may be. For Frank, unfakeable expressions of emotions associated with moral sentiments are what allow other agents to tell this (here, the index and commitment accounts intersect).

Of course, commitment devices need not be internal to agents and signaled in some way to influence the partner choices of others. Commitment devices themselves may be discernible to others. Noë's example of purple martin landlords preferring male tenants who are "sexually handicapped" by juvenile plumage might fit here. Commitments bind agents; they foreclose some future option(s). Assuming that a male displaying juvenile plumage at the start of a breeding season cannot change his appearance rapidly enough to seduce tenant females that same season, the male has bound himself (at least in the short term) to being a relatively sexually nonthreatening tenant for the dominant landlord male. Then again, we may want to reserve the term "commitment device" for factors that foreclose options indefinitely. And in any case, it is unclear whether males displaying juvenile plumage really are juveniles (making plumage a cue), or whether they are mature birds that have retained juvenile plumage as a breeding strategy (making plumage, potentially at least, a signal).

A clear case of commitment playing a role in partner choice-mediated cooperation comes from the case of ritual scarification, tattooing, and other forms of highly visibly body modification (Fessler & Quintelier, this vol.). Such modifications can serve to mark the modified individual as a member of a particular group. Depending on the wider social dynamics, such marking may strongly prejudice an individual's partner choices, even to the extent of precluding some choices, such as the choice to defect to a different group. If so, then marked individuals may well be more attractive than unmarked ones as partners in group-beneficial cooperative endeavors, precisely because such individuals' fates are tied to the fate of the group.

To sum up, there are numerous ways in which signaling might work in the context of partner choice. Nothing in this section has been groundbreakingly new. Even so, it is worthwhile to organize the scattered discussions of false advertising extant in the partner choice literature, and to explicitly bring together work in signaling theory with the market perspective on cooperation.

Punishment and Partner Choice

I want to turn now to cooperation and partner choice in humans, and consider the role of punishment. I think punishment is a neglected option in the partner choice literature. Noë, for instance, is skeptical about the possibility that punishment of false advertising might serve to maintain signal reliability in the context of partner choice (see, e.g., Noë & Hammerstein, 1995, p. 337). This is partly because he underestimates the ways in which punishment can be efficacious. He says that

> Punishment as revenge for past behaviour without future fitness advantages cannot be "evolutionarily stable." "Punishment" can only work in long-lasting relationships in which the aggression of the punisher moulds the behaviour of the punished individual in a manner beneficial to the punisher. (Noë, 2001, pp. 101, 105)

Noë assumes here that the only way that punishment could benefit the punisher is if it rehabilitates the punishee. That is one way in which punishment might benefit the punisher, but it is not the only way. Importantly (and importantly by Noë's own lights as a proponent of the partner choice approach), punishment might benefit the punisher by influencing the partner choices of observers in the punisher's favor.

Experimental economists have studied the effect of costly punishment on partner choice in humans. Rob Nelissen investigated "how the costs invested in an altruistic act influence its interpersonal consequences" (Nelissen, 2008, p. 243). By "altruism," Nelissen meant moralistic punishment, specifically, the paying of a cost to punish unfairness. He predicted that "people [would] confer social benefits (both in terms of enhanced preference and financial rewards) on altruistic punishers proportionally to the cost they incurred in punishing" (ibid., pp. 243–244).

Nelissen's subjects were given a sum of money with which to play a "trust game," which worked as follows: Each player was given a sum of money and the option of sending some or all of that money to another player. Any money sent to a trustee would be tripled, and the trustee would then have the option of returning some, all, or none of that amount to the sender. Subjects had to choose a partner for the trust game from among the participants of a previous experiment. Subjects were told that their three potential partners—labeled *A*, *B*, and *C*—had observed a "dictator game" in which the dictator split $10 unevenly with the receiver, keeping $8 and giving only $2. Subjects were also told that *A*, *B*, and *C* had had the opportunity to spend some of their own money to take money away from the dictator: Giving up $1 would reduce the dictator's total by $2. Finally, subjects were told that *A* chose to spend $0 out of $5 on punishment, *B* chose to spend $1.50 out of $5 on punishment, and *C* chose to spend $1.50 out of $10 on punishment. In one condition, subjects were randomly matched with *A*, *B*, or *C* and were then asked how much they would entrust to that partner. In the other condition, subjects were asked which of *A*, *B*, and *C* they wanted to play the trust game with and were then asked how much they would entrust to that partner.

Nelissen found that subjects chose *B* over *A* and *C* (ibid., p. 244). He also found that, when pairing was random, subjects paired with *B* sent the most money in the trust game (ibid., p. 246). As Nelissen interprets the findings:

> [T]he costs incurred in altruistic punishment were perceived as signalling the extent to which punishers value fairness. . . . [P]eople prefer punishers more [as trust-game partners] if they invest more to punish unfairness but only if the invested amount can be perceived as a reliable index of fairness concerns. (Ibid., pp. 244, 246)

The difference between *B* and *C* lies in the relative cost each paid in order to punish unfairness. Punishment was, relatively speaking, twice as costly for *B* as for *C*. Subjects thus seem to take the cost of punishment into account when deciding who to interact with or how to behave toward partners who are forced upon them. Although this does not show that more costly acts of punishment are more reliable signals that the punisher values fairness, it at least suggests that observers *judge* them to be such.[6]

The evidence for a signaling role for punishment in the context of human cooperative partner choice is admittedly rather thin at this stage. The influence of costly moralistic punishment on partner choice is yet to be fully described. Recent work by Horita (2010) indicates that being a punisher influences others' partner choice to one's own advantage in some cases, but works to one's disadvantage in others. Specifically, punishing unfairness is good for one's prospects of being chosen as a partner when one will play the role of provider of resources (i.e., when the punisher will have control over distribution of resources in later interactions). However, punishers seem to be chosen less frequently than nonpunishers when, in the coming interaction, the chooser has control over the distribution of resources.

At least in the case of humans, punishment may play an important role in partner choice and cooperation, both by imposing costs on noncooperation and by influencing decisions about whom to pair with for mutually beneficial interactions. Proponents of the partner choice approach should therefore not be too quick to dismiss punishment as a potential means of solving the false advertising problem, at least in the context of human cooperation and partner choice.

Conclusion

There is a cheater problem for the partner choice approach: the problem of false advertising. That problem is currently understudied. Addressing it will require importing some signaling theory into the partner choice approach. I have begun that process here, but much remains to be done. Hopefully, my contribution will help inform future work on partner choice and cooperation.

Notes

1. I set aside here group selectionist explanations (see Sober & Wilson, 1998), which show how, under certain conditions, genuinely fitness-sacrificing behavior can be maintained in populations under selection.

2. One reason that has been suggested for this is that the cognitive demands of managing reciprocal interactions are rarely met (Stevens & Hauser, 2004; Stevens, Cushman & Hauser, 2005). But, the cognitive demands of reciprocity need not be high—even if most reciprocating organisms are in fact quite cognitively sophisticated—since even plants and paramecia can trade costs and benefits in a way that counts as reciprocity as defined above. Another reason for the dearth of clear empirical examples of reciprocity is that excluding alternative hypotheses is difficult. The classic case of (supposed) reciprocity-based cooperation is blood sharing in vampire bats (Wilkinson, 1984). Subsequent work has challenged this as an example of reciprocity, since the data fail to rule out the possibility that blood sharing results from kin selection plus the occasional misidentification of kin, for instance, or is a form of tolerated theft (Clutton-Brock, 2009). Even demonstrating Tit-for-Tat reciprocity in action in the laboratory is problematic, given the difficulty of designing experiments that faithfully replicate an iterated prisoner's dilemma (Noe, 2006, p. 11).

3. There is yet another stage that must also be considered, namely, the *generation of benefit* stage, at which the cooperating individuals actually produce whatever good is then distributed. Both the partner control and the partner choice model have tended to neglect this stage, and I must do so here simply for reasons of space. For detailed discussion, see Calcott (2008; this vol.).

4. Thanks to one of the editors—Kim Sterelny—for pointing this out.

5. Host stabilization is typically used to soothe clients after a conflict, or to induce waiting fish to remain in the area when the cleaning station is crowded and busy.

6. Nelissen's work on moralistic punishment dovetails with Barclay's (2006) work on the topic. Barclay ran an experiment in which accepting a cost to punish free riding during a public goods game benefited other players. He found that individuals who paid to punish were subsequently rated as more trustworthy and more worthy of respect than nonpunishers and were chosen over nonpunishers as partners in subsequent trust games (Barclay, 2006, p. 330).

References

Axelrod, R. (1984). *The evolution of cooperation*. New York: Basic Books.

Barclay, P. (2006). Reputational benefits for altruistic punishment. *Evolution and Human Behavior, 27*, 325–344.

Bshary, R. (2002). Biting cleaner fish use altruism to deceive image-scoring client reef fish. *Proceedings of the Royal Society, Series B: Biological Sciences, 269*, 2087–2093.

Bshary, R., & Grutter, A. (2003). Cleaner wrasse prefer client mucus. *Proceedings of the Royal Society, Series B: Biological Sciences, 270*, S242–S244.

Bshary, R., & Noë, R. (2003). Biological markets: The ubiquitous influence of partner choice on the dynamics of cleaner fish–client reef fish interactions. In P. Hammerstein (Ed.), *Genetic and cultural evolution of cooperation* (pp. 167–184). Cambridge, MA: MIT Press.

Bshary, R., & Wurth, M. (2001). Cleaner fish *Labroides dimidiatus* manipulate client reef fish by providing tactile stimulation. *Proceedings of the Royal Society, Series B: Biological Sciences, 268,* 1495–1501.

Bull, J., & Rice, W. (1991). Distinguishing mechanisms for the evolution of cooperation. *Journal of Theoretical Biology, 149,* 63–74.

Calcott, B. (2008). The other cooperation problem: Generating benefit. *Biology and Philosophy, 23,* 179–203.

Clutton-Brock, T. (2009). Cooperation between non-kin in animal societies. *Nature, 462,* 51–57.

Dugatkin, L. (1995). Partner choice, game theory, and social behaviour. *Journal of Quantitative Anthropology, 5,* 3–14.

Dugatkin, L. (1997). *Cooperation among animals.* Oxford: Oxford University Press.

Edwards, D., Hassal, M., Sutherland, W., & Yu, D. (2006). Selection for protection in an ant-plant mutualism: Host sanctions, host modularity, and the principal-agent game. *Proceedings of the Royal Society, Series B: Biological Sciences, 273,* 595–602.

Frank, R. (1988). *Passions within reason.* New York: W. W. Norton.

Hamilton, W. (1964). The genetical evolution of social behavior (I and II). *Journal of Theoretical Biology, 7,* 1–52.

Horita, Y. (2010). Punishers may be chosen as providers but not as recipients. *Letters on Evolutionary Behavioural Science, 1,* 6–9.

Kotiaho, J. (2001). Costs of sexual traits. *Biological Reviews of the Cambridge Philosophical Society, 76,* 365–376.

Maynard Smith, J., & Harper, D. (2003). *Animal signals.* Oxford: Oxford University Press.

Miller, G. (2007). Sexual selection for moral virtues. *Quarterly Review of Biology, 82,* 97–121.

Nelissen, R. (2008). The price you pay: Cost-dependent reputation effects of altruistic punishment. *Evolution and Human Behavior, 29,* 242–248.

Nesse, R. (2007). Runaway social selection for displays of partner value and altruism. *Biological Theory, 2,* 143–155.

Noë, R. (1990). A Veto game played by baboons: A challenge to the use of the Prisoner's Dilemma as a paradigm for reciprocity and cooperation. *Animal Behaviour, 39,* 78–90.

Noë, R. (2001). Biological markets: Partner choice as the driving force behind the evolution of mutualisms. In R. Noë, J. van Hoof, & P. Hammerstein (Eds.), *Economics in nature: Social dilemmas, mate choice, and biological markets* (pp. 93–118). Cambridge: Cambridge University Press.

Noë, R. (2006). Cooperation experiments: Coordination through communication versus acting apart together. *Animal Behaviour, 71,* 1–18.

Noë, R., & Hammerstein, P. (1994). Biological markets: Supply and demand determine the effect of partner choice in cooperation, mutualism and mating. *Behavioral Ecology and Sociobiology, 35*, 1–11.

Noë, R., & Hammerstein, P. (1995). Biological markets. *Trends in Ecology & Evolution, 10*, 336–339.

Nowak, M., & Sigmund, K. (1998). Evolution of indirect reciprocity by image scoring. *Nature, 393*, 673–677.

Petrie, M., & Halliday, T. (1994). Experimental and natural changes in the peacock's (*Pavo cristatus*) train can affect mating success. *Behavioral Ecology and Sociobiology, 35*, 213–217.

Potts, G. (1973). The ethology of *Labroides dimidiatus* on Aldabra. *Animal Behaviour, 21*, 250–291.

Sachs, J., Mueller, U., Wilcox, T., & Bull, J. (2004). The evolution of cooperation. *Quarterly Review of Biology, 79*, 135–160.

Searcy, W., & Nowicki, S. (2005). *The evolution of animal communication*. Princeton, NJ: Princeton University Press.

Smith, E., & Bliege-Bird, R. (2000). Turtle hunting and tombstone opening. *Evolution and Human Behavior, 21*, 245–261.

Sober, E., & Wilson, D. (1998). *Unto others*. Cambridge, MA: Harvard University Press.

Stevens, J., & Hauser, M. (2004). Why be nice? Psychological constraints on the evolution of cooperation. *Trends in Cognitive Sciences, 8*, 60–65.

Stevens, J., Cushman, F., & Hauser, M. (2005). Evolving the psychological mechanisms for cooperation. *Annual Review of Ecology Evolution and Systematics, 36*, 499–518.

Trivers, R. (1971). The evolution of reciprocal altruism. *Quarterly Review of Biology, 46*, 35–57.

West, S., Griffin, A., & Gardner, A. (2007). Social semantics. *Journal of Evolutionary Biology, 20*, 415–432.

Wilkinson, G. (1984). Reciprocal food sharing in the vampire bat. *Nature, 308*, 181–184.

Zahavi, A. (1975). Mate selection: A selection for handicap. *Journal of Theoretical Biology, 53*, 205–214.

Zahavi, A. (1977). The cost of honesty: Further remarks on the handicap principle. *Journal of Theoretical Biology, 67*, 603–605.

9 MHC-Mediated Benefits of Trade: A Biomolecular Approach to Cooperation in the Marketplace

Haim Ofek

1 Introduction

The discussion in this chapter is an attempt to reconcile two observations: (1) that vertebrates are generally the least cooperative form of life in the animal kingdom and, conversely, (2) that humans are the most cooperative form of life in the same kingdom (at least in terms of cooperation with nonkin). The main challenge is to reconcile such a diametric departure on the part of humans with the fact that humans are perfect vertebrates in all other respects. The second section of this chapter explains why, excepting our own species, the vertebrates are all but universally defective or nearly defective in terms of cooperation among conspecifics at large, if not among kin. The "defect" responsible for this failure can be traced to the inescapable reliance on sexual reproduction combined with further reliance on MHC diversity—the latter being a special need associated with the major histocompatibility complex, a site in the genome peculiar to the vertebrates. The third section explains how humans have managed (in an evolutionary sense) to circumvent this defect by turning it over on its head and using it as a facilitating rather than a limiting factor of cooperation.

For the purpose of the present discussion cooperation is best conceived in the form of division of labor. Division of labor (as measured, approximately, by the volume of trade) is the most obvious economic measure of cooperation among nonkin in the ordinary business of human life. Next to economics, the field best poised to conceive the essence of division of labor as key to cooperation is entomology—the scientific study of insects.[1] Notwithstanding the fundamental differences in the division of labor between humans and insects, it should be clear that when it comes to the sphere of subsistence, division of labor describes cooperation exceedingly well in both. However, going to spheres of life beyond the prevailing system of subsistence, it should be equally clear that division of labor is not necessarily the best account of cooperation nor the most common form of it, either in humans or in insects.[2] One exception in humans is the sphere of reproduction. Indeed, when it comes to *reproductive* division of labor (the litmus test of eusociality), humans are utterly unimpressive compared to the social insects. By the same token, the social insects are for their part equally unimpressive, compared with humans, when it comes to division of labor among nonkin. Taken together, reproductive division of labor in insects and conspecific division of labor (including nonkin) in humans come out as two separate cooperative poles that set the two groups far apart.

Such a disparity suggests distinct origins not only in descent but also in more recent adaptation. The basic adaptations that facilitate division of labor in social insects (such as *haplodiploidy* in the *hymenoptera*: wasps, bees, and ants) are obviously unshared by humans. So, one can ask, what is the counterpart adaptation that facilitates the division of labor in human society? An attempt to answer this question is undertaken in section 3. In the meantime, leaving the invertebrates behind for a while, it will shortly become apparent that one place where we can look only in vain for such an adaptation is closer to home, in the vertebrates.

2 The Vertebrate Cooperative Disadvantage

It is true that the degree of cooperation that regularly occurs among kin, and occasionally even among nonkin, is quite extensive in certain species of vertebrates (see Cockburn's chapter, this vol., for examples). However, with the exception of our own species and, perhaps, with the possible exception of some of its predecessors, the kind of division of labor that transpires among vertebrate conspecifics falls nowhere near the levels of refinement and elaboration observed in insect division of labor or, for that matter, in human division of labor (let alone cellular division of labor)—and this is precisely the failure for which we seek an explanation in this section. With better than half a billion years in the making (the age of the vertebrates) and some 60,000 extant species on Earth, the inability to establish a fully fledged elaborate system of division of labor in all but one vertebrate species (humans) is a failure that demands explanation. Ordinarily, such a long-lasting failure can be endured under natural selection only if an overriding counteracting selective force had been at work the whole time. At issue is a force as old as sex itself.

2.1 The Twofold-Plus Cost of Sex

The twofold cost of sex arising from the "needless" existence of males is (phenotypically) best evaluated from the viewpoint of the female who provides a resource in short supply: typically, a relatively large and immotile egg, compared with minute and highly motile sperm that are offered in great abundance by males. An egg produced by an asexual female will contain a full complement of chromosomes and genes, whereas an egg produced by a sexual female will contain only half of each. The asexual female seems therefore to propagate her genome, or any part of it, twice as efficiently as the sexual female. Alternatively, the dilemma can be viewed from the standpoint of a gene that can switch meiosis on or off. If such a gene suppresses meiosis it is certain to reach all the eggs produced by a female, whereas a gene that tolerates meiosis will reach only half. "So isn't it just daft," asked Richard Dawkins in a rhetorical manner, "to throw half your genes away with every egg or sperm you make, in order to mix the other half with the genes of somebody else?" (Dawkins, 2005, p. 430). This is essentially the evolutionary paradox of sexuality.

To the lost opportunity of doubling the number of offspring (due to the fact that asex is twice as fecund as sex), an economist would be tempted to add the lost opportunities of division of labor due to the uncooperative or otherwise selfish progeny produced by meiotic recombination. Indeed, so long as meiosis is kept switched off, barring mutations, all siblings produced by an asexual mother are genetically identical to her, and to each other.[3] In other words, all the members of a family are related to each other like identical twins, scoring a perfect $r = 1$ in terms of Hamilton's coefficient of relatedness. Moreover, unlike identical twins, here perfect relatedness applies not only to concurrent siblings, but also to parents (mothers) and offspring. Obviously, there is plenty of altruism to go around among family members and, by implication, sufficient opportunities for uninhibited cooperation and division of labor to take their full course. Relatedness among members of a sexual family pales in comparison. With the exception of identical twins, it will never exceed $r = ½$, and occasionally, depending only on the prevalence of half-siblings, it may actually drop to $r = ¼$. Consequently, the opportunities for division of labor seem far more compelling under asex than under sex. Hence, the "plus" component in the "twofold-plus" cost of sexuality.

The paradox of sexuality is considered to be the paramount problem in evolutionary biology. For more than a century it has occupied some of the best minds in the field, from August Weismann (1834–1914) to William D. Hamilton (1936–2000). A partial list of surviving (though challenged) theories and explanations as outlined, for instance, by Graham Bell (1982, pp. 91–157), includes the following titles: *The Vicar of Bray*, *The Best-Man*, *The Hitch-Hiker*, *The Tangled Bank*, *Muller's Ratchet*, and *The Red Queen*.[4] Though each of the works under these titles was viewed in its time as a serious breakthrough, and some even today still hold their own ground, most have been powerfully challenged by two paradigm shifts that occurred in the last quarter of the twentieth century. The first shift was driven by the introduction of the "selfish gene" paradigm. The growing influence of this paradigm helped to discount most of the earlier explanations as group selectionist in the sense of yielding adaptive benefits only at the level of the population or species rather than at the operative level of the individual gene. The second paradigm shift was driven by one of the theories within the list itself: the red queen principle. Indeed, judging from the literature, by the mid-1980s only two contenders were left in the arena of discourse: Muller's ratchet mechanism and, of course, the red queen principle itself.

2.2 The Red Queen Principle

The red queen principle was originally triggered by a macroevolutionary observation suggesting that the probability of a species going extinct is independent of how long the species has already existed (Van Valen, 1973). The queen herself is a surreal character out of Lewis Carroll's *Through the Looking Glass*, where she informs us (and Alice) that "it takes all the running you can do, to keep in the same place." The red queen principle has a range of applications in biology that exceeds macroevolution, going beyond evolution to the borderline with economics.[5] The significance of the red queen principle to evolution in general, and to the evolution of sexuality in particular, is threefold: (1) the recognition of a fundamental distinction between the roles of biotic factors and physical factors in the formation of adaptations to the environment; (2) that it is the biotic factors that are crucial to the evolution of sex; and (3) that among biotic factors, it is the response to parasites, rather than the response to predators or to any other coevolving form of life, that is most crucial to the evolution of sex (Glesener & Tilman, 1978; Bell, 1982; Hamilton, 1988). These three essential implications obviously depend on each other in the order listed and, in the same order, derive from one another (and from corresponding empirical observations). Let us take them one by one.

The fundamental difference between adaptations in response to the physical environment and adaptations in response to the biotic environment is best understood in light of the general distinction between response in the form of *reaction* and response in the form of *interaction*, respectively (Ofek, 2001, pp. 89–90). To see why, note that the relationship between an organism and the biotic factors of its environment (availability of food, predation, parasitic pressures, symbiotic relationships, etc.) is essentially *interactive*: natural selection represents (at least) two "customers" and consequently operates on their behalf on two opposite sides of the ecological equation.

In contrast, the relationship between an organism and the physical factors of its environment (climate, topography, availability of moisture, the composition of soils, salinity of water, etc.) is essentially *reactive*: natural selection operates here only on one side of the ecological equation, much like a tennis player who hits against a concrete wall. Unlike a game against a real opponent, alive and thinking, the strategic combinations are few and simple, and they tend to repeat themselves almost in the same manner for every player (and every wall). Much like that solitary player, organisms do not play games against their physical environment in any sense that can get the attention of game theorists, as such. On the other hand, the games organisms play against their biotic environment (including members of their own species) are of the highest interest to game theorists. The key component present in games against biotic factors, but completely missing from games against physical factors, is the strategic challenge facing the players.

Note that the essence of the one-sided play or, equivalently, the absence of strategic challenge, applies not only to organisms in relation to a fixed physical environment, but also to organisms that modify their physical environment—like the earthworm in relation to soil. It is true that such organisms set up dynamic interactions with physical factors in their environment. However, dynamic interactions do not amount to strategic interactions in the sense of a two-sided conflict between optimizing forces. So long as natural selection (an optimizing force) operates on behalf of the worm but not on behalf of the soil, the essence of the one-sided play is in effect, and the final outcome remains fairly predictable at least in terms of statistical expectations.

Adaptations to the physical environment, however harsh, are patently rationalizable, if only by virtue of being subject to highly predictable constraints. Natural selection (or the human mind) can easily economize under such constraints by optimal choice among well-defined alternatives. The same privilege is no longer available to adaptations in response to the biotic environment, where, by the very nature of the struggle between living things, all constraints are in constant motion and all alternatives behave like elusive moving targets. Moreover, note that unlike most biotic factors, physical factors typically affect a wide range of species simultaneously, friend and foe in nearly the same manner (Ofek, 2001, p. 89). Consequently, when it comes to the physical environment, the red queen seems to run backward: the harsher the environment, the lesser the competition—at great savings in cost. In the process of acquiring adaptations to the physical environment, the odds of incurring costs high enough to justify the twofold cost of sexual reproduction are, therefore, quite implausible.

On the other hand, the cost of adaptations acquired in response to biotic factors is highly unpredictable and on many occasions will admit no upper bound short of extinction. The reason for this is the tendency of antagonistic relationships between living things to slide into runaway evolutionary arms races (Dawkins & Krebs, 1979). The only ecological way to avoid the staggering cost of such (interspecific) races is to branch off (or "niche apart") in the familiar pattern attributed by Darwin's principle of divergence to all species that compete for the same resources. "Niching apart" is no longer an option, however, for species that do not compete for any vital resource precisely because one serves as such for the other—as between predators and prey, or between parasites and their hosts.

The staggering costs associated with such inescapable arms races typical of many biotic relationships seem to provide tentative support to the hypothesis that biotic factors rather than physical factors are primarily responsible for the expensive evolution of sex.[6] However, in itself, high cost is no proof of any hypothesis, however expensive. An attempt to provide further support and corroborating evidence for this hypothesis is undertaken in the following discussion.

Regardless of cost, it can be argued that the applicability of an adaptation across different states of the environment bears testimony to the specific state of the environment under which it has evolved. Meiosis, the main adaptation associated with the evolution of sex, seems to have clear applications in the biotic environment but no obvious application, as far as the eye can see, in the physical environment. Meiosis produces ever new unpredictable structures at rates that far exceed the typical rate of change in the physical elements of the environment. However, for most organisms larger than bacteria, the rate of change in meiotic products hardly catches up with the typical rate of change in the biotic environment, where almost everyone is constantly adopting to their fastest evolving predator or prey, parasite or host—as the case may be. Ample evidence in support of this conclusion (at least for parasites) is provided in the field of epidemiology. It can be argued, therefore, that evolution at the rate of meiotic recombination is essential for adaptation to the biotic environment. On the other hand, the needs of adaptation to the physical environment can be met by evolution at a much slower rate, say, at the rate of mutation—for which there is no need for sex. It is hard therefore to see, even without the argument of cost, how the physical factors of the environment could have exerted any influence over the evolution of sex.

That biotic factors, rather than physical factors, are crucial to the evolution of sex is an extrapolation supported by a substantial body of data documented in the literature dealing with the geographic distribution of asexual populations compared to that of their closely related sexual counterparts (see, e.g., Suomalainen, 1950; Ghiselin, 1974; and particularly Glesener & Tilman, 1978). Overall, the evidence indicates that among animals that live on land, sexual forms tend to occupy regions of greater biotic stress: lower altitudes and latitudes, mesic rather than xeric, mainlands rather than islands, and undisturbed rather than disturbed (e.g., by agriculture). On the other hand, asexual forms are unusually common in regions of greater physical stress. Reduced biotic stress in such harsh physical environments apparently causes reduced need for sexuality (Hamilton, 1988, p. 80). With a reinforced focus on biotic factors, and physical factors largely out of the way, it is tempting to think that the task of tracing the evolution of sex to its root causes—supposedly, predation and parasitism—has reached its proper closure.

However, on closer examination, as Hamilton was the first to insist, the process of narrowing the quest for the culprit responsible for sex is not quite over. Noting that all biotic factors are not the same in terms of the stress that they exert on the evolution of sexuality, he was quick to draw attention to one of the most striking differences between predators and parasites.

A predator is typically larger than its victim, but a parasite is far smaller than its host. Consequently, as Seger and Hamilton pointed out, "Predators usually have generation times as long or even longer than those of their prey, while parasites may have generation times many orders of magnitude shorter than those of their hosts" (1988, p. 176). This provides parasites, microscopic parasites in particular, with a deadly edge in the form of exceedingly rapid rates of mutations. "The bacteria in your gut," reckons Matt Ridley, "pass through six times as many generations during your lifetime as people have passed through since they were apes" (1993, p. 66). For lack of mutation in the host, the battle against microscopic parasites is relegated primarily to recombination (and hence largely, though not exclusively, to sexual reproduction). At issue is recombination across genes highly specialized for the role of resistance against parasites that seek access to the internal environments of bodies and cells.[7] The best testimony to the role of such specialized genes is provided by the existence and function of the MHC site in the vertebrate genome.

2.3 The Major Histocompatibility Complex (MHC)

The MHC, or human leukocyte antigen (HLA) as it is sometimes referred to, is a region in the genome of many vertebrates that is responsible for control of cellular immune responses against foreign antigens and nonself material, such as transplanted foreign organs. Indeed, the MHC first drew widespread attention in the late 1950s under the looming concern about tissue compatibility between donor and recipient in organ transplants. It has since become a screening device widely used to improve donor-recipient matching and reduce the risk of organ rejection. The main function of the MHC, however, is not resistance to foreign organs as such, but resistance to foreign material in general—and to invading parasites, in particular. This task is accomplished primarily by immune recognition: encoding cell-surface proteins that the immune system constantly monitors in order to distinguish between self and nonself entities, maintaining the former and destroying the latter.

Residing on the short arm of chromosome 6 in humans (but on chromosome 17 in mice), two peculiarities of the MHC make it especially suitable for its main immune function outlined above. First, each MHC locus is exceedingly polymorphic (typically, with 100 or more possible alleles). Second, each allele is expressed in codominant form (in which the heterozygote exhibits the phenotypic traits of both alleles). Taken together, a heterozygote MHC locus is 55 times more diverse than the homozygote and nearly 5,000 times more diverse than a two-allele classical Mendelian counterpart. As such, the probability of two unrelated individuals sharing the same MHC genotype is negligible for all practical reasons, though, in the case of close relatives, it keeps rapidly rising as the relationships get closer, reaching ¼ for full siblings (and 1 for identical twins). The main idea is to keep the parasites "guessing" as they move from one host to the next by creating a moving target for their adaptation. The advantage of MHC polymorphism and of heterozygosity in the struggle against infectious agents is demonstrated by the observation that HIV-infected MHC heterozygotes progress to AIDS more slowly than homozygotes (Carrington et al., 1999).

2.4 MHC-Mediated Behavior

Growing evidence indicates that the MHC is a primary genetic site involved not only in immune responses at the level of molecules and cells, but also in kin recognition and mate selection, influencing patterns of breeding and cooperation at the organismic level (Grafen, 1992). At issue is a system of behaviors partly mediated through the MHC site in the vertebrate genome.

MHC-mediated behavior is a mechanism made up of three components that operate consecutively. First, its unprecedented level of diversity makes the MHC genotype an exceedingly useful tool as a kin recognition marker. Second, in response to that marker, the MHC genes encode genetic products that are thought to be in the form of activating olfactory signals or odors. Finally, the signals or odors influence behavior according to predetermined adaptations. In principle, such a mechanism can serve any number of adaptations, as long as they all rely on kinship. In the present framework, the mechanism serves three such adaptations: it lends disease-resistance to offspring, it helps to avoid inbreeding, and it supports a certain minimal level of kin cooperation. The only problem is that the first two are in conflict with the third. Indeed, since both cooperation (in nonhumans) and offspring health depend largely on kinship—albeit in opposite directions—it follows that any improvement in cooperation in the present generation puts a drag on health in the next, and vice versa. Stated differently, cooperation is kin-assortative, whereas mating is essentially kin-disassortative, taken together we end up with a putative conflict.

The mechanism just outlined, including the associated conflict, has been experimentally demonstrated or observed primarily in mice and, to a lesser extent, in other species (e.g., rats, salmon, sticklebacks), including humans. Thus, for instance, mice of congenic strains have been shown to prefer mating with the mice that differ from them at MHC, even when this is the only genetic difference present (Yamazaki et al., 1980). Similar conclusions have been drawn from the high rates of natural miscarriage that are found in human couples who are too similar at MHC (Beer, Gagnon & Quebbeman, 1981).

Grossly underestimated in the past, the kin-disassortative force of mating seems indeed to go well beyond the mere incest taboo. Extended beyond close relatives, the kin-disassortative force of mating may shed some light on the poorly understood reason for the exceedingly low rates of reproductive success in captivity of otherwise free-ranging endangered species—like the giant panda, for instance. Apparently, the MHC trigger fails to take its normal course given the limited choice of sufficiently unrelated mates under typical zoo confinement. Yet, at least in the case of mice, females are seen not only to prefer MHC-dissimilar mates but, conversely, are seen also to prefer MHC-similar communal nesting partners (Potts, Manning & Wakeland, 1991; Manning, Wakeland & Potts, 1992).

2.5 The Predicament of the Female Mouse

The clash between the kin-disassortative force of mating and the kin-assortative force of cooperation results in a dilemma best illustrated, perhaps, by the predicament of a female mouse. House mice (*Mus musculus domesticus*) raise their young in communal nests where they seem to nurse each other's pups more or less indiscriminately (Manning, Wakeland & Potts, 1992). To minimize exploitation by free riders, female mice prefer to form communal nests with relatives. Their attraction to kin, however, does not go so far as to prevent a female at the brink of conception from seeking consort with nonkin mating partners—with the apparent aim of chasing, as noted by Hamilton, after genes beneficial to the health of descendants. With the onset of pregnancy and the approach of birth, preferences revert back to an undiminished attraction to kin—the most trustworthy source of helpers to a parent at the nest.

The key to success in such a mixed cooperative-reproductive strategy is an agile sense of kin recognition. Based on the ability for olfactory detection of genetic products of MHC, the mouse shares this sense of recognition with other vertebrate species, including, to some degree, humans (Wedekind et al., 1995). The human ability to perceive MHC (or HLA as the human version is called) is largely confined to females, and especially to females at the fertile stage of their menstrual cycle. On the other hand, the ability for MHC detection declines and the accompanied preference for nonkin even slightly reverses for females taking the contraceptive pill, which is known to mimic the influence of pregnancy in women. Drawing parallels with the mouse MHC model, some authors suggest the existence of a switch in female preferences from an attraction toward nonkin when a conception is in prospect to an attraction toward kin at onset of pregnancy.

2.6 The Post-Cambrian Tragedy of Cooperation

One of the most noticeable features that generally sets vertebrates apart from other animals (i.e., from invertebrates) is large body size, though there are occasional exceptions.[8] The main challenge of large body size is on land rather than in water, although even in water special adaptations are necessary. Though the largest known land vertebrates (dinosaurs) are long extinct, the basic set of adaptations that enable large bodies to survive on land is present in vertebrates of all sizes. These include the skeletal support for a large body and the underpinning for the muscles needed to power it, special organs that compensate for reduced surface-to-volume ratio (e.g., lungs), and, above all, the necessity to maintain a fully fledged system of sexual reproduction.

The reliance on sexual reproduction (i.e., meiotic recombination) is, of course, a phenomenon neither exclusive nor original to the vertebrates. It evolved with the first appearance of eukaryotic cells at least a billion years before vertebrates, and is important to all forms of eukaryotic life today. To the vertebrates, however, it is of paramount importance for at least two reasons. First, because of their larger body size (relative to other animals), and consequently their prolonged generation times, they are easy targets for infectious agents and parasites. It was probably this reason that led Hamilton to make the following observation: "corresponding with theory, the smaller and the more short-lived is the species, the more likely the species is to be asexual (although there are many exceptions)" (Hamilton, 1988, p. 77). The second reason that sexual reproduction is important to vertebrates, with the possible exception of some amphibians, is the lack of chemical defenses of the type otherwise common among plants, fungi, and even some insects. Plants routinely produce chemicals that are toxic to their pests and parasites and thereby allow, among other things, giant trees to reach their gigantic stature and exceedingly prolonged life span with surprisingly little sex. More accomplished, in this respect, are the fungi that produce antibiotics (essentially, chemicals that kill bacteria) on demand. Overall, in the absence of such advanced chemical defenses, given their typical body size, vertebrates are left with little choice but to rely on an immune response augmented by an uncompromised system of sexual reproduction.

The sweeping proposition made by E. O. Wilson that "sex is an antisocial force in evolution" was largely based, as far as I am aware, on the following observation:

> The vertebrates are all but universally sexual in their mode of reproduction. . . . And with the exception of man, vertebrates have assembled societies that are only crudely and loosely organized in comparison with those of insects and other invertebrates. Sex is a constraint overcome only with difficulty within the vertebrates. (Wilson, 1975, pp. 314–315)

Vertebrate sex is, in fact, a twofold constraint on cooperation owing to the forces of (1) meiotic recombination, and (2) MHC-driven mate selection. Under the influence of the former it rules out the possible existence of organisms most suitable for kin-cooperation and, under the influence of the latter, it places them in the company of partners least suitable for such cooperation. Most suitable for cooperation driven by kinship are organisms related to each other by a factor of $r = 1$ on the scale of Hamilton's coefficient of relatedness. With the exception of identical twins (i.e., monozygotic siblings), no pair of organisms produced under regular meiotic recombination can possibly reach anywhere close to this ideal of kinship. With a coefficient of relatedness that in the case of vertebrates can almost never exceed $r = ½$, the prospect of kin mediated cooperation is not promising to begin with.[9] The MHC genes make matters only worse, since the pervasive chase after MHC-dissimilar mating partners leads the chasers to share close quarters with remotely related companions, if not total strangers, thus driving relatedness arbitrarily close to $r = 0$.

The dismal state of cooperation in the vertebrates should therefore come as no surprise. Compared with the invertebrates and other forms of life, the scope and depth of the division of labor in nonhuman vertebrate societies are utterly unimpressive. The two pillars of division of labor, differentiation of form and specialization of function, are largely confined in such societies to little or nothing more than sexual dimorphism (an adaptation that serves primarily the function of reproduction but occasionally the functions of subsistence and antipredator safeguard, as well).[10] Indeed, the vertebrates seem to represent half a billion years' worth of lost opportunities for division of labor or, so to speak, a post-Cambrian tragedy in the evolution of cooperation.

3 Trade: The Human Cooperative Advantage

The only vertebrate species known to escape the "tragedy" of cooperation are anatomically modern humans like us and, arguably, some of our more or less recent ancestors.[11] The transition was smooth and complete, if not spectacular. The best indication to this effect is the fact that the division of labor in human society (viewed on the face of it as a superorganism) is almost as elaborate as the division of labor within the internal environment of a multicellular body. Full credit for this achievement, it will be argued, can be reasonably ascribed to the existence of the market as a central organizing principle in the ordinary business of human life. While the final outcome speaks for itself, the precise market mechanism that facilitated the breakthrough is still poorly understood.

The mechanism in question is perhaps best approached in light of insights gained long ago by Adam Smith. His insights, in this respect, are fourfold: (1) that the benefits of division of labor tend to exceed by far all expectations by casual observers; (2) that the existence of division of labor is invariably dependent on a coexisting system of redistribution; (3) that, in the case of humans, the necessary system of redistribution is largely provided by market exchange; and (4) that the scope of the division of labor (in humans) is determined by the extent of the market. To this list we can add in the spirit of Adam Smith one additional insight that he himself apparently failed to mention, namely (5) that the extent of the market itself is determined in large measure by the prevailing means of exchange. A brief review of these insights is outlined below.

Our tendency to underestimate the potential benefits of the division of labor (insight 1, above) is well demonstrated by the observations made by Adam Smith in a tiny Scottish pin factory (of ten workers) where output exceeded the output of ten workers operating apart at least by a factor of 240 to one(!). The nexus between division of labor and redistribution (insight 2) is based on the simple fact that narrowly specialized producers are never completely self-reliant. They mostly work for the necessities of others, and they rely on others for most of their own necessities. Consequently, they cannot survive without redistribution in one form or another. Redistribution in the case of humans takes the unique and highly peculiar form of market exchange (insight 3). Adam Smith ascribed this form of redistribution to the human propensity to exchange. He considered this propensity to be responsible not only for the capacity of bilateral transfer of one thing for another between two traders but also for the unique human capacity to settle such a transaction in the framework of a market—"by treaty, by barter," and above all, "by purchase"—to use his own words (Smith, 1776/1976, p. 19). As such, exchange fulfills a necessary condition for division of labor by providing a system of redistribution of goods and services among specialists that are not necessarily related to each other in any other way.

Assuming that the general condition for the overall existence of division of labor (i.e., a coexisting system of redistribution) is in place, the remaining question is what determines the limit on the scope and intensity of such a division in a specific setting. Adam Smith's answer to this question (insight 4) was unequivocal: the limit is the extent of the market. Economically considered, the extent of the market is essentially the volume of trade passing through it. By that measure, there is little doubt that the single most consequential event in the history of the market and, by implication, in the history of the division of labor, took place with the introduction of money as a medium of exchange (insight 5). It should be noted, however, that a necessary and sufficient condition for the existence of money as a meaningful means of exchange is the existence of the price mechanism.

Indeed, as long as exchange is confined to barter, it does not impart a sufficient condition for division of labor to reach its fullest potential as we know it, or anywhere even remotely close to it. In the absence of money, the main impediment to the division of labor is the "double coincidence of wants"—that is, the improbability that two persons whose disposable possessions mutually suit one another's wants will somehow meet each other. The inhibiting reliance on this coincidence is largely alleviated, however, under the price mechanism (that is almost always accompanied by the existence of money in one form or another). Since prices in conjunction with money serve to separate the act of sale from the act of purchase, the double coincidence of wants is no longer a prerequisite for exchange. This eliminates the major impediment on the division of labor. Finally, it should be noted that the price mechanism provides not only a highly improved system of redistribution, but also a system of incentives that can induce potential participants, however selfish, to take part in a collective enterprise where the benefits originally occur at the level of groups or otherwise to unrelated parties, rather than to self or kin.

To anticipate a possible source of confusion associated with the use of the term "market," it should be noted that when economists use the term without a qualifier they generally refer to the business of buying and selling by means of money. Alternatively, when they refer to exchange by means other than money, they usually modify the term market with the appropriate qualifier, as in: *the marriage market* (Becker, 1991; Ofek & Merrill, 1997), *the market for ideas* (Cukier, 2005), or for that matter, *biological markets* (Noë & Hammerstein, 1995; see also the discussion in ch. 7, this vol.).

However, for the purpose of the discussion in the present chapter, it will be useful to focus on the market only in the strongest unqualified sense of the word, for that is where the division of labor in human society best unfolds, and that is where the mechanism most responsible for it—the price mechanism—is invariably situated. Similarly, the term "trade" as used in the following discussion will refer only to forms of exchange that take place under the price mechanism.

3.1 The Peculiarities of Trade under the Price Mechanism

Trade, as taken here, is a form of exchange that relies on an adaptation far more elaborate than a simple capacity for reciprocal transfer of one item for another between two traders. Trade relies on the capacity for exchange by means of money. That is, to begin with, the capacity of conceiving the material world in terms of prices. At issue is a cognitive ability of transforming a multidimensional universe of matter, energy, and information into a one-dimensional money-metric abstraction, without loss in essential information useful for decision making.[12] The existence of such a cognitive capability, and the fact that it repeats itself in nearly the same manner and without conscious effort in all members of the species, seems to suggest the presence of some sort of a hardwired mental algorithm.

The use of prices as guidelines that facilitate action, and interaction, is so pervasive in economic affairs that the foundation of economic theory (also known as microeconomics) is often labeled "price theory." As a form of exchange that relies on prices, trade exhibits five major peculiarities:

1. Trade is a patently impersonal form of exchange.

Since prices are assigned to commodities but not to persons, trade is a market activity completely determined by the merit of the commodities traded (transaction costs included) rather than—fully or partly—by the identity of the trading partners. By virtue of this peculiarity humans apparently acquired the unique capacity and propensity to cooperate and trade with complete strangers—including remote MHC carriers.

2. Trade is a nonkin form of cooperation.

The fact that prices rarely come into play in transfers among relatives, and almost never in other family affairs, seems to suggest that the price mechanism is designed not so much for kin and nepotistic exchange but more nearly, if not exclusively, for formal trade among unrelated members of the species. Moreover, for reasons of comparative advantage—a major source of benefits from trade—agents in the market may on many occasions opt to trade or otherwise cooperate with complete strangers rather than with relatives or close compatriots. It follows that humans, in the business part of their life, not only tolerate remote MHC carriers, but may actually single them out for preferential interaction.

3. Trade facilitates the formation of division of labor, not only beyond the boundaries of kinship, but also across the boundaries of populations.

Benefits of trade derived from comparative advantage and enjoyed by trading partners that belong to the same population apply, with added force, to trading partners that belong to different populations. The more remote the trader the better. The same argument applies, of course, to the benefits derived from MHC diversity. What is true of trading partners is true of mating partners.

4. Trade largely mitigates the ill effects of selfishness as well as the ill effects of group selection failure.

A system of division of labor unsupported by trade is at the mercy of altruism. Altruism is a noble endeavor, but in the absence of trade the vast majority of the necessities that we take for granted would never have reached us. In fact, they would had never been produced or even invented. On the other hand, trade provides a sufficient condition in the form of proper incentives for division of labor to take its course to the fullest extent. Key to these incentives is the reliance of trade on the price mechanism. Prices possess built-in incentives that lead trading partners to privatize collective effects and internalize external affects, thus keeping the interaction relatively safe from the ill effects of free ridership, group selection failure, as well as from the very selfishness of the participants toward each other (Ofek, 2001). The pervasive human engagement in division of labor with strangers seems to have no alternative explanation.

5. Trade is in its element only within the realm of subsistence.

It is true that the concept of trade can be extended (by analogy) to almost all walks of life, but its explanatory and predictive powers rapidly decline beyond the boundaries of human subsistence broadly defined (the notion of "subsistence" should be understood here to include not only the provision of food, but the provision of all forms of matter, energy, and information necessary to sustain life). On most occasions it would be sensible to reserve explanations based on the mechanism of trade strictly for market applications and, however tempting, keep them away from nonmarket applications.[13] Extramarket episodes of cooperation associated, for instance, with kin selection, mate selection, or situations at the brink of risk to life or limb—from alarm calls to predator mobbing to warfare—are on many occasions better explained, even in humans, by mechanisms other than trade. Owing in large part to the study of cooperation in animal societies, where commercial markets and formal trade do not exist, we have at our disposal a growing body of explanations for cooperative structures derived, if not from first principles of natural selection, then from related fields (e.g., game theory): *relatedness* (Hamilton, 1964), *reciprocity* (Trivers, 1971), *eusociality* (E. O. Wilson, 1975), *Tit-for-Tat* (Axelrod, 1984), *trait-group selection* (D. S. Wilson, 1975, 1992), *biological markets* (Noë & Hammerstein, 1995)—to name a few. With the possible exception of eusociality, such structures are occasionally applicable to human affairs as well.

3.2 A Biological Approach to the Benefits of Trade

The existence and incidence of benefits of trade is a topic of great importance to the study of economics and, accordingly, has always received considerable attention in the literature. Humans reap the benefits of trade at three distinct levels of organization: the private level, the public (or market) level, and the level of the species at large. The benefits that occur at the private level are under our direct awareness and, as such, are easily traceable to their sources (allocative improvements through redistribution facilitated by trade, productive improvements through division of labor facilitated by trade, etc.). The benefits that occur at the market level (typically, in some form of consumer surplus or producer surplus) are not as evident to the eye of an untrained observer, but are fully accounted for conceptually if not empirically by the analytical tools available to any student of economics.

On the other hand, the benefits of trade that occur at the level of the species have thus far received little to no attention by economists and biologists alike. Without a clear idea of the existence of benefits in this category, there is probably no meaningful way to add an evolutionary dimension to the study of economics nor, vice versa, to add an economic dimension to the study of evolution.

Insights gained throughout the discussion in previous parts of this chapter suggest at least four sources of benefits associated with trade that occur at the level of the species:

1. improvement in health;
2. improvement in productivity;
3. reduction in the cost of sexuality; and
4. elimination or mitigation of ill effects associated with selfishness and the failure of group selection.

The seemingly unrelated improvements in health and productivity (the first two sources on this list) are inferred from the fact that cooperation and division of labor have been set free, by the price mechanism, from an otherwise inescapable dependence on kinship. By adaptation an organism would always seek to maximize the health of descendants as well as the gains from cooperation. In the absence of trade, the two objectives seem to work against each other, as we learned from the predicament of the female mouse described earlier. Indeed, the (nonmarket) key to cooperation is interaction with kin. Conversely, the (MHC) key to good health throughout generations is interaction with strangers or outbreeding. One or both objectives, in the absence of trade, must be compromised at least in part, reaching a balance that is less than desirable from the separate viewpoints of each objective. On the other hand, under trade, the key to cooperation is no longer interaction with kin, thus allowing each of the two objectives to be independently optimized under unconstrained maximization. The differences in the total gain can be considerable. If only for the extra benefits thus gained from MHC diversity, the total gain from trade may be arbitrarily larger than previously expected.[14]

The reduction in the cost of sexuality (the third source of benefits, as listed above) is obtained by eliminating the "plus" component from the "twofold-plus" cost of sex discussed earlier (section 2.1). Recall that this added component was intended, in the absence of trade, to account for the lost opportunities of division of labor due to the "noncooperative" or otherwise selfish nature of the progeny produced by meiotic recombination. Under trade, however, nobody can possibly be selfish enough to resist the incentives for cooperation offered by prices.

In the same fundamental sense, the benefits from mitigating the failure of group selection and altruism (the last source of benefits as listed above) can also be traced to the human capacity of conducting business, especially exchange, by means of prices and to the equilibrium properties of prices themselves. Among these properties, as we already saw, are built-in incentives that lead trading partners to internalize external affects and privatize collective effects, thus shielding a transaction from the ill effects of group selection failure and from the inborn selfishness of the parties themselves. However selfish, and without being aware of it, traders are led to act as perfect altruists—an undertaking that is not necessarily part of their original intention. In fact, with the habit of using credit cards, whereby visible money is no longer changing hands, a visiting Martian would be pretty sure that all market transactions are a form of pure altruistic transfer and, in that, the Martian would not be entirely wrong.

4 Conclusion

The existence of a mechanism geared to facilitate cooperation among otherwise selfish agents is inferred from the unprecedented incidence of division of labor in human society. Division of labor occurs within and across human populations at levels of organization well above the single kinship group—that is, well above the unit of selection, and potentially at the level of the species at large. Traced to its economic foundations, the facilitating mechanism has been shown here to rely on the human propensity to exchange, on the human capacity to conduct business by means of prices, and on the properties of prices themselves. Overall, an explanation for the existence and the uniqueness of such a system of division of labor has been provided, as well as an explanation for the benefits gained from it.

Overall, the discussion in this chapter draws attention to at least two new sources of benefits previously unknown or unaccounted for in relation to trade: (1) improvement in human health associated with MHC diversity, and (2) reduction in the cost of sexuality. It adds therefore a new (epidemiological) dimension to the economic analysis of benefits of trade. In addition, it puts into a slightly new perspective the long standing evolutionary discourse on the paradox of sexuality.

Notes

1. To the entomologist, the division of labor is a fundamental aspect in the organization of insect societies (Robinson, 1992). Thus, for instance, *The Insect Societies* by Edward O. Wilson (1971) has references to division of labor on at least 47 separate pages (out of some 500 total). By comparison, *The Wealth of Nations* by Adam Smith (1776/1976) refers to the division of labor "only" on 29 pages (out of nearly 1,000 total).

2. Situations in which division of labor is not necessarily the best or the most common form of cooperation, as well as alternative forms of cooperation more suitable in such situations, are outlined later in the discussion (e.g., toward the end of subsection 3.1).

3. This is true not only in the obvious cases of somatic or vegetative reproduction, but also in the case where offspring are produced from unfertilized eggs (parthenogenesis) so long as meiosis is either suppressed or, otherwise, is preempted by a form of preceding mitosis (i.e., apomixis and endomitosis, respectively). The only exception is the case where meiosis is fully retained (by a process known as automixis) but fertilization is replaced by fusion of two products of meiosis that restores diploidy in the offspring (for further details, see Maynard Smith, 1989, pp. 235–239).

4. Aimed at the general reader, a slightly more recent account covering most of the items in this list has been provided by Matt Ridley (1993, pp. 23–87). Compared with Bell, Ridley's account reflects the growing influence of the notion of resistance to parasites as a possible explanation for sex.

5. One of the best-known implications from the study of economics suggests that, under perfect competition, maximum economic profit gained by profit-maximizing firms (driven either by rational choice or by selection), no matter how hard they work, is inescapably zero in the long run.

6. All "costs" (or "benefits") in the present discussion are understood to be in terms of lost (or gained) reproductive opportunities.

7. The main function of such "resistance" genes is to guarantee ceaseless and never-ending replacement of, so to speak, molecular locks and passwords with the idea of keeping the parasites guessing anew in each generation and in each individual host (save identical twins).

8. For instance, exceptions such as vertebrates as small as the tiny carp fish *Paedocypris* (0.8 cm long) or, conversely, invertebrates as large as squid (20 m).

9. Alternative forms of cooperation are outlined later in the discussion (section 3.1, below).

10. For instance, in the case of raptors females are typically significantly larger than males, and this is likely to be a form of niche division, allowing males and females to exploit differently sized prey classes.

11. Judging from the antiquity of *hunting-gathering*: a system of human subsistence based on division of labor that predated not only *Homo sapiens* but also *Homos erectus*. However, it did not, apparently, go as far back as *Australopithecus* (Milton, 1987; Isaac, 1989; Ofek, 2001).

12. Under consideration are competitive prices (i.e., market-clearing prices determined at equilibrium rather than arbitrarily fixed prices). Such prices are known to serve as signals that carry almost all the information necessary for optimal market decision making, along with the right amount of incentives to act on the decisions themselves.

13. Sentiments along these lines have been prudently expressed by Alfred Marshall, the most influential neoclassical economist at the turn of the nineteenth century. "Economics," he noted, "is a study of men as they live and move and think in the ordinary business of life. But it concerns itself chiefly with . . . man's conduct in the business part of his life" (Marshall, 1961, p. 14).

14. Arguably, the massive potential for benefits of trade creates an equally massive temptation to expropriate them. There is certainly a kernel of truth in that argument, but its overall impact is largely exaggerated, for a number of reasons. For one, the benefits of trade are largely received in the form of flows (e.g., wages, rents, revenues, and for that matter the entire GDP), rather than in the form of stocks. Such flows are fairly well protected against one-shot acts of expropriation (e.g., by a foreign invader) because the yield to their perpetrators in loot is a disappointingly small fraction of the wealth of a nation that they are trying to victimize. Similarly, economic agents (e.g., workers) are largely protected against domestic expropriation through exploitation by the mere existence of a sufficient number of alternative trading partners (e.g., employers) in a competitive market. Protection against exploitation by free riders and other defectors is on many occasions provided by the repetitive (Tit-for-Tat) nature of the transactions, and the list goes on. On the other hand, there is at least one form of expropriation against which the benefits of trade seem to be ill-protected: taxation.

References

Axelrod, R. (1984). *The evolution of cooperation.* New York: Basic Books.

Becker, G. S. (1991). *A treatise on the family.* Cambridge, MA: Harvard University Press.

Beer, A. E., Gagnon, M., & Quebbeman, J. F. (1981). Immunologically induced reproductive disorders. In P. G. Crosigniani & B. L. Ruhin (Eds.), *Endocrinology of human infertility: New aspects* (pp. 419–439). London: Academic Press.

Bell, G. (1982). *The masterpiece of nature: The evolution and genetics of sexuality.* Berkeley: University of California Press.

Carrington, M., Nelson, G., Martin, M., Kissner, T., Vlahov, D., Goedert, J., et al. (1999). HLA and HIV-1: Heterozygote advantage and B*35-Cw*04 disadvantage. *Science, 283,* 1748–1752.

Cukier, K. 2005. A market for ideas (interview). *Economist* (special report, October 20).

Dawkins, R., & Krebs, J. R. (1979). Arms races between and within species. *Proceedings of the Royal Society of London, Series B: Biological Sciences, 205,* 489–511.

Dawkins, R. (2005). *The ancestor's tale: A pilgrimage to the dawn of evolution.* New York: Mariner Books.

Ghiselin, M. T. (1974). *The economy of nature and the evolution of sex*. Berkeley: University of California Press.

Glesener, R. R., & Tilman, D. (1978). Sexuality and the components of environmental uncertainty: Clues from geographic parthenogenesis in terrestrial animals. *American Naturalist, 112*, 659–673.

Grafen, A. (1992). Of mice and the MHC. *Nature, 360*, 530–531.

Hamilton, W. D. (1964). The genetical evolution of social behaviour, I & II. *Journal of Theoretical Biology, 7*, 1–52.

Hamilton, W. D. 1988. Sex and disease. In G. Stevens & R. Beling (Eds.), *Nobel Conference XXIII: The Evolution of Sex* (pp. 65–95). San Francisco: Harper & Row.

Isaac, G. 1989. *The archaeology of human origins: Papers by Glynn Isaac*. Ed. B. Isaac. Cambridge: Cambridge University Press.

Manning, C. J., Wakeland, E. K., & Potts, W. K. (1992). Communal nesting patterns in mice implicate MHC genes in kin recognition. *Nature, 360*, 581–583.

Marshall, A. (1961). *Principles of economics*. London: Macmillan.

Maynard Smith, J. (1989). *Evolutionary genetics*. Oxford: Oxford University Press.

Milton, K. (1987). Primate diets and gut morphology: Implications for hominid evolution. In M. Harris & E. B. Boss (Eds.), *Food and evolution: Toward a theory of human food habits* (pp. 96–116). Philadelphia: Temple University Press.

Noë, R., & Hammerstein, P. (1995). Biological markets. *Trends in Ecology & Evolution, 10*, 336–339.

Ofek, H., & Merrill, Y. (1997). Labor mobility and the formation of gender wage gaps in local markets. *Economic Inquiry, 35*(1), 28–47.

Ofek, H. (2001). *Second nature: Economic origins of human evolution*. Cambridge: Cambridge University Press.

Potts, W. K., Manning, C. J., & Wakeland, E. K. (1991). Mating patterns in semi-natural populations of mice influenced by MHC genotype. *Nature, 352*, 619–621.

Potts, W. K., Manning, C. J., & Wakeland, E. K. (1994). The role of infectious disease, inbreeding, and mating preferences in maintaining MHC diversity: An experimental test. *Philosophical Transactions of the Royal Society of London, Series B: Biological Sciences, 346*, 369–378.

Ridley, M. (1993). *The red queen: Sex and the evolution of human nature*. London: Viking.

Robinson, G. E. (1992). Regulation of division of labor in insect societies. *Annual Review of Entomology, 37*, 637–665.

Seger, J., & Hamilton, W. D. (1988). Parasites and sex. In R. E. Michod & B. R. Levine (Eds.), *The evolution of sex: An examination of current ideas* (pp. 176–193). Sunderland, MA: Sinauer.

Smith, A. (1776/1976). *The wealth of nations*. Chicago: University of Chicago Press.

Suomalainen, E. (1950). Parthenogenesis in animals. *Advances in Genetics, 3*, 193–253.

Trivers, R. (1971). The evolution of reciprocal altruism. *Quarterly Review of Biology, 46*, 35–57.

Van Valen, L. (1973). A new evolutionary law. *Evolutionary Theory, 1*, 1–30.

Wedekind, C., Seebeck, T., Bettens, F., & Paepke, A. J. (1995). MHC-dependent mate preference in humans. *Proceedings of the Royal Society, Series B: Biological Sciences, 260*, 245–249.

Wilson, D. S. (1975). A general theory of group selection. *Proceedings of the National Academy of Sciences of the United States of America, 72*, 143–146.

Wilson, D. S. (1992). Group selection. In E. F. Keller & E. A. Lloyd (Eds.), *Keywords in evolutionary biology*. Cambridge, MA: Harvard University Press.

Wilson, E. O. (1971). *The insect societies*. Cambridge, MA: Harvard University Press.

Wilson, E. O. (1975). *Sociobiology: The new synthesis*. Cambridge, MA: Harvard University Press.

Yamazaki, K., Yamaguchi, M., Boyse, E. A., & Thomas, L. (1980). The major histocompatibility complex as a source of odors imparting individuality among mice. In D. Muller-Schwarze & R. M. Silverstein (Eds.), *Chemical signals* (pp. 267–273). New York: Plenum.

10 What We Don't Know about the Evolution of Cooperation in Animals

Deborah M. Gordon

The starting point for many studies of the evolution of cooperation is to explain why cooperation ever happens at all. Beginning with the premise that each individual's actions should serve that individual's interests, the question is why would anyone ever act in the interest of someone else? Models in evolutionary biology set up this question as a quantitative problem, in which the interests of the cooperating actor must be measured in the currency of reproductive success. How can we add up the benefits of cooperation so that it turns out that acting cooperatively serves the interests of the actor? The problem is to find an account that gives a net benefit, in reproductive success, to an individual that cooperates.

Recent discussion about the role of genetic relatedness in the evolution of cooperation centers around disagreement about how best to add up the benefits of social life. Some say that the best way is to count the reproductive success of the cooperating actor and add a measure of the reproductive success of relatives affected by those actions (Abbot et al., 2011). Others say this procedure does not necessarily give the best answer (Nowak, Tarnita & Wilson, 2010). Still others say that the argument is moot since all the available ways of counting are instances of the same general procedure (Sober & Wilson, 1998).

In any case, adding up benefits to see how evolution will proceed is a merely theoretical exercise unless we can relate actual behavior to ecological outcomes. To measure the benefits of cooperation, we need to know how particular kinds of cooperative behavior function, and why acting cooperatively matters to the individuals involved: what one animal does for another, and what difference this makes to each animal's livelihood and its relations with the rest of its environment. As Calcott (2008) points out, we need to understand what produces benefits before we can measure how the benefits are distributed.

How cooperative behavior benefits individuals is an empirical question, with unique answers related to the ecology of any particular social system. Any attempt to find out how individuals benefit from participating in a cooperative society leads directly to the conceptual problem of how to integrate the study of organization at the individual and group levels. Empirical studies of cooperation in animal groups reveal webs of interactions that do not easily come apart to provide a measure of what each individual is getting. Theoretical attempts to demonstrate how cooperation could evolve from self-interest often lead to models that depict cooperation as uncooperative, with cheaters, spiters, defectors, and so on, who adjust their actions according to self-interest. But cooperative group processes are not easily understood this way.

Cooperative social relations, whose evolution we are trying to explain, are by nature resistant to the usual procedures for constructing evolutionary models, because such models require us to consider the individual separately from the group. But to specify the benefits to an individual of its life within a group, we have to grapple with the problem of characterizing how the welfare of the individual and that of the group are related.

Social Insects

Social insects provide a canonical example of cooperation. A colony consists of one or more queens, who lay eggs, and workers, who usually do not. Being a worker and not reproducing is called "altruism," defined as having both some benefit to others in the colony and some cost to the altruist in its loss of reproduction. The assumption is that by not laying eggs, the worker is instead providing some benefits to others through her work. In this way, not laying eggs has come to be considered to be equivalent to cooperation.

Thus the form of cooperation in social insects that kin selection was originally invoked to explain is defined negatively. In eusocial social insect colonies, there are individuals that do not reproduce. The evolutionary question was to explain how not reproducing could possibly be inherited. As originally formulated by Hamilton (1970), the problem is that once genes for not reproducing appear, they should quickly be lost in future generations because they are imprisoned in individuals that do not reproduce. Hamilton's solution was to notice that probably the relatives of workers would have the same genes as the workers do, so the genes could be passed on when the relatives reproduce.

Hamilton's argument about the role of kin selection in the evolution of worker sterility rested on the peculiar haplodiploid genetic system of the ants, bees, and wasps. This system has the consequence that a worker is more closely related to her sisters than she would be to her daughters. However, this is true only if the mother of the workers, the queen, mates only once. Since Hamilton introduced this idea, it has become clear that in many ant species the queens mate many times, so his argument does not apply. Others have argued for using kin selection in different ways to explain the evolution of worker sterility (Queller & Strassman, 1998).

The discussion about kin selection and the evolution of worker sterility sidesteps broader questions about the evolution of cooperation in social insects. Not laying eggs is not equivalent to cooperation. Workers sometimes lay eggs and still work, and they often do not work even when they don't lay eggs. This means that identifying who lays the eggs is not sufficient to produce an accurate count of the benefits of cooperation. To explain how cooperation in social insects evolved, we also have to count up the benefits of a worker's work. So we need to understand what workers do, and then figure out how to measure the extent to which each worker's actions contribute to the colony's ability to thrive and reproduce.

To determine how an ant's work benefits the colony, the first step is to find out how the colony operates and its ecological relations. A worker goes out and gets a seed, and brings it back to the nest. That seed is taken deep into the nest and stored. Months later, it is fed to a larva. We do not know how to calculate the benefit to the other ants of that worker's foraging trip. If it had stayed in the nest instead, how much worse off would the colony be, and how quickly would another ant have replaced it? The answer varies with current conditions: how much food the colony has stored, how many larvae need to be fed, the amount of each nutrient available in that seed, and so on.

In social insects, colonies can be considered the reproductive individuals. Colonies produce reproductive males and females (the queens); the males and females of different colonies mate, and the newly mated queens found new colonies. Each new colony is thus the offspring of the colonies that produced the founding queen and those that produced her mate or mates.

Since the colony reproduces as a single unit, producing more colonies, the benefits to any individual depend on the reproductive success of the colony, which in turn depends on how all the individuals in the colony function together. What we are learning about the organization of social insect colonies suggests new perspectives on what it means for a colony to operate well. A social insect colony regulates its behavior as a result of a network of interactions among workers (Gordon, 2010). For example, in a harvester ant colony, at any time there are many workers doing nothing. In the standard kin-selection story, these inactive workers are viewed as cheaters, serving their own interest at others' expense, who need to be policed. But inactive workers may in fact serve to buffer the interaction rate, keeping response down and dampening positive feedback from interactions (Pacala, Gordon & Godfray, 1996). Thus a worker, by doing nothing, may contribute to the ability of the colony to respond appropriately to changing conditions.

Harvester ants regulate foraging using a simple feedback system. They live in the desert and eat seeds. While a forager is out in the hot sun searching for a seed, it loses water. The ants get water from metabolizing the fats in the seeds they eat. So a colony must spend water to obtain water, as well as food. A colony also has to deal with its neighbors, with whom it competes for food. If a colony does not use its foraging area for several days, its neighbors will collect seeds there instead (Gordon, 1992). Thus, for a given current food supply and current amount of brood to feed, the colony has to manage the trade-off between loss of water when it forages and loss of foraging area to neighbors where it does not forage.

From moment to moment, the colony adjusts the numbers foraging to the current food supply. Each forager spends most of its trip searching (Beverly et al., 2009), so the more food is available, the more quickly it finds food and the more quickly it returns to the nest. The rate at which foragers return with food is the feedback that regulates how quickly foragers leave the nest on the next trip (Gordon, Holmes & Nacu, 2008). The more food, the faster the foragers come back and the sooner more ants go out.

Colonies differ in how closely they regulate foraging (Gordon et al., 2011). In experiments where we manipulate how many foragers return, some colonies adjust within minutes, while others are less likely to change foraging activity. It seems that what differs among colonies is an outgoing forager's responsiveness to returning foragers. When foragers are less responsive, then foraging activity is steadier. When foragers are more responsive, then foraging activity is fine-tuned to the current level of food availability.

How closely a colony regulates foraging determines how it manages water loss and competition with neighbors for foraging area. The ecological question is: How important is the regulation of foraging for a colony's survival, growth, and reproductive success? If individual responsiveness to returning foragers is inherited from parent to offspring colonies, and it matters to a colony how well it regulates foraging, then natural selection can shape the regulation of foraging.

The regulation of foraging by an ant colony is a facet of its cooperative behavior. Working together, the ants help the colony to manage its food and water supply. Each individual forager's contribution consists of waiting for the right number of foragers with food to touch its antennae, going out to find a seed and bring it back, meeting the outgoing foragers along the way back inside, and then, after enough interactions, going out again. All of this is stochastic; the ant does the right thing, in a certain range of conditions, most of the time.

Suppose that natural selection is shaping this cooperative behavior by favoring the colonies that regulate foraging more closely. If the reactions of foragers are heritable from parent to offspring colonies, such selection could occur when the colonies with foragers who are a little more likely to rush out on a trip when a few foragers come back with food have higher fitness than the colonies whose foragers require more persuasion. The only way to measure the effect of any individual's contribution is to compare the reproductive success of its colony to that of other colonies in the population.

Colonies, and thus the ants in them, differ in the values used in the algorithm that links forager-return rate to the probability of going out on the next foraging trip (Prabhakar, Dektar & Gordon, 2012). The benefits to the colony of an individual's particular version of the forager-response algorithm depend on how that version contributes to the colony's water and food stores and relations with its neighbors. The benefits to the individual arise from how much the resulting ecology allows the colony to reproduce. To understand how benefit is generated requires understanding the colony-level behavior that regulates foraging.

Group Behavior

Many studies of cooperative behavior in animals, like models of its evolution, start from the same kind of question that Hamilton asked about social insects. For Hamilton, the problem was why some individuals do not reproduce. More generally, the problem is usually framed as why animals live or hunt in groups. This is seen as a sacrifice, analogous to worker sterility, of all of the resources or food that the animal could obtain on its own. Living in a group means the animal has to share, and sometimes has to put up with being bossed around by others. What does it get in return, and does it get more, the larger the group?

African wild dogs hunt in groups. Fanshawe and Fitzgibbon (1993) watched hunts and measured the relation between the size of the hunting group and its success in killing the prey. They found that large groups are no more successful than small ones in killing gazelles, because they are usually able to catch only young or sick gazelles, and it doesn't take many dogs to bring down a small gazelle. Larger groups are more successful in killing wildebeest, however, because the wildebeest is large enough that one dog has to hold it down while the others tear it apart. After the kill, hyenas often show up, and the higher the ratio of dogs to hyenas, the more likely are the dogs to keep control of the carcass.

It is not easy to see how to add up the costs and benefits of group hunting in wild dogs. A dog that hunts in a group of five eats more meat, relative to a dog that hunts alone, when the wildebeest are passing through the area, or when it's a good year for hyenas, but not when the most available prey are gazelles.

Any attempt to count up the benefits of group hunting has to make assumptions about the answers to a host of interesting questions, and making those assumptions distracts us from the fact that we don't know the answers. How do the dogs work together? How fluid are the groups and what is the network structure; do the same individuals tend to hunt together over and over? Does group structure change according to conditions, such as the availability of wildebeest and gazelles, and the group behavior of the local hyenas? The dogs' group size influences their interactions with hyenas, and so there is another set of questions about how the relative size of each kind of group affects the outcome of a conflict, and how the network structure of the hyena group interacts with that of the dogs.

Many species of fish form schools that move and forage together, forming temporary coordinated groups. A school of fish moves through the water, turning, sometimes breaking apart around an obstacle and then fusing again. To decide whether this is cooperation, and to provide an account for how it could evolve, we'd have to describe the costs and benefits for the fish of being with the others instead of swimming around on its own. The larger the group of fish, the less likely is any individual to be the fish captured by a predator—but a large group is more conspicuous than a fish on its own. A fish in a group is likely to find a food source that a neighboring fish found—but it has to share its discoveries with the others. Both predator hunger and the distribution of food must vary in time and space. To specify the benefits added by each additional fish, we would have to know how often the predators appear, and how many fish they eat, and how patchy are the resources that the fish are searching for.

A separate question is how the fish coordinate their movements. There is no leader or central control. Each fish uses a simple algorithm, based on what it sees of neighboring fish, and on how the neighbors' movements alter the flow of the water (Couzin & Krause, 2003). In the aggregate, this produces the movement of the school.

How does evolution shape this algorithm? Natural selection acts on heritable variation and requires that the variation in behavior is associated with variation in reproductive success. If fish vary slightly in how they respond to each other, and these differences are inherited, and the coherence of a fish's school affects that fish's survival and reproductive success, then over time, in certain conditions, particular algorithms would increase in frequency.

To learn about the evolution of fish schooling behavior, we would have to know why, and in what situations, it matters to a fish how well coordinated is its school. The spectacular maneuvers of fish schools may be unusual. How often do the fish travel together, and under what conditions? What evolves is the fishes' capacity to move in and out of schools, to work as a group sometimes and as individuals at other times. To understand the evolution of this behavior, as of any other instance of cooperation, we have to begin by asking how the behavior is organized and how it functions ecologically.

Individuals and Groups

Evolutionary questions about cooperation are framed by contrast with an imaginary alternative in which each individual is out for itself. But for most animal groups, living in groups probably preceded the particular behavior we are studying. Ants evolved from wasps, who form temporary associations in which some females do not reproduce some of the time. Group living in canids is ancestral to the particular hunting strategies of African wild dogs. Evolution has shaped an enormous diversity of nuanced interactions in groups, each arising through particular histories and conditions. Considering the benefits of cooperation relative to acting alone invites us to ignore all the ecological detail that could explain how particular kinds of behavior evolved.

It is clear that the more we learn about any cooperative system, the more we grapple with trying to evaluate benefits at different levels—what is good for the individual and what is good for the group. The study of cooperation has always ricocheted back and forth between individual and group. Debates about individual, kin, and group selection are one current manifestation of the basic problem of how to reconcile the individual and group levels when thinking about the advantages of cooperation. These debates can't be resolved in principle. Questions about how animals actually interact, and how individuals function within groups, will help to formulate realistic assessments about how individuals benefit from living together and cooperating.

How individual behavior is related to group behavior is the focus of our debates about how best to count the benefits of cooperation. More fundamentally, how individuals construct groups is what cooperation *is*. We need to ask first how the actions of individuals affect others, and how this produces the development, ecology, and history of the group. Evolutionary models of cooperation can be only as sophisticated as their measures of its benefits. The challenge is to understand the group organization and ecological relations that generate those benefits.

Acknowledgments

This work benefited greatly from comments from Brett Calcott, Ben Fraser, and Richard Joyce, from the Dispersed Authority project funded by the Stanford University President's Fund for Innovation in the Humanites, and from the Stanford Emergence of Cooperation project supported by a Templeton Foundation planning grant.

References

Abbot, P., Abe, J., Alcock, J., Alizon, S., et al. (2011). Inclusive fitness theory and eusociality. *Nature, 471*, E1–E3.

Beverly, B., McLendon, H., Nacu, S., Holmes, S., & Gordon, D. M. (2009). How site fidelity leads to individual differences in the foraging activity of harvester ants. *Behavioral Ecology, 20*, 633–638.

Calcott, B. (2008). The other cooperation problem: Generating benefit. *Biology and Philosophy, 23*, 179–203.

Couzin, I. D., & Krause, J. (2003). Self-organization and collective behavior in vertebrates. *Advances in the Study of Behavior, 32*, 1–75.

Fanshawe, J. H., & Fitzgibbon, C. D. (1993). Factors affecting hunting success of an African wild dog pack. *Animal Behaviour, 45*, 479–490.

Gordon, D. M. (1992). How colony growth affects forager intrusion in neighboring harvester ant colonies. *Behavioral Ecology and Sociobiology, 31*, 417–427.

Gordon, D. M. (2010). *Ant encounters: Interaction networks and colony behavior.* Princeton, NJ: Princeton University Press.

Gordon, D. M., Guetz, A., Greene, M. J., & Holmes, S. (2011). Colony variation in the collective regulation of foraging by harvester ants. *Behavioral Ecology, 22*, 429–435.

Gordon, D. M., Holmes, S., & Nacu, S. (2008). The short-term regulation of foraging in harvester ants. *Behavioral Ecology, 19*, 217–222.

Hamilton, W. D. (1970). Selfish and spiteful behaviour in an evolutionary model. *Nature, 228*, 1218–1220.

Nowak, M. A., Tarnita, C. E., & Wilson, E. O. (2010). The evolution of eusociality. *Nature, 446*, 1057–1062.

Pacala, S. W., Gordon, D. M., & Godfray, H. C. J. (1996). Effects of social group size on information transfer and task allocation. *Evolutionary Ecology, 10*, 127–165.

Prabhakar, B., Dektar, K. N., & Gordon, D. M. (2012). The regulation of ant colony foraging activity without spatial information. *PLoS Computional Biology, 8*(8): e1002670. DOI: 10.1371/journal.pcbi.

Queller, D. C., & Strassman, J. E. (1998). Kin selection and social insects. *Bioscience, 48*, 165–175. jouu

Sober, E., & Wilson, D. S. (1998). *Unto others: The evolution and psychology of unselfish behavior.* Cambridge, MA: Harvard University Press.

11 Task Partitioning: Is It a Useful Concept?

Adam G. Hart

Introduction

Division of labor allows for specialization (Smith, 1776). Division of labor in human groups has long been an important organizational principle; and, as societies expand and become more complex, so division of labor has become both more complex and more essential. Insect societies, with highly interrelated tasks, such as foraging, caring for young, and nest building, also depend to a great extent on the effective and efficient organization of the workforce through division of labor. Studies of social (more precisely, eusocial) insect organization have tended to lead us toward an increasingly sophisticated understanding of the functions, causes, and implications of division of labor (e.g., Smith et al., 2008; Jandt & Dornhaus, 2009; Robinson, Feinerman & Franks, 2009). As our understanding of insect societies has become more nuanced, however, so has our appreciation of how colony work is structured and organized. Division of labor was, and remains, the primary principle underpinning studies of social insect organization, but, in 1986, Jeanne added a further organizational principle—that of task partitioning (Jeanne, 1986a).

The essential feature of task partitioning is the subdivision of a single task into two or more subtasks. In the examples described in social insects these subtasks are usually connected into a single process by a transfer of material (Jeanne, 1986a; Ratnieks & Anderson, 1999a). Thus, honeybee nectar foragers returning to the nest with nectar in their crops do not regurgitate their load directly into a convenient storage cell in the honeycomb. Instead, these returning foragers proceed to an "unloading area" inside the nest entrance where they wait until a "receiver" bee (typically slightly younger than the foragers) becomes available (Seeley, 1995; Hart & Ratnieks, 2001b). The forager then transfers the nectar directly to the receiver via trophallaxis, with the receiver extending her mouthparts to take up the liquid from the forager's mouth. The receiver bee may unload several foragers in one receiving bout; likewise, foragers may have to unload to more than one receiver (for data on transfers, see Hart & Ratnieks, 2001b, and references therein). Once the forager is empty she is free to leave the nest to forage while the receiver can either store the nectar in a cell or, before that, commence a bout of evaporation, where she manipulates the nectar in her mouthparts so as to concentrate it and hasten the honey-forming process (Park, 1925; Maurizio, 1975). The task of nectar foraging is thus subdivided (partitioned) into nectar *collection* and nectar *storage*. Of course, this could also be considered division of labor since the workforce is divided up among the tasks that need to be undertaken, and the receivers and foragers are, at least in the short term, task specialists. The difference between task partitioning and division of labor is not always obvious, but a useful scheme is provided by Ratnieks and Anderson (1999a); division of labor is the division of the workforce into different tasks, whereas task partitioning is the division of the tasks themselves. Thus, task partitioning is an additional "level of organization" that can act in concert with (and indeed encourage) division of labor.

Advantages and Disadvantages of Task Partitioning

In honeybees, nectar foraging is a partitioned task. The partitioning of the notional single task of "nectar foraging" into its core components—collection and storage—provides a diversity of benefits for a honeybee colony. Nectar is an often ephemeral resource, with nectar availability shifting spatially and temporally even within the period of a single day's foraging (Seeley, 1995). Successful nectar foragers, with fresh knowledge of rewarding patches, are better employed advertising these resources (via the forager-recruiting waggle dance) or returning directly to them, than storing nectar within the nest. By transferring nectar to receivers, foragers reduce the time they spend away from currently rewarding foraging sites and increase the influx of nectar into the hive (discussed in Ratnieks & Anderson, 1999a). The decoupling of collection and storage also enables receivers to process the nectar through evaporation (a time-consuming task that would be wasteful of foragers' time were they to perform it), which makes far more efficient use of colony storage space and speeds up the production of honey. Receivers are able to unload several bees in one bout (foragers may not always be full for a number of practical and ergonomic reasons), and foragers are kept within the unloading area, which prevents them from interfering with each other as they search for storage cells (the location of which can be known by receivers and therefore found quickly). In this case, task partitioning (which at its core is simply the subdivision of tasks into smaller subtasks) is allowing for additional specialization (division of labor), providing benefits to the colony.

However, partitioning also introduces costs into the foraging system. The arrival of foragers and receivers to the unloading area is an unpredictable process that inevitably introduces delays, meaning that queues can become lengthy if the numbers of foragers and receivers are badly unbalanced (Anderson & Ratnieks, 1999a). These delays include searching for a transfer partner, waiting for a partner (a queuing delay), and the time involved in the transfer itself (Ratnieks & Anderson, 1999a). Interestingly, the delays that foragers experience in finding a receiver are closely related to the probability of them performing certain recruitment dances characteristic of honeybees. Longer delays provoke receiver-recruiting tremble dances, a complex dance performed by forgers involving simultaneous shaking of the body, rotation of the body axis, and a slow walk across the comb (Seeley, 1992). Shorter delays provoke forager-recruiting waggle dances that serve both to recruit foragers and to direct them to rewarding patches by conveying information on distance and direction (von Frisch, 1967; see Seeley, 1995, for a highly readable account of honeybee dances). However, it is important to realize that the ability to balance the workforce in this sophisticated manner is not an advantage of task partitioning, but rather a cost-reducing reaction to a feature of foraging that does not exist without partitioning (Ratnieks & Anderson, 1999a). Were it not for dividing the task in the first place, thereby introducing receivers and subsequent queuing delays, there would be no need for a colony to act to reduce those queuing delays.

Task partitioning as a concept has been reviewed by Ratnieks and Anderson (1999a) and, as an organizational feature of leafcutting ants, by Hart, Anderson, and Ratnieks (2002). These reviews provide more comprehensive details of the concept as well as an overview of its occurrence both in terms of species and tasks. Should further specific information be required, the reader is directed toward these sources and the references therein. The purpose here is not to review the field but rather to discuss theoretical aspects of task partitioning as an organizational principle. Consequently, following a brief consideration of its advantages and disadvantages, the discussion will be focused on the following questions:

1. Does task partitioning have to involve material transfer?
2. Is it a useful concept; including
   i. How widely can it be applied in terms of task types?
   ii. How taxonomically widespread is it?
3. Is it useful terminology?
   i. Has it been applied correctly in reported instances?
   ii. Does it add anything to our understanding of organization over and above that provided by division of labor?

1 Does Task Partitioning Have to Involve Material Transfer?

Reported examples of task partitioning within social insects involve more than just the subdivision of a single task into two or more subtasks. There is also a connection of these subtasks through the processing and transfer of *material*, exactly as we find in a human factory production line. This material transfer may be direct (for example, via trophallaxis, as in honeybee nectar foraging, or via mandible-to-mandible transfer) or indirect (typically via a cache, a temporary storage "structure"). Indirect transfer via caches introduces another benefit of task partitioning: the ability to buffer supply and demand. Foraging in leafcutting ants provides an excellent example of this buffering effect (Hart, Anderson & Ratnieks, 2002). Leafcutting ants cut and collect leaf fragments, which are carried back to the nest and processed to feed fungus gardens that the ants use for food. Leaf-fragment piles are commonly associated with nest entrances, and it appears, at least in some cases, that these caches are adaptive structures. At times, the supply of fragments will outstrip the colony's ability to process them within the fungus gardens, leading to a processing bottleneck (Hart & Ratnieks, 2001a). By caching leaf fragments temporarily, foragers can undertake further foraging trips without waiting to transfer their load, as would occur if direct transfer were employed. Caching is not without cost, though. Cached materials may not be recovered, wasting the resources used in their initial collection, and caching may make material liable to infection with harmful microorganisms (Hart & Ratnieks, 2000, 2001a). Thus, as in human production lines, caching can be possible only when the material processing is not time-sensitive or at risk of contamination; components can wait on a car assembly line, but not on a sandwich production line, where caching would be unhygienic and result in spoiling.

Given that all reported examples of task partitioning in social insects involve the transfer of material between the subtasks, it is tempting to include material transfer as a key component of the definition of task partitioning, as indeed was done by Hart, Anderson, and Ratnieks (2002). However, the initial definition of task partitioning (Jeanne, 1986a,b) and some further development of the theory underpinning it as an organizational concept (Ratnieks & Anderson, 1999a) do not explicitly include material transfer. To be partitioned, a task merely has to have "two or more individuals [contributing] sequentially to a particular task or piece of work" or be when a "piece of work is divided among two or more workers, such as the partitioning of the collection of a load of forage between a forager and a storer or transporter" (Ratnieks & Anderson, 1999a). Thus, if the task of feeding larvae involved two or more workers sequentially feeding the same larvae (as indeed it usually does), then larval feeding fits the definition of task partitioning but does not involve material transfer in anything other than an abstract manner; the larvae act as the "material" element of the task and the transfer is effected by workers moving to the task location.

The term "task partitioning" has also been used to describe hygienic behavior in honeybees, whereby sealed brood cells containing diseased or dead pupae are uncapped and the contents removed (e.g., Arathi, Ho & Spivak, 2006). This single task can be, and often is, subdivided into the two subtasks of "uncapping" and "removal." However, this does not include the connection of subtasks via a material transfer, unless one considers the unsealed brood cell to be the "material" (rather like in the notional example of larval feeding above), which seems neither accurate nor helpful. Using this interpretation of task partitioning enables practically any two or more tasks within any colony that are in any way connected to be considered "task partitioning." For example, ant foragers bring food back to the nest that is subsequently consumed by nurse ants who then feed larvae with secretions made from that material. Thus the task of "feeding larvae" might, in an extreme interpretation, be considered "task partitioning," with the subtasks (collection of food, processing of food, nursing of larvae with glandular secretions) being linked, potentially through a single adult. This interpretation certainly emphasizes the often interrelated and connected nature of division of labor within a social insect colony, but it is not helpful to call it task partitioning.

While these semantic issues are stimulating, "task partitioning" like the larval feeding example above is clearly different from the transfer of nectar between foragers and receivers, or the transfer of leaves between workers in leafcutting ants. Furthermore, the example of larval feeding involving two or more individuals sequentially feeding the same larva does not involve the divisions of the task (feeding) into subtasks. Ratnieks and Anderson (1999a) state that "task partitioning requires explicit linking of related tasks, i.e., sequential stages in the handling and processing of material," a definition that implies both material handling and division of a task into subtasks. Although the initial definitions of task partitioning do not explicitly include material handling, this further development of the concept reveals it to be a core component. A further crucial aspect to the definition of task partitioning involves the definition of tasks and subtasks, which will be considered in more detail below.

2 Is Task Partitioning a Useful Concept?

(a) Limited Occurrence among Tasks

In their initial review of task partitioning, Ratnieks and Anderson (1999a) focused on foraging, indicating that "despite [our] failure to find examples we feel it is probable that examples of task partitioning do occur outside foraging and [we] encourage others to seek and publish them." In fact, their review contains two examples of task partitioning other than in foraging, both of which are in nest construction. The neotropical wasp *Polybia occidentalis* has a complex three-way partitioning of two groups: "foragers" who collect wood pulp and water, and "receivers" back at the nest who use these foraged materials to build the nest carton (Jeanne, 1986b). A similar system is employed by *Lasius fuliginosus*, with foragers collecting honeydew from aphids and particles of soil or organic matter (Hölldobler & Wilson, 1990). Honeydew is transferred directly to builders back at the nest, as might be expected with a perishable liquid, whereas the particles are transferred indirectly after being deposited within the nest cavity, enabling the colony to buffer supply and demand of this building material. In the ten years following the publication of this review, task partitioning examples outside of foraging have been described, but these have mostly been limited to leafcutting ants, and then generally within the genus *Atta*, for example, in waste management, leaf processing (although this is a problematic example as is discussed below), and colony emigration (a historical record from the naturalist Thomas Belt, 1928) (Hart, Anderson & Ratnieks, 2002).

If we allow that task partitioning is limited to tasks that both involve "material" and that have a well-defined transfer of that material (as discussed above) then this relative exclusivity perhaps explains why, to date, task partitioning has been reported only in a very small subset of colony tasks, namely foraging, waste disposal, leaf processing in leafcutting ants (though bear in mind the caveats below for this application), nest construction, and, in one case, colony emigration. Although other tasks within a colony can involve handling material (for example, feeding larvae), these tasks are conceptually "monolithic" in nature, not allowing for any sensible subdivision into subtasks that can be connected by a transfer of the material involved in those subtasks. Many tasks within a colony come under this banner; for example, it is hard to see how guarding or grooming could be partitioned in any sensible fashion.

Given the extensive behavioral repertoire of most social insect colonies it would be easy to dismiss task partitioning as a relatively idiosyncratic organizational principle limited primarily to foraging, with a few other examples found in nest building and waste management. To do this would be to dramatically underestimate the importance of task partitioning in those species that perform it. Foraging is a critical task within any social insect colony and it typically involves a relatively high proportion of workers. In the leafcutting ant *Atta colombica*, the vast majority of workers working outside the nest are either involved with foraging or with waste management, both of which involve task partitioning (Hart & Ratnieks, 2002a). Many workers inside the nest are involved with processing leaves to incorporate into the fungus gardens, which also involves task partitioning (albeit an arguable case of it—see below). Overall, a large proportion of the total workforce may be involved in partitioned tasks, despite these tasks being only a small subset of the total colony task repertoire.

(b) Limited Occurrence across Species

Division of labor, at least at the fundamental level of reproduction, is, by definition, a universal feature of eusocial insects. Whereas more complex aspects of division of labor, such as the evolution of physical castes among the workforce (where individuals become morphologically "trapped" in certain roles) are more unusual, division of labor whereby individuals perform, at any given time, different tasks on an age-based schedule (so-called age polyethism or temporal polyethism) is nearly universal (Oster & Wilson, 1978). Task partitioning, on the other hand, is, at least in terms of the number of species and types of tasks reported to employ it, more unusual, with most published work focusing on the honeybee *Apis mellifera* and leafcutting ants of the genus *Atta*. It is fair to say that examples of task partitioning in novel species have not been especially forthcoming in the ten years following Ratnieks and Anderson's (1999a) review of the topic.[1]

Despite a relatively small number of examples, and a very small number of novel studies in the recent literature, task partitioning has been reported in a broad range of social insect taxa. It occurs in termites (e.g., *Hodotermes mossambicus* [Leuthold, Bruinsma & van Huis, 1976]), bees (stingless bees [Hart & Ratnieks, 2002b] and honeybees [Seeley, 1995] but not bumblebees [Michener, 1974]), wasps (including examples of foraging in Polistinae and Vespinae [Jeanne, 1991]), and ants (e.g., *Oecophylla longinoda* [Hölldobler, 1984], *Messor* [Sudd, 1965], and *Ectatomma ruidum* [Schatz, Lachuad & Beugnon, 1996]). However, the broad taxonomic presence of task partitioning masks the fact that it does not occur in the majority of species (a fact alluded to by Ratnieks & Anderson, 1999a). Furthermore, the costs (delays and waiting times) involved in mismatching the arrival of transfer partners to a transfer site where direct transfer occurs indicate that colony size is an important consideration, with task partitioning involving direct transfer being more likely to evolve in larger colonies (Anderson & Ratnieks, 1999b,c; Ratnieks & Anderson, 1999b). Indirect transfer, with its buffering effect, is less affected by the dynamics of queuing delays, but as is discussed below, some examples of task partitioning with indirect transfer can prove to be problematic when analyzed within the framework provided by the definition of task partitioning.

The relationship between colony size and queuing delays in particular also seems to favor swarm-founding species in the evolution of task partitioning, with swarm founders always having relatively large colonies (Anderson & Ratnieks, 1999a,b; Ratnieks & Anderson, 1999b). Support for this hypothesis is found in the pattern of task partitioning in the corbiculate bees and in the *Vespinae* wasps. Task partitioning is absent in the independently founding *Bombus* but present in closely related swarm founding species such as *Melipona* and *Apis*. A similar pattern of absence (independently founding *Vespa*) and presence (swarm-founding *Polybia*) is also found in the *Vespinae* (surveyed in Hart & Ratnieks, 2002b). However, this limited correlative evidence is far from compelling, and since the richest occurrence of task partitioning examples is found in the independently founding leafcutting ants (Hart, Anderson & Ratnieks, 2002), it is likely that large eventual colony size can overcome the initial size limitations imposed by independent founding.

The evolution and maintenance of task partitioning in some species clearly shows that the benefits can outweigh the costs. Given its wide taxonomic occurrence, and the pattern of absence and presence observed even among closely related species, we can conclude that it has evolved more than once. However, the absence of partitioning in so many species, and in so many situations where it *could* occur, indicates that the cost-benefit ratio is not favorable in many cases. It would seem to be limited to species with relatively large colonies and relatively complex materials-handling tasks.

However, as with the limited occurrence of task partitioning among task types, this need not be considered a weakness in the concept in task partitioning. By connecting tasks together, it provides an added level of sophistication to colony organization that provides a suite of clear advantages to those colonies large enough and complex enough to make use of it.

3 Is "Task Partitioning" Useful Terminology?

A theme in the discussion above has been the definition of task partitioning particularly with respect to the connection of subtasks via a transfer of material. However, the definition and categorization of "task" and "subtask" is another fundamental issue that impinges on task partitioning as a useful concept, namely, how we can define "units of work" within a colony? Anderson and Franks (2001) provide a useful synthesis. Briefly, a task is defined as "a set of behaviours that must be performed to achieve some purpose of the colony" (Oster & Wilson, 1978) and that makes some "positive contribution" to colony fitness, while a subtask is a unit of work that "may appear as a discrete unit but will not enhance fitness unless other individuals complete their own additional work units" (Anderson & Franks, 2001). These definitions have a compelling simplicity and logic to them. However, they must be used with caution. Technically, fitness in these cases concerns the production of sexual offspring or new colony units (by swarming or budding), and so the only way any "task" not directly related to the production of sexual offspring or colony growth can enhance fitness is through a series of connections with a multitude of other "tasks" involved in producing sexual offspring or colony growth. Continuing this to its logical conclusion, virtually all "tasks" within a colony are really "subtasks," since the only way they can enhance fitness is through their conjunction with other tasks. Clearly, it is possible to become enmeshed in semantics with this topic, and equally clearly the definition of fundamental units of work within a social insect colony is not as straightforward as might first be thought. However, the definitions of tasks and subtasks proposed by Anderson and Franks (2001) continue to provide a useful framework in which to consider colony work.

Consider the task of nectar foraging in honeybees—a canonical example of task partitioning, with "foragers" and "receivers." We define this task as "partitioned" because the notional single task of foraging (comprising collection of nectar from flowers and its subsequent storage in the wax combs of the nest) has been subdivided such that collection and storage are performed by different individuals that are connected by a material transfer (in this case a direct transfer of nectar). This is all very well, but it relies absolutely on the notion that foraging *is* a single task and that this single task can be broken down into subtasks. Following the definitions of Anderson and Franks (2001), this would seem to be the case. However, an alternative way of considering this situation is that "nectar collection" and "nectar storage" are two discrete tasks in their own right (simple division of labor) that are connected by the transfer of nectar between the two groups performing them. The term "task partitioning" implies a level of organization overlying conventional division of labor and thereby adding an extra level of sophistication that might not be justified. In this case it could be argued that what is occurring is merely "task connection": Two discrete ergonomic tasks organized through division of labor are linked together. Were pollen and water collection also subject to "task partitioning" (in fact both materials are stored by the forager without transfer to a receiving partner), then perhaps we would view "foraging" in honeybees as a two-stage process and speak instead of division of labor ("collection" versus "storage") with "task connection" or some similar term. In fact, many tasks within in a colony are linked in some way, and integrated division of labor is a ubiquitous feature of colony organization (see the argument above regarding the fitness definition of tasks). However, the compelling logic behind the key factor in defining a subtask (according to Anderson & Franks, 2001) is that a subtask cannot enhance colony fitness unless it is linked with another task. Providing we take this in the spirit in which it was intended, rather than becoming too fixated on a strict definition of fitness, then this definition gives us a solid foundation underpinning task partitioning and allows (at least theoretically) for any given "work unit" to be defined as a task or a subtask.

In the example of nectar foraging in honeybees, task partitioning as a concept and as a useful term can in fact be further justified along two lines. In related species performing similar tasks, such as bumblebees, *Bombus*, nectar foraging is not partitioned between foragers and receivers, and a single forager performs both roles (Michener, 1974). Extending the comparative approach, we find that the stingless bees (of the corbiculate tribe *Meliponini*) partition nectar foraging, as seen in honeybees (Hart & Ratnieks, 2002b). However, in both partitioning taxa, some foragers can and do treat the tasks of "collection" and "storage" as subtasks of a single overall task, storing their nectar directly into the combs without transfer to a receiver (although what little published data there are indicate that this is extremely unusual [e.g., 1.5% in *Apis mellifera*; Hart & Ratnieks, 2001b]). So, taking the social corbiculate bees overall, it seems reasonable to assume that (1) the basal condition is that nectar foraging is not partitioned and that (2) the current situation, whereby foraging comprises the two roles of foragers ("collectors") and receivers ("storers"), satisfies the first part of the definition of task partitioning since foraging and receiving can properly and usefully be considered as subtasks. Finally, material transfer occurs between the now well-defined subtasks, thereby satisfying fully the extended definition of task partitioning proposed above. As discussed in the example of hygienic behavior in honeybees, though, it not sufficient to just have well-defined subtasks. To qualify as a useful application of task partitioning, it is also necessary to connect these subtasks by material transfer. The reverse of this situation—the existence of material transfer in the absence of well-defined subtasks—has also, wrongly, been considered task partitioning. This is discussed below.

The concept of material transfer has always been central to discussions of task partitioning—the transfer serves as the connection between the subtasks. However, in some instances the individuals performing each subtask are clearly specialists (at least for a period of time) with a discrete role profile. For example, foragers collect material from the field while receivers process and store that material. Consequently, in these examples, whatever "task partitioning" can be said to be occurring is clearly acting in concert with division of labor. This is conceptually quite different from other uses of "task partitioning" that appear in the literature. An excellent example to illustrate this difference in the use of terminology is provided by leaf foraging in neo-tropical leaf-cutting ants, particularly those in the genus *Atta*.

Task Partitioning in Leafcutting Ants

The leafcutting ants (*Atta* and *Acromyrmex*) of Central and South America use leaf fragments cut from trees and plants to grow a mutualistic fungus within their underground nests (Weber, 1972). This fungus is cultivated and used by the colony as food, especially to nourish the larvae. *Atta* typically have very large nests, both spatially and with respect to numbers of individuals (reaching perhaps 8 million or more workers; Hölldobler & Wilson, 1990), and the rather specialized biology of this genus in particular has provided an unusually rich array of task partitioning examples (Hart, Anderson & Ratnieks, 2002). The propensity for task partitioning in *Atta* is highlighted by the large number of transfers that can occur between plant and fungus garden and fungus garden and waste heap. Transfers, both direct and indirect, can occur at the source tree, at any point along the foraging trail, at the nest entrance, within the fungus gardens, and as waste is conveyed from the fungus gardens to the waste heaps (reviewed in Hart, Anderson & Ratnieks, 2002). Task partitioning has also been reported in colony emigration (Belt, 1928). In most cases an adaptive explanation can be offered to explain any given transfer in terms of ergonomic efficiency (e.g., transfers on the trails, leaf cutting and dropping, and transfers in the fungus gardens) (Hart & Ratnieks, 2000, 2001a; Anderson & Jadin, 2001), efficacy (e.g., leaf caching at the entrance in response to processing bottlenecks), or colony hygiene (transfers associated with waste management) (Hart & Ratnieks, 2001c, 2002a). However, in many cases, the actual adaptive benefit of task partitioning has yet to be demonstrated, and a number of untested hypotheses remain.

Indirect Transfer of Leaves

A prominent feature of task partitioning in leaf foraging is the relatively high incidence of indirect transfer, usually mediated via caches of leaves, at various points along the trails leading from forage sites back to the nest. Several forces are likely to be driving cache formation on foraging trails in these species, and in some cases these forces may not be acting independently. Certainly bottlenecks in leaf processing within the fungus gardens can drive cache formation near the nest entrance. Allowing leaves to stand for some time outside the nest may help to reduce the toxicity of plant-produced fungicides that are harmful to the mutualistic fungus, and this may also be a factor in cache formation, particularly at the nest entrance. However, in some instances "cached" leaves at the nest entrance are never retrieved by ants (something that can be seen readily both in field and laboratory colonies). If leaves are not retrieved, then, at least in these instances, it is incorrect to refer to this caching behavior as a component of partitioned foraging. Leaves are simply being abandoned.

Caches away from the nest entrance form readily on trails wherever "obstacles" are encountered (such as flooded sections and small branches) and can also form if the trail is disturbed (for example by human footfall) or where the trail gradient changes abruptly (Hart & Ratnieks, 2001a). These caches have been shown to build up through positive feedback (Hart & Ratnieks, 2000) and in many (but by no means all) cases, unladen ants on the foraging trail will collect these leaf fragments and convey them back to the nest. In these instances, "droppers" and "collectors" are connected by a transfer of material, but this only satisfies one criterion in the definition of task partitioning. The second criterion, a single task subdivided into two or more subtasks, is more problematic in this case.

The cache-mediated indirect transfer found in *Atta* foraging is said to be part of the partitioned task of foraging, but it is clearly quite different from the situation found in honeybees. Caches build up through the action of positive feedback (Hart & Ratnieks, 2000), and at some point, when the obstacle is removed, unladen foragers pick up cached fragments and convey them back to the nest. The "task partitioning" observed is merely a by-product of ants dropping leaf fragments and other ants subsequently being attracted to leaf fragments on the ground. This attraction can be readily demonstrated in the laboratory and is often exploited to study ant foraging choices and other aspects of behavior; for example, ants will readily pick up uniform leaf sections cut using a cork borer. To our knowledge, individual ants do not specialize in being "cachers" and "collectors" (unlike the foragers and receivers found in honeybees), and so the "dropping and retrieving" behavior found in *Atta* is quite different from the elegant system task partitioning linked with division of labor and specialization found in honeybee nectar foraging. Furthermore, given the fact that the vast majority of leaf fragments are cut and conveyed back to the nest in a single, uninterrupted trip by a single foraging ant without any transfer (e.g., Anderson & Jadin, 2001), the task of "leaf collecting" should properly be thought of as a single task. This again contrasts with nectar foraging in honeybees, where virtually all foragers transfer nectar to receivers and the task is therefore virtually always partitioned. It is also instructive to compare this type of organization with hygienic behavior in honeybees. The latter is reported as task partitioning by some authors, but although it has well-defined subtasks, it lacks a well-defined transfer of material. Leaf caching in leafcutting ants, on the other hand, has transfer of material but lacks well-defined subtasks.

Task Partitioning and "Bucket Brigades"

In the leafcutting ants, it makes little sense to talk of the single task of leaf carrying being "partitioned" between different ants. The putative "subtasks" are identical (carrying a leaf fragment an undefined distance along the trail before dropping it), and in many cases the ants involved are dropping leaves because of some obstacle or another and other ants are opportunistically picking up the dropped fragments. Undoubtedly, this is an interesting and potentially important component of the leafcutting ant foraging system, especially in certain circumstances. However, using "task partitioning" as a guiding organizational principle does little to increase our understanding of this very specialized foraging system, and it offers little toward a more general understanding of social insect organization. Indeed, it might be said to be jargon for jargon's sake. A more intuitive term for this type of organization might be a "bucket brigade," and indeed this term has been used to refer to similar "sequential load transport" in *Anoplolepis gracilipes* (Hsu, Yang & Horng, 2006) and "prey chain transport behaviour" reported in *Pachycondyla tarsata* (López, Agbogba & Ndiaye, 2000). Anderson, Boomsma, and Bartholdi (2002) argue that "bucket brigade" should be restricted to situations where only direct transfer occurs between the component parts of the bucket brigade. While this undoubtedly makes modeling the situation more straightforward (and relates favorably to some field observations of sequential transfers [Anderson & Jadin, 2001]), correctly we should further restrict the use of the term since bucket brigades usually involve linear, unidirectional transfer along a stationary line of individuals. Although the term "bucket brigade" is far more intuitive than "task partitioning," it is perhaps too restricted in its definition to be useful for social insect studies, involving, as they do, direct and indirect transfers, nonlinear transfers, considerable variation in the length of the "links" in the transfer "chain," and variation in the behavior of individuals acting as those links.

Leaf Cutting, Dropping, and Carrying—Task Partitioning?

A common behavior observed in field colonies of leafcutting ants involves the partitioning of leaf foraging (in general performed by a single ant) into "cutting" and "carrying," whereby individuals in the tree canopy ("cutters") cut the petioles of leaves, which then fall to the ground where other individuals ("carriers") pick them up and convey them back to the nest (reviewed in Hart, Anderson & Ratnieks, 2002). Although the cutters in the trees have not been studied to any great extent, it seems likely that they specialize in this role for at least a short time. Certainly individuals foraging "normally" in the canopy cut fragments rather than focusing on the petiole, indicating at least some role specialization in the cutters. This could be a temporary specialization or, more intriguingly, might be underpinned by caste polymorphism. Certainly it would make sense for larger ants with stronger mandibles better suited to slicing through leaf petioles to assume the role of cutters. It is less clear whether carriers are specialized "carriers" or whether they are simply foragers attracted to leaf material on the ground; but nonetheless, this seems a case where "task partitioning" is useful terminology. A single task, foraging, can be clearly identified, and in some cases that single task is subdivided into subtasks (again clearly identifiable). Furthermore, these tasks are connected through a well-defined material transfer, in this case an indirect transfer.

Leaf Processing—A Problematic Example

Leafcutting ants process leaf fragments within their fungus gardens (Weber, 1972). The sequential processing of leaves involves workers within the fungus garden initially cutting and chewing leaf fragments after they are brought into the garden by foragers. This activity produces pieces of 2 mm or less that then enter the second stage where workers chew and lick the fragments, further reducing them in size before they are inserted into the fungus garden (Wilson, 1980a). If "leaf processing" is considered to be the task, then this begins when leaf fragments enter the fungus garden and ends at their point of insertion into the fungal substrate. This task can be considered to be partitioned into at least four subtasks (movement into nest, degradation I, degradation II, insertion) that are linked by a flow of material between workers—therefore fitting the definition of task partitioning discussed above. Wilson (1980b) showed that each stage is performed by sequentially smaller workers (*Atta* are polymorphic), a good example of size-based division of labor. Observation of the processing of leaves, however, suggests that the situation may not be a straightforward "production line." Ants within the fungus garden undertake Wilson's stages 1 and 2 concurrently, with ants joining and leaving the leaf piece as it is advances through the process. Furthermore, small workers may lick and chew a leaf at the same time that larger workers cut the leaf up into smaller fragments. Given this concurrent processing, leaf processing has been identified by Hart, Anderson, and Ratnieks (2002) as "a problematic example of task partitioning." As they state, the process is without a well-defined material transfer from worker to worker, and in fact the material remains at a central place while the "sequential" subtasks (as implied by Wilson, 1980a) are performed concurrently at this location. In this respect, leaf processing is organizationally similar to hygienic behavior in honeybees. In both cases, there is a well-defined set of "connected" tasks (capping and removal of brood; fragmenting and processing leaf fragments), but the material transfers are ill-defined, relying on concurrent working at a common location rather than an explicit handing on of material.

Division of Labor and Task Partitioning—Blurring the Boundaries

The difference between "division of labor" (the division of the work force between the tasks performed by a colony) and "task partitioning" (the division of a task into linked subtasks connected by material transfer) is, at least semantically, very clear. Also, it is clear that division of labor can act together with task partitioning—the two are by no means mutually exclusive. This is well illustrated by perhaps the clearest example of task partitioning: honeybee nectar foraging. Receivers are typically younger than foragers, and both task-groups specialize in their role for a period of time (an example of age polyethism in work organization). In cases where work organization falls under the rigid definition of task partitioning suggested above (encompassing both subdivision of tasks and material transfer) (e.g., nectar foraging in honeybees and stingless bees, harvesting of honeydew from aphids in *Oecophylla longinoda*, waste management in *Atta cephalotes*), there is also clear division of labor. In waste management in *Atta cephalotes*, for example, waste is carried from fungus chambers to just outside waste chambers by "carriers," where it is cached ready for "waste heap workers" to collect and deposit on the waste heap (Hart & Ratnieks, 2001c). On occasions, waste is transferred directly to the heap by carriers without caching, indicating that this task could be organized without partitioning. It is hypothesized that partitioning enables colonies to isolate the heap and to prevent back contamination of the fungus gardens that would occur if contaminated waste heap ants had to collect waste from the gardens rather than the cache. Waste heap workers and carriers are clearly defined roles, and this is a good example of division of labor. Indeed it seems likely that carriers progress to waste heap working in accordance with Schmid-Hempel's (1998) conveyor belt model of colony organization.

Given the important underlying principle of division of labor in many putative examples of task partitioning and the importance of the connection of tasks, does the term "task partitioning" really add anything to our understanding? To put it another way, is it necessary to invent additional terminology (which is used to mean different things in other disciplines such as computer science and engineering) where "connected division of labor" might be more understandable and more useful? To help untangle this, we can identify three types of situations that have been identified as "task partitioning" in the literature, only one of which is actually valid.

1. Clearly defined subtasks connected by a well-defined transfer

The use of the term "task partitioning" in these situations, including nectar foraging in honeybees and stingless bees and waste management in *Atta spp*, emphasizes the additional level of organization involved in subdividing and connecting tasks and is therefore useful terminology.

2. Clearly defined subtasks but poorly defined transfers

Task partitioning has been used to describe hygienic behavior in honeybees and leaf processing in *Atta*, but in both these cases satisfying the material transfer component of the definition outlined above in any useful way is difficult. If a description is needed, then these cases are better described as "connected (or codependent or contingent) tasks in division of labor," since task partitioning is neither strictly correct nor especially helpful in describing the level of organization observed.

3. Poorly defined subtasks but clearly defined transfers

In the case of sequential leaf transfer (either directly, or indirectly via caches) in leaf-cutting ant foraging, sequential load transport in *Anoplolepsis gracilipes*, and other examples, the transfer of material between individuals is central to the organization observed, but there is no clear subdivision of the overall task into subtasks. Leaf caching in *Atta* may be a flexible and useful response to changing conditions, but it is not a "task divided into subtasks connected by material transfer" in any meaningful way. The task of "leaf carrying" does not subdivide into N clearly different carrying portions (the notional subtasks in this situation), completed by P different ants (where $P \leq N$). In cases like this, the use of the term "bucket brigade" has been suggested. Anderson, Boomsma, and Bartholdi (2002) consider bucket brigades to be (1) an example of task partitioning and (2) defined only for cases of direct transfer of material between individuals within the bucket brigade. I propose that these situations are not task partitioning under the strict definition (there are no clearly defined subtasks), and that the term "bucket brigade" has a specific definition outside of social insect studies—the sequential, unidirectional direct transfer of material between stationary individuals to convey that material between locations. "Sequential load transport" (suggested by Hsu, Yang & Horng, 2006) is a more suitable term in these cases and should be defined as the direct and/or indirect transfer of material between individuals within a transport chain.

Bucket brigades and similar dynamics introduce another aspect to task partitioning that has been largely overlooked. Transfers between individuals, whether direct or indirect, may not be adaptive. In *Atta*, caching leaves at a blockage on the trail or at the nest entrance during a processing bottleneck provides a clear benefit if these leaves are later recovered. However, it would be a very Panglossian approach to assume a priori that any dropped leaves, or any direct transfer between individuals, is part of an adaptive foraging system. Anderson and Jadin (2001) showed that leaves traveled significantly faster post-transfer, which may indicate that transfers increase the rate of delivery of leaf fragments to the nest. However, Burd and Howard (2008) found that the highest colony-wide rate of processing was achieved when leaf-fragment sizes were suboptimal for individual delivery rate by foragers, indicating that getting resources back to the nest as rapidly as possible may not be the best way to organize foraging. In the laboratory, *Atta* foragers can often be seen pulling on the same leaf fragment (foragers are attracted to fragments, sometimes even when they are being carried by another ant). A larger ant is likely to win this tussle, and by virtue of a longer stride length will convey the leaf fragment to the nest quicker than the smaller ant could. However, given the time spent "transferring" the fragment, and the fact that absolute rate of leaf delivery is perhaps not an important factor in this system, the adaptive value of such "transfers" is open to question. I argue above that bucket brigades should not be treated as example of task partitioning, but I would also argue that until transfers at different points can be shown to have an adaptive value, they should not a priori be considered adaptive.

Conclusion

Task partitioning in social insects was defined in the early literature as two or more individuals contributing sequentially to a particular task, or the situation when a piece of work is divided among two or more workers. These definitions are problematic since they potentially identify many tasks within the colony as being "partitioned" while adding little to our understanding. A far more useful (and, in practice, more used) definition of this form of work organization (typified by the nectar transfer system of the honeybee) involves the subdivision of a task into clearly defined subtasks, linked together by the transfer of material. Under this definition, the differences between task partitioning and division of labor become much clearer, as do the advantages of the terminology in increasing understanding. When task partitioning "proper" occurs, the organization of the task itself (and its partitioning into subtasks) can be seen as an additional level of organizational complexity within the colony. What is more, focusing on the subtasks and on the transfer of material emphasizes biologically important and organizationally interesting features, such as caching. However, employing a more rigid definition (which was in any case being used implicitly in most early work) makes task partitioning as a concept limited with respect to both task types and species. Overall, although widespread in terms of general taxonomic occurrence, it is not a common feature across social insect species. It has proven to be useful in larger and more complex colonies, particularly the leafcutting ants; and, although the types of tasks that may involve partitioning are limited, task partitioning may nonetheless be involved in a substantial proportion of the total work undertaken by species where it is present.

Though some reported instances of task partitioning are correctly and usefully described as such, some reported instances are best thought of in other ways. Tasks with clear subtasks but poorly defined transfers (such as hygienic behavior in the honeybee) are perhaps better thought within the framework provided by division of labor, whereas tasks with well-defined transfers but unclear subtasks are in general better referred to as "sequential load transport." Task partitioning is a valid and useful concept, but its use should be restricted to those situations where a material-handling task is subdivided into clear subtasks that are linked by the transfer of the material involved. Its relative rarity is not a failure of the concept but a reflection of the relative rarity of the type of colonies performing the kind of complex material-handling tasks that together can benefit from this "extra" level of organization.

Note

1. A literature search on ISI Web of Knowledge was conducted (November 2009) based on those papers citing Ratnieks and Anderson (1999a). Simply searching for task partitioning brings up well over 1,000 hits because the term is widely used in engineering and computing contexts; a cited reference search of Ratnieks and Anderson (1999a) would be most likely to reveal reports of novel task partitioning examples. The search yielded 59 citations, of which 45 involved task partitioning as an organizational term in a social insect context. Fourteen papers (31%) were concerned with theoretical aspects of social insect organization, including review articles or robotics applications, 11 (24%) were directly involved with foraging in honeybees (solely *A. mellifera*), and 13 (29%) focused on leafcutting ants (11 on *Atta*, 2 on *Acromyrmex*). In total, 53 percent of papers were focused on honeybees (again, only *A. mellifera*) and leafcutting ants, and a total of 84 percent were focused solely on these two taxa, theoretical studies, reviews, or robotics. A further two papers (4%) used "task partitioning" in reference to bucket brigades (see below), two (4%) focused on colony size and organization, and two (4%) discussed foraging examples in taxa already recorded by Ratnieks and Anderson (1999a) (*Ectatomma ruidum* and *Messor*). Only one paper (2%) introduced task partitioning in a previously unrecorded species (*Aphaenogaster rudis*) (Banschbach et al., 2006).

References

Anderson, C., & Franks, N. R. (2001). Teams in animal societies. *Behavioral Ecology, 12*, 534–540.

Anderson, C., & Jadin, J. L. (2001). The adaptive value of leaf transfer in *Atta colombica*. *Insectes Sociaux, 48*, 404–405.

Anderson, C., & Ratnieks, F. L. W. (1999a). Worker allocation in insect societies: Coordination of nectar foragers and nectar receivers in honey bee (*Apis mellifera*) colonies. *Behavioral Ecology and Sociobiology, 46*, 73–81.

Anderson, C., & Ratnieks, F. L. W. (1999b). Task partitioning in insect societies (I): Effect of colony size on queueing delay and colony ergonomic efficiency. *American Naturalist, 154*, 521–535.

Anderson, C., & Ratnieks, F. L. W. (1999c). Task partitioning in foraging: General principles, efficiency and information reliability of queueing delays. In C. Detrain, J. L. Deneubourg, & J. M. Pasteels (Eds.), *Information processing in social insects* (pp. 31–50). Basel: Birkhäuser.

Anderson, C., Boomsma, J. J., & Bartholdi, J. J. (2002). Task partitioning in insect societies: Bucket brigades. *Insectes Sociaux, 49*, 171–180.

Arathi, H. S., Ho, G., & Spivak, M. (2006). Inefficient task partitioning among nonhygienic honeybees, *Apis mellifera L.*, and implications for disease transmission. *Animal Behaviour, 72*, 431–438.

Banschbach, V., Brunelle, A., Bartlett, K., Grivetti, J., & Yeamans, R. (2006). Tool use by the forest ant *Aphaenogaster rudis*: Ecology and task allocation. *Insectes Sociaux, 53*, 463–471.

Belt, T. (1928). *The naturalist in Nicaragua*. London: Everyman Library, Dent.

Burd, M., & Howard, J. J. (2008). Optimality in a partitioned task performed by social insects. *Biology Letters, 4*, 627–629.

Hart, A. G., & Ratnieks, F. L. W. (2000). Leaf caching in leafcutting ants *Atta*: Discrete cache formation through positive feedback. *Animal Behaviour, 59*, 587–591.

Hart, A. G., & Ratnieks, F. L. W. (2001a). Leaf caching in the leafcutting ant *Atta colombica*: Organizational shift, task partitioning and making the best of a bad job. *Animal Behaviour, 62*, 227–234.

Hart, A. G., & Ratnieks, F. L. W. (2001b). Why do honey-bee (*Apis mellifera*) foragers transfer nectar to several receivers? Information improvement through multiple sampling in a biological system. *Behavioral Ecology and Sociobiology, 49*, 244–250.

Hart, A. G., & Ratnieks, F. L. W. (2001c). Task partitioning, division of labor, and nest compartmentalisation collectively isolate hazardous waste in the leafcutting ant *Atta cephalotes*. *Behavioral Ecology and Sociobiology, 48*, 387–392.

Hart, A. G., & Ratnieks, F. L. W. (2002a). Waste management in the leaf-cutting ant *Atta colombica*. *Behavioral Ecology, 13*, 224–231.

Hart, A. G., & Ratnieks, F. L. W. (2002b). Task partitioned nectar transfer in stingless bees: Work organisation in a phylogenetic context. *Ecological Entomology, 27*, 163–168.

Hart, A. G., Anderson, C., & Ratnieks, F. L. W. (2002). Task partitioning in leafcutting ants. *Acta Ethologica, 5*, 1–11.

Hölldobler, B. (1984). The wonderfully diverse ways of the ant. *National Geographic, 165*, 778–813.

Hölldobler, B., & Wilson, E. O. (1990). *The ants*. Cambridge, MA: Harvard University Press.

Hsu, H. Y., Yang, R. L., & Horng, S. B. (2006). Sequential load transport in *Anoplolepis gracilipes* (Hymenoptera: Formicidae): A novel case of non-cooperation. *Sociobiology, 48*, 571–584.

Jandt, J. M., & Dornhaus, A. (2009). Spatial organization and division of labor in the bumblebee *Bombus impatiens*. *Animal Behaviour, 77*, 641–651.

Jeanne, R. L. (1986a). The evolution of the organization of work in social insects. *Monitore Zoologico Italiano, 20*, 119–133.

Jeanne, R. L. (1986b). The organization of work in *Polybia occidentalis*: The costs and benefits of specialization in a social wasp. *Behavioral Ecology and Sociobiology, 19*, 333–341.

Jeanne, R. L. (1991). Polytheism. In K. G. Ross & R. W. Matthews (Eds.), *The social biology of wasps* (pp. 389–425). Ithaca, NY: Cornell University Press.

Leuthold, R. H., Bruinsma, O., & van Huis, A. (1976). Optical and pheromonal orientation and memory for homing distance in the harvester termite *Hodotermes mossambicus* (Hagen) (Isopt., Hodotermitidae). *Behavioral Ecology and Sociobiology, 1*, 127–139.

López, F., Agbogba, C., & Ndiaye, I. (2000). Prey chain transfer behaviour in the African stink ant, *Pachycondyla tarsata* Fabr. *Insectes Sociaux, 47*, 337–342.

Maurizio, A. (1975). How bees make honey. In E. Crane (Ed.), *Honey: A comprehensive survey* (pp. 77–105). London: International Bee Research Association.

Michener, C. D. (1974). *The social behaviour of bees: A comparative study*. Cambridge, MA: Harvard University Press.

Oster, G. F., & Wilson, E. O. (1978). *Caste and ecology in the social insects*. Princeton, NJ: Princeton University Press.

Park, W. (1925). The storing and ripening of honey by honeybees. *Journal of Economic Entomology, 18*, 405–410.

Ratnieks, F. L. W., & Anderson, C. (1999a). Task partitioning in insect societies. *Insectes Sociaux, 46*, 95–108.

Ratnieks, F. L. W., & Anderson, C. (1999b). Task partitioning in insect societies (II): Use of queueing delay information in recruitment. *American Naturalist, 154*, 536–548.

Robinson, E. J. H., Feinerman, O., & Franks, N. R. (2009). Flexible task allocation and the organization of work in ants. *Proceedings: Biological Sciences, 276*, 4373–4380.

Schatz, B., Lachaud, J. P., & Beugnon, G. (1996). Polyethism within hunters of the ponerine ant, *Ectatomma ruidum* Roger (Formicidae, Ponerinae). *Insectes Sociaux, 43*, 111–118.

Schmid-Hempel, P. (1998). *Parasites in social insects*. Princeton, NJ: Princeton University Press.

Seeley, T. D. (1992). The tremble dance of the honeybee: Message and meanings. *Behavioral Ecology and Sociobiology, 31*, 375–383.

Seeley, T. D. (1995). *The wisdom of the hive*. Cambridge, MA: Harvard University Press.

Smith, A. (1776). *An Inquiry into the Nature and Causes of the Wealth of Nations*. The Adam Smith Institute, London: http://www.econlib.org/library/Smith/smWN.html.

Smith, C. R., Toth, A. L., Suarez, A. V., & Robinson, G. E. (2008). Genetic and genomic analyses of the division of labor in insect societies. *Nature Reviews. Genetics, 9*, 735–748.

Sudd, J. H. (1965). The transport of prey by ants. *Behaviour, 25*, 234–271.

von Frisch, K. (1967). *The dance language and orientation of bees*. Cambridge, MA: Harvard University Press.

Weber, N. A. (1972). *Gardening ants: The attines*. Philadelphia, PA: American Philosophical Society.

Wilson, E. O. (1980a). Caste and division of labor in leaf-cutter ants (Hymenoptera: Formicidae: *Atta*) I. The overall pattern in *A. sexdens*. *Behavioral Ecology and Sociobiology, 7*, 143–156.

Wilson, E. O. (1980b). Caste and division of labor in leaf-cutter ants (*Hymenoptera: Formicidae: Atta*). II. The ergonomic optimization of leaf cutting. *Behavioral Ecology and Sociobiology, 7*, 157–165.

12 Cooperative Breeding in Birds: Toward a Richer Conceptual Framework

Andrew Cockburn

In most organisms, care for offspring stops once the egg is provisioned, or else is dominated by just one parent. However, in some taxa, cooperation occurs between both the male and female parent, and parental care is shared. True cooperative breeding, where more than two individuals combine to rear a single brood of offspring, has also evolved many times, though its phylogenetic distribution is patchy, being particularly common in some groups of insects and in birds. By simple arithmetic, some of the individual carers in cooperative groups are assisting young that are not their own. As parental care is costly, this behavior is altruistic and requires special explanation within evolutionary theory, which predicts that behavior will be motivated by self-interest.

Cooperative breeding is troubling to theory precisely because of the stark simplicity of the situation. Consider eclectus parrots, where females are utterly dependent on male provisioning to feed themselves and their offspring. Females mate with many males each mating period and receive food for as much as ten months of the year as a consequence (Heinsohn et al., 2007). Yet females lay just two eggs, so most of the males cannot have sired the brood they provision.

Attempts to understand cooperative breeding have given rise to sophisticated theory (e.g., Gardner, West & Wild, 2011), increasingly clever experiments (e.g., Baglione et al., 2002), and some of the most intricate long-term studies of free-living organisms ever attempted, particularly among birds (Stacey & Koenig, 1990), where about 9 percent of species breed cooperatively (Cockburn, 2006). My purpose is not to review that literature, but to analyze how preconceptions and assumptions may have shaped the study of cooperative breeding. I will argue that the original questions that have guided the study of cooperative breeding in birds are declining sharply in heuristic value, and are increasingly hindering attention to a new set of questions, solutions to which will afford an opportunity for conceptual and theoretical synthesis, yet allow recognition of the incredible diversity of cooperative systems. In this review, I focus on the empirical literature on cooperative breeding in birds, though many of arguments recall debates that are well developed in the literature of social insects.

I also unashamedly concentrate in some detail on empirical examples from my own research group, as it is in these that I understand from bitter experience where conceptual straitjackets can be most misleading.

The Standard Framework: A Problem in Two Steps

The primary question of why some bird species breed cooperatively has traditionally been structured around two questions (Emlen, 1978; Brown, 1987). First, why in some species do young remain into adulthood on the territory on which they were born (natal philopatry) rather than dispersing to pursue independent reproduction? Second, given philopatry, why do they assist the adults on their natal territory to rear additional broods?

Philopatry

The problem of philopatry helps illuminate two problems that will recur throughout this discussion. The first difficulty is one of untested assumption. It makes sense to focus on philopatry as the critical phenomenon to be resolved only if cooperative breeding is inevitably (or at least largely) associated with individuals remaining with their parents on the natal territory, which leads to the formation of family groups (Emlen, 1995; Covas & Griesser, 2007; Ekman, 2006). Although it is certainly true that most *well-studied* species conform to this model, the issue of prevalence remains unresolved. Biased sampling can arise because species that live in family groups on year-round territories are unusually easy to study, for example by allowing easy detection of whether the size of the breeding group is greater than two. It also seems sensible to try and understand philopatry in taxa where it occurs, but assuming that philopatry and cooperative breeding are intimately intertwined can constrain the choice of study organisms and the questions addressed in analysis. Indeed, it is now clear that cooperative breeding occurs in many other circumstances, including family groups, completely unrelated individuals, and bizarre mixtures of the two (Cockburn, 1998). Without a great deal of additional work, it will be impossible to assign most cooperative breeders to these three categories.

It is also true that remaining philopatric to the natal territory is not a prerequisite for cooperative breeding, as we have seen in the case of eclectus parrots that move throughout the forest and feed females at more than one nest. For example, a recent compilation shows the prevalence of cooperative breeding is much higher than has previously been supposed (Cockburn, 2006). Many potential or actual cooperative breeders such as Asian babblers and neotropical tanagers do not live their lives on year-round territories, but spend a great part of their lives foraging in mobile multispecies feeding flocks (bird waves), only breaking away from the flocks to breed. None of these species have ever been the subject of detailed study, which is scarcely surprising in view of the formidable difficulty of studying animals whose breeding site is unpredictable and whose nest may be hidden in the rainforest canopy.

However, an indication of the possibilities comes from weavers in the genus *Malimbus*. These birds were among the first tropical species whose nesting behavior was filmed, confirming the startling suspicion that whereas several birds cooperate to weave an impressive nest, only a single pair appeared to contribute to rearing the young (Brosset, 1978). Alexander Skutch (1987) went on to use this half-hearted assistance as the lowest possible level of cooperative behavior in his proposed hierarchy based on group selectionist philosophy, in which he asserted that helping other species was the pinnacle of the evolution of cooperation. However, recent study of the endangered *Malimbus ballmanni* has shown that the situation is even more curious than originally supposed. Although the supernumerary birds that do not initially provision the young follow the flock away from the nest that they have helped build, the mixed species flock may return many days later to the vicinity of the same nest. The supernumeraries then start provisioning, but only until the flock moves on, so they may help with rearing chicks for just a few hours in the entire nestling period (Gatter & Gardner, 1993). It remains unclear whether this help or the assistance with construction of the nest affects the fitness of the pair they assist.

The second difficulty is that while the evolution of natal philopatry is profoundly fascinating and a rewarding and insufficiently developed area of research, its relevance to understanding cooperative breeding may have been overestimated. As for many aspects of cooperative breeding (and studies of mating systems in general), investigation of philopatry in different cooperative species provides reasonably unequivocal support for all the competing hypotheses that have been proposed to explain its prevalence (e.g., Komdeur, 1992). In many cases this means that there is robust evidence that all these hypotheses are also untrue, in certain species in certain circumstances. Hence understanding philopatry is less critical than recognizing that it occurs.

To illustrate how confusion can arise, consider the hypothesis that philopatry arises and leads to cooperative breeding because opportunities to breed elsewhere are unavailable (the habitat saturation or ecological constraints hypothesis). There are two classical textbook examples of such constraints. First, in the Seychelles warbler, philopatry of daughters only arises once a certain population density is exceeded, implicating crowding in the occurrence of cooperative breeding (Komdeur, 1992). Second, in superb fairy-wrens, experimental removal of breeding pairs from territories fails to elicit any interest in the vacant territory from neighboring supernumerary males, but return of the female immediately prompts a neighbor to move in and pair with her, suggesting that the unavailability of unpaired females is the constraint on dispersal (Pruett-Jones & Lewis, 1990).

However, where one only one sex remains philopatric, as is true for both these species, the assumption of constraints can often produce logical inconsistency. In

Seychelles warblers, supernumerary males bud off small territories at the boundary of their natal territory to establish a vacancy (Komdeur & Edelaar, 2001a), raising the question of why females cannot use similar tactics. A likely explanation is that philopatric females often contribute eggs to the nest they help rear (Richardson, Burke & Komdeur, 2002), providing them with direct benefits that are more important than constraints in influencing the decision to remain at home. In superb fairy-wrens, the shortfall of females is itself largely a consequence of the obligate dispersal of young females (Cockburn et al., 2003), as most of these females die during dispersal as the result of the failure to obtain a vacancy (Cockburn et al., 2008b). We are therefore left with the dilemma that dispersal by one sex causes the shortfall in the other, which is hardly consistent with an insoluble constraint. The inconsistency is heightened by the red-winged fairy-wren, *Malurus elegans*, in which supernumerary females are also philopatric, and the incidence of cooperative breeding is much higher than in other *Malurus* species (Russell & Rowley, 2000), suggesting that in the genus the shortfall of one sex cannot be the primary influence shaping cooperation. Further clear evidence that prolonged philopatry can be beneficial rather than enforced through a constraint comes from species where there are limits to the numbers of brood-mates that are tolerated on the territory, and the philopatric individuals enjoy advantages over their dispersive brood-mates. In the best-studied example, direct lifetime advantages of philopatry have been convincingly shown in the Siberian jay (Ekman et al., 1999), and appear to be linked to the ability of the parents to procure safe access to food for their offspring (Ekman et al., 2000). Notably, this species does not breed cooperatively, though one of its congeners has been shown to do so (Jing, Sun & Fang, 2003).

Hence we are left with little insight other than the straightforward observation that in species with prolonged natal philopatry, there is the possibility that supernumerary family members will be available to help their parents raise offspring. However, there are also species with prolonged natal philopatry where help does not occur, and cooperative breeders where there is no natal philopatry; and we are poor at predicting why this is so.

Help

Numerous hypotheses have been suggested to explain why more than two individuals can care for offspring, but as I have reviewed this topic elsewhere (Cockburn, 1998), I will summarize these only briefly. Recent interest has surrounded five possibilities.

First, helping may provide future benefits, such as practice in rearing young, or production of a future workforce. Helping behavior may represent "rent" that allows courtship of future mates, inheritance of all or part of the territory, or a safe haven to wait for neighboring vacancies. Understanding how to assess these multigenerational advantages is in its infancy (Kingma, Hall & Peters, 2011), and this is an area that will unquestionably come under greater scrutiny.

Second, subordinates can gain direct access to at least some reproduction, subject to the important constraint that incestuous mating is rare or absent in most cooperative breeders (Cockburn, 2004; Koenig & Haydock, 2004). Where this constraint does not apply, it has long been known that in some species direct access to reproduction is shared in a reasonably egalitarian manner, though in species with low fecundity, reproduction needs to be summed over several broods to understand the extent of sharing (Haydock & Koenig, 2002; Millar et al., 1994). In other species, dominants gain most success, though in the majority of well-studied species, subordinates do gain significant access to reproduction unless constrained by incest avoidance (Cockburn, 2004). Why some species remain completely monogamous remains poorly understood. What is most bewildering about the contest between group members for reproduction is that it occurs within very many highly distinctive mating systems, and there is much less convergence between unrelated taxa than might have been anticipated. In an earlier review I showed that it was possible to recognize 22 distinct mating systems among the 30 species that had been subject to detailed molecular dissection of mating success (Cockburn, 2004), and that where different species had the same system, they were generally close relatives. Indeed, the sheer variety of cooperative systems, some of which will be illustrated below, represents a fundamental failure of our ability to understand the evolution and maintenance of cooperative breeding.

Third, there can be indirect (but immediate) access to fitness, by helping relatives to rear young. There is an obvious nexus between this hypothesis and the ability of philopatry to promote kin association. However, it is increasingly clear that kin association can arise from diverse mechanisms in addition to philopatry (Hatchwell, 2009, 2010). Kin selection has played a crucial role in the development of theory pertaining to cooperative breeding, and the eusociality found in other taxa (Hamilton, 1964a,b). This theory and its empirical basis have recently been strongly attacked (Nowak, Tarnita & Wilson, 2010), but such attacks appear to be based on a straw-man trivialization of theory, and a dramatic underestimation of the power of inclusive fitness theory to make predictions that can and have been tested successfully (Gardner, West & Wild, 2011).

Indeed, studies of bird species have yielded sophisticated contextual and experimental evidence for kin-selected helping behavior. Remarkably, the best support has been from societies quite different from the model where philopatric young assist their parents to rear offspring. The most compelling results are from research by Hatchwell and his colleagues on long-tailed tits, where breeders that have failed in their own independent attempts seek out a relative with an intact nest and help with that breeding effort (Hatchwell & Sharp, 2006). The active choice confirms the crucial role of kinship. Although it has been argued that such failed breeder systems represent a transitional state to the complex state where helpers never attempt independent reproduction (Ligon & Burt, 2004), I know of no case where that is strongly indicated.

By contrast, there are many cases where such a transition is unlikely, primarily in species that produce several broods each season, so the failure of their first nest does not remove options for future reproduction.

Good evidence for kin effects also occurs in migratory species (Lessells, Avery & Krebs, 1994), and birds that breed in colonies (Covas et al., 2006). In another remarkable study from carrion crow breeding groups comprised of both philopatric and immigrant supernumeraries, the best evidence comes from immigrants, which preferred to help young to which they were likely to be most closely related, suggesting that they had actively chosen groups comprising relatives (Baglione et al., 2003).

It is true that in these cases active choice increases the probability of unambiguous detection of kinship effects. Indeed, definitive tests are difficult where only philopatric young have the opportunity to help, as there is no context to evaluate helping in circumstances where relatedness is low. It is nonetheless important to realize that kin selection can still work without any kin discrimination, provided that relatedness on average is reasonably high and that the actions of the subordinates enhance the output of the relatives they are assisting. However, it can be difficult to demonstrate such enhancement, because experiments often perturb far more than the amount of provisioning (Dunn & Cockburn, 1996; Jamieson & Quinn, 1997), and correlative studies are plagued by numerous confounding variables, most notably the difficulty of demonstrating unambiguously that extra provisioners improve productivity (Cockburn et al., 2008c). In philopatric species, high group productivity in the presence of help could indicate that supernumeraries provide a benefit, but it could also suggest that productive territories or parents are more likely to acquire supernumeraries. Alternatively, an absence of a correlation between help and productivity could arise because helpers only assist when help is beneficial (Reyer & Westerterp, 1985). Some progress had been made in developing statistical methods to overcome these problems, but the data requirements are formidable, and the methods remain rarely applied (Cockburn et al., 2008c).

Finally, despite claims to the contrary, kinship is irrelevant or of uncertain importance in the contemporary social systems of other avian cooperative breeders, most obviously where unrelated individuals collaborate (e.g., greater anis, Riehl, 2011), but also in many cases of "family" groups, where helping is provided even where relatedness is unlikely, such as the fairy-wrens (Dunn, Cockburn & Mulder, 1995), or in an extreme case, helpers unrelated to the dominants are most likely to provision (white-browed scrubwrens, Magrath & Whittingham, 1997). These examples come from families where cooperative breeding is universal (Crotophagidae, Maluridae) or fairly common (Acanthizidae), suggesting that either kinship was unimportant in the evolution of complex social organization in some of the main clades where it now occurs, or perhaps more plausibly, that it has declined in importance as social evolution proceeded.

Summary

In summary, available evidence suggests that cooperative societies have diverse origins and evolve along idiosyncratic trajectories. Explaining this variation is unlikely to be achieved by focusing exclusively on what is an important but nonetheless specific case: natal philopatry leading to the formation of family groups on year-round territories. Even within these species, testing any of the links in the causal chain (constraint leads to philopatry leads to kin association leads to kin-selected cooperation) is technically challenging and rarely accomplished.

Can We Unpack the Diversity Using Comparative Analysis?

The usual approach to dissection of diversity is broad-scale comparative analysis. However, avian cooperative breeding has been quite resistant to such approaches. The difficulties can be illustrated by reviewing a number of recent analyses that have claimed to discern general patterns in the distribution or evolution of cooperative breeding.

First, in attempting to test particular hypotheses, it is possible to construct the analysis in ways that are biased toward supporting the conventional framework. Cornwallis and his colleagues (2010) recently analyzed what has become known as the monogamy hypothesis. According to this view, which originated in studies of cooperatively breeding insects, kinship between potential helpers and the broods they provision is higher when parents mate faithfully, as subordinates are likely to be helping rear their full sibs, whereas if the female mates promiscuously, the subordinates will often be caring just for half-sibs (Boomsma, 2007; Boomsma et al., 2011). Cooperation is therefore expected to arise more commonly in the case of true fidelity. Two of the results obtained by Cornwallis et al. (2010) provide support for this hypothesis. First, as predicted, evolutionary transitions to cooperative breeding are most likely when females mate monogamously, and loss of cooperative breeding is most likely when females are unfaithful to their mates. Second, in a more subtle result, they showed that there was little evidence for sophisticated kin discrimination when females are faithful, which is expected when a rule of thumb that presumes relatedness is reliable, not when fidelity was very low, when presumption of relatedness is unjustified. However, kin discrimination occurred at intermediate levels of fidelity. This study is fascinating, and will provoke further work, but in order to construct these contrasts it was necessary to define cooperative breeding as being confined to family groups. Hence, the construction of the question is heavily biased to support kin-based explanations of cooperation (Cockburn, 2010).

Second, there is the difficulty of how to treat the underlying variation in the trait. The simple and most commonly adopted approach is to treat cooperative breeding as a binary (yes/no) trait. For example, Jetz and Rubenstein (2011) recently used a global analysis to argue that cooperative breeding is associated with interannual variation in rainfall, at least in the passerine birds that dominate avian diversity and include many transitions to and from cooperative breeding. In constructing this test all other parental care systems are lumped together as a single alternative state, so brood parasitism of the nests of other species and uniparental care by females or males are treated as equivalent to biparental care, which, as we have just seen, does not seem particularly sensible, as transition to group breeding from uniparental care is not expected because the pattern of relatedness in this case is less clear than when young are helping rear the young of their parents.

Third, particularly when sample sizes are extremely high, it is possible to obtain highly significant associations that provide negligible additional biological information. In the Jetz and Rubenstein (2011) study, the amount of extra variation explained over and above the well-known but poorly understood phylogenetic concentration of cooperative breeding is very slight (Cockburn & Russell, 2011).

Finally, choice of explanatory variables is itself problematic. Jetz and Rubenstein (2011) attempted a global analysis extrapolated from a more focused study on African starlings (Rubenstein & Lovette, 2007), which had implicated interannual variability as the best predictor of which species are cooperative. One of the difficulties in a global analysis is that most climatic and many other ecological variables will be strongly correlated with the best-known geographic predictor of cooperative breeding, a sharp decline in frequency at far northern latitudes (Cockburn, 2003; Ekman & Ericson, 2006). This not only makes interpretation of causation from correlation even more obscure than usual, but it fails to deal with the unexplained variation within the tropical and subtropical latitudes within which most birds live.

Finally, even if the correlation is the correct one, its interpretation would not be straightforward. Jetz and Rubenstein (2011) speculate that the benefits associated with help will be most pronounced in harsh conditions. However, although there is some empirical support for the contention that help is most likely and beneficial under harsh conditions (Covas, du Plessis & Doutrelant, 2008; Reyer & Westerterp, 1985), it has more often been shown that philopatry and helping are more likely under favorable conditions, perhaps because younger birds are more likely to have excess resources they can afford to donate to others (Baglione et al., 2006; Cullen, Heinsohn & Cockburn, 1996). Hence, the importance and theoretical implications of these observations remain uncertain.

These comments are not meant to indicate that comparative analysis will not help unpack the causes of cooperative breeding. However, progress may be greatest with a series of structured analyses that recognize the suite of different ways that prevalence of a trait can be influenced. For example, I have argued that cooperative taxa where at least one sex is highly philopatric may have different speciation dynamics than species that breed as pairs, because colonization of new habitats such as oceanic islands, or via migration over hostile habitat, is unlikely unless both sexes disperse (Cockburn, 2003). This leads to macroevolutionary consequences, by restricting the diversity of cooperative clades, but also macroecological effects, as assemblages in recently colonized habitats will be dominated by pair-breeders.

A New Set of Questions

What can we do about this ongoing uncertainty and lack of progress? I believe we need to introduce new questions to focus research, and abandon the straitjacket that the traditional model has imposed. Progress will be furthered the most by acknowledging variation and trying to explain it. I believe that further investigation of five intimately related questions is likely to be extremely valuable.

First, why do the mating systems of cooperative breeders diverge so dramatically? Second, why does the role of kinship in cooperative breeding seem so obvious on the twigs of the evolutionary tree (e.g., among failed breeders and migrants), whereas the contemporary importance of kinship appears to be much less obvious in some of the clades where cooperative breeding is best developed? Third, how do societies proceed from facultative cooperative breeding to an inability to breed successfully as unassisted pairs? Fourth, why does cooperative breeding persist in social systems that are riven by conflict between the separate participants? Fifth, how do the underpinnings of cooperation in species that live in family groups compare to those in groups comprising unrelated individuals, and is transition from one state to the other possible?

Here I will argue that all of these questions can be answered within a common conceptual framework. In essence, once low levels of cooperative breeding arise, new evolutionary pressures arise for the participants. Solutions to these pressures will push lineages along idiosyncratic trajectories, and potentially obscure the conditions under which cooperation first developed. Under some circumstances, these trajectories trap the lineage on a peak in the adaptive landscape, meaning that the mating system is resistant to change, so that it persists despite changes in environment and speciation.

Obligate Cooperative Breeding: The Social Cradle

Some of the new evolutionary pressures are largely positive, offering benefits to all the participants. For example, prolonged kin association allows new types of investment by parents in their offspring. This in turn provides new opportunities for young. For example, learning may be facilitated as young are under less pressure to develop the skills they need to live independently.

The family Corcoracidae is an Australian passerine group with a long evolutionary history. Despite this, the family contains just two species, both of which occur over very large geographical ranges across long latitudinal gradients. Apostlebirds occur in drier inland habitats, and white-winged choughs live in more mesic woodland habitats closer to the coast, though the range of both species overlaps extensively. Both species are primarily insectivorous, though apostlebirds have a more generalist diet, in particular with a greater dependence on seed. Despite ecological differentiation, the social organization and mating system of both species are nearly identical. They are obligate cooperative breeders, generally requiring at least four birds to rear young, as only a workforce of at least this size is capable of delivering sufficient food to the chicks (Heinsohn, 1992; Woxvold, 2004). Neither species is conventionally territorial, though the young are initially philopatric to the group in which they are born. Breeding success increases across all group sizes, creating the dynamic that small groups dwindle while large groups grow rapidly. However, conflict eventually develops within large groups, leading to formation of factions that attempt to disperse and breed.

These dynamics raise a series of problems. If the group dwindles, or drought or hostile conditions increase death rates among older birds, groups can be reduced to just juvenile birds that are incapable of breeding, or to a group incapable of successfully rearing chicks. In addition, when the size of large groups prompts dispersal, it is difficult for individuals or pairs of birds to disperse on their own, as independent reproduction would not be possible. Hence, the dispersing faction must exceed the critical group size, or the fragments must coalesce with other factions to exceed the critical group size. Genetic and behavioral studies suggest how these difficulties are resolved. In groups that have been stable for a long period of time, reproduction is dominated by just the dominant pair, but new groups that have been assembled from separate smaller groups share reproduction, though only one member of each faction ever gained success, and then only when they were supported by relatives (Woxvold & Mulder, 2008; Heinsohn et al., 2000). Garnering support therefore becomes critical, and it has been hypothesized that helping behavior itself becomes a signal of suitability as a faction partner. Young birds can scarcely afford to give up hard-won food. However, they ostentatiously carry food to the nest but then swallow it themselves, sometimes after lowering it into the bills of the nestlings. This suggests that they are signaling their suitability as a coalition member even when they cannot afford to lose food (Boland, Heinsohn & Cockburn, 1997).

The critical question is how does such extreme dependency on cooperative breeding arise in the first place? Arguments concerning the harshness of the local environment seem inadequate, given the extensive range of the birds and the stability of the system over millions of years, indicated by persistence through a speciation event after which considerable divergence in foraging habit occurred. At least in choughs, young do not learn adult foraging skills until they are four years of age (Heinsohn, Cockburn & Cunningham, 1988), and are fed for an extremely long period after fledging (Heinsohn, 1991). Prolonged learning affords many advantages, including a more nuanced exploitation of the resources in the local environment. However, it is easiest to sustain when a large workforce supports the young. This relaxed social cradle allows even slower development to an extent that is extreme among passerine birds, but also ensures a workforce for parents attempting to rear young (Heinsohn, 1991). I suspect that this trajectory has stranded the birds on an evolutionary peak in which independent reproduction is no longer possible.

Winners Are Grinners: What Stabilizes Cooperation When Interests Differ?

In contrast to the favorable opportunities afforded by such a social cradle, it is also possible for costly antagonistic relationships to develop among the members of cooperative groups. There are three distinct disadvantages of group living, the first two of which are most pronounced among family groups. First, Hamilton (1964b) argued strongly that just as population viscosity can set the scene for kin cooperation, close association also leads to competition over resources and mating opportunities that limits the advantages of associating with kin (West, Pen & Griffin, 2002). The collapse of chough groups beyond a certain size illustrates these pressures. Second, a particular problem is posed by the limits to mating opportunities where incest avoidance prevails, or the cost of inbreeding if incest is not avoided. These difficulties and their idiosyncratic consequences have been reviewed by Koenig and Haydock (2004). Finally, although studies of cooperative breeders sometimes treat gender as irrelevant, there can be strong conflict of interest between males and females, and selection for one to exploit the other (Cockburn, 2004).

Cases of sexual conflict are rife. The extreme example where conflict leads to extraordinarily complicated relationships between group members (including kin) has been uncovered through my own work on superb fairy-wrens. These birds have played a central if ultimately completely misplaced role in the classic models of cooperation emerging from natal philopatry and kin selection (Brown, 1975; Grafen, 1984). Dominant males and females form a socially monogamous relationship to defend a year-round territory. All females disperse from their natal territory to breed, but males are more faithful to their natal territory than any other species thus far studied, with dispersal by and large confined to movements to immediately adjacent territories where all the males have died (Cockburn et al., 2008b, 2003). Many females fail to gain vacancies, so the sex ratio of breeding adults is strongly male biased, and as many as four subordinate adults can live with the dominant pair (Cockburn et al., 2008b). All males contribute to defense and provisioning of the young produced by the female on the natal territory (Dunn & Cockburn, 1996).

All males also become sexually active in their first and subsequent breeding seasons (Mulder & Cockburn, 1993). However, despite having access to as many as five sexually active males on her territory, the female always mates with a male from outside the group, whom she visits on a predawn foray three days before she lays the first egg (Cockburn et al., 2009; Double & Cockburn, 2000). Although she mates with her own partner (and subordinate males unless they are her sons) when she returns from the foray, the extra-group male is most likely to sire chicks, so most young are reared by males other than their sire (Mulder et al., 1994). The female allocates a greater proportion of paternity to her mate when there are no helpers, probably because there is no alternative source of care (Mulder et al., 1994). Once she has helpers, nearly all paternity is allocated to the extra-group sires. What little paternity is made available is shared with helpers unrelated to the female, who, when present, gain about one quarter of all such success (Cockburn et al., 2008b,c).

Clearly this poses two disadvantages for the dominant males, which should undermine their willingness to breed cooperatively. Less paternity is available at home, and what is available sometimes has to be shared. However, helpers also pose an additional burden. Dominant males advertise to foraying females by singing in the dawn chorus (Dalziell & Cockburn, 2008). Helpers of attractive dominants join the chorus, and sing as close to the dominant as they are able, even though they are subject to attacks by him every single morning of the breeding season (Cockburn et al., 2009). Females arriving to mate with the dominant male are often confused by this parasitism, and helpers of attractive dominants are actually more successful at gaining extra-group parentage than average or low -quality dominants (Double & Cockburn, 2003). Helpers therefore undermine both the extra-group and within-group paths to reproductive success available to dominant males. Remarkably, there do not appear to be any compensating advantages. Helpers do not increase the productivity of the territory, which might have increased their chance of siring at least some offspring within the brood (Cockburn et al., 2008c). Males with helpers are able to feed less (Dunn & Cockburn, 1996), and do convert some of that advantage into more time courting neighboring females (Green et al., 1995), which they do by a dramatic dance display. However, there is strong evidence that female preference is finalized before the breeding season starts, so such additional display is unlikely to enhance male attractiveness (Green et al., 2000). Nor are there future benefits, as males with helpers are not more likely to survive until the following breeding season (Cockburn et al., 2008c).

By contrast, the breeding female gains considerable advantages from breeding cooperatively. Not only is the need to allocate the dominant some paternity removed (Mulder et al., 1994), but female survival to breed again is greatly enhanced (Cockburn et al., 2008c). This is initially surprising because while females with helpers reduce feeding rates, they do so to a lesser extent than males, so would be expected to achieve less benefit. This paradox has been resolved by the discovery that females achieve the survival advantage by laying smaller eggs (Russell et al., 2007). Their young are not disadvantaged, because the reduced investment in eggs is fully compensated by higher provisioning rates at nests where more than two individuals provision. Hence, females completely monopolize the advantages of cooperative breeding.

Initially such monopolization seems completely unstable, yet all but one of the comparatively old and species-rich genus *Malurus* share a mating system based on the curious combination of extra-group mating and cooperative breeding, despite being found in almost every habitat in Australia and New Guinea. The mating system is thus extremely resistant to evolutionary change.

The strongest direct evidence that these tensions eat at the heart of fairy-wren society come from experiments that trigger behavior never seen in unmanipulated populations. In the first, helpers were removed for the week leading up to egg-laying (Dunn & Cockburn, 1996). This manipulation was expected to increase confidence of paternity for the dominant males, and hence induce increased provisioning. Instead, females refused to proceed with the nesting attempt, either resorbing the eggs, abandoning them once they were laid, or divorcing the male and moving elsewhere (Dunn & Cockburn, 1996). In the second experiment, helpers were removed for 24 hours at different times of the year and then returned to the group. It was expected that removal during the nestling phase would prevent load-lightening by the adults, and force them to feed at the rate of unassisted dominants. Dominant males by and large ignored the removal outside the provisioning period, but did react when there were nestlings to be fed. However, they did not change their provisioning rate at all, but attacked the helper repeatedly on his return, and chased and harassed him for the next couple of days (Mulder & Langmore, 1993). The response was so severe that the experiment was terminated for ethical reasons. How can these results be interpreted? It seems that the dominant pair have clear expectations of the resources they can obtain from other group members. Females expect a level of provisioning from dominant and subordinate males, and dominants expect provisioning from subordinates.

Disentangling the path to evolution a system of this intricacy will always involve speculation, but I believe the following is the most plausible sequence, driven by the ploy and counterploy that arises from the conflict between males and females.

1. Females face fierce competition for vacancies and are selected to settle with a minimum of prior assessment (Cockburn et al., 2003).
2. They compensate for lack of mate choice by seeking extra-pair fertilizations (Møller, 1992).
3. Males increase courtship of extra-group females in response and are able to do this by coercing care from subordinates (Mulder & Langmore, 1993), which allows the dominants to lower their own provisioning rate (Dunn & Cockburn, 1996).
4. Females take advantage of the coercion to increase the cuckoldry rate to a level that males would not usually tolerate (Mulder et al., 1994), and, by imposing progressively stronger sexual selection on males (Dunn & Cockburn, 1999), tip the balance in male fitness toward extra-group mating success.
5. The chief target of selection changes from the brilliance of plumage and courtship to the amount of time males spend displaying, so only the small number of males that can display continuously for months before the breeding season are attractive (Cockburn et al., 2008a; Dunn & Cockburn, 1999).

6. Females need to be able to find these males when they become fertile and seek them on their territories while they are advertising at dawn (Double & Cockburn, 2000). This locks males into needing a territory in order to be successful.
7. Females appease their own mates by copulating with them on their return (Cockburn, unpublished data).
8. Sperm competition arises between the extra-group and home male, leading to production of massive ejaculates that take time to replenish (Mulder & Cockburn, 1993; Tuttle & Pruett-Jones, 2004, 2007).
9. Females mate earlier and earlier to avoid sperm depletion of the attractive extra-group males (Double & Cockburn, 2000).
10. Subordinates (and unattractive dominants) exploit the uncertainty faced by females seeking attractive males in the dark by displaying parasitically in close proximity to him (the hidden lek hypothesis), increasing their own benefits of participating in the system (Cockburn et al., 2009).

This system achieves resilience to the collapse that would ensue if dominant males refused to provision chicks, because of the implications of high variance in reproductive success. Although in fairy-wrens this heterogeneity is affected to an unusual extent among cooperative breeders by sexual selection, virtually all studies of avian cooperative breeders report high variance among individuals that leads to proliferation of particular lineages (Stacey & Koenig, 1990). Regardless of the source, the variation generates real winners and, as a corollary, and much more commonly, losers. In fairy-wrens, most males are unattractive and perform poorly, so virtually all the reproductive success they obtain comes from mating with their own partner, as it is only subordinates of attractive dominants that are likely to achieve substantial fitness gains via the parasitic route. Although within-group mating is reduced in the presence of subordinates, the dominant male is always oldest, and it is feasible that the subordinates are his only sons. Low-quality birds are thus obliged to make the best of a bad job and cooperate with the female in rearing young, allowing her to monopolize the benefits of the cooperative system.

However, it is the extreme success of the winners that drives the evolution of the society, as most offspring will be progeny of these high-quality birds. For these birds, the convoluted evolution of the society constrains their behavior in a variety of ways. First, attractive dominants require a territory on which they can be found, and can gain reproductive benefits there, even though within-pair fertilizations are only a small component of their success. Therefore, they might as well have their cake and eat it too.

This line of argument has been challenged by the discovery that purple-crowned fairy-wrens, while remaining cooperative breeders, have secondarily reverted from a society based on extra-group mating to become genetically as well as socially monogamous (Kingma et al., 2009). The only clear distinction between this species and its congeners is the configuration of its habitat, which is linear rather than spread across the landscape. The linearity arises because the species specializes on the riparian habitat that occurs along rivers and streams in northern Australia, many of which are filled with water only seasonally. Birds have no neighbors along most of the length of their territories, encountering other birds only at the ends of their territories. Kingma et al. (2009) speculate that genetic benefits from extra-group mating are unlikely to be available, as the probability that a female will have a high-quality male increases linearly instead of exponentially with the number of territories traversed, in contrast to the case of species living in a more complex landscape. The difficulty with this hypothesis is that most extra-group mating in well-studied species of fairy-wrens is over just a short distance, where this effect of the difference between the two landscape configurations would remain modest (Double & Cockburn, 2000, 2003).

An alternative hypothesis is that the system collapses because signaling by high-quality males is compromised in linear habitats, undermining the benefits to both males and females. Males advertising from beyond immediately neighboring territories can only be accessed if females fly past the display sites of males on intervening territories, sharply increasing the risk of parasitism to the males, and diminishing the likely returns to the female. The system is thus jeopardized for both parties in the mating transaction and is undermined by a condition at the top of the hierarchy of male-female conflict (selection for parasitism) rather than a lower-level condition that has been obscured by subsequent coevolution.

This model has been worked through in such length in part to illustrate the extraordinary complexity of some cooperative systems, but also to demonstrate how the endpoints of evolution can be fundamentally distinct from the founding conditions.

Can Mating Systems Transition from Families to Unrelated Groups?

What of cooperative breeding among unrelated individuals? It seems likely that conflict of interest between males and females over mating opportunities may once again be the explanation for most of these cases. In the best-studied species, the dunnock, the female incites supernumerary males to mate with her in order to persuade them to feed at her nest, which in turn increases her productivity (Davies, 1992). Males are sensitive to the risk of cuckoldry and apportion their care according to the proportion of time they have monopolized access to the female (Davies, 1992).

It has been suggested that such conflicts impose an upper limit on the number of males that could be induced to participate (Hartley & Davies, 1994). Unfortunately, the theory pertaining to this issue remains in its infancy (Johnstone, 2011), but there is empirical evidence that sometimes large coalitions of males may participate. In two of these cases (brown skuas and Galapagos hawks), there has been remarkable convergence on a system where multiple males spend a long apprenticeship in bachelor flocks on land unsuitable for breeding, and form coalitions to overthrow the males breeding on prime territories elsewhere. On gaining a territory, these birds copulate with the female in an egalitarian fashion, and reproductive dominance is only gained when the other coalition members die. However, this also exposes the long-lived victor to a risk of overthrow by a new coalition of younger birds (Faaborg et al., 1995; Millar et al., 1994). By contrast, in eclectus parrots, large numbers of males can attend an individual female, but they do so facultatively rather than by becoming a member of a permanent group (Heinsohn & Legge, 2003; Heinsohn, 2008). Two subtleties explain this unusual system. First, males can actually attend several nests. Second, there is extreme heterogeneity in female reproductive success, driven largely by failure of nesting attempts due to flooding and predation, prompting males to direct care to high-quality females and their nests (Heinsohn & Legge, 2003; Heinsohn, Legge & Endler, 2005). Perhaps the most curious feature of this system is that in this case the females have evolved to become utterly dependent on the males. In order to defend quality nest sites, they occupy the nest cavity for as much as ten months of the year and are dependent on males for their own food throughout that time.

A mitigating circumstance that unites these three species is that the males in each case are potentially very long-lived (Faaborg & Bednarz, 1990; Millar et al., 1994; Heinsohn & Legge, 2003). Thus, despite the low clutch size in each species, males are likely to gain access to parentage over time, and molecular studies reveal that success by a male in one brood does not predict his success in later broods, so when integrated over many years, all males may gain direct reproduction from attending females (Faaborg et al., 1995; Millar et al., 1994; Heinsohn et al., 2007). We can therefore predict that large cooperatively polyandrous groups are likely to be found only in long-lived species.

Occasionally, cooperative breeding among unrelated individuals becomes obligatory. Riehl (2011) has shown that greater ani pairs cooperate to lay in and provision and defend a joint nest, despite considerable tensions involving initially tossing the eggs from other pairs from the nest. Despite these costs, reproductive success increases as the number of pairs contributing increases because of enhanced defense against nest predators. Once again, success at one nest does not predict success at subsequent nests, suggesting that reproduction is shared over time.

However, the transition to this state is even more difficult to explain than other obligate cooperative societies. One possibility I believe requires further investigation is the possibility that kin association could provide a founding condition for social evolution but be completely obscured by subsequent evolution. I know of no particularly convincing demonstration that this is the case, but there are numerous cases where the signature of kinship appears to have been diminished by a progressively greater role for unrelated helpers. For example, in *Manorina* miners, as in eclectus parrots, males within colonies can help at many nests, and nests are provisioned by very large numbers of individuals. While there is an underlying signature of kinship that predicts provisioning rate, a great deal of the assistance is from unrelated individuals (McDonald, Kazem & Wright, 2009; Wright et al., 2010), which possibly gain benefits from subsequent mate acquisition. The transition from a condition where kinship is all important to one where mate acquisition becomes most important and a role for kinship disappears is certainly feasible.

Conclusions

In summary, I contend that discussions of cooperative breeding in birds have overemphasized ecological constraints and philopatry, forcing a restricted view of the extraordinary diversity of cooperative systems within birds. An alternative view focuses on the diversity and extraordinary evolutionary trajectories of cooperative lineages and recognizes that the evolutionary consequences of cooperative breeding may often obscure the founding conditions under which cooperation evolved. For example, kin selection may be a crucial founding condition but may diminish in importance within lineages. These evolutionary consequences of cooperation can be both mutualistic, allowing obligate cooperative breeding to evolve, and antagonistic, forcing complex mating arrangements. Both forces can strand lineages on an adaptive peak from which movement is difficult, stabilizing complex systems over evolutionary time. Attention to these possibilities will allow both theoretical and empirical progress.

References

Baglione, V., Canestrari, D., Marcos, J. M., & Ekman, J. (2003). Kin selection in cooperative alliances of carrion crows. *Science, 300*(5627), 1947–1949.

Baglione, V., Canestrari, D., Marcos, J. M., & Ekman, J. (2006). Experimentally increased food resources in the natal territory promote offspring philopatry and helping in cooperatively breeding carrion crows. *Proceedings of the Royal Society of London, Series B: Biological Sciences, 273*(1593), 1529–1535.

Baglione, V., Canestrari, D., Marcos, J. M., Griesser, M., & Ekman, J. (2002). History, environment, and social behaviour: Experimentally induced cooperative breeding in the carrion crow. *Proceedings of the Royal Society of London, Series B: Biological Sciences, 269*(1497), 1247–1251.

Boland, C. R. J., Heinsohn, R., & Cockburn, A. (1997). Deception by helpers in cooperatively breeding white-winged choughs and its experimental manipulation. *Behavioral Ecology and Sociobiology, 41*(4), 251–256.

Boomsma, J. J. (2007). Kin selection versus sexual selection: Why the ends do not meet. *Current Biology, 17*(16), R673–R83.

Boomsma, J. J., Beekman, M., Cornwallis, C. K., Griffin, A. S., Holman, L., Hughes, W. O. H., et al. (2011). Only full-sibling families evolved eusociality. *Nature, 471*(7339), E4–E5.

Brosset, A. (1978). Social organization and nest-building in the forest weaver birds of the genus *Malimbus (Ploceinae)*. *Ibis, 120*, 27–37.

Brown, J. L. (1975). *The evolution of behavior*. New York: W. W. Norton.

Brown, J. L. (1987). *Helping and communal breeding in birds: Ecology and evolution*. Princeton: Princeton University Press.

Cockburn, A. (1998). Evolution of helping behavior in cooperatively breeding birds. *Annual Review of Ecology and Systematics, 29*, 141–177.

Cockburn, A. (2003). Cooperative breeding in oscine passerines: Does sociality inhibit speciation? *Proceedings of the Royal Society of London, Series B: Biological Sciences, 270*(1530), 2207–2214.

Cockburn, A. (2004). Mating systems and sexual conflict. In W. D. Koenig & J. L. Dickinson (Eds.), *Ecology and evolution of cooperative breeding in birds* (pp. 81–101). Cambridge: Cambridge University Press.

Cockburn, A. (2006). Prevalence of different modes of parental care in birds. *Proceedings of the Royal Society of London, Series B: Biological Sciences, 273*(1592), 1375–1383.

Cockburn, A. (2010). Oh sibling, who art thou? *Nature, 466*(7309), 930–931.

Cockburn, A., Dalziell, A. H., Blackmore, C. J., Double, M. C., Kokko, H., Osmond, H. L., et al. (2009). Superb fairy-wren males aggregate into hidden leks to solicit extragroup fertilizations before dawn. *Behavioral Ecology, 20*(3), 501–510.

Cockburn, A., Osmond, H. L., & Double, M. C. (2008a). Swingin' in the rain: Condition dependence and sexual selection in a capricious world. *Proceedings of the Royal Society of London, Series B: Biological Sciences, 275*(1635), 605–612.

Cockburn, A., Osmond, H. L., Mulder, R. A., Double, M. C., & Green, D. J. (2008b). Demography of male reproductive queues in cooperatively breeding superb fairy-wrens *Malurus cyaneus*. *Journal of Animal Ecology, 77*(2), 297–304.

Cockburn, A., Osmond, H. L., Mulder, R. A., Green, D. J., & Double, M. C. (2003). Divorce, dispersal, and incest avoidance in the cooperatively breeding superb fairy-wren *Malurus cyaneus*. *Journal of Animal Ecology, 72*(2), 189–202.

Cockburn, A., & Russell, A. F. (2011). Cooperative breeding: A question of climate? *Current Biology, 21*(5), R195–R97.

Cockburn, A., Sims, R. A., Osmond, H. L., Green, D. J., Double, M. C., & Mulder, R. A. (2008c). Can we measure the benefits of help in cooperatively breeding birds? The case of superb fairy-wrens *Malurus cyaneus*. *Journal of Animal Ecology, 77*, 430–438.

Cornwallis, C. K., West, S. A., Davis, K. E., & Griffin, A. S. (2010). Promiscuity and the evolutionary transition to complex societies. *Nature, 466*(7309), 969–991.

Covas, R., Dalecky, A., Caizergues, A., & Doutrelant, C. (2006). Kin associations and direct vs. indirect fitness benefits in colonial cooperatively breeding sociable weavers *Philetairus socius*. *Behavioral Ecology and Sociobiology, 60*(3), 323–331.

Covas, R., du Plessis, M. A., & Doutrelant, C. (2008). Helpers in colonial cooperatively breeding sociable weavers *Philetairus socius* contribute to buffer the effects of adverse breeding conditions. *Behavioral Ecology and Sociobiology, 63*(1), 103–112.

Covas, R., & Griesser, M. (2007). Life history and the evolution of family living in birds. *Proceedings of the Royal Society of London, Series B: Biological Sciences, 274*(1616), 1349–1357.

Cullen, N. J., Heinsohn, R., & Cockburn, A. (1996). Food supplementation induces provisioning of young in cooperatively breeding white-winged choughs. *Journal of Avian Biology, 27*(1), 92–94.

Dalziell, A. H., & Cockburn, A. (2008). Dawn song in superb fairy-wrens: A bird that seeks extra-pair copulations during the dawn chorus. *Animal Behaviour, 75*, 489–500.

Davies, N. B. (1992). *Dunnock behaviour and social evolution*. Oxford: Oxford University Press.

Double, M. C., & Cockburn, A. (2003). Subordinate superb fairy-wrens (*Malurus cyaneus*) parasitize the reproductive success of attractive dominant males. *Proceedings of the Royal Society of London, Series B: Biological Sciences, 270*(1513), 379–384.

Double, M., & Cockburn, A. (2000). Pre-dawn infidelity: Females control extra-pair mating in superb fairy-wrens. *Proceedings of the Royal Society of London, Series B: Biological Sciences, 267*(1442), 465–470.

Dunn, P. O., & Cockburn, A. (1996). Evolution of male parental care in a bird with almost complete cuckoldry. *Evolution: International Journal of Organic Evolution, 50*(6), 2542–2548.

Dunn, P. O., & Cockburn, A. (1999). Extrapair mate choice and honest signaling in cooperatively breeding superb fairy-wrens. *Evolution: International Journal of Organic Evolution, 53*(3), 938–946.

Dunn, P. O., Cockburn, A., & Mulder, R. A. (1995). Fairy-wren helpers often care for young to which they are unrelated. *Proceedings of the Royal Society of London, Series B: Biological Sciences, 259*(1356), 339–343.

Ekman, J. (2006). Family living among birds. *Journal of Avian Biology, 37*(4), 289–298.

Ekman, J., Bylin, A., & Tegelstrom, H. (1999). Increased lifetime reproductive success for Siberian jay (*Perisoreus infaustus*) males with delayed dispersal. *Proceedings of the Royal Society of London, Series B: Biological Sciences, 266*(1422), 911–915.

Ekman, J., Bylin, A., & Tegelstrom, H. (2000). Parental nepotism enhances survival of retained offspring in the Siberian jay. *Behavioral Ecology, 11*(4), 416–420.

Ekman, J., & Ericson, P. G. P. (2006). Out of Gondwanaland: The evolutionary history of cooperative breeding and social behaviour among crows, magpies, jays, and allies. *Proceedings of the Royal Society of London, Series B: Biological Sciences, 273*(1590), 1117–1125.

Emlen, S. T. (1978). The evolution of cooperative breeding in birds. In J. R. Krebs & N. B. Davies (Eds.), *Behavioural ecology: An evolutionary approach* (pp. 245–281). Oxford: Blackwell.

Emlen, S. T. (1995). An evolutionary theory of the family. *Proceedings of the National Academy of Sciences of the United States of America, 92*, 8092–8099.

Faaborg, J., & Bednarz, J. C. (1990). Galápagos and Harris' hawks: divergent causes of sociality in two raptors. In P. B. Stacey & W. D. Koenig (Eds.), *Cooperative breeding in birds* (pp. 359–383). Cambridge: Cambridge University Press.

Faaborg, J., Parker, P. G., Delay, L., DeVries, T., Bednarz, J. C., Paz, S. M., et al. (1995). Confirmation of cooperative polyandry in the Galápagos hawk (*Buteo galapagoensis*). *Behavioral Ecology and Sociobiology, 36*, 83–90.

Gardner, A., West, S. A., & Wild, G. (2011). The genetical theory of kin selection. *Journal of Evolutionary Biology, 24*(5), 1020–1043.

Gatter, W., & Gardner, R. (1993). The biology of the Gola malimbe *Malimbus ballmanni* Wolters 1974. *Bird Conservation International, 3*(2), 87–103.

Grafen, A. (1984). Natural selection, kin selection, and group selection. In J. R. Krebs & N. B. Davies (Eds.), *Behavioural ecology: An evolutionary approach* (2nd ed., pp. 62–84). Oxford: Blackwell Scientific Publications.

Green, D. J., Cockburn, A., Hall, M. L., Osmond, H., & Dunn, P. O. (1995). Increased opportunities for cuckoldry may be why dominant male fairy-wrens tolerate helpers. *Proceedings of the Royal Society of London, Series B: Biological Sciences, 262*(1365), 297–303.

Green, D. J., Osmond, H. L., Double, M. C., & Cockburn, A. (2000). Display rate by male fairy-wrens (*Malurus cyaneus*) during the fertile period of females has little influence on extra-pair mate choice. *Behavioral Ecology and Sociobiology, 48*(6), 438–446.

Hamilton, W. D. (1964a). The genetical evolution of social behaviour. I. *Journal of Theoretical Biology, 7*, 1–16.

Hamilton, W. D. (1964b). The genetical evolution of social behaviour. II. *Journal of Theoretical Biology, 7*, 17–52.

Hartley, I. R., & Davies, N. B. (1994). Limits to cooperative polyandry in birds. *Proceedings of the Royal Society of London, Series B: Biological Sciences, 257*, 67–73.

Hatchwell, B. J. (2009). The evolution of cooperative breeding in birds: kinship, dispersal and life history. *Philosophical Transactions of the Royal Society of London, Series B: Biological Sciences, 364*(1533), 3217–3227.

Hatchwell, B. J. (2010). Cryptic kin selection: Kin structure in vertebrate populations and opportunities for kin-directed cooperation. *Ethology, 116*(3), 203–216.

Hatchwell, B. J., & Sharp, S. P. (2006). Kin selection, constraints, and the evolution of cooperative breeding in long-tailed tits. *Advances in the Study of Behavior, 36*(36), 355–395.

Haydock, J., & Koenig, W. D. (2002). Reproductive skew in the polygynandrous acorn woodpecker. *Proceedings of the National Academy of Sciences of the United States of America, 99*, 7178–7183.

Heinsohn, R. (1991). Slow learning of foraging skills and extended parental care in cooperatively breeding white-winged choughs. *American Naturalist, 137*(6), 864–881.

Heinsohn, R. (1992). Cooperative enhancement of reproductive success in white-winged choughs. *Evolutionary Ecology, 6*(2), 97–114.

Heinsohn, R. (2008). The ecological basis of unusual sex roles in reverse-dichromatic Eclectus parrots. *Animal Behaviour, 76*, 97–103.

Heinsohn, R., Cockburn, A., & Cunningham, R. B. (1988). Foraging, delayed maturation, and advantages of cooperative breeding in white-winged choughs, *Corcorax melanorhamphos. Ethology, 77*(3), 177–186.

Heinsohn, R., Dunn, P., Legge, S., & Double, M. (2000). Coalitions of relatives and reproductive skew in cooperatively breeding white-winged choughs. *Proceedings of the Royal Society of London, Series B: Biological Sciences, 267*(1440), 243–249.

Heinsohn, R., Ebert, D., Legge, S., & Peakall, R. (2007). Genetic evidence for cooperative polyandry in reverse dichromatic eclectus parrots. *Animal Behaviour, 74*, 1047–1054.

Heinsohn, R., & Legge, S. (2003). Breeding biology of the reverse-dichromatic, co-operative parrot Eclectus roratus. *Journal of Zoology, 259*, 197–208.

Heinsohn, R., Legge, S., & Endler, J. A. (2005). Extreme reversed sexual dichromatism in a bird without sex role reversal. *Science, 309*(5734), 617–619.

Jamieson, I. G., & Quinn, J. S. (1997). Problems with removal experiments designed to test the relationship between paternity and parental effort in a socially polyandrous bird. *Auk, 114*(2), 291–295.

Jetz, W., & Rubenstein, D. R. (2011). Environmental uncertainty and the global biogeography of cooperative breeding in birds. *Current Biology, 21*(1), 72–78.

Jing, Y., Sun, Y.-H., & Fang, Y. (2003). Notes on the natural history of the Sichuan jay (*Perisoreus internigrans*). *Chinese Journal of Zoology, 38*(3), 91–92.

Johnstone, R. A. (2011). Load lightening and negotiation over offspring care in cooperative breeders. *Behavioral Ecology, 22*(2), 436–444.

Kingma, S. A., Hall, M. L., & Peters, A. (2011). Multiple benefits drive helping behavior in a cooperatively breeding bird: An integrated analysis. *American Naturalist, 177*(4), 486–495.

Kingma, S. A., Hall, M. L., Segelbacher, G., & Peters, A. (2009). Radical loss of an extreme extra-pair mating system. *BMC Ecology, 9*(15), 1–11.

Koenig, W. D., & Haydock, J. (2004). Incest and incest avoidance. In W. D. Koenig & J. L. Dickinson (Eds.), *Ecology and evolution of cooperative breeding in birds* (pp. 142–156). Cambridge: Cambridge University Press.

Komdeur, J. (1992). Importance of habitat saturation and territory quality for evolution of cooperative breeding in the Seychelles warbler. *Nature, 358*, 493–495.

Komdeur, J., & Edelaar, P. (2001). Male Seychelles warblers use territory budding to maximize lifetime fitness in a saturated environment. *Behavioral Ecology, 12*(6), 706–715.

Lessells, C. M., Avery, M. I., & Krebs, J. R. (1994). Nonrandom dispersal of kin: Why do European bee-eater (*Merops apiaster*) brothers nest close together? *Behavioral Ecology, 5*(1), 105–113.

Ligon, J. D., & Burt, D. B. (2004). Evolutionary origins. In W. D. Koenig & J. L. Dickinson (Eds.), *Ecology and evolution of cooperative breeding in birds* (pp. 5–34). Cambridge: Cambridge University Press.

Magrath, R. D., & Whittingham, L. A. (1997). Subordinate males are more likely to help if unrelated to the breeding female in cooperatively breeding white-browed scrubwrens. *Behavioral Ecology and Sociobiology, 41*(3), 185–192.

McDonald, P. G., Kazem, A. J. N., & Wright, J. (2009). Cooperative provisioning dynamics: Fathers and unrelated helpers show similar responses to manipulations of begging. *Animal Behaviour, 77*, 369–376.

Millar, C. D., Anthony, I., Lambert, D. M., Stapleton, P. M., Bergmann, C. C., Bellamy, A. R., & Young, E. C. (1994). Patterns of reproductive success determined by DNA fingerprinting in a communally breeding oceanic bird. *Biological Journal of the Linnaean Society, 52*, 31–48.

Møller, A. P. (1992). Frequency of female copulations with multiple males and sexual selection. *American Naturalist, 139*, 1089–1101.

Mulder, R. A., & Cockburn, A. (1993). Sperm competition and the reproductive anatomy of superb fairy-wrens. *Auk, 110*(3), 588–593.

Mulder, R. A., Dunn, P. O., Cockburn, A., Lazenby-Cohen, K. A., & Howell, M. J. (1994). Helpers liberate female fairy-wrens from constraints on extra-pair mate choice. *Proceedings of the Royal Society of London, Series B: Biological Sciences, 255*(1344), 223–229.

Mulder, R. A., & Langmore, N. E. (1993). Dominant males punish helpers for temporary defection in superb fairy-wrens. *Animal Behaviour, 45*(4), 830–833.

Nowak, M. A., Tarnita, C. E., & Wilson, E. O. (2010). The evolution of eusociality. *Nature, 466*(7310), 1057–1062.

Pruett-Jones, S. G., & Lewis, M. J. (1990). Sex ratio and habitat limitation promote delayed dispersal in superb fairy-wrens. *Nature, 348*, 541–542.

Reyer, H.-U., & Westerterp, K. (1985). Parental energy expenditure: A proximate cause of helper recruitment in the pied kingfisher (*Ceryle rudis*). *Behavioral Ecology and Sociobiology, 17*, 363–369.

Richardson, D. S., Burke, T., & Komdeur, J. (2002). Direct benefits and the evolution of female-biased cooperative breeding in Seychelles warblers. *Evolution: International Journal of Organic Evolution, 56*, 2313–2321.

Riehl, C. (2011). Living with strangers: Direct benefits favour non-kin cooperation in a communally nesting bird. *Proceedings of the Royal Society of London, Series B: Biological Sciences, 278*, 1728–1735.

Rubenstein, D. R., & Lovette, I. J. (2007). Temporal environmental variability drives the evolution of cooperative breeding in birds. *Current Biology*, *17*(16), 1414–1419.

Russell, A. F., Langmore, N. E., Cockburn, A., Astheimer, L. B., & Kilner, R. M. (2007). Reduced egg investment can conceal helper effects in cooperatively breeding birds. *Science*, *317*(5840), 941–944.

Russell, E., & Rowley, I. (2000). Demography and social organisation of the red-winged fairy-wren, *Malurus elegans*. *Australian Journal of Zoology*, *48*(2), 161–200.

Skutch, A. F. (1987). *Helpers at birds' nests: A worldwide survey of cooperative breeding and related behavior*. Iowa City: University of Iowa Press.

Stacey, P. B., & Koenig, W. D. (Eds.). (1990). *Cooperative breeding in birds: Long-term studies of ecology and behavior*. Cambridge: Cambridge University Press.

Tuttle, E. M., & Pruett-Jones, S. (2004). Estimates of extreme sperm production: Morphological and experimental evidence from reproductively promiscuous fairy-wrens (*Malurus*). *Animal Behaviour*, *68*, 541–550.

Tuttle, E. M., & Pruett-Jones, S. (2007). Estimates of extreme sperm production: Morphological and experimental evidence from reproductively promiscuous fairy-wrens (*Malurus*) (vol. 68, pg. 541, 2004). *Animal Behaviour*, *73*, 555–556.

West, S. A., Pen, I., & Griffin, A. S. (2002). Cooperation and competition between relatives. *Science*, *296*(5565), 72–75.

Woxvold, I. A. (2004). Breeding ecology and group dynamics of the apostlebird. *Australian Journal of Zoology*, *52*(6), 561–581.

Woxvold, I. A., & Mulder, R. A. (2008). Mixed mating strategies in cooperatively breeding apostlebirds *Struthidea cinerea*. *Journal of Avian Biology*, *39*(1), 50–56.

Wright, J., McDonald, P. G., te Marvelde, L., Kazem, A. J. N., & Bishop, C. M. (2010). Helping effort increases with relatedness in bell miners, but "unrelated" helpers of both sexes still provide substantial care. *Proceedings of the Royal Society of London, Series B: Biological Sciences*, *277*, 437–445.

II Agents and Mechanisms

13 Why the Proximate–Ultimate Distinction Is Misleading, and Why It Matters for Understanding the Evolution of Cooperation

Brett Calcott

Introduction

In a review article on the evolution of cooperation, Stuart West and colleagues suggested that future work on cooperation could profit from "an emphasis on the distinction and interplay between mechanistic (proximate) and evolutionary (ultimate or selective value) approaches" (West, Griffin & Gardner 2007, p. 16). I've previously argued that a variety of explanatory approaches to cooperation is desirable (Calcott, 2008, 2011), so I'm on board—at least in principle. But Mayr's proximate–ultimate distinction (Mayr 1961), once examined, turns out to be a rather muddled affair. I'm not the first to point this out,[1] but here I give an analysis of why this muddle matters when the subject is the evolution of cooperation.

I show that Mayr's distinction conflates two quite separate issues, and as a result, completely obscures a distinctive puzzle about how cooperation evolves. Recognizing this connects the evolution of cooperation to two other traditionally puzzling areas in evolution: the evolution of novelty and evolution of complexity.

The Proximate and the Ultimate in Cooperation

In principle, the idea of dividing up explanatory labor in biology is a good one.[2] It is often misguided to insist that there is just one thing to explain about some biological phenomenon, or just a single way to explain it. Cooperation is no exception, and the proximate–ultimate distinction does seem to capture two distinct issues. Consider two amateur naturalists—one an economist, the other an engineer—coming across an ant nest:

Economist "Their cooperative behavior is truly puzzling. The workers toil endlessly, with little concern for their own well-being, all their efforts directed to the success of the nest as a whole. How could such selfless behavior have evolved?"

Engineer "I agree their cooperative behavior is puzzling, but for very different reasons. The workers take on many different roles, coordinating their behavior with others in fascinating ways, yet nobody seems to be in charge. How do they achieve such complex cooperative organization?"

Both are both struck by the cooperation of the ants, but they approach the issue from two very different perspectives. The economist is puzzled by the apparent fitness costs that the worker ants pay, without gaining any obvious benefit. The worker ants appear to be sacrificing their reproductive future for the benefit of others. Why does such behavior persist? Surely natural selection would favors "defectors"—those that take the benefits derived from group interaction but do nothing to earn them. The engineer, in contrast, is puzzled by something quite different: how the ants manage to organize their complex tasks. This is puzzling because, although complex organized behavior is commonplace for humans, the ants appear to accomplish it without hierarchical organization or complex language. In contrast to humans, ants coordinate their behavior impressively without either individual intelligence or top-down control.

This contrast between the fictional economist and engineer captures a difference in explanatory interests, and these two perspectives look ripe for the application of the proximate–ultimate distinction. So, for example, Thomas Seeley, in his book on the social physiology of beehives, deploys the distinction when identifying these two puzzles in his introduction on the social organization of bees:

> The first [puzzle] deals with ultimate causation: why exactly is there strong cooperation among of the lower-level entities? . . . The second puzzle lies in the realm of proximate causation: how exactly do the lower-level entities work together to form the high-level entity? (Seeley 1995, p. 6)

In Mayr's terminology, the economist's puzzle is about ultimate causation, and the engineer's puzzle is about proximate causation.

Notice that these two perspectives idealize the situation in very different ways. The engineer treats the group as a mechanism, and the individuals within it as functionally integrated parts that, together, enable the whole to perform some adaptive behavior. For the engineer, the internal complexity of the group—the specialization of individuals and their allocation to different tasks, and the cues and signals that regulate this behavior—plays a central role in the explanation. The details tell us how the individuals are organized to achieve a group-level task. They tell us something about benefits of cooperating too, for complex organization can lead to increased efficiency, or an ability to do something that is simply not possible for a solitary individual (Calcott, 2008). This is an important issue to understand, for if cooperating couldn't generate some benefit over and above acting alone, then problems of defection wouldn't even arise.

In contrast to this mechanistic approach, the economist's perspective treats the group as a social collective containing individuals with divergent reproductive interests, something completely ignored by the engineer. Typically, the economist abstracts away from the complex details of interest to the engineer, for what matters to the economist is the final fitness payoffs for individuals, rather than the complex organization that generates these payoffs. The focus on payoffs is not just true of theoretical studies of cooperation; it is also largely the case in experimental work. So, for example, in public goods games, popular in experimental economics, the profit of cooperation is given by experimental fiat—perhaps by simple doubling the "pot"—yet the agents do nothing to generate that profit.

A real example can show both kinds of idealization at work. *Dictyostelium discoideum* is a single-celled organism (an amoeba) that, when short of food, assembles together with others of its kind into a multicellular body that looks and acts remarkably like a small worm. Here is Richard Kessin, talking of this multicellular phase:

> Once they have aggregated, aggregates have many of the properties of an embryo—they have polarity, they have exquisite proportioning, they regulate, and they have an organising centre, the anterior tip. (Kessin, 2001, p. 5)

Kessin tries to explain how complex group behavior is produced by the individual cellular behavior—he is our engineer.

We can take an economists' perspective on *D. discoideum* too. The worm, once assembled, makes its way to the top of the leaf litter, where it forms a base and grows a thin stalk with a blob on top. The blob eventually sends forth spores to create the next generation of amoebae—with luck, they'll land far enough way to find a fresh source of food. But only those in the blob get to reproduce; those stuck in the stalk have no evolutionary future, even though they aided the worm in making it to the top of the leaf litter. The stranded amoebae are like worker ants, doing a job with no direct reward.

Kessin's book-long treatment of *D. discoideum* contains a few paragraphs on altruism and cooperation (Kessin, 2001, pp. 29–30), but the subject is largely ignored. Strassman and her colleagues, however, advocate an entirely different view on *D. discoideum*—they are our economists:

> The social amoeba, Dictyostelium discoideum, is widely used as a simple model organism for multicellular development, but its multicellular fruiting stage is really a society. . . . if D. discoideum aggregates include more than one clone, they may be more analogous to the societies of social insects, which have sterile castes, but also have multiple genotypes and genetic conflicts of interests. (Strassmann, Zhu & Queller, 2000)

Strassman et al. focus on whether relatedness among the amoebae has measurable effects on the worm and stalk stage in the life cycle, rather than understanding the particular mechanisms that enable the single cells to assemble and coordinate their behavior. If they look for mechanisms at all, they are interested in those that facilitate or prevent cheating, such as mechanisms for kin recognition.

It is important to acknowledge these idealizations for what they are. Neither denies the existence of the other puzzle. Each perspective slices off different bits of the world for examination. We see Strassman et al. claiming that the assemblage "really is a society," suggesting they think their perspective is to be preferred to Kessin's. But if we acknowledge that each is using a different idealization of the biological phenomenon because they have different explanatory goals, then we can avoid arguments about what "really is" the right view.

These idealizations are partly a pragmatic issue—they are required simply because of the complexity of the biological world; but there is also an epistemological payoff. Abstracting away from certain issues enables us to apply our insights to a broader set of cases, making our theories more general. For example, the mathematical models of the evolution of cooperation (such as kin selection) apply regardless of the details of complex organization within a group. Likewise, models of self-organization that putatively explain the behavior of eusocial insects have been applied to a broad range of engineering problems, where issues of internal conflict are irrelevant.

The Dichotomy That Wasn't

Thus far, the proximate–ultimate distinction appears to carve up the different explanatory interests within cooperation quite well. If we look more closely, however, we'll see that this apparent dichotomy actually squashes together two quite different distinctions. Once we separate them, a third kind of explanation becomes apparent, and with it, and third puzzle worthy of our attention.

In order to explore these issues, I'm going to put aside cooperation for the moment, and just focus on the more general usage of the distinction. Once we have a firm grasp of the problem, I'll show how it matters for cooperation.

The first way to interpret Mayr's dichotomy is like this: The proximate concerns the present, whereas the ultimate concerns the past. This is often how the distinction is interpreted:

> One apparently plausible interpretation of this dichotomy is that proximate causes concern processes occurring during the life of an organism while ultimate causes refer to those processes (particularly natural selection) that shaped its genome. (Francis, 1990)

> In essence, the proximate–ultimate distinction is that between the causes of behaviour that occur within an animal's lifetime and those that preceded its life, i.e. evolutionary factors. (Dewsbury, 1999)

The distinction between past and present is also apparent in a common way of stating the distinction between "how" and "why" questions: If we want to know how something works now, we ask a proximate question, and if we want to why how something came be the way it is, we ask an ultimate question. Proximate, then, relates to the current operation, whereas ultimate has something to do with the evolutionary origins of a trait.

The second way to interpret the distinction is this: Instead of thinking in terms of past or present, we contrast explanations involving individual mechanisms to those involving population dynamics.

Prime examples of mechanistic explanations appear in the field of biomechanics. So, for example, we might explain how a flea jumps by describing the relationships and operations of various parts of the flea's legs and body (Rothschild et al., 1975). What we seek to explain in these cases is the operation of some mechanism at the individual level, and we explain it by referring to the parts of the mechanism, and their various interactions.

In contrast, many questions in biology focus on dynamics of population-level properties—on how the frequency of certain traits, or the population structure, affect the change (or lack of change) over time in a population. This includes most theoretical work on the evolution of cooperation. For example, we might explain the stability of cooperation in a population by referring to the relatedness of the individuals that interact. What we seek to explain in these cases is something that can be stated only in population terms: the frequency of a particular trait and whether it is maintained in a population.[3]

These two distinctions—between past and present, and between mechanisms and populations—are collapsed in the proximate–ultimate distinction: Mechanistic explanations tell us about the present, and population dynamics tell us about the past. This is a problem, because we might be led to believe that an interest in mechanisms must focus on present behavior only, and any story about the history or origins of a trait is always about population dynamics.

This cannot be true, however, for there is a further puzzle we can ask about mechanisms, one that doesn't just focus on how some complex system works at a time, but looks at how it is possible for such a system to change over time. This puzzle is particularly salient when the evolved mechanism being explained is something complex or novel. Although it is assumed that evolution proceeds by incremental change, in these cases it is often far from obvious how such change arose piecemeal. When puzzles like these are addressed in the literature, we see a kind of historical explanation being used that idealizes away from population dynamics and instead, focuses on understanding how a series of small differences can explain the changing behavior of a mechanism. I've called this kind of historical-mechanical explanation a "Lineage Explanation" (Calcott, 2009).

Table 13.1
The proximate–ultimate distinction conflates two separate distinctions, and obscures an important third option: Lineage explanations (explained in the text).

	Current Operation	Historical Origins
Individual Mechanisms	Proximate explanation	*Lineage explanation*
Population Dynamics		Ultimate explanation

Lineage Explanations

It is instructive to see one of these explanations at work. A classic example of complexity is the vertebrate eye. Here is how Darwin posed the question:

> if numerous gradations from a perfect and complex eye to one very imperfect and simple, each grade being useful to its possessor, can be shown to exist . . . then the difficulty of believing that a perfect and complex eye could be formed by natural selection, though insuperable by our imagination, can hardly be considered real. (Darwin 1872, pp. 143–144)

Notice that the population process of natural selection plays the role of a background assumption here. What is actually in question whether there is a path of incremental variations that connects a simple eye to one that is much more complex. Without such a path, natural selection would have nothing to get a grip on.

In response to Darwin's question, Nilsson and Pelger produced a computer model of the evolution of an eye. The model showed how, beginning with a simple eye-spot, a plausible series of small variations could produce a complex camera eye. Each stage in Nilsson and Pelger's model has the same parts: a light-sensitive patch of cells sandwiched between a transparent layer and a dark pigmented layer. The parts undergo various modifications throughout the series from a simple light-sensitive spot to a camera-lens eye. And at each subsequent stage, the acuity of the eye—its ability to project a detailed image on the retina—increases.

Nilsson and Pelger's model tells us something about the origins of the eye, but the explanation does not include any populations. Rather, it shows a plausible trajectory connecting a simple eye-spot to a complex camera-eye—the "numerous gradations" that Darwin requested. Each successive stage, and its accompanying increase in acuity, is attained through a small change to the prior stage. Moreover, the increase in acuity in each subsequent stage is not simply stated, but actually explained by the changes in the underlying parts and organization of the mechanism. The decomposition

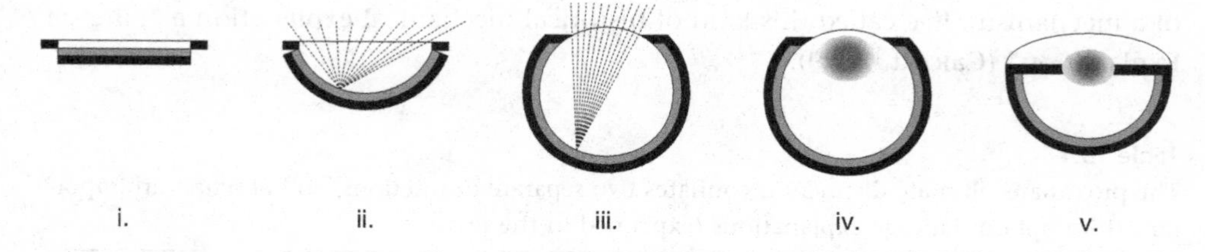

Figure 13.1
Nilsson and Pelger's model of the evolution of vertebrate eye. (i) A simple eye-spot. (ii) A depression forms making it sensitive to direction. (iii) A "pin-hole camera" eye. (iv) Variable density in the retina improves acuity. (v) Eye shape changes to parabola, again improving acuity.

of the mechanism into its parts thus serves two simultaneous roles. It shows how the eye works at each stage, and it supplies the grain of change between each successive stage.[4]

This kind of explanation is commonplace in evolutionary developmental biology, where it is often accompanied by a cartoon-strip-like representation of the changes present at each stage. So, for example, the evolution of feather shape is explained by appealing to changes in the mechanism in the follicle from which it grows (Prum, 1999, 2005), or the evolution of metamorphosis is explained by appealing to changes in the timing of juvenile hormone release (Truman & Riddiford, 1999), or the evolution of segmentation is explained by appealing to changes in genes that control the cyclic expression responsible for producing segment boundaries (Peel & Akam, 2003).

In each of these cases, the explanation tells us about the origins of some novel form or capacity. Again, these explanations do not include population dynamics. These cases are even more removed from accounts that include population dynamics for, unlike the case of the eye, there is no attempt to even assess what is adaptive about such changes. That is because what matters for these explanations is providing some plausible foundations for how these new variants arose. Again, it is important to emphasize that doing so is an idealization. No one denies that selection played a role in each case, and that this selection required populations. But we put these issues aside to focus on the origins of the requisite variation. These origins are explained by showing how some preexisting developmental components could be redeployed in a way that produced the novel features.

In the case of feathers, for example, the appearance of hierarchical treelike structure of flight feathers depended on a number of complex changes in the way the feather grows. Understanding how the changes in feather form could be produced gradually is only possible by understanding the mechanistic details of how a feather actually grows. The explanation, like the one for the eye, both shows a mechanism at each stage—in this case, one detailing how proto-feathers develop—and demonstrates what changes in the mechanism were sufficient to produce the subsequent mechanism.

Although these explanations focus on evolutionary change, they are very different from explanations involving population dynamics. Lineage explanations show how some prior mechanism capable of one task can, through only minor variations, be modified into a new mechanism capable of a different task. The explanation plausibly demonstrates the existence of a continuous trajectory of change over time, rather than explaining the selective forces that directed and maintained this change.

Power versus Path

An analogy can help bring out the differences between these two types of explanation. A highlight for any tourist visiting Paris (France) is Montmartre, a hill in the 18th arrondissement with a great view, a striking church, the last remaining vineyard in Paris, and many overpriced cafés. Getting to the top of Montmartre presents the tourist with two problems. Problem 1 is simply that Montmartre lies at the top of a hill, one hundred and thirty meters high (imagine climbing it in fifty years' time if you don't see this as a problem). Problem 2 is that, although the top the hill is easily visible from much of Paris, there is no simple direct route to the top. There are several routes to get to the top, but each winds its way through many streets and stairways.

Imagine meeting two tourists at the top of Montmartre—a man with one leg and woman without a map of Paris. We can ask each of them how they got up here, but we're really asking two different questions. For the man with one leg, Problem 1 is pertinent; we want to what how someone with one leg could raise himself 130 meters (did he hop, use crutches, or perhaps rely on a partner?). For the woman without the map, Problem 2 is pertinent; with no map, what particular trajectory did she discover that led her to the top (how circuitous was the route, did she have to reverse her path at any stage, were there any obvious clues that guided her, or did she proceed randomly?).

Any successful expedition to the heights of Montmartre requires both a way to power the uphill climb and a navigable path from the bottom to the top. Whether it is the power or the path that requires explanation will depend on our interests, and these are dictated by the particular context in which the question is asked, just as they were with the one-legged man and the map-averse woman.

A similar contrast is apparent when we want to explain the evolutionary origins of some phenotypic feature. Explanations using population dynamics show how natural selection (among other things) could drive a population to a particular equilibrium. Lineage explanations, in contrast, shows that a path of incremental change is possible. Which of these explanations is pertinent will depend on the context, and the answer given will idealize the situation in a different way.

From Organs to Organizations

It's time to return to see how these issues pertain to cooperation. Thus far, I've been talking about complex organs, such as eyes, or novel morphology, such as feathers. Yet there is a straightforward parallel between complex organs and complex organized groups. They both consist of heterogeneous parts interacting in very specific ways to produce an adaptive outcome. Recall that, from an engineering perspective, this is exactly how a cooperative group is idealized: Individuals are interacting parts of a mechanism, and we use this to explain how the group does its job.

Given that cooperative organization can be viewed as a mechanism, it follows that we can give a lineage explanation for its origins, much like we did for the origins of eyes or feathers. A lineage explanation for cooperation would show how the redeployment of the antecedent capacities of the individuals and their interactions could lead to changes in group-level abilities. In other words, we could explain how the behavior of the group changed over time in virtue of the changes to the individuals and interactions that made it up.

I say we could give a lineage explanation, but is one needed? Remember that the eyes and feathers were puzzling because it was not obvious how such features could come about through incremental change. Does cooperation, viewed as a mechanism, present this same sort of puzzle? I outline three situations where cooperation might call for a lineage explanation. In each case, the explanatory goal is to demonstrate the trajectory behind the origin of some complex cooperative capacity.

Situation 1: Complex Interdependent Cooperation

The bacterial flagellum, one of the few known evolved rotary mechanisms, is a poster-child for creationism. The flagellum is, according to Michael Behe, "a single system which is composed of several well-matched, interacting parts that contribute to the basic function, and where the removal of any one of the parts causes the system to effectively cease functioning" (Behe, 1996, p. 96). This kind of system is irreducibly complex—or so Behe would have us believe. Sahotra Sarkar (2007), among others, does an excellent job of showing just how bankrupt this argument is.

What Sarkar does not deny, however, is that explaining the origins of such mechanisms is a worthwhile and sometimes difficult task. Sarkar suggests two generic pathways that can produce mechanisms of the kind that Behe describes. The first involves an incremental loss of functional redundancy and the second the co-option of secondary function from the same structure. Noticeably, both of these pathways take the form of lineage explanations—showing how some previous mechanisms could be gradually modified over time, producing a complex mechanism with interdependent parts.

These kinds of complex integrated mechanisms can occur in a cooperative context as well. Consider green tree ants (*Oecophylla smaragdina*), which form nests by gluing together large leaves. The leaves are pulled together by chains of ants, attached head-to-toe, and stuck together by other ants wielding larvae-like tubes of glue, persuading them to emit a sticky silk to hold the leaves together. We have three separate tasks here: pulling the leaves together, applying the glue, and producing the glue. And making a nest is not simply a matter of efficiently dividing labor—close coordination between each of these three tasks is essential. The three tasks appear to be highly interdependent. Removal of any one of them wouldn't simply make the nest building less efficient; the nests would simply not get made at all. Viewed from an engineering perspective, such cooperative behavior constitutes a mechanism of the kind Behe describes.

Any claim that this interdependence in the mechanism is somehow irreducible is, of course, barking mad. Yet there is a genuine puzzle here, for, as Sarkar acknowledges, it is not obvious what series of simple steps could produce such a mechanism. One thing is certain, however: Kin selection, or any other population-level framework aimed at explaining the evolution of cooperation, is simply not the right tool for approaching this puzzle about evolutionary change. Nor would a mechanistic understanding of how nest construction is currently accomplished be sufficient to answer the question, for we need to understand not just how it works now, but how it has changed over time. This is what lineage explanations do.

Situation 2: The Origins of Individual Capacities That Support Cooperation

Imagine finding the following two cooperative practices in different social mammals. In the first mammal, each individual in the group hunts for food, bringing it back to a central location. Everyone in the group then helps themselves to the centrally located food. Hunting is itself a solitary task, however, and an individual's hunting success is independent of what others do. The benefit of cooperation, in this case, is simple insurance: If you are unlucky today, you can still eat, and tomorrow, your success may compensate another's bad luck.

In the second case, the group hunts together as a team. Individuals take on different roles during the hunt, and they signal one another to effectively surround and take down the prey. This kind of cooperation is more complex, for it requires individuals to tune their behavior to one another, taking on different roles and timing their interactions with one another. The final result depends on precisely coordinated parts, just as a complex mechanism relies on the individual parts being highly tuned to one another.

In the first case, it seems there is little new going on; individuals hunt just as they might have done when they were solitary. We certainly need to explain why this sharing behavior remains stable (the economist's puzzle), but the individual hunting behavior require no new explanation; individuals are the same as they were when they were solitary. In the second case, however, the hunting behavior of the individuals—the division of labor, and the plastic response to the actions of other individuals within the group—only makes sense within a cooperative task. There would be no point in having these abilities before cooperation began, yet for the cooperative behavior to pay off (for the hunt to be successful) individuals require these very abilities. So how do we explain the origins of these individual capacities, if they weren't required before cooperation?

Consider, for example, how we might explain the origins of cellular differentiation—a sexual division of labor—in the green algae *Volvox carteri*. Some of the cells in this simple multicellular organism are somatic, performing the important role of swimming to keep the algae afloat, while others are specialized for reproduction. According to Richard Michod, we can explain this division of labor as a co-option of two separate life stages (Michod, 2011). As individuals, the ancestors of *V. carteri* went through two distinct life stages: first swimming, then reproducing. These life stages are now expressed differentially in the cells making up the organism. So these capacities did not arise de novo but were co-opted from a different context.

Carl Schlichting (2003) takes a different and more generic approach to explaining differentiation. During normal multicellular growth, cells will be exposed to different environments, some on the inside, and others on the outside. If the cells are sensitive to the local environment, then the asymmetry in environmental conditions will be sufficient to produce some differences in their behavior. Certain combinations of these differentiated cells will result in a fitter group, as the cells can take on different roles. This theory relies on the prior plasticity of the cells, but it is important that this prior plasticity itself is not in need of explanation. As Schlichting rightly points out, plasticity is the primitive condition of biological processes—by default, things are sensitive to the context in which they occur. In general, it is adaptation that channels, controls, or prevents plasticity—it does not create it. Schlichting's explanation requires co-option, as did Michod's, but in this case the capacity being co-opted is previously hidden plasticity, rather than previously adaptive behavior.

Obviously, the puzzles I'm drawing attention to here have been examined, and some solutions have been suggested. But it is important to recognize that the explanations given for these problems don't neatly fit into the proximate–ultimate dichotomy. Each of them explains the origins of some new capacity by referring to an older capacity, redeployed in a new context. These are lineage explanations: mechanistic explanations that address change over time.

Situation 3: Novel Group-Level Cooperative Capacities

Explaining how feathers evolve is puzzling because the morphology of feathers is novel: It has no homologous precursor. The hierarchical treelike structure is not formed by a gradual modification of scales, as was previously thought (Maderson, 1972). Rather, the structure is produced from small changes in the growth pattern within the follicle, which effectively extrudes the feather, rather like toothpaste being squeezed out (Prum, 1999, 2005). Though the morphology of feathers presents a significant jump from any predecessor, this jump can be explained by small changes in the growth pattern, once we understand how these affect change in the resulting feather. In other words, by understanding how feathers develop, we can explain how small changes in developmental mechanisms can produce large jumps in morphology.

The general message is this: It is crucial to understand the organizational processes that link small changes at a lower level of organization to large changes at higher levels of organization. Without understanding these processes, we cannot explain gradual change over time at the higher level. In the case of the feather, it is the details of feather growth that matter. But in the context of cooperation, there are cases where novel group-level advantages cannot be explained without understanding the details of processes underlying them. We need to understand not only how they work, but also how they change.

Again, the evolution of multicellularity can provide an example. Newman and coauthors have suggested that many core morphological features of multicellular organisms are produced as a direct consequence of physical interactions between simple cellular capacities (Newman, Forgacs & Müller, 2006; Newman & Bhat, 2009; Newman & Müller, 2000). For example, the sorting of different cell types into layers can be achieved if there are cell types with differential strength in adhesion. That is, the first appearance of cell layering does not require a story of gradual selection, as it can arise purely as an upshot of the differential capacities of individual differentiated cells.

From an engineering perspective, this cell layering is a novel capacity that putatively provides a cooperative benefit. It is not simply the differentiation of cells that makes multicellularity pay off, but also the robust, repeatable organization of the differentiated cells into a useful structure. Such internal organization is a feature of complex cooperation. But to explain this we need to understand how this self-organizing process works, and how it is both enabled and stabilized by small changes at the level of the cell. This is a different task than in the previous two examples, as we're not trying to explain the origins of the abilities of the individuals making up the group, nor are we puzzled by the interdependence of parts. Instead, we're trying to show how novel group-level ability can be produced from the interactions of lower-level individuals.

Summary

I began the chapter by suggesting that the proximate–ultimate distinction failed to capture some important aspects of the evolution of cooperation. My main objection is that the distinction can lead us to overlook mechanistic explanations for the origins of complex cooperative interactions, in the form of what I've called "lineage explanations."

One response to this view is to suggest that I'm simply focusing on issues that are not central to the evolution of cooperation. The complaint might go like this: Explaining such things as the origins of differentiation in multicellular creatures is simply a different issue than explaining the origins of cooperation; the real issue at the heart of cooperation is explaining why it is stable in the face of defection. Yet such a view fails to capture much of what we mean by cooperation. Ants and humans are often heralded as examples of highly cooperative organisms, but such heights refer to more than just a willingness or capacity to sacrifice individual fitness; it also captures the complexity of cooperative interaction and the benefits that are derived from it. Moreover, the issue of the stability of cooperation only arises because cooperative ventures hold the promise of increasing fitness for all those involved. Surely understanding the mechanisms that enable this, and how they might have evolved, is part of understanding cooperation.

More importantly, the simple split between proximate and ultimate focuses all issues about the origins of cooperation within a framework that idealizes away from the organizational complexity that produces the benefits. One effect of this is to see all barriers to cooperation defined in terms of the economists' puzzle. For example, Strassman et al. suggest that one reason *D. dictyostelium* has remained "simple" may be because it lacks an "effective means to suppress conflict" (Strassmann, Zhu & Queller, 2000). Perhaps they are right. Yet it might also be the case that there is no simple direct path from its mode of organization to something more complex. Strassman et al. tell us that *D. dictyostelium* possesses "most of the kinds of molecular mechanisms required to evolve more complex forms of multicellularity." But the idea that merely possessing the requisite cell-level capabilities is sufficient to produce group-level complexity fails to consider the problems of putting those capabilities together, and doing so in an incremental fashion. It is exactly this kind of possibility that is obscured when we adhere to a simple dichotomy of proximate–ultimate.

Acknowledgments

Many thanks to my fellow editors for their helpful comments and especially for their patience. I gratefully acknowledge support from the Philosophy Department in the Research School of the Social Sciences at the Australian National University, the Australian Research Council, and the Konrad Lorenz Institute for Evolution and Cognition Research, Altenberg, Austria.

Notes

1. See, e.g., West-Eberhard (2003) and Francis (1990).

2. Mayr's proximate–ultimate distinction isn't our only option here. Another way to capture these differing perspectives would be with Tinbergen's four kinds of explanation (Tinbergen, 1963). Tinbergen's types offer a more fine-grained set of distinctions than Mayr (indeed, Tinbergen's distinctions are often presented as subcategories within Mayr's). But I'm going to stick with Mayr's distinction in this chapter, for two reasons: introducing further distinctions will cause more confusion than clarity, and none of Tinbergen's finer-grained categories can aid us with the issues I identify later in the chapter.

3. A good example of such population dynamics approaches to cooperation appears in Lehmann and Keller (2006) and the subsequent commentaries.

4. See Calcott (2009) for a more detailed explanation and some further distinctions.

References

Behe, M. J. (1996). *Darwin's black box: The biochemical challenge to evolution.* New York: Free Press.

Calcott, B. (2008). The other cooperation problem: Generating benefit. *Biology and Philosophy, 23*(2), 179–203.

Calcott, B. (2009). Lineage explanations: Explaining how biological mechanisms change. *British Journal for the Philosophy of Science, 60*(1), 51–78.

Calcott, B. (2011). Alternative patterns of explanation for major transitions. In B. Calcott & K. Sterelny (Eds.), *The major transitions in evolution revisited.* Cambridge, MA: MIT Press.

Darwin, C. (1872). *The origin of species.* London: Murray.

Dewsbury, D. (1999). The proximate and the ultimate: Past, present, and future. *Behavioural Processes, 46*(3), 189–199.

Francis, R. (1990). Causes, proximate and ultimate. *Biology and Philosophy, 5*(4), 401–415.

Kessin, R. H. (2001). *Dictyostelium: Evolution, cell biology, and the development of multicellularity.* Developmental and Cell Biology Series. Cambridge: Cambridge University Press.

Lehmann, L., & Keller, L. (2006). The evolution of cooperation and altruism—a general framework and a classification of models. *Journal of Evolutionary Biology, 19*(5), 1365–1376.

Maderson, P. F. A. (1972). On how an archosaurian scale might have given rise to an avian feather. *American Naturalist, 106*(949), 424–428.

Mayr, E. (1961). Cause and effect in biology. *Science, 134*, 1501–1506.

Michod, R. 2011. Evolutionary transitions in individuality: multicellularity and sex. In B. Calcott & K. Sterelny (Eds.), *Major transitions in evolution revisited.* Cambridge, MA: MIT Press.

Newman, S. A., & Bhat, R. (2009). Dynamical patterning modules: A "pattern language" for development and evolution of multicellular form. *International Journal of Developmental Biology, 53*(5–6), 693–705.

Newman, S. A., Forgacs, G., & Müller, G. B. (2006). Before programs: The physical origination of multicellular forms. *International Journal of Developmental Biology, 50*(2/3), 289.

Newman, S. A., & Müller, G. B. (2000). Epigenetic mechanisms of character origination. *Journal of Experimental Zoology, 288*(4), 304–317.

Peel, A., & Akam, M. (2003). Evolution of segmentation: Rolling back the clock. *Current Biology, 13*(18), R708–R710. doi:10.1016/j.cub.2003.08.045.

Prum, R. O. (1999). Development and evolutionary origin of feathers. *Journal of Experimental Zoology, Part B: Molecular and Developmental Evolution, 285*(4), 291–306.

Prum, R. O. (2005). Evolution of the morphological innovations of feathers. *Journal of Experimental Zoology, Part B: Molecular and Developmental Evolution, 304*(6), 570–579.

Rothschild, M., Schlein, J., Parker, K., & Neville, C. (1975). The jumping mechanism of Xenopsylla Cheopis III. Execution of the jump and activity. *Philosophical Transactions of the Royal Society of London, Series B: Biological Sciences, 271*(914), 499–515.

Sarkar, Sahotra. (2007). *Doubting Darwin: Creationist designs on evolution*. London: Wiley-Blackwell.

Schlichting, C. D. (2003). Origins of differentiation via phenotypic plasticity. *Evolution & Development, 5*(1), 98–105.

Seeley, T. D. (1995). *The wisdom of the hive: The social physiology of honey bee colonies*. Cambridge, MA: Harvard University Press.

Strassmann, J. E., Zhu, Y., & Queller, D. C. (2000). Altruism and social cheating in the social amoeba Dictyostelium discoideum. *Nature, 408*(6815), 965–967.

Tinbergen, N. (1963). On aims and methods of ethology. *Zeitschrift für Tierpsychologie, 20*, 410–433.

Truman, J. W., & Riddiford, L. M. (1999). The origins of insect metamorphosis. *Nature, 401*(6752), 447–452.

West, S. A., Griffin, A. S., & Gardner, A. (2007). Evolutionary explanations for cooperation. *Current Biology, 17*(16), R661–R672.

West-Eberhard, M. J. (2003). *Developmental plasticity and evolution*. New York: Oxford University Press.

14 Emergence of a Signaling Network with *Probe and Adjust*

Brian Skyrms and Simon M. Huttegger

1 Introduction

Once individuals have developed a system of signals, in one way or another, they may spontaneously assemble into a signaling network. The network structure achieved may depend both on the payoffs involved and on the kind of adaptive dynamics driving the evolution of the network. We focus here on one example due to Bala and Goyal (2000), in which a ring network, where information flows cheaply, fully, and without degradation but only in one direction, has strong distinguishing properties. In such a ring network, each participant has exactly one connection to another participant such that the overall configuration of connections constitutes a circle. This implies that every member of the ring receives all of the information flowing between the participants at a minimum cost paid for establishing one connection. In game-theoretic terms, the ring structure is both optimal for all involved and the unique structure of strict Nash equilibria in the associated network formation game. A famous example of such a ring structure is provided by the Kula Ring of the Trobriand islands, which was first described by Bronislaw Malinowski nearly a hundred years ago (Malinowski, 1920, 1922). In the Kula Ring, two articles of value, necklaces and bracelets, circulate between islands in reverse direction. The circles so established also facilitate the exchange of goods and information. The ring is also one of the fundamental configurations for local computer area networks.

Bala and Goyal prove that a myopically rational dynamics of *Best Response with Inertia* leads to this ring structure (in a sense to be made precise later). On this rule, agents are conservative, but when they do change their informational connections, they do so optimally relative to what others did last time. This dynamics requires individuals who best respond to know the whole network configuration in the previous round, as well as the payoff structure of the game. These requirements for the application of best response may not be too onerous in small groups but—at least in certain cases—may presume too much in larger settings. Even in small groups in the economics laboratory, where the experimenter makes sure that this information is available, *Best Response with Inertia* does not seem to fit the empirical data well (Berninghaus et al., 2007). Nevertheless, individuals do tend to learn the ring structure. It is therefore interesting for two reasons to explore alternative adaptive dynamics—to see whether dynamics with a lower informational requirement can learn the ring, and to generate candidates for empirical testing.

Here we address the first question. Is there a plausible low-information adaptive dynamics that learns the ring? We focus on a low-rationality, payoff-based dynamics of *Probe and Adjust*, introduced by Skyrms (2010). This learning rule is conservative, but agents do experimentally change their informational connections. It does not presuppose that agents know how others are connected, assuming only that agents can compare current benefit against immediate past benefit. This dynamics thus only requires individuals to know their own actions and payoffs, and to remember their actions and payoffs from the last round. Nevertheless, individuals using this dynamics also learn the ring configuration, that is, they settle into ring configurations and usually stay there. These results have broader application, both to other network structures and to broader classes of games. These broader classes of games will be similar to the Bala–Goyal game described below in certain respects. Most importantly, there are decision paths leading from any nonring configuration to a ring configuration such that no player's payoff is lowered by a player changing connections.

2 The Bala–Goyal Ring Game

We proceed by specifying the game that we introduced informally. Individuals get private information by observing the world. Each gets a different piece of information. Information is valuable. An individual can pay to connect to another and get her information. The individual who pays does not give any information; it only goes from payee to payer. The payer gets not only the information from private observations of those whom she pays, but also that which they have gotten from subscribing to others for their information. Information flows freely and without degradation along the links so established. It flows in one direction, from payee to payer. We assume that information flow is fast, relative to any adjustment of the network structure.

More precisely, each of a finite number of individuals can make links to any set of other individuals. Making a connection has a cost, c. Each individual has information with a value, v. If individual 1 has a link to another individual 2, individual 1 gets all the information that 2 has including all information gotten from links with other players. That is, a player does not only get the information from another player she is immediately connected to, but also from all of the other player's (direct or indirect) connections. More formally, let x *be connected to* y if there are players $z_1, \ldots, z_n$ such that x is identical to z_1, z_i is linked to z_{i+1} (i=1, ..., n), and z_n is identical to y. Then a player has all her own original information and all the information of players to whom she is connected.

If the cost of subscribing to someone's information is too high, then it won't pay for anyone to do it. But let's suppose that that the cost of establishing a connection is less than the value of each piece of information, $c<v$. Then connections certainly make sense. We assume that any individual can make as many connections as she wishes. This model can be viewed as a game, with an individual's strategy being a decision of what connections to make. It could be none, all, or some. The game has multiple equilibria, but one is special. This is the ring (a.k.a. the circle, the wheel). The ring structure in this game is special in two ways. The first is that it is *strict*; the second that it is *efficient*. It is a strict equilibrium in that someone who unilaterally deviates from such a structure finds herself worse off. It is *Pareto efficient* in that there is no way to change it to make someone better off without making someone worse off. In fact, there is no way at all to make anyone better off. Everyone has the highest possible payoff that they could get in any network structure. The key to both these properties is that information flows freely around the ring, so that for the price of one connection a player gets all the information that there is.

Consider a player in such a ring who changes her strategy. She could establish additional links, in which case she pays more and gets no more information. She could break her link, in which case she would forgo the cost but get no information. She could break the link and establish one or more new ones, but every way to do that would deliver less than total information. Every deviation leaves her worse off. That is to say that the ring is a *strict* Nash equilibrium of the game. Now suppose that, starting from the ring, there is some lucky guy that everyone else would like to make better off. There is nothing they can do! He is already getting all the information at the cost of one link. They can't alter their links so as to give him more information, since he is already getting it all. Only he can avoid the cost of the link by breaking it—that is, not visiting anyone—but then he gets no information at all. The ring is *efficient*.

These rather special properties of the ring depend on rather special properties of the model—one-way flow of information, flow without informational decay or degradation, and the subscriber pays all the costs. Various modifications of these assumptions lead to a rich array of situations with different properties. For the moment, however, we will focus on the ring.

3 *Best Response with Inertia*

Bala and Goyal propose a myopically rational adaptive dynamics that learns the ring. Most of the time an agent just keeps the network connections that she had the last time—this is *inertia*. But with some small probability, e, she wakes up and chooses a set of connections that is optimal against the network configuration of her associates the previous time—this is *best response*. The best response probabilities are independent between trials and between players. (Thus, for small *e*, simultaneous or subsequent best responses are highly unlikely.) If there are ties for best response, the individual chooses between the tied set of connections at random.

Since, for small *e*, simultaneous or subsequent best responses are highly unlikely, Bala and Goyal analyze a simpler process in which nature, with small probability, chooses one player to best respond. First, since the ring is a strict Nash equilibrium configuration, no best response will lead a player to depart from a ring. By definition, deviation by a single player would leave him worse off. Next, since the ring is the unique strict Nash equilibrium, it is the *only* network configuration with this property. Players will exit any other configuration with positive probability. This is true even of Nash equilibria that are not strict, because tie-breaking for best response can lead to an exit.

The whole process, consisting of all players using best response with inertia, is thus a Markov chain with ring configurations as the unique absorbing states.[1] Next, Bala and Goyal show that there is a best response path from any state of the system to a ring. That is, there is a sequence of network configurations starting from the original state and ending in a ring, such that each change is the result of a best response by a player. It follows from the definition of best response with inertia that every such path has a positive probability. In this simplified theory, it then follows from standard Markov chain theory that from any starting point, players will reach a ring structure with probability one. Notice that we are still assuming that only one player switches strategies at a time. This may seem to be very restrictive, but considering the fact that choosing a best response has a low probability that is independent between players, it is a useful first approximation.

4 *Probe and Adjust*

We now assume that players only know their payoffs and can remember them for one period. They need not know the previous configuration of the network or the structure of the game. We retain *inertia*. Most of the time players just keep doing the same thing. But with some small probability, *e*, a player *probes*. A probe consists of choosing some set of connections at random and trying them out. If a player probes, she compares the payoff obtained on the probe with the payoff obtained on the previous move. If it is higher, she sticks with the connections tried on the probe. If it is lower, she goes back to the previous connections. If there is a tie, she chooses between the alternatives with equal probability. Probe probabilities are independent, just like best-response probabilities in the previous dynamics. Notice that there is no constraint of having only one player probing at a time.

5 Analysis of *Probe and Adjust*

If the probe probability is small, simultaneous or subsequent probes are very unlikely. So (as before) we start by analyzing a simpler process. With small probability, nature chooses a player at random to probe and adjust. That player probes and then adjusts according to the results of the probe, while all other individuals keep doing the same thing. Since nothing happens most of the time, we can just look at the cases where there are probes. So we consider this even simpler *process S*:

Nature chooses a player at random to probe. This player probes in one round and reacts in another. Repeat.

This process contains an embedded Markov chain, consisting of the even times. Probes are omitted but reactions to them are included. As before, rings are absorbing states because they are strict Nash equilibria, and the only absorbing states because they are the unique strict Nash equilibria. Probes may lead away from nonstrict Nash configurations, but they cannot lead away from strict Nash configurations.[2]

That there is a positive probability of transition from any state to a ring is a consequence of Bala and Goyal's proof that there is a best response from any state to a ring, together with the fact that if there is a best response a probe will—with positive probability—lead to its being taken. The embedded Markov chain leads to a ring with probability one, just as before. That means that process S spends most of its time in a ring configuration, except for occasional unproductive probes that can discover nothing better.

In the original *Probe and Adjust* dynamics, simultaneous or subsequent probes, although unlikely, are not impossible. We now remove our simplifying assumptions and address the effect of such unlikely perturbations on the stability of the ring. Suppose that from a ring, many players probe simultaneously. Unless all probe and hit another ring structure, probes will lead to worse payoffs, and all will go back to the same ring. These simultaneous probes don't change anything. But suppose from a ring, player 1 probes, thus lowering the payoff of player 2, and immediately player 2 probes and gets a higher payoff than in the previous round (although lower than her payoff in the ring). This can lead away from the ring.

Strings of subsequent probes can, with small probability, lead away from the ring. The question for the real *Probe and Adjust* dynamics is whether getting away from the ring to another state in this way is more or less probable than getting back from that state to the ring. Consider the scenario for departure from the ring just discussed. At t_1 player 1 probes, gets a worse payoff, and causes a worse payoff for player 2. At t_2 player 1 goes back to her original connect, and player 2 probes and gets a better payoff. At t_3 player 2 adopts her probe strategy. This is a scenario of probability order of e^2. (This is just the probability that player 1 probes at t_1 *and* player 2 probes at t_2.)

To get back to the ring, it suffices that player 2 then probes her original strategy. This has a probability of order e. For small e, the probability of getting from the alternative state to the ring is much greater than the transition in the opposite direction. Basically the same argument works in general: At t_0 we have a ring. The player or players that probe at t_1 return to their original strategies at t_2. Then we suppose that there are subsequent probes, either sequential or simultaneous or some combination of them—it doesn't matter. To get m players to switch from their original circle strategies then takes *at least* m+1 probes (probability of order e^(m+1)). To get back only requires the m players to simultaneously probe their original strategies (probability of order e^m). We may conclude that (i) *Probe and Adjust* dynamics learns the ring, and (ii) for infrequent probes, it then spends most of its time in a ring configuration.

6 *Probe and Adjust* in the Short Run

To get a sense of how long it takes *Probe and Adjust* to learn the ring in the Bala–Goyal game, we conducted some numerical simulations with three players. The results of these simulations are shown in figure 14.1. In these simulations, the cost for establish-

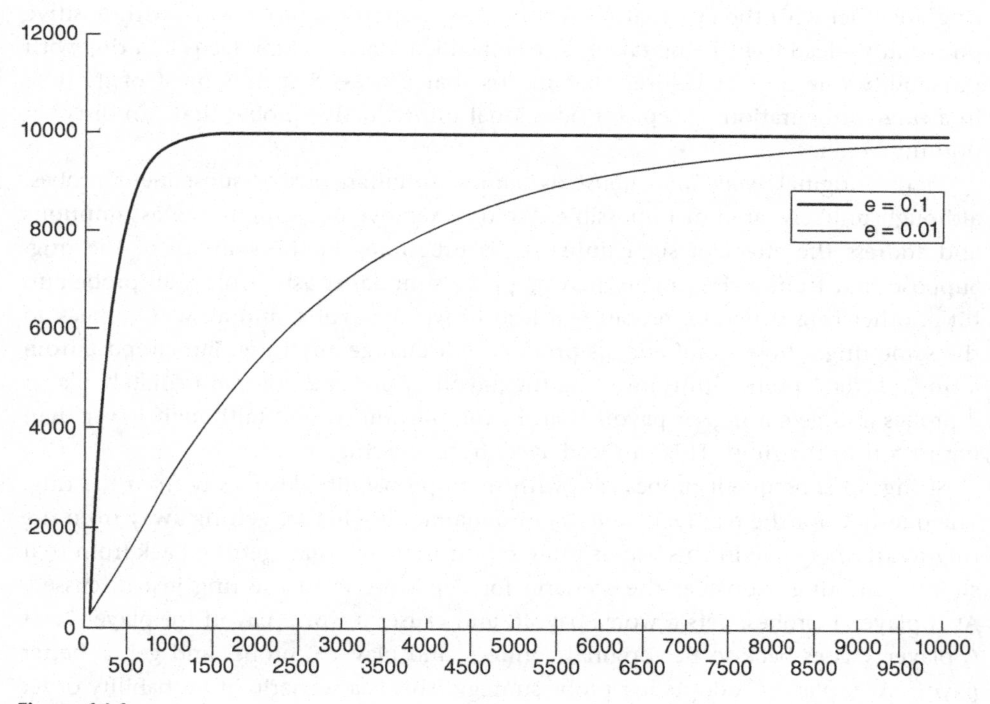

Figure 14.1
Simulations: probe rates .1 and .01.

ing a link was set to 0.6. The cost parameter could be varied and still lead to the same simulations as long as establishing links does not get too costly or too cheap ($0 < c < 1$), for adjustment exclusively rests on the ordinal relationships of the payoffs. When faced with a tie, players are assumed to choose equivalent strategies with equal probabilities.

Figure 14.1 shows two simulations, one with a probe rate of 0.1 and the other with a probe rate of 0.01. There were 10,000 trials each. The data points show the first time of hitting the ring configuration. If each player probes every tenth round on average, convergence to the ring appears to be very rapid. About 87 percent of all trials hit the ring before round 500. By round 1,000, almost all trials did hit the ring at least once. If each player on average probes only once in a hundred rounds, convergence appears to be somewhat slower, although by 10,000 rounds all trials have hit the circle. Looking at individual trials, once players choose according to the ring configuration, switching stably to a different configuration requires at least two players to probe consecutively and thus is a rare event. Even if it happens, the players typically switch back to the ring fairly rapidly.

7 Related Literature

On the *Best Response with Inertia* dynamics for network formation, see also Watts (2001), Jackson and Watts (2002), Goyal (2007), Jackson (2008). An alternative, low-rationality model of network formation based on *reinforcement learning* was introduced in Skyrms and Pemantle (2000). See also Pemantle and Skyrms (2004a,b), Skyrms and Pemantle (2010). Extensive simulations show that this dynamics does not learn the ring in the game discussed in this chapter. That slightly modified *Fictitious Play* dynamics reliably learns the ring is shown in Huttegger and Skyrms (2008). There we ask whether there is a low-rationality, low-knowledge, payoff-based dynamics that learns the ring—a question that is given an affirmative answer here. The *Probe and Adjust* dynamics discussed here is related to a class of payoff-based learning dynamics studied in Marden et al. (2009) and Young (2009). It is somewhat simpler than these in that it only requires agents to remember the past round.

There is a growing experimental literature on network formation. See Berninghaus et al. (2007), Callander and Plott (2005), Kosfeld (2004), and Falk and Kosfeld (2003). Experiments find that subjects do learn the ring in the game discussed here. They also tend to show that subjects are not using rules like *Best Response with Inertia* or *Fictitious Play*. Rather, there is a noisier learning process in which individuals at a strict equilibrium may temporarily leave it and then come back (Berninghaus et al., 2007). It would be interesting to see which learning rules best fit the empirical data.

Alternative network games with different payoff structures support star network configurations as strict equilibria. See Bala and Goyal (2000), Berninghaus et al. (2007), and Hojman and Szeidl (2008). *Probe and Adjust* dynamics also has applications in these cases, as well as well as in more general classes of games. We plan to explore this elsewhere.

Notes

1. A Markov chain is a stochastic process whose probabilities of transiting from one state to another depend only on the former state, and not on the previous history of the process. In a Markov process a state is absorbing if the stochastic system stays there with probability one once it has entered this state.

2. A Nash configuration is not the same as a Nash equilibrium. A Nash configuration is a set of Nash equilibria that are equivalent up to permutations of the players (such as rings).

References

Bala, V., & Goyal, S. (2000). A noncooperative model of network formation. *Econometrica: Journal of the Econometric Society, 68*, 1181–1231.

Berninghaus, S., Ehrhart, K.-M., Ott, M., & Vogt, B. (2007). Evolution of networks—an experimental analysis. *Journal of Evolutionary Economics, 17*, 317–347.

Callander, S., & Plott, C. R. (2005). Principles of network development and evolution: An experimental study. *Journal of Public Economics, 89*, 1469–1495.

Falk, A., & Kosfeld, M. (2003). It's all about connections: Evidence on network formation. IEW Working Paper 146. University of Zurich.

Goyal, S. (2007). *Connections: An introduction to the economics of networks*. Princeton: Princeton University Press.

Hojman, D. A., & Szeidl, A. (2008). Core and periphery in networks. *Journal of Economic Theory, 139*, 295–309.

Huttegger, S. M., & Skyrms, B. (2008). Emergence of information transfer by inductive learning. *Studia Logica, 89*, 237–256.

Jackson, M. (2008). *Social and economic networks*. Princeton: Princeton University Press.

Jackson, M., & Watts, A. (2002). On the formation of interaction networks in social coordination games. *Games and Economic Behavior, 41*, 265–291.

Kosfeld, M. (2004). Economic networks in the laboratory. *Review of Network Economics, 3*, 20–41.

Malinowski, B. (1920). Kula: The circulating exchange of valuables in the archipelagoes of eastern New Guinea. *Man, 20*, 97–105.

Malinowski, B. (1922). *Argonauts of the Western Pacific*. New York: Dutton.

Marden, J. P., Young, H. P., Arslan, G., & Shamma, J. S. (2009). Payoff-based dynamics for multiplayer weakly acyclic games. *SIAM Journal on Control and Optimization, 48*, 373–396.

Pemantle, R., & Skyrms, B. (2004a). Network formation by reinforcement learning: The long and the medium run. *Mathematical Social Sciences, 48*, 315–327.

Pemantle, R., & Skyrms, B. (2004b). Time to absorption in discounted reinforcement models. *Stochastic Processes and Their Applications, 109*, 1–12.

Skyrms, B. (2010). *Signals: Evolution, learning, and information.* Oxford: Oxford University Press.

Skyrms, B., & Pemantle, R. (2000). A dynamic model of social network formation. *Proceedings of the National Academy of Sciences of the USA, 97*, 9340–9346. (Reprinted with a postscript in T. Gross & H. Sayama [Eds.], Adaptive networks [Springer, 2009], pp. 231–251.)

Skyrms, B., & Pemantle, R. (2010). Learning to network. In E. Eells & J. Fetzer (Eds.), *Probability in science*. Berlin: Springer.

Watts, A. (2001). A dynamic model of network formation. *Games and Economic Behavior, 34*, 331–341.

Young, H. P. (2009). Learning by trial and error. *Games and Economic Behavior, 65*, 626–643.

15 Bacterial Social Life: Information Processing Characteristics and Cooperation Coevolve

Livio Riboli-Sasco, François Taddei, and Sam Brown

Introduction

Bacteria are fast becoming a new paradigm for social evolution, owing to their widespread investment in costly collective traits, such as the secretion of externally active proteins (Crespi, 2001; West, Diggle et al., 2007). The experimental and theoretical study of bacterial systems generates new perspectives on cooperation by drawing attention to the particular importance of information transfer in shaping bacterial cooperation. Information transfers are achieved in bacterial systems through very different processes, such as quorum sensing, conjugation, and transformation, and these channels can be established even between bacteria of different species. Interestingly, recent research has shown that genes coding for cooperative traits are more often transferred on infectious elements known as plasmids and temperate phages than are noncooperative traits (Nogueira et al., 2009).

In this chapter, we explore the links between an individual's ability to process information, population scale information dynamics, and cooperation. We argue that the study of bacterial systems opens new perspectives about how information dynamics may affect cooperation. We also argue that an individual's ability to process information constrains such dynamics. However, information processing characteristics that could be used for the analysis of a wide range of biological systems have not yet been defined. We will propose a few possible definitions. Finally, we will argue that the evolution of these generic processing characteristics is important not only in bacterial evolution but across a range of social species, including humans.

Information Is Necessary to Perform Cooperation; Its Dynamics May Thus Affect Cooperation

Recent research (Nogueira et al., 2009) has shown that in bacterial genomes, genes contributing to public goods production (genes coding for secreted proteins) are found in a higher proportion on plasmids. Secreted proteins are considered as potential public goods, as the benefits are not secured to the individual who has secreted the protein but can potentially be accessed by any other agent present in the environment. This result triggers questions about the potential causal links between the *mobility* of genetic elements involved in cooperative traits and the potential *emergence and maintenance of cooperation* in populations. In other words, we ask what are the dynamical links between gene mobility and the level of cooperation in a population. Nogueira et al. (2009, p. 1683) consider that "more mobile loci generate stronger among-individual genetic correlations at these loci (higher relatedness) and thereby allow the maintenance of more cooperative traits via kin selection." In other words, their model predicts that given a high level of mobility (for instance, due to the parasitic, horizontally transferred lifestyle of plasmids), more cooperative traits will evolve on these mobile loci. What we shall question in this chapter is more globally how cooperation and mobility coevolve. In particular, we also ask the reverse question of the one raised by Nogueira et al. (2009), which is how the mobility of specific genes already involved in cooperation evolves.

Research on cooperation is currently mostly driven by two perspectives. The first one is related to the way costs and benefits impact agents involved in an interaction (Doebeli & Hauert, 2005). Many such studies have benefited from theoretical developments inspired by economics (e.g., Rapoport, 1966). The second main perspective explores how population structure influences potential cooperation. It questions how individuals are genetically related to each other and tries to describe the networks of interaction (e.g., Taylor & Nowak, 2007; Grafen, 2007; Queller, 2003). These two perspectives can be translated into Hamilton's (1964, 1970) rule, which explains that altruism can be selected for if the expected benefit via genetically related individuals is higher than the cost of performing an altruistic action (Frank, 1998; Sachs et al., 2004; Lehmann & Keller, 2006; West, Griffin & Gardner, 2007). The rule is most of the time stated as $rB > C$ (where r represents the relatedness, B the expected benefit received from the altruist, and C the cost for the altruist performing the action). With a gene's-eye view, this can be also understood as an action that yields benefits to individuals that are holding the same genes as the altruist. The altruistic action thus allows these genes to spread. In terms of Hamilton's rule, the first perspective we have presented focuses on the cost-benefit ratio. The second perspective focuses on computing relatedness among individuals of specific populations in specific environments. These two lines of exploration have been fruitful and have contributed a lot to our understanding of cooperation. Population structure as well as the cost-benefit ratio resulting from the interaction can be deduced from the study of physical constraints of the environment, from the network of social interaction, and from the characterization of behaviors at stake.

In this chapter we focus on a key element that is needed for the establishment of cooperation, namely, on the basic and rather obvious fact that information is needed for any cooperative action to be performed. All biological processes, and not only cooperative ones, are specified, at least in part, by information (genetic, epigenetic, neuronal, hormonal, etc.). We acknowledge that the influence of genes on social behavior has been well integrated into current theoretical frameworks (Wenseleers, Gardner & Foster, 2010). Nonetheless, our perspective will focus more generically on information dynamics, in the light of bacterial examples. Bacterial systems highlight that information coding for social traits can be transferred by conjugation, within the life time of individuals. Coming back to the results presented by Nogueira et al. (2009), the genes found at a higher rate on plasmids are those needed for the production of public goods and thus those needed for cooperation to be established. We argue that the dynamics of information that specifies cooperative behaviors and the dynamics of cooperation in a population can be better understood together. This should be particularly relevant whenever the dynamics of information can reshape the information that individuals hold at a faster time scale than the generation time.

Accordingly, we argue that such a dynamical perspective may be suitable to explore social evolution when culture is at stake. Indeed, in humans, cultural elements, just as plasmids, can be horizontally transferred between individuals. Let's consider the following situation. You are driving, coming back from a vacation on the seaside. You really are in no hurry to get back to work at the university. On the road you see someone who seems to be having trouble with their car. You stop and offer some help. You do not mind losing a few minutes to help someone who's in trouble. Yet you quickly realize that you cannot do much, as you have so little mechanical knowledge (imagine you are a theoretical biologist). No matter how you offer to cooperate, you are totally useless—unless you know, for example, where to find the nearest mechanic open on a Sunday evening in August. This little story tells us something about the potential links between cooperation and information. However, only a few experimental studies have questioned human cooperation in a setting where people were using a significant set of cultural information. Milinski et al. (2006) showed, for example, that the more people were informed about climate change, the more they would financially contribute to a public good. This contribution to the public good was intended to pay for a communication campaign to raise awareness in the local population about the consequences of climate change.

Looking at Hamilton's rule from a dynamical perspective is relevant and widespread (Wenseleers, Gardner & Foster, 2010). We will specifically discuss how information dynamics may affect the relatedness term in the Hamilton equation. We propose to explore the consequences of an individual's information processing characteristics on information dynamics within a population and eventually on relatedness. Relatedness becomes thus a function of the information processing abilities of individuals in a population.

First, we need to clarify the concept of relatedness, as we consider it to be a way to integrate the impact of dynamical information into Hamilton's rule. The concept of kinship refers specifically to the relatedness of individuals generated via common descent. However, even in the simple example of bacteria, things can already become more complex. Take two sister bacteria, coming from the division of one bacterium. They will have the same chromosome (except for a few potential mutations), but they could have different plasmids. They are kin, as their chromosomal relatedness is high; however, their "plasmidic relatedness" may be low. On the opposite scale, two bacteria from different species and thus with different chromosomes may have the same plasmid. The classical way to overcome any ambiguity regarding issues of kinship and cooperation is to focus on information (i.e., genes) that directly generates the cooperative action and to compute relatedness at the precise level of this information, as illustrated by Nogueira et al.'s (2009) calculation of locus-specific relatedness coefficients. When defined as such, relatedness can vary even for periods of time shorter than the generation time of the species at stake, and according to specific mechanisms that allow the transfer of the specific information.

We shall distinguish this intraindividual, coding perspective from the one that questions the influence of the availability of *social* information within a population of individuals. We define "social information" as information about the current state of a population of interacting individuals. As shall be discussed later on, this social information may afterward trigger cooperative behavior. By contrast, the kind of information we were just discussing—for example, a gene for a shared, secreted protein—specifies cooperative behavior. It is not social information, describing the state of the social world.

We should also delimit how we will use the concept of cooperation. We consider cooperation as any action that is beneficial to others and has been selected for this beneficial function, thus encompassing definitions of altruism and mutualism (West, Griffin & Gardner, 2007). On this perspective, information sharing could be seen as a special case of cooperation, whenever sharing the information comes at some cost for some individuals and benefits others. However, this simple perspective skips the question of the use of the information. We thus consider that eventually making use of information allows cooperation to take place. We thus prefer to consider information sharing as a requisite for cooperation rather than as cooperation itself.

We have seen that genes contributing to public good production are found at a higher rate in plasmids. Experimental research in microbiology (Dionisio et al., 2002) has also shown very high variation (10^8-fold variation) in abilities to transfer or receive plasmids among different strains of bacteria. No evolutionary explanation has yet been offered to explain such variability. What could favor or disfavor the emission of information abilities of a given system? What could favor or disfavor the reception? We propose that information processing abilities may be evolving in particular to allow more or less "sharing" of information, depending on specific selective pressures. We also argue that it is difficult to develop a good understanding of these issues without a clarification of an individual's information processing characteristics.

Characterizing Information Processes

If we want to be able to characterize how information dynamics may affect relatedness, and eventually cooperation, we first have to understand what "information" is and what are the processing characteristics of the biological systems that are at stake.

The concept of information was imported into biology just after World War II, from developments in communication theory and cybernetics. Research in these fields had been fueled by military needs during the war (unveiling enemy coded communication, secure communication within the military, and secure communication between humans and machines, used for example in dropping bombs). The concept of information has been replacing the concept of specificity. Information became a key element in the early theoretical developments of molecular biology and later genetics. It also lies at the heart of the "central dogma" defined by Crick in 1958. It is very interesting to note that most scientists willing to define information refer to Shannon's self-information definition, which is associated with the unpredictability of an event. It is mathematically defined as log(probability of the outcome of a random variable). Weaver, introducing Shannon's book *Mathematical Theory of Communication* (Shannon & Weaver, 1949, p. 8), said "the word 'information' in this theory is used in a special sense that must not be confused with its ordinary usage. In particular, information must not be confused with meaning." Defining information as a concept in biology is a very complex issue. As biological information has an "impact" on biological systems, a potential definition may be closer to our ordinary-language concept of meaning than to Shannon's quantification. We shall not go into details here about defining information in biology; rather, we focus on clarifying information processing characteristics that we argue are key to understand relatedness dynamics. For example, this may help us define precisely what a microbiologist means when stating that bacteria that produce quorum-sensing molecules or transferring plasmids "talk to each other" (Bassler, 2004). Clear definitions and comparable characteristics are lacking and should be proposed to facilitate interspecific and even interkingdom comparisons of information processing. Eventually this may even lead to potentially useful comparisons between, for example, information processing via the nervous system of a human being, information processing via the immune system, and so on (Riboli-Sasco, 2010).

Information Processing Characteristics

We propose an analysis along three main sets of characteristics: content-holding characteristics, interfacing, and transferring characteristics (Riboli-Sasco, 2010). *Content-holding* characteristics relate to the ability of systems to hold information within themselves. "Holding" information thus goes along with structural identity. Information is enclosed within the system at stake (cell, brain, book, DVD, etc.). Content-holding is first of all about the nature of the content. By nature we refer to the primary physical and chemical dimensions of the object at stake. Is it a chemical molecule? An electrical wave? The second main aspect of these systems is that there are interfaces between one layer of information and a resulting process. We call an *interface* a process that directly interacts with a piece of information to convert it into a subsequent information content or into an action. Encoding–decoding processes lie at the heart of any analysis of information processes. The last main dimension of our analysis is about *transferring* processes. The easiest approach is to consider the emission and reception characteristics separately, as these are not symmetrical. Types of emission vary depending on the network of agents involved. First of all we suggest analyzing emission in terms of *frequency* of emitted content. Two other key elements are whether the emission is *targeted* and whether the process is *context-dependent*. By targeting we refer to any process where information is emitted toward other agents in a nonrandom manner. Reception can be analyzed in terms of *rate* of reception and in terms of possible *acceptability* mechanisms whenever the reception process is nonrandom or context-dependent (Riboli-Sasco, 2010).

Aggregates of Information Processing Characteristics

We suggest that it might be useful to build concepts referring to aggregates of information characteristics (Riboli-Sasco, 2010). We argue that these may be a relevant level of organization upon which selection is acting. These aggregates can thus be viewed as so-called emergent properties. We will refer more simply to these aggregates as second-order information processing properties, to distinguish them from the first-order characteristics outlined in the previous section. We will present only one of such aggregate, namely the one we consider useful for the understanding of the evolution of cooperation.

We define *interoperability* as measure of the range of systems that are able to operate mobile information (Riboli-Sasco, 2010). By "operating information," we mean that the system is able to run on this mobile information a set of processes that we have characterized as being of the first order. We thus consider that interoperability is an aggregate of first-order characteristics that will allow a piece of information to be effectively transferred and used in various biological systems. For example, the content-holding characteristics should be such that the information, when transferred, is not altered. The interfacing processes should be based on encoding and decoding processes that allow various biological systems to interface the same information. Eventually, transferring characteristics will modulate the extension of interoperability in terms of the number of systems reached by a piece of information.

Plasmids, and in particular "broad host range plasmids," are a paradigmatic example of interoperable information. We analyze in table 15.1 how plasmid transfer can be interpreted according to our previous characteristics.

Table 15.1

Content-holding	Nature	DNA: chemical nature
Interfacing	Encoding/decoding	Through the so-called genetic-code
Transferring	Type	One to many (copies of a plasmid can reach multiple bacteria)
	Emission frequency	Variable
	Targeting	Yes, through pili adhesion
	Context-dependence	Not well known
	Reception rate	Variable
	Acceptability	Possible mechanisms to prevent plasmid expression after reception: degradation by restriction enzymes

We know that bacteria from different strains are able to express and thus interoperate the same plasmids. This situation is made possible by a commonality of interfacing processes among different strains of bacteria. Whenever this is associated with intense mobilization of plasmids (owing to the interfacing characteristics) we have highly interoperable plasmids.

Coevolution of Interoperability and Cooperation

According to Hamilton's rule, any modification of focal relatedness of genes specifying cooperative behaviors may have an effect on the stability of cooperative behaviors between interacting individuals. Thus, changes in the interoperability of information specifying cooperative action could have an impact on the overall level and stability of cooperation within a population.

As mentioned in the previous section, Nogueira et al. (2009, p. 1683) consider that "more mobile loci generate stronger among-individual genetic correlations at these loci (higher relatedness) and thereby allow the maintenance of more cooperative traits via kin selection." Our first suggestion is to enlarge this consideration from mobility to a more general interoperability approach. This move allows us to stress the need to operate the mobile information, and not only to receive it. Second, while retaining the validity of this prediction, we question whether the coevolution of interoperability and cooperation can also happen the other way round, that is, what is the impact of cooperation on selection for gene mobility and interoperability?

Such a question can benefit from the more general niche construction theoretical framework (Odling-Smee, Laland & Feldman, 2003), which accounts for environmental alteration by organisms that affects the evolutionary pressures acting on them. By definition, interoperability allows large-scale sharing of action, as the information is not only shared but can also be operated. Thus, within the niche construction framework, higher levels of interoperability may correspond to, and coevolve with, important shared niche-construction abilities.

The importance of gene mobility in maintaining cooperation was first identified by Smith (2001), who proposed a model whereby populations of bacteria vulnerable to cheats (nonproducers) could be rescued by making the cooperative gene infectious (plasmid borne), and so able to convert or reprogram cheats into cooperators. Yet as McGinty, Rankin, and Brown (2011) argue, as soon as competition with noncooperative mobile variants is introduced (e.g., plasmids without the cooperative gene), cooperators are no longer maintained in an unstructured environment. Spatial structure can then rescue mobile cooperative alleles, by allowing for the development of stronger genetic correlations (higher relatedness) at mobile loci (Nogueira et al., 2009). A significant assumption of this work is that mobile genes are fully interoperable, and there is little difference in transmission or interoperability properties with increasing investment in cooperation. We argue that it is crucial to enhance our understanding of information-processing characteristics to predict how they can favor or disfavor interoperability. This will eventually help us understand and question more globally the way information dynamics and cooperation abilities coevolve, in systems such as bacterial communities transferring plasmids.

Until now we have mostly presented examples in the bacterial domain, in large part because of the clear mechanistic basis of social traits and gene mobility in bacteria. We shall now explore a biological process—teaching behavior in animals—where sufficient knowledge has been acquired to raise questions about the coevolution of information interoperability and cooperation. Teaching has been defined by Caro and Hauser (1992, p. 153) as follows:

> an individual actor A can be said to teach if it modifies its behavior only in the presence of a naïve observer B at some cost or at least without obtaining an immediate benefit for itself. A's behavior thereby encourages or punishes B's behavior, or provides B with experience, or sets an example for B. As a result, B acquires knowledge or learns a skill earlier in life or more rapidly or efficiently than it might otherwise do, or that it would not learn at all.

This definition allows extending the study of teaching processes well beyond humans as it implies that teaching need not involve complex cognitive abilities. Also following the definition, teaching can be considered an altruistic behavior. If such a definition is accepted then one can consider that there is teaching in ants (Franks & Richardson, 2006) and meerkats (Thornton & McAuliffe, 2006, 2008), as presented in table 15.2. Teaching has been suggested in other species but proper data are lacking to fit to the definition. For example, bees have behaviors similar to those highlighted in table 15.2 for ants. They can transfer information about food location. However, the cost and benefits of such information-transfer behaviors have not been yet computed.

The current literature on teaching does propose evolutionary explanations for such altruistic behavior. These are mainly based on kin selection effects in structured populations (Hoppitt et al., 2008). We do not reject such explanations, but we think that one could frame the explananda differently: In the case of meerkats and ants, the

Table 15.2

Teaching in ants and meerkats

	"Teacher"	"Pupil"	Process at stake	Cost for teacher	Benefit for "pupil"
Ants	Any ant that recently found the location of a resource	Any ant passing close by	Guiding another ant on the way toward resources	Time not devoted to gathering resources for the colony	Quicker access to scattered resource than by exploration alone
Meerkats	Adults	Young ones (not necessarily related ones)	Capturing scorpions as a food resource	Time not devoted to feeding. Risky behavior	Quicker and safer increase in the ability to access a dangerous resource

environment offers resources that one individual cannot access on its own. In the case of ants, the individuals lack the knowledge about the location of food. If they were to search alone and independently for food they may well starve before finding anything. Access to information about food location is thus easier and cheaper than accessing food directly. In the case of meerkats, accessing food is highly dangerous. The probability of getting killed is high. Accessing information about procedures to deal with dangerous food is thus easier and cheaper than trying to independently and naively access food.

In both examples, maintaining and sharing within the population information about resources can be viewed as a niche construction process. Knowledge about location or about how to process the resource allows for the modification of the selective pressures initially imposed by the environment. Abilities to interoperate information could thus result from selection favoring individual niche construction abilities. This level of interoperability is made possible by specific information processing characteristics as outlined in table 3.3.

Individuals within bacterial populations or ant colonies do not interact with each other at random. We shall now investigate how issues related to population structure and social information can be understood in connection with the coevolution of interoperability and cooperative abilities.

Dynamical Interactions between Relatedness and Population Structure in Bacterial Systems

It has been shown that bacterial conjugation is an important feature of biofilms, and it has been suggested (Ghigo, 2001) that conjugation can favor biofilm formation.

Table 15.3

Information processing characteristics allowing interoperability of information between ants and between meerkats. We consider in these tables the information at the transfer stage.

Species		Ants	Meerkats
Content-holding	Nature	Physical contact	Visual (observation)
Interfacing	Encoding/decoding	Through the neuronal system	Through the brain
Transferring	Emission frequency	Unknown	Many events each day
	Targeting	Yes	Yes
	Context-dependence	When food resource is available, when location is known	When food (scorpio) is available
	Reception rate	Unknown	Can be inferred from success rate of "pupils" in the task at stake
	Acceptability	Unknown mechanisms if any	Unknown mechanisms if any

Indeed, plasmid transfer is made possible by a physical connection through a pilus. These pilus connections participate to increase the cohesion of the biofilm, and in the end create a more rigid structure for the population. Biofilms can also be seen as a niche resulting from a niche construction process. Indeed, biofilms offer better collective access to nutrients such as oxygen, as they allow bacteria to float at the surface of liquid media. The example of bacterial biofilms, where conjugative processes induce physical structuration, shows that information-transfer processes cannot be considered independently from issues of population structure. Population structure plays in itself an important role to allow or prevent cooperation as it constrains interactions with neighboring partners.

Interestingly, we can also question what the role might be of potential *social* information about the current state of the population structure. Bacteria are able, for example, to produce molecules that allow them to "sense" the amount of other bacteria in the environment. These molecules have been called "quorum-sensing molecules." There is some debate over whether these molecules instead should be called "density sensing" or "efficiency sensing" (Hense et al., 2007). Still, these processes allow individuals to get "information about the current state of a population." In particular, these systems may allow the triggering of a cooperative behavior such as virulence factor production. For example, various proteases can be secreted by *Pseudomonas aeruginoa* in the external environment to exploit its hosts. The potential benefits will be shared by the whole population. Models have explored how such a process can be reliable while costly (Brown & Johnstone, 2001, empirically supported by Diggle et al., 2007). Among humans, cooperative actions are also under the dependence of social information. Knowing that a person is supposedly "generous" will increase the probability of initiating a cooperative action toward her (Milinski, Semmann & Krambeck, 2002).

Following our discussion, we need to ask how such "social signals" evolved together with cooperative abilities and what the difference is between dynamics related to information that directly specifies the cooperative behavior and dynamics about social information. Will cooperative settings select for increased interoperability of social signals as well? The few examples we have given point to the advantage that individuals may have if they are able to trigger cooperative behavior only when the environmental and social conditions are such that the behavior will turn out to be beneficial. One may also expect the emergence of cheating behaviors that induce a bias in the inferred social information (Brown & Johnstone, 2001; Diggle et al., 2007). We expect that the information processing ability that allows one to distinguish between "relevant" versus "useless," as well as "reliable" versus "unreliable" social information, is under selection. We might also predict a red queen dynamics to rescue cooperation by restoring the reliability of social signals. This can be a way to interpret much of the social tag literature (Jansen & van Baalen, 2006): Social tags that cheaters have started to use are replaced by new tags that are not yet "contaminated" by cheaters. When talking about social information rather than procedural information, issues of "reliability" become even more important here than issues of pure interoperability. It might be important not to operate received information that is "manipulating" the agent, leading to behaviors that in the end reduce one's fitness. In our conceptual framework, the "reliability" of information should be highly dependent on interfacing characteristics of the systems at stake. A better understanding of information processing characteristics might once again prove to be helpful to our understanding of the emergence and maintenance of cooperation.

Perspectives

We would like to point first to a key notion that should be further explored, namely the fact that the characteristics of information processing are such that any modification to them may have a large-scale impact. For example, the evolution of a focal social trait may select for information processing characteristics that enhance the interoperability of a larger amount of information processed by the individuals at stake. We suggest, for example, that so-called major evolutionary transitions could perhaps be better understood in the light of selection over information properties with large-scale effects.

We also call for experimental research and simulations to explore the extent of selection for interoperability in distinct biological and cultural systems. We should also refine our conceptual tools to analyze more deeply and quantify the way information is shared, stored, and balanced between interoperable and noninteroperable settings in living systems. The level of exploration of the information properties we are calling for is still to be achieved. We have presented only a glimpse of what this approach can bring. Information properties should be thoroughly analyzed, quantified, and compared. Even if there is little of such data on living systems, we predict that acquiring it may lead to a better understanding of living systems. Such an approach has indeed been fruitful in computer sciences. Why should it be different in biology? We have tried to argue that the analyses of specific information properties may at least help with our understanding of cooperation.

Conclusion

Cooperation is an amazing property present in many living systems. It is amazing because it is so puzzling to scientists. It is also amazing because it is so fundamental to our daily human lives. How could we live without cooperating with our colleagues, with our friends, with our loved ones? We have tried here to explore how agents are engaged in cooperation. We argue that we can enhance our understanding of cooperation from an understanding of one of the core properties of life, namely that it is based on information processing. Information can potentially be interoperated between different agents or systems. We suggest that selection can "play" with this interoperability, just as selection plays with what genes specify in terms of phenotype. We propose that cooperation coevolves with information processing characteristics, and in particular with interoperability. Such a view, linking cooperation and information transfer, does not contradict most of the existing theories about cooperation. We feel that, on the contrary, such a perspective is coherent with all explanations related to Hamilton's rule, as interoperability can impact focal relatedness.

References

Bassler, B. L. (2004). Cell-to-cell communication in bacteria: A chemical discourse. *Harvey Lectures, 100*, 123–142.

Brown, S. P., & Johnstone, R. A. (2001). Cooperation in the dark: Signalling and collective action in quorum-sensing bacteria. *Proceedings of the Royal Society of London, Series B: Biological Sciences, 268*(1470), 961–965.

Caro, T. M., & Hauser, M. D. (1992). Is there teaching in nonhuman animals? *Quarterly Review of Biology, 67*(2), 151–174.

Crespi, B. J. (2001). The evolution of social behavior in microorganisms. *Trends in Ecology & Evolution, 16*(4), 178–183.

Crick, F. H. C. (1958). On protein synthesis. *Symposia of the Society for Experimental Biology, 12*, 139–163.

Dawkins, R. (1976). *The selfish gene*. Oxford: Oxford University Press.

Diggle, S. P., Griffin, A. S., Campbell, G. S., & West, S. A. (2007). Cooperation and conflict in quorum-sensing bacterial populations. *Nature, 450*(7168), 411–414.

Dionisio, F., Matic, I., Radman, M., Rodrigues, O. R., & Taddei, F. (2002). Plasmids spread very fast in heterogeneous bacterial communities. *Genetics, 162*(4), 1525–1532.

Doebeli, M., Hauert, C., & Killingback, T. (2004). The evolutionary origin of cooperators and defectors. *Science, 306*(5697), 859–862.

Doebeli, M., & Hauert, C. (2005). Models of cooperation based on the prisoner's dilemma and the snowdrift game. *Ecology Letters*, (8), 748–766.

Frank, S. A. (1998). *Foundations of social evolution*. Princeton, NJ: Princeton University Press.

Franks, N. R., & Richardson, T. (2006). Teaching in tandem-running ants. *Nature, 439*(7073), 153.

Ghigo, J. M. (2001). Natural conjugative plasmids induce bacterial biofilm development. *Nature, 412*(6845), 442–445.

Grafen, A. (2007). An inclusive fitness analysis of altruism on a cyclical network. *Journal of Evolutionary Biology, 20*(6), 2278–2283.

Hamilton, W. D. (1964). The genetical evolution of social behaviour. I & II. *Journal of Theoretical Biology, 7*(1):1–52.

Hamilton, W. D. (1970). Selfish and spiteful behaviour in an evolutionary model. *Nature, 228* (5277), 1218–1220.

Hasman, H., Chakraborty, T., & Klemm, P. (1999). Antigen-43-mediated autoaggregation of Escherichia coli is blocked by fimbriation. *Journal of Bacteriology, 181*(16), 4834–4841.

Hense, B. A., Kuttler, C., Muller, J., Rothballer, M., Hartmann, A., & Kreft, J. U. (2007). Does efficiency sensing unify diffusion and quorum sensing? *Nature Reviews Microbiology, 5*(3), 230–239.

Hoppitt, W. J., Brown, G. R., Kendal, R., Rendell, L., Thornton, A., Webster, M. M., et al. (2008). Lessons from animal teaching. *Trends in Ecology & Evolution, 23*(9), 486–493.

Jansen, V. A., & van Baalen, M. (2006). Altruism through beard chromodynamics. *Nature, 440*(7084), 663–666.

Lehmann, L., & Keller, L. (2006). The evolution of cooperation and altruism—a general framework and a classification of models. *Journal of Evolutionary Biology, 19*(5), 1365–1376.

McGinty, S. E., Rankin, D. J., & Brown, S. P. (2011). Horizontal gene transfer and the evolution of bacterial cooperation. *Evolution: International Journal of Organic Evolution 65*(1), 21–32.

Milinski, M., Semmann, D., & Krambeck, H. J. (2002). Donors to charity gain in both indirect reciprocity and political reputation. *Proceedings of the Royal Society of London, Series B: Biological Sciences, 269*(1494), 881–883.

Milinski, M., Semmann, D., Krambeck, H. J., & Marotzke, J. (2006). Stabilizing the earth's climate is not a losing game: Supporting evidence from public goods experiments. *Proceedings of the National Academy of Sciences of the United States of America, 103*(11), 3994–3998.

Nogueira, T., Rankin, D. J., Touchon, M., Taddei, F., Brown, S. P., & Rocha, E. P. (2009). Horizontal gene transfer of the secretome drives the evolution of bacterial cooperation and virulence. *Current Biology, 19*(20), 1683–1691.

Odling-Smee, F. John, Laland, Kevin N., & Feldman, Marcus W. (2003). *Niche construction: The neglected process in evolution*. Princeton, NJ: Princeton University Press.

Queller, D. C. (2003). Theory of genomic imprinting conflict in social insects. *BMC Evolutionary Biology, 3,* 15.

Rapoport, A. (1966). *Two-person game theory: The essential ideas*. Ann Arbor: University of Michigan Press.

Riboli-Sasco, L., & Brown, S. P., Taddei, F. (2008). Why teach? The evolutionary origin and ecological consequences of costly information transfer. In P. d'Ettorre & D. P. Hughes (Eds.), *Sociobiology of communication: An interdisciplinary perspective*. Oxford: Oxford University Press.

Riboli-Sasco, L. (2010). Evolving information in living systems: A pathway for the understanding of cooperation and major transitions. Frontiers of Life Graduate School, Paris Descartes University, Paris.

Sachs, J. L., Mueller, U. G., Wilcox, T. P., & Bull, J. J. (2004). The evolution of cooperation. *Quarterly Review of Biology, 79*(2), 135–160.

Shannon, C. E., & Weaver, W. (1949). *The mathematical theory of communication*. Urbana: University of Illinois Press.

Smith, J. (2001). The social evolution of bacterial pathogenesis. *Proceedings of the Royal Society of London, Series B: Biological Sciences, 268*(1462), 61–69.

Taylor, C., & Nowak, M. A. (2007). Transforming the dilemma. *Evolution: International Journal of Organic Evolution, 61*(10), 2281–2292.

Thornton, A. (2008). Variation in contributions to teaching by meerkats. *Proceedings of the Royal Society of London, Series B: Biological Sciences, 275*(1644), 1745–1751.

Thornton, A., & McAuliffe, K. (2006). Teaching in wild meerkats. *Science, 313*(5784), 227–229.

Wenseleers, T., Gardner, A., & Foster, Kevin, R. (2010). Social evolution theory: A review of methods and approaches. In T. Székely, A. J. Moore & J. Komdeur (Eds.), *Social behaviour: Genes, ecology, and evolution*. Cambridge: Cambridge University Press.

West, S. A., & Buckling, A. (2003). Cooperation, virulence, and siderophore production in bacterial parasites. *Proceedings of the Royal Society of London, Series B: Biological Sciences, 270*(1510), 37–44.

West, S. A., Diggle, S. P., Buckling, A., Gardner, A., & Griffin, A. S. (2007). The social lives of microbes. *Annual Review of Ecology Evolution and Systematics, 38,* 53–77.

West, S. A., Griffin, A. S., & Gardner, A. (2007). Social semantics: Altruism, cooperation, mutualism, strong reciprocity, and group selection. *Journal of Evolutionary Biology, 20*(2), 415–432.

16 Two Modes of Transgenerational Information Transmission

Nicholas Shea

1 Introduction

Much of the organized complexity found in living things depends on communication. Many differences between forms of social organization can be traced to differences in the way information is communicated between individuals. The complexity of an organism depends on communication between cells, organs, and other component parts. These forms of horizontal communication rightly receive considerable attention. However, transgenerational communication is equally significant. Differences in the way that information is communicated down the generations also play an important role in explaining differences among living things. In particular, major innovations in transgenerational signaling have probably been especially important in recent hominin evolution.

It has long been recognized that teaching and learning transmits information between generations in humans, and that more limited forms of cultural inheritance are also found in other species. Now the burgeoning literature on epigenetic and other parental effects shows that there may be very many routes for communication between generations that go well beyond the bounds of cultural inheritance. We gain a useful perspective on these results by asking how information is being transmitted between generations in each case.

This chapter argues for a distinction between two ways that adaptively significant information may be transmitted between generations. The distinction is roughly between parental organisms blindly transmitting information and their finding out information for themselves. Both generate *correlational information* to which offspring can react adaptively. Correlational information is found whenever some entity's being in a particular state changes (usually raises) the probability that some other entity is in another particular state. Organisms can make use of correlational information because they can react to the information-carrier as they would if they had access directly to the thing with which it correlates. With detection-based transgenerational effects, the parent detects a fact and transmits a correlate of it to its offspring. By contrast, selection-based effects depend on parents reliably transmitting variants without doing any detection for themselves. Natural selection on such variants then generates the correlational information on which offspring organisms can rely.

We have two tasks: to draw the distinction and to justify that it is a matter of information transmission. The philosophical material on channels of semantic information is logically prior, but much more comprehensible if the distinction is first laid out and justified empirically. So section 2 will set out the distinction and section 3 will argue that to understand the evolutionary consequences of transgenerational effects it is important to categorize mechanisms in terms of how they form channels of information transmission. Section 4 completes the task of motivating the utility of the distinction by showing how it is a useful way of thinking about hominin cultural inheritance. Sections 5 and 6 turn to the second, more obviously philosophical task. Section 5 justifies the assumption that selection-based and detection-based effects are channels of information in a substantive sense that differs from other kinds of causal interaction by showing how they carry semantic information. Section 6 argues that appealing to semantic information offers a kind of explanatory purchase that is missing from a purely causal-mechanical take on transgenerational effects.

2 Detection-Based and Selection-Based Information

This section sets out the distinction between detection-based and selection-based effects and motivates it on the basis of empirical findings. An important class of our detection-based effects is studied in the empirical literature on transgenerational phenotypic plasticity. In such cases, a range of phenotypic outcomes is open to the developing organism, and which one it takes depends on an effect from a parent. For example, the herb *Campanulastrum americanum* can follow one of two life history strategies. It can either germinate in the autumn and flower the following summer (annual), or it can germinate in the spring and flower in its second summer (biennial). Which strategy is adaptive depends on the local environment in which the plant grows. Plants that grow in shady woodland understory have higher overall fitness if they adopt the biennial strategy. Those that grow in a light gap do better if they grow as annuals, making use of the extra light to reproduce more quickly. Interestingly, experiments have shown that the strategy adopted by a seedling is affected by whether its mother grew in understory or light gap (Galloway & Etterson, 2007), and that this strategy is adaptive in the wild.

This form of transgenerational plasticity makes sense since a seedling is likely to be in the same kind of light environment as its mother, given limited seed dispersal. The choice between germinating in spring or autumn has to be made at an early stage, when it is presumably difficult or impossible for the germinating seed to detect directly which form of light environment it is in. Instead, the observed plasticity seems to be driven by some form of nongenetic maternal effect that correlates with the maternal light environment. Seedlings rely on this natural sign of light environment and produce an adaptive developmental response. In this respect, the case is no different from within-generation examples of adaptive phenotypic plasticity, where the developing organism picks up on correlational information in its environment in order to fix on an adaptive developmental outcome—for example, when the water flea *Daphnia pulex* reacts to chemical traces of predators by growing a protective outer shell.

Wherever there is causal interaction, there is information in the thin, correlational sense. However, in these cases there is information in a stronger, semantic sense, as argued for in section 5. Out of all the ways that a parent organism causally affects its offspring, there is a principled reason for singling out these cases of transgenerational adaptive plasticity as involving information transmission in a substantive sense. The offspring acts on detection-based information transmitted by the parent.

In addition to detection-based effects, some epigenetic effects are stable over very many generations. For example, a single treatment of the ciliate *Tetrahymena* with insulin leads to a significant increase in insulin binding in progeny for at least 664 generations (Csaba, 2008; see Jablonka & Raz, 2009, for further examples). Natural selection can act on stably transmitted epigenetic effects just as it can on genes. Such cases should not be assimilated to the category of transgenerational plasticity. Instead, if such epigenetic effects carry information, it will be much more like the information carried by genes (Shea, 2007a). Their effects may be context sensitive, but these are cases where differences in phenotypic outcome are due not to epigenetic variation, but rather to variation in the environment. In a standard heritability analysis, these stably transmitted epialleles will be indistinguishable from genes (Helanterä & Uller, 2010).

An initial reaction is that stably transmitted epigenetic factors are not transmitting information at all. After all, they are unvarying, unlike the usual cases where a variable signal transmits information about a varying matter of fact. But that is a mistake. If an epigenetic factor is selected as a result of some phenotypic difference it makes, then the frequency of that factor in the population will increase. At the outset, the epigenetic factor will correlate with the phenotypic result it produces. Assuming it arises at random, it will initially carry no information about the external environment. However, as a result of selection, it will come to carry information about the external environment—it will correlate with whatever environmental parameter or parameters are conducive to the phenotypic effect with which it in turn correlates. The same process occurs when a genetic variant is selected. The process of selection leads genetic types to carry correlational information about features of the environment in which they were selected. To the extent that the environment has not changed in relevant respects, this information will still be useful to the developing organism.

With genes, it is not just fortuitous that the mechanisms of DNA expression and replication are acted on by natural selection. DNA has been designed by evolution to

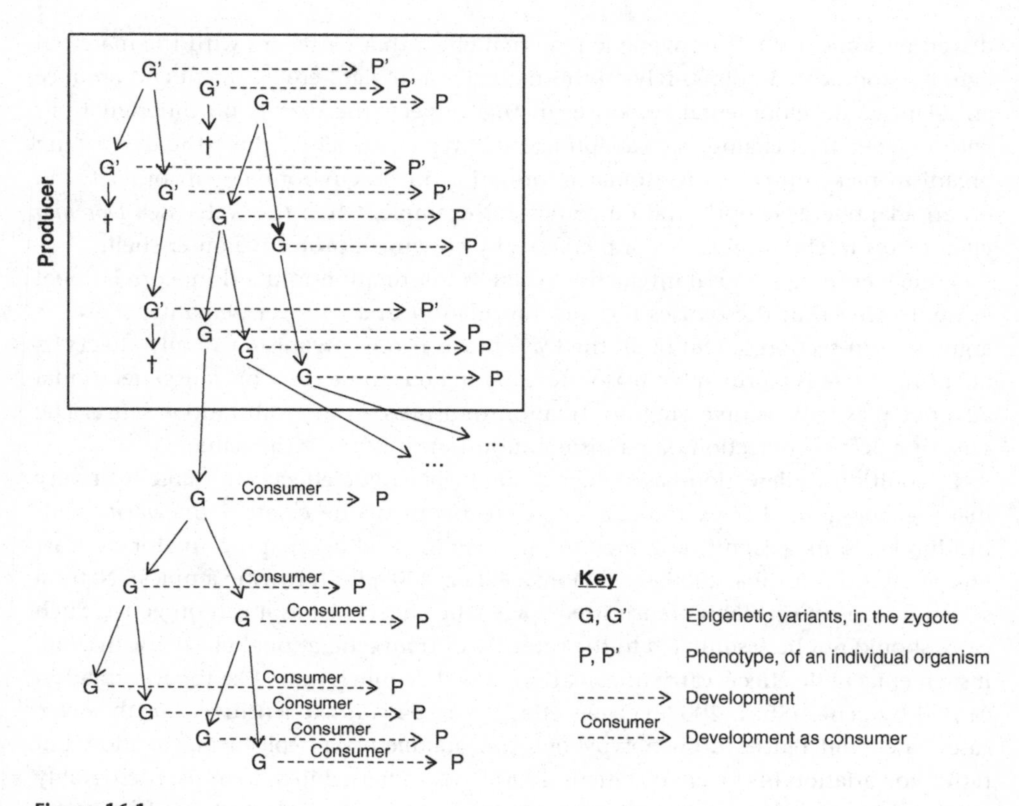

Figure 16.1

Epigenetic selection-based information. Information *production* takes place over many generations and involves selection. During this time, development (of phenotype P in response to epigenetic type G) is part of the process of representation production. At the outset, G arises by a random process (e.g., mutation) and leads to, and correlates with, phenotype P; at the end of representation production, G correlates both with P and with some environmental factor(s) E in virtue of which G was selected (E not shown). *Consumption* of the selection-based information carried by G takes place in each subsequent generation, in every individual carrying G in the zygote and developing P as a result. (Content of G is: E is the case; develop P—see section 5.)

perform the function of transmitting phenotypes to future generations (Freeland & Hurst, 1998; Haig & Hurst, 1991). So the consumption of such DNA-environment correlations by the mechanisms of development, in giving rise to environmentally appropriate phenotypes, is a matter of design (Shea, 2007a). The same may be true of some mechanisms of epigenetic inheritance. They, too, may have been selected to perform the function of transmitting phenotypes down the generations. If so, such epigenetic factors will count as transmitting information in the semantic sense argued for in section 5 below. Unlike detection-based effects, the information they carry is due to selection. They are *selection-based effects* carrying *selection-based information.*

Sometimes an epigenetic variant will be selected because of an effect that it has within the organism, a physiological effect, say, without altering the way the organism interacts with its external environment. In those cases the "environment" about which the epigenetic variant carries information (as a result of selection) is the existing physiology of the organism. If the epigenetic effect is selected because of some effect that occurs within the organism, then there is some feature or features of the organism with which that effect interacts so as to increase fitness. The epigenetic factor comes to carry information about those features.

Selection-based effects can be context sensitive. This is still information generated by selection, but information of a more complex sort. The selected epigenetic variant instructs development to produce a range of outcomes and carries the information that the environment is likely to be conducive to phenotypes within that range. Where each of the phenotypes induced in different environmental contexts is adaptive, we can say something more specific. Then the information is roughly: The environment is likely to be E1 or E2; in E1 produce phenotype P1 and in E2 produce P2. Selection on stably transmitted epigenetic factors could be the basis of such adaptive plasticity, which would be epigenetic accommodation of a plastic phenotype, just as there can be genetic accommodation of a plastic phenotype (Gilbert & Epel, 2009). A particular adaptive phenotype (P1) would then result from a combination of selection-based information about the likely range of environments and detection-based information about the specific environment (E1) within which a particular organism is developing. The latter could be detected by the developing organism itself or, as with *Campanulastrum,* could be detected by a parent and transmitted to the developing offspring.

With the basic distinction between detection-based and selection-based information in hand, we can revisit *Campanulastrum.* Why does its form of developmental plasticity rely on a maternal effect rather than a genetic cue? Modeling work by Leimar, Hammerstein, and Van Dooren (2006) suggests that that question should be answered by examining the quality of the information available through those different routes. Genes would be unable to build up correlational information at the available timescale. Paternal DNA is dispersed by the wind and so is drawn from a much larger area than the spatial scale on which the variation between light gap and understory is found, making paternal information about the light environment unreliable. And lineages of offspring do not remain in the same light environment as their mothers for sufficiently many generations for natural selection to act and build up maternally inherited information about being in a light gap, say (either in genes, or in long-term maternal epigenetic effects). In short, the structure of the problem suggests that detection-based information from the mother will be the most reliable source of correlational information about the local light environment, if a mechanism for conveying such signals is available—as it seems to be. The literature on the evolution of individual and social learning also focuses on the circumstances in which detection-based information (of various sorts) is superior to selection-based information (Boyd & Richerson, 1985; Feldman, Aoki & Kumm, 1996; Sterelny, 2003, pp. 162–171).

In addition to selection-based information that produces particular adaptive phenotypes, there are other cases where epigenetic mechanisms serve to produce undirected variation in offspring phenotypes. Typically environmental stress can trigger epigenetically caused variability in offspring. For example, oxidative stress or heat shock can directly alter epigenetic marks in the germ line, leading to random variations in offspring phenotypes. Such variability can be adaptive for the parent, and while there may be cases where it is an adaptation (a bet-hedging response to a change in environment), in many cases the epigenetic variability is just a direct causal effect of the environmental stress and not the result of selection. Such cases would not then count as cases of the parent transmitting adaptively relevant semantic information to the offspring.

Intergenerational Conflict

Detection-based epigenetic information transmission raises the same issues about the possibility of honest signaling as are canvassed in the existing literature on animal signaling. If the interests of sender and receiver diverge, what prevents the sender from sending dishonest signals? But if the signals are unreliable, then receivers would not have a reason to react to the signal with an informationally relevant response. How then could a signaling system ever evolve? This problem is overcome if the signaling occurs in a cooperative context, so that the interests of the sender and receiver are aligned. For example, the famous honeybee waggle dance is underpinned by the fact that nonreproductive worker bees all have the same reproductive interest in their shared hive. These obstacles are of course overcome in many cases of animal signaling that are not fully cooperative, as explained by factors such as costly signaling of fitness, or production of signals that are otherwise hard to fake, or where dishonest signaling imposes large developmental costs. Should we expect similar obstacles to the evolution of transgenerational detection-based information transmission?

The fitness interests of parents and offspring are usually closely aligned in most respects. That provides the basis for cooperative signaling about many aspects of the environment. However, they do come apart in nonsignaling contexts, for example where the parent has limited resources to distribute among its progeny. Parental fitness will often be best served by distributing resources equally among several progeny, whereas each one of those individuals has a stronger fitness interest in getting an unequally large share of the resources.

Something similar may occur with respect to offspring of offspring. Suppose that, in resource-poor environments, it is adaptive for a mother to limit the number of offspring produced in turn by each of her progeny. When resources are in short supply, it might be in the mother's best interests for each daughter to have on average m offspring, whereas each daughter's individual fitness interest would be to have $n > m$ offspring. The mother could send a signal to its offspring that correlates with whether the environment is likely to be resource poor or resource rich. But would the offspring react to the signal by producing the behavior that is adaptive for the parent? Modeling work strongly suggests that they would not, unless the mother has a means to take control of the relevant aspect of her offspring's development and canalize it toward her preferred result (Uller and Pen, 2011). It is easy to see why. Offspring will evolve to react to correlational information in a way that is adaptive for themselves (cf. the noninformational treatment in Hinde, Johnstone & Kilner, 2010). The result is that, as with regular animal signaling, transgenerational detection-based epigenetic information transmission will not evolve in noncooperative contexts unless there is some special mechanism in place to deal with the conflict between how the receiver can best act in the light of a piece of correlational information, on the one hand, and how the sender would like the receiver to act on that information, on the other.[1]

3 Different Channels of Transgenerational Information Transmission

Selection-based effects and detection-based effects are two ways of transmitting information between generations. This way of categorizing cross-cuts standard ways of dividing up epigenetic effects based on the types of mechanism involved. In this section, we argue that to understand the different channels by which information flows between generations, and their evolutionary consequences, it is important to consider the way information is generated and transmitted as well as the mechanisms that are involved.

The transgenerational information flow is fundamental to the evolution of complexity. The process of differential survival and reproduction—selection—generates information about which phenotypic variants are better suited to the environment in which they exist. Life can evolve only if that information is somehow transmitted to subsequent generations. DNA, with its associated machinery of transcription, translation, and replication, is a wonderful solution to that problem. But it is becoming clear that there are several other mechanisms of heredity (Jablonka & Lamb, 1995, 2005). Jablonka and Lamb argue that information should be used as the common denominator by which different systems of heredity are compared (Jablonka & Lamb, 2007, p. 382).

One of the themes of this volume is that communication and signaling is an evolutionary challenge to all, not only to cognitively sophisticated creatures like humans. Even the simplest forms of multicellularity depend on efficient signaling between cells to solve coordination problems. A parallel point can be made about transgenerational information transmission. It used to be thought that genetic inheritance was the only mechanism of transgenerational information transmission in all but the most cognitively sophisticated creatures. It is now becoming clear that there are many more mechanisms of transgenerational information transmission than just DNA, on the one hand, and human culture, on the other. Following Jablonka and Lamb's suggestion, an appreciation of the different ways in which information is transmitted across the generations can help us to understand aspects of various branches of the tree of life.

As we saw in the last section, influences that are unified mechanistically under the label "parental effects" can generate and transmit information in different ways. Reliably transmitted epigenetic factors can generate selection-based information that works like genes, acting as a channel of inherited information generated through a process of selection. Epigenetic factors probably operate on a different timescale to genes, owing to different mutation rates and levels of transmission fidelity. For example, some forms of antibiotic resistance in *E. coli* work by epigenetically based selection (Adam et al., 2008), suggesting that some epigenetic mechanisms are well suited to the rapid evolution of adaptation.

When detection-based information is transmitted from parents to offspring, parental effects operate at the timescale of a single lifetime, with the parent detecting some fitness-relevant environmental parameter and the offspring relying on a signal from the parent to fix a developmental parameter for life. (Of course, for such effects to be adaptations, there must also have been natural selection on the intergenerational disposition; but the epigenetic signals themselves vary generation by generation.) So epigenetic mechanisms bring in information channels that operate in parallel to genes over at least two different timescales.

The animal signaling literature fills in the story at the shortest timescale, where organisms signal to one another about transient matters of fact, with receivers producing not a fixed response that lasts a lifetime, but a transient response that is appropriate to the particular fact signaled (e.g., fleeing in response to the information that there is a predator overhead now). Of course, signals at this timescale can be sent between generations as well. Parents signal significant pieces of transient information to their offspring. Given their highly overlapping reproductive interests, there are many cases of cooperative communication of transient matters of fact between generations.

4 The Distinction Applied to Hominin Cultural Transmission

Section 2 distinguished between two ways that information is transmitted between generations by epigenetic factors. In the last section we argued that it is useful in respect of all kinds of inheritance mechanisms, genetic and nongenetic, to analyze the way information is generated and transmitted between generations. In this section we will follow that advice in examining hominin cultural transmission.

Although it is now clear that there are many nongenetic channels of transgenerational information transmission throughout the natural world, it remains true that humans are particularly well endowed with means for sending information between generations. Humans have far richer cultural traditions than any other species, showing that transgenerational transmission of phenotypes has an especially central role for us. In this section we argue that the distinction we drew above between selection-based and detection-based information is particularly germane to understanding human cultural transmission.

It is tempting to think that human cultural learning all falls in the category of detection-based information. Individuals learn about some aspect of their environment (how to exploit some resource, how to perform some skill, how to turn some raw material into a useful tool), and then pass that knowledge on to conspecifics, including those in the next generation. If those individuals in turn pass on their knowledge, then this pattern of teaching and learning will produce cultural traditions. But the utility of the behavioral phenotypes is usually thought to be detectable by the individuals who learn them. Unlike the information carried by genes, or by long-term epigenetic effects, the generation of useful information does not itself depend on a history of selection. The standard idea is that, at some point, an individual learns for itself something useful. That information can then be passed on and preserved by cultural inheritance, so that it does not need to be relearned anew by the individuals who come after. When the first individual uncovers a useful fact or hits on a clever trick, perhaps by trial-and-error learning, that individual's behavioral phenotype immediately carries correlational information about a fitness-relevant aspect of the environment. A process of individual learning has generated that information.

However, not all cultural traditions work like that. Suppose offspring were to copy the behavioral patterns of their parents blindly, without ever assessing the utility of those forms of behavior. Then behavioral phenotypes would follow reproductive lineages. If there were fitness differences between individuals with different behavioral phenotypes, then those with higher relative fitness would be selected. Even if novel behavioral phenotypes were introduced at random, such a process would give rise to selection, and so generate correlational information about the selective environment. In short, given reasonably high-fidelity vertical or oblique transmission, selection on behavioral phenotypes would work, informationally, just like genetic transmission and long-term epigenetic effects. That is, behavioral phenotypes would carry selection-based information down the generations.

The distinction between these two broad ways that cultural learning may operate, which are often elided, emerges clearly when we take Jablonka and Lamb's advice and use information as a common denominator. If individuals generate exploitable correlational information through individual learning, then the process of cultural inheritance is Lamarkian, with useful characteristics acquired in individual ontogeny then passed down the generations. The role played by individual learning also makes the transmission process inherently noisy, as individuals in subsequent generations go on learning for themselves. If the exploitable correlational information depends on a history of selection, then the process loses its directed character. A potential for higher-fidelity transmission is balanced by a reduction in the speed with which the system can respond to selection pressures. Nevertheless, even behavioral traditions that depend on selection—those that carry selection-based information—can likely be selected on a much more rapid timescale than gene-based selection. (Compare the different timescales over which selection generates information for genetic versus epigenetic inherited information, as we saw above.)

Selection-Based Information: Overimitation

There is evidence that some forms of human imitation-based learning do operate so as to carry selection-based information. The baseline is that both humans (Meltzoff, 1988) and other primates (Voelkl & Huber, 2000) are capable of copying the means by which a demonstrated action is performed, and of doing so rationally, appreciating the difference between means to an intended goal and irrelevant actions (Horner & Whiten, 2005; Buttlemann et al., 2007, Wood & Hauser, 2008; Tennie, Call & Tomasello, 2009; cf. Povinelli, 2000).

Curiously, humans have a disposition to imitate even action-steps that seem to them to be irrelevant to achieving the demonstrated goal. When chimpanzees are shown a series of actions performed on an opaque box they copy them all so as to obtain a reward, as do children. A transparent version of the same apparatus makes it obvious that some of the demonstrated steps are irrelevant to achieving the reward. Chimps omit unnecessary steps, but children copy them (Horner & Whiten, 2005, Lyons, Young & Keil, 2007). There seems to be a developmental trajectory in which infants (12 and 18 months old) are poor at copying specific actions and are more focused on outcomes (Nielsen, 2006), 3-year-olds are capable of copying specific actions (Williamson, Jaswal & Meltzoff, 2010) but less inclined to engage in overimitation, and 5-year-olds imitate obviously irrelevant actions in a wide range of conditions

(McGuigan et al., 2007; McGuigan & Whiten, 2009; Whiten et al., 2009; but cf. Nielsen & Tomaselli, 2010). Overimitation appears to be even more pronounced in adults (McGuigan, Makinson & Whiten, 2011).

Five-year-olds can tell the difference between the necessary and apparently unnecessary actions, but seem to think that the latter have an unspecified purpose (Kenward, Karlsson & Persson, 2010; cf. Hernik & Csibra, 2009). Their tendency nevertheless to copy these steps may just be the most rational hypothesis about the underlying causal structure (Buchsbaum, Gopnik & Griffiths, 2010), but the developmental results suggest that the tendency may be more pervasive than causal rationality would dictate.

It is counterintuitive that humans should engage in such "overimitation" when other apes do not. Humans seem to be copying in the less intelligent way. Overimitation makes more sense when we realize that it may be the basis on which selection-based information is transmitted. The tendency to imitate actions that are registered as intentional but causally irrelevant to the outcome may be an adaptation to increase the fidelity with which behavioral phenotypes are transmitted down the generations. Arguments for and against that hypothesis are detailed in Shea (2009) and only summarized here.

Since there is a fitness cost of transmitting useful phenotypes to nonkin, our hypothesis that overimitation may be the basis of a selection-based inheritance system is only plausible if vertical transmission is somehow favored. If it is, selection on phenotypic variants would have the power to accumulate behavioral adaptations. If human children selectively imitate in "natural pedagogical contexts" (Gergely & Csibra, 2006; Csibra & Gergely, 2009) that may encourage vertical transmission, lending support to the inheritance system hypothesis.

However, the evidence is by no means one way. The importance of imitation may instead be to allow the transmission of behaviors whose utility has been discovered through individual learning—detection-based information. Studies of diffusion chains reach different conclusions about the reliability of child-to-child copying of irrelevant actions (McGuigan & Graham, 2009; Whiten & Flynn, 2010), although they do seem to show fidelity to a particular method of achieving an outcome (Flynn, 2008). Social transmission of a particular method of performing an action is not distinctive of hominin cultural transmission (Hobaiter & Byrne, 2010), so some of these results count against the claim that there is a distinctively hominin form of imitation-based transmission of selection-based information. Instead they suggest that individual children are still behaving on the basis of an assessment of the consequences of their actions. Considered as an inheritance system, such modifications would reduce the fidelity of long-term transmission. On the other hand, these studies do not replicate the chains of parent-child dyads that are essential for vertical inheritance. For now, the importance of overimitation for transmitting selection-based information remains an open empirical question.

Detection-Based Information and Group Selection

Detection-based behavioral effects raise their own theoretical problems. Once individual learning is allowed free reign, the transmission process becomes noisy, making it hard to see how new behaviors, starting at low frequency, would be selected and then maintained in a population. Boyd and Richerson (1985, 2005) have shown how various kinds of learning biases can help to overcome these problems, for example by enabling learners to observe several different techniques and copy the best.

The problem of honest signaling from the animal communication literature also raises concerns. If useful skills and information spread horizontally to nonkin, that will often reduce the relative fitness of the sender (vis-à-vis those competitors). Even if spreading the information is good for both senders and receivers (in the everyday sense of "good for"), senders would have an incentive not to transmit useful information if keeping it secret is more detrimental to others than it is to themselves. For example, even if the sender would be healthier, on average, if everyone in the group were to adopt his clever hygiene practice, he will increase his relative fitness by keeping it to himself. There will be more disease going around for everyone, but he has the relative advantage of a behavior that protects him a bit more than others.

So there is a puzzle about how there can be the right kinds of cooperative contexts for detection-based information transmission to evolve. A restriction to vertical transmission would help here, too; but unlike in the case of blind imitation, once agents are reasoning for themselves about how to behave it is unclear that a disposition to copy only the behaviors of kin forebears would be evolutionarily stable. A population of agents who copied detection-based information only from their parents would be invaded by individuals with a disposition to copy useful detection-based information from nonkin as well.

The problem of why information should be shared with nonkin can be overcome if there is strong group selection. Then the relative fitness advantage of keeping information secret is outweighed by the relative fitness advantage of being in a group of agents disposed to transmit detection-based information. Several of the mechanisms pointed to by Boyd and Richerson for improving the maintenance of adaptive behaviors in social groups work best given group selection. Although there is widespread skepticism about the importance of group selection in other species, it may have been particularly important in human evolution. The increasing richness of human culture in our recent evolutionary past may have gone hand in hand with the operation of a powerful form of group selection between small tribes.

We saw above that vertical or oblique overimitation of behavioral phenotypes by kin could have formed a channel of selection-based information in hominin evolutionary history. In this section we have argued that this is not the exclusive channel of cultural inheritance. Socially mediated learning may also transmit detection-based information about behavioral phenotypes, if the right kind of social structure is in place to underpin group selection.

So it seems likely that recent hominin evolution has involved both detection-based and selection-based information transmission. Their respective contributions are still unclear. The answer has implications for the timescale on which cultural changes can respond to selection pressures, and the kinds of mechanisms that may be in place to support the transmission of information down the generations. Asking Jablonka and Lamb's question about how information is generated and transmitted in cultural inheritance produces the insight that phenomena that are standardly grouped together as being based on social learning might actually operate in ways that have significantly different evolutionary consequences. Much work has implicitly assumed that social learning transmits detection-based information. Our focus on selection-based culturally mediated information is a useful counterweight. It may make sense of empirical phenomena like the human tendency toward overimitation. Periods of apparent stasis in the evolutionary history of hominin tools or behavioral practices may also be due to their being based on the relatively high-fidelity transmission of selection-based information.[2] By contrast, where we find behavioral phenotypes and cultural practices that seem to vary widely between populations but spread rapidly within a population, it is more likely that detection-based information has mediated their transmission.

5 A Theory of Content Applicable to Transgenerational Information Transmission

The sections above have set out the argument that a distinction between selection-based and detection-based information transmission can do useful explanatory work. It is now time to pay out on the promissory note—to show that it is legitimate to treat these as channels of information transmission in a substantive sense, over and above the mere existence of causal correlations. In this section we argue that both cases involve the transmission of semantic information. That invites the question: "But what does semantic information add to a full causal understanding of these interactions?" To answer that we must first set out the framework that gives rise to semantic information. That is the task of this section, before turning in the final section to show how this semantic information plays an explanatory role over and above the explanations afforded by underlying causal structure.

If the causal intercourse between parents and offspring described above comprised nothing more than chains of causal interactions, then it would remain true that we could describe these interactions using the tools of information theory, but there would be no justification for thinking that there was any way of distinguishing between channels of information transmission and other kinds of causal interactions. All the myriad ways in which parents affect offspring carry information in the thin correlational sense, whether or not they are adaptive or even heritable, as do causal interactions between an offspring and all other aspects of its environment. Our claim that it is useful to distinguish between two ways that information is generated and transmitted in inheritance would be correspondingly thin were this not information in some more substantial sense.

The stronger, substantial sense is furnished by infotel semantics. Infotel semantics is a philosophical framework that characterizes systems that are adaptations for information transmission. It shows how some information-carrying elements in the system are representations with semantic content. Having semantic content is a matter of there being correctness conditions or satisfaction conditions attached to each carrier of information. This is significant for philosophical purposes because it differentiates representations from mere correlations. For our purposes, the importance of the framework is to delineate those cases where there is an adaptation to transmit information in certain ways, which gives substance to the claim that these are channels of information transmission. So we argue for the claim that detection-based effects and selection-based effects are based on information channels by showing how they involve representations with semantic contents.

Detection-Based Information

The framework of infotel semantics is well suited to capturing the semantic content of low-level varieties of representation, like those found in animal signaling and in inheritance systems (Shea, 2007b). It is a development of the teleosemantics pioneered by Millikan (1984) and Papineau (1987), but with an additional requirement that representations should correlate with the facts that they represent (i.e., that they should also carry information in the "thin" sense of information theory).[3] The framework applies when there is some producer system that produces a range of different natural signs and a consumer system that responds to those natural signs, behaving in a characteristic way or ways with respect to each.

An intermediate carries correlational information if its occurrence affects the probability that some further state of affairs obtains. Correlational information is cheap,

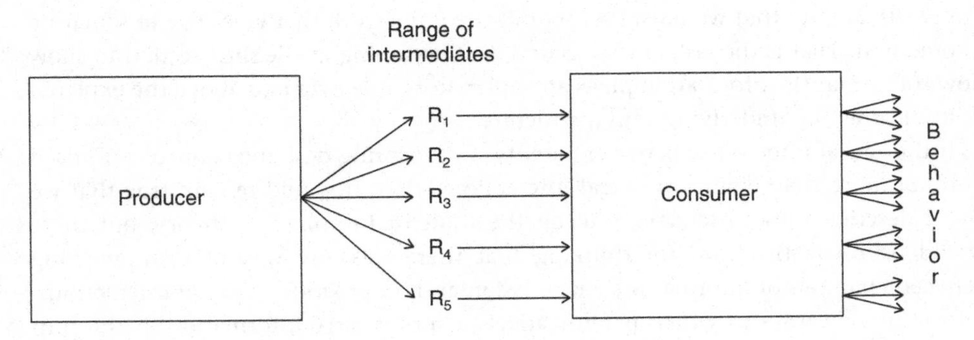

Figure 16.2
The basic representationalist model.

so the natural signs sent by a producer system will typically carry correlational information about very many states of affairs. An intermediate has content, however, only if the consumer system reacts to it with a form of behavior that is appropriate given the obtaining of one of the states of affairs with which it correlates. In the cases we are interested in, the idea of the consumer system producing an output that is appropriate given the information carried by an intermediate should be understood in evolutionary terms. Behaving in that way is one of the evolutionary functions of the consumer system; for that function to lead to survival and reproduction of the consumer in an evolutionarily normal way a specific condition has to obtain, and the intermediate carries correlational information about that condition. In short, the content of the intermediate is that condition the production of the intermediate correlated with during a history of selection and the obtaining of which explains why the specific forms of behavior prompted by that intermediate led to the survival and reproduction of the consumer system.

Less abstractly, consider how this applies to systems of animal communication. Suppose female frogs preferentially mate with males that produce low-pitch calls. In the animal communication literature, researchers substantiate the idea that communication is occurring by showing:

(i) "that call frequency is well correlated with caller size";
(ii) "that females show a behavioral preference for calls of lower frequency";
(iii) "that size is what matters to the receivers" [e.g.,] "that female frogs benefit from mating with larger males but not from mating with older ones" (Searcy & Nowicki, 2005, p. 3).

According to infotel semantics, the content of a signal is fixed by a set of conditions that match the factors listed above: correlation on the input side and consumption for content on the output side—consumption for content being a matter of the condition under which the behavioral response to the signal will be evolutionarily beneficial.

The infotel framework carries over straightforwardly to cases of detection-based information. Consider the epigenetic signal of light environment sent by maternal *Campanulastrum*. It correlates with whether the mother plant was in an understory or light gap environment, and hence whether the progeny are likely to be in understory or a light gap. The developing seedling responds to that signal by germinating in the spring or in the autumn, with the respective outputs showing relative fitness benefits if spring germination occurs in understory and if autumn germination occurs in a light gap. So we have correlation on the input side (maternal signal of light gap) and consumption for content on the output side (offspring behavior that is adaptive on condition that it is in a light gap), with the input correlation matching the output condition. In addition to ascribing indicative contents—*that male frog is large* or *the local environment is sunny*—infotel semantics also ascribes imperative contents to these kinds of signals, corresponding to the functions that they serve. For example, the epigenetic signal in *Campanulastrum* carries the imperative content *grow as an annual* or *grow as a biennial*.

The cultural cases may also include adaptations for information transmission. Where individuals are learning for themselves, then they are using a learning mechanism that is an adaptation for picking up on correlational information. Where that correlational information happens to have been generated by their parents, then it will fall into our category of transgenerational detection-based information. In some cases the connection between parent and offspring may be tighter, for example if there have been adaptations for teaching and learning (a "pedagogy complex"; see Sterelny, 2012).

For exposition, it is simplest to think in terms of simple correlations between natural signs and the external affairs they signify: Some epigenetic effect is produced when and only when the maternal plant is in the understory and a different epigenetic effect is produced when and only when the mother is in a light gap. The framework does not depend on this simplification. The correlational information carried by a sign emitted by a sender will often consist in a joint probability distribution between a range of properties the sign can have and a range of possible external states of affairs. These correlations can be captured by Shannon information theory, but also by other means, such as by the Kullback–Leibler divergence (Cover & Thomas, 2006), which measures how informative the range of states of one system (with their probabilities) are about the range and probability distribution of states of another system. The umbrella term "correlational information" is intended to be neutral as between the various ways of capturing such correlations.

The treatment in terms of a single correctness condition or satisfaction condition is also a dispensable idealization, albeit one that is innocuous in very many applications. In some cases there may be no single condition the obtaining of which explains the fitness advantage of the behavior produced in response to some natural sign. Just as a particular sign is informative about the probabilities of a range of states of affairs, the behavioral output produced in response to the sign may have aided survival and reproduction in a normal way in a range of different conditions, each of which is among the states of affairs with which the sign correlates. We should give the content of such signals not in terms of a single correctness condition, but in terms of a set of conditions and a degree of probability or credence that the system attaches to each.

Selection-Based Information

Application of the infotel framework to selection-based information involves an additional subtlety. In these cases, correlational information is generated through a process of selection on reliably transmitted epigenetic variants over many generations. Once natural selection has generated correlational information, that information is reliably transmitted down the generations in the germ line of chains of individuals. The con-

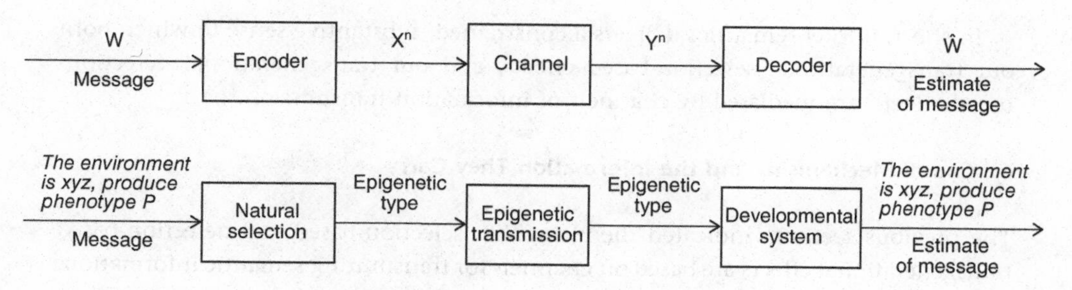

Figure 16.3
Transmission of selection-based epigenetic information.

sumer is the individual developing organism, and the transmission chain also consists of a series of individuals. However, the producer in the infotel model is realized by the process of selection over many generations.

Infotel semantics delivers contents when there is a producer system that sends a variety of signals and a consumer system designed to react to the various signals with different outputs. Correlations are produced by the process of selection over many generations, but consumed by the process of development in a single generation. With reliably transmitted epigenetic variants there is a genuine consumer designed to respond to such transgenerational effects only if it is an evolutionary function of that family of epigenetic variants that they should be reliably transmitted down the generations. It is not enough that they lead to heritable effects so that selection can in fact build up correlational information.

It is reasonably clear that DNA has the relevant function (Shea, 2007a)—a metafunction, in addition to its myriad object-level functions of producing particular selected phenotypes. The system of DNA transcription, translation, and replication has evolved in order effectively to fulfill the function of producing heritable phenotypes. It is less clear whether any nongenetic mechanisms have evolved in order to produce heritable phenotypes (rather than just happening to be a basis on which some heritable phenotypes arise). The evidence discussed above that some features of human imitation appear to be adaptations for the high-fidelity transmission of behavioral phenotypes is an indication that human blind overimitation, at least, may also meet the criterion.

Given the metafunction, the rest of the framework applies directly. A range of different messages is produced by episodes of selection in different historical epochs. The consumer (the developing individual) responds to each with a particular behavior or set of behaviors. That behavior is adaptive in the environmental conditions in which it was selected; and the result of selection is that the signal sent between generations carries correlational information about that environmental condition.

In short, infotel semantics delivers a constrained, substantive sense in which both our transgenerational selection-based effects, and our transgenerational detection-based effects, are mediated by channels of information transmission.

6 Causal Mechanisms and the Information They Carry

The previous section vindicated the claim that selection-based and detection-based transgenerational effects are based on channels for transmitting semantic information. This section highlights the kind of explanatory payoff that can be achieved by focusing on semantic information, over and above causal interactions and correlational information. The point applies equally to detection-based and selection-based information, so to keep things simple we'll focus on detection-based information.

With any kind of explanation that appeals to representational contents there is always a question about the explanatory work that is done by adverting to the fact that various states bear contents. What does semantic content add to a causal explanation in terms of noncontentful properties of those states? Returning to the example of the mate choice of female frogs, it looks as if researchers had a complete causal story once they discovered that females respond to low-pitch calls with mating behavior. The causal sequences are:

(a) Male emits lower-pitch call → female hears lower-pitch call → female mates with male.
(b) Male emits higher-pitch call → female hears higher-pitch call → female doesn't mate with male.

Even if it's true that, according to a defensible theory of content, the low-pitched call has the representational content *my body size is large*, what additional explanatory purchase does that provide?

The answer is that the content-based story addresses a further explanandum: Why is the system set up to have that causal structure? Why do females choose mates based on the pitch of male songs? Of course, the answer is that pitch correlates with body size and, in the evolutionary past, females who mated with larger males had a higher fitness. So we can understand why the signaling system is set up as it is when we see the female responding to the male call in virtue of its content. We also get a corresponding explanation of a particular episode of behavior. Why did *that* female engage in mating behavior with *that* male? Because she was responding to a representation of his body size. We get *an* explanation of her behavior by citing the causal regularities (a) and (b) above and noting that this bit of female behavior exemplified causal regularity (a). But we get a deeper explanation when we advert to the representational content of the call.

Since we have offered a naturalistic reduction of content in these simple systems, content is nothing over and above a particular conjunction of producer, consumer, correlational information, and evolutionary function. But that conjunction is a real property that is important in nature. Explanations in terms of the conjunction identify real patterns that generalize across a wide range of cases. The reduced property (semantic information) is ontologically real and explanatorily useful.

The same dialectic applies to detection-based transgenerational effects. The experimental literature on epigenetic effects focuses on demonstrating the existence of such effects and documenting their robustness. So we observe that progeny of *Campanulastrum americanum* that grow in light gaps tend, in their turn, to grow as annuals (Galloway & Etterson, 2007). That causal knowledge offers us some explanatory purchase. Why did *that* plant grow as an annual? Because it grew from a seed from a plant in a light gap. However, the semantic treatment offers additional explanatory purchase. It fits with the other half of Galloway and Etterson's research on *Campanulastrum*, which investigates the adaptive significance of that strategy. The causal story is supplemented when we understand that the annual strategy is an adaptive response to an epigenetic signal that correlates with a fitness-relevant fact about the environment.

The point can be made most cleanly by reference to abstract models of signaling systems. Brian Skyrms (2010) has adapted the sender-receiver games discussed by Lewis (1969) and modeled them in terms of evolutionary game theory. The sender has private access to some fitness-relevant facts about the environment and the receiver can perform some actions whose fitness may depend on the environment (see table 16.1). Senders can send a range of messages which the receiver can react to. A sender strategy is a rule for sending messages, which can be conditional on the environment. A receiver strategy is a rule for action, which can be conditional on the message. Senders and receivers jointly receive the payoffs from an interaction. Agents are endowed with strategies consisting of a pair of sender and receiver strategies, and the system is then allowed to evolve using replicator dynamics.

A *signaling system* is a complementary set of strategies where players always achieve a payoff of 1. For example, we can see that the combination of the sender and receiver strategies illustrated above constitutes a signaling system:

In E1, sender sends signal S1, receiver acts on S1 to do action A1. Payoff: 1.
In E2, sender sends signal S2, receiver acts on S2 to do action A2. Payoff: 1.

Skyrms's modeling work shows that, under standard assumptions, the signaling systems are attractors whose basins of attraction cover almost all the possible initial

Table 16.1

Example of a sender strategy	
E1	S1
E2	S2

Example of a receiver strategy	
S1	A1
S2	A2

Payoffs for both sender and receiver
A1: 1 in E1, 0 in E2
A2: 0 in E1, 1 in E2

distributions of strategies. There are many other combinations of strategies that are less adaptive; for example, the sender can send the same signal irrespective of the environment, or the receiver can perform the same action irrespective of the signal.

The important result from the modeling work is that when sender-receiver strategies are embedded in a selective regime where there are payoffs for actions, with payoffs conditional on the state of the environment, a certain kind of communication system is likely to evolve. Not only will the strategies evolve to maximize the transmission of information, in the sense that the sender will correlate its signal to the state of the environment and the receiver will correlate its action to the state of the signal; they will also correlate in a particular way. Correlations are efficiently transferred in the pure antisignaling case, too, where the sender correlates its signals to the state of the environment and the receiver acts on them conditionally to perform the action that is not rewarded in those circumstances:

Antisignaling

In E1, sender sends signal S1, receiver does action A2. Payoff: 0.
In E2, sender sends signal S2, receiver does action A1. Payoff: 0.

From the point of view of information theory, for example of the Shannon variety, there is maximum information flow in both cases. Representational content marks out the case where there is transmission of correlation in the service of maximizing payoffs. The payoff structure allows us to see why the evolutionary dynamics tend to result in pairs of sender-receiver strategies that both (i) transmit correlational information, and (ii) do so in ways that help the pair.

In the kinds of cases considered here, representational properties arise where there is a certain internal structure—transmission of correlation—that is embedded in the context of success conditions for its outputs. Therefore, when we explain some behavior in terms of the content of a representation that caused it, we get a different, and in some ways deeper, explanation than we would have were we just to appeal to causal regularities about the operation of the system. Content depends on the mechanism in question being embedded in a structure of payoffs, fitness, success, or some such. Correspondingly, we can explain the success of particular pieces of behavior by the fact that a system is responding to some true content.

7 Conclusion

Within the broad category of adaptive transgenerational effects, an important distinction should be drawn between selection-based and detection-based effects. With detection-based effects, an individual in one generation detects some fitness-relevant parameter and conveys correlational information about it to individuals in the next generation, who react with a fitness-relevant output (phenotype or behavior). With selection-based effects, the fitness-relevant information is instead generated by a process of natural selection. The distinction is particularly illuminating when applied to information transmission in human cultural traditions.

Acknowledgments

I would like to thank audiences at the Australian National University (RSSS), Oxford Brookes University, and Bristol University; and Brett Calcott, Ben Fraser, Richard Joyce, Kim Sterelny, and Tobias Uller for helpful comments. This research was supported by the Oxford University Press John Fell Research Fund, the Oxford Martin School, Somerville College (Mary Somerville Junior Research Fellowship), and the Wellcome Trust (grant 086041 to the Oxford Centre for Neuroethics).

Notes

1. The case where the mother takes control of an aspect of the offspring's development in her own interests would not be an example of semantic information transmission of the type considered here. Those cases are better treated as manipulations (Dawkins & Krebs, 1978).

2. Haldane (1956, p. 9) suggested that the stability of the Acheulean biface could have been due to its being transmitted by selection-based information. He had in mind a genetically transmitted instinct, but culturally transmitted selection-based information could have a similar effect.

3. Standard teleosemantics bases content purely on the way that a consumer system is disposed to react to a putative representation and so is indifferent to whether representations transmit information of the correlational sort.

References

Adam, M., Murali, B., Glenn, N., & Potter, S. S. (2008). Epigenetic inheritance based evolution of antibiotic resistance in bacteria. *BMC Evolutionary Biology 8*, 52.

Boyd, R., & Richerson, P. J. (1985). *Culture and the evolutionary process*. Chicago: University of Chicago Press.

Boyd, R., & Richerson, P. J. (2005). *The origin and evolution of cultures*. Oxford: Oxford University Press.

Buchsbaum, D., Gopnik, A., & Griffiths, T. L. (2010). Learning from actions and their consequences: Inferring causal variables from continuous sequences of human action. In *Proceedings of the 32nd Annual Conference of the Cognitive Science Society*. Austin: Cognitive Science Society.

Buttelmann, D., Carpenter, M., Call, J., & Tomasello, M. (2007). Enculturated chimpanzees imitate rationally. *Developmental Science, 10*, F31–F38.

Cover, T. M., & Thomas, J. A.. (2006). *Elements of information theory* (2nd Ed.). Hoboken, NJ: John Wiley & Sons.

Csaba, G. (2008). Hormonal imprinting: Phylogeny, ontogeny, diseases, and possible role in present-day human evolution. *Cell Biochemistry and Function, 26* (1), 1–10.

Csibra, G., & Gergely, G. (2009). Natural pedagogy. *Trends in Cognitive Sciences, 13*, 148–153.

Dawkins, R., & Krebs, J. R. (1978). Animal signals: Information or manipulation. In J. R. Krebs & N. B. Davies (Eds.), *Behavioural ecology: An evolutionary approach*. Oxford: Blackwell.

Feldman, M. W., Aoki, K., & Kumm, J. (1996). Individual versus social learning: Evolutionary analysis in a fluctuating environment. *Anthropological Science, 104*, 209–231.

Flynn, E. (2008). Investigating children as cultural magnets: Do young children transmit redundant information along diffusion chains? *Philosophical Transactions of the Royal Society of London, Series B: Biological Sciences, 363*, 3541–3551.

Freeland, S. J., & Hurst, L. D. (1998). The genetic code is one in a million. *Journal of Molecular Evolution, 47*, 238–248.

Galloway, L. F., & Etterson, J. R. (2007). Transgenerational plasticity is adaptive in the wild. *Science, 318*, 1134–1136.

Gergely, G., & Csibra, G. (2006). Sylvia's recipe: The role of imitation and pedagogy in the transmission of cultural knowledge. In S. Levenson & N. Enfield (Eds.), *Roots of human sociality: Culture, cognition, and human interaction*. Oxford: Berg.

Gilbert, S. F., & Epel, D. (2009). *Ecological developmental biology. Integrating epigenetics, medicine, and evolution*. Sunderland, MA: Sinauer.

Haig, D., & Hurst, L. D.. (1991). A quantitative measure of error minimization in the genetic code. *Journal of Molecular Evolution, 33*, 412–417.

Haldane, J. B. S. (1956). The argument from animals to men: An examination of its validity for anthropology. *Journal of the Royal Anthropological Institute of Great Britain and Ireland, 86* (2), 1–14.

Helanterä, H., & Uller, T. (2010). The Price equation and extended inheritance. *Philosophy and Theory in Biology, 2*, e101.

Hernik, M., & Csibra, G. (2009). Functional understanding facilitates learning about tools in human children. *Current Opinion in Neurobiology, 19*, 34–38.

Hinde, C. A., Johnstone, R. A., & Kilner, R. M. (2010). Parent-offspring conflict and coadaptation. *Science, 327*, 1373–1376.

Hobaiter, C., & Byrne, R. W. (2010). Able-bodied wild chimpanzees imitate a motor procedure used by a disabled individual to overcome handicap. *PLoS ONE, 5* (8), e11959. doi:10.1371/journal.pone.0011959.

Horner, V. K., & Whiten, A. (2005). Causal knowledge and imitation/emulation switching in chimpanzees (*Pan troglodytes*) and children. *Animal Cognition, 8,* 164–181.

Jablonka, E., & Lamb, M. J. (1995). *Epigenetic inheritance and evolution: The Lamarkian dimension.* Oxford: Oxford University Press.

Jablonka, E., & Lamb, M. J. (2005). *Evolution in four dimensions: Genetic, epigenetic, behavioral, and symbolic variation in the history of life.* Cambridge, MA: MIT Press.

Jablonka, E., & Lamb, M. J. (2007). Précis of *Evolution in four dimensions. Behavioral and Brain Sciences, 30,* 353–392.

Jablonka, E., & Raz, G. (2009). Transgenerational epigenetic inheritance: Prevalence, mechanisms, and implications for the study of heredity and evolution. *Quarterly Review of Biology, 84* (2), 131–176.

Kenward, B., Karlsson, M., and Persson, J. (2010). Over-imitation is better explained by norm learning than by distorted causal learning. *Proceedings of the Royal Society of London, Series B: Biological Sciences,* published online October 13, 2010. doi: 10.1098/rspb.2010.1399.

Leimar, O., Hammerstein, P., & Van Dooren, T. J. M. (2006). A new perspective on developmental plasticity and the principles of adaptive morph determination. *American Naturalist, 167* (3), 367–376.

Lewis, D. (1969). *Convention: A philosophical study.* Oxford: Blackwell.

Lyons, D. E., Young, A. G., & Keil, F. C. (2007). The hidden structure of overimitation. *Proceedings of the National Academy of Sciences of the United States of America, 104,* 19751–19756.

McGuigan, N., & Graham, M. (2009). Cultural transmission of irrelevant tool actions in diffusion chains of 3- and 5-year-old children. *European Journal of Developmental Psychology.* doi:10.1080/17405620902858125.

McGuigan, N., Makinson, J., & Whiten, A. (2011). From over-imitation to super-copying: Adults imitate causally irrelevant aspects of tool use with higher fidelity than young children. *British Journal of Psychology, 102*(1), 1–18.

McGuigan, N., & Whiten, A. (2009). Emulation and "overemulation" in the social learning of causally opaque versus causally transparent tool use by 23- and 30-month-olds. *Journal of Experimental Child Psychology, 104,* 367–381.

McGuigan, N., Whiten, A., Flynn, E., & Horner, V. (2007). Imitation of causally-opaque versus causally transparent tool use by 3- and 5-year-old children. *Cognitive Development, 22,* 353–364.

Meltzoff, A. (1988). Infant imitation after a 1-week delay: Long-term memory for novel acts and multiple stimuli. *Developmental Psychology, 24,* 470–476.

Millikan, R. G. (1984). *Language, thought, and other biological categories.* Cambridge, MA: MIT Press.

Nielsen, M. (2006). Copying actions and copying outcomes: Social learning through the second year. *Developmental Psychology, 42,* 555–565.

Nielsen, M., & Tomaselli, K. (2010). Overimitation in Kalahari Bushman children and the origins of human cultural cognition. *Psychological Science, 21*, 729–736.

Papineau, D. (1987). *Reality and representation*. Oxford: Blackwell.

Povinelli, D. (2000). *Folk physics for apes: The chimpanzee's theory of how the world works*. Oxford: Oxford University Press.

Searcy, W. A., & Nowicki, S. (2005). *The evolution of animal communication*. Princeton: Princeton University Press.

Shea, N. (2007a). Representation in the genome, and in other inheritance systems. *Biology and Philosophy, 22*, 313–331.

Shea, N. (2007b). Consumers need information: Supplementing teleosemantics with an input condition. *Philosophy and Phenomenological Research, 75* (2), 404–435.

Shea, N. (2009). Imitation as an inheritance system. *Philosophical Transactions of the Royal Society of London, Series B: Biological Sciences, 364*, 2429–2443.

Skyrms, B. (2010). *Signals*. Oxford: Oxford University Press.

Sterelny, K. (2003). *Thought in a hostile world*. Oxford: Blackwell.

Sterelny, K. (2012). *The evolved apprentice: How evolution made humans unique*. Cambridge, MA: MIT Press.

Tennie, C., Call, J., & Tomasello, M. (2009). Ratcheting up the ratchet effect: On the evolution of cumulative culture. *Philosophical Transactions of the Royal Society of London, Series B: Biological Sciences, 364*, 2405–2415.

Uller, T. & Pen, I. (2011). A theoretical model of the evolution of maternal effects under parent-offspring conflict. *Evolution, 65*(7), 2075–2084.

Voelkl, B., & Huber, L. (2000). True imitation in marmosets. *Animal Behaviour, 60*, 195–202.

Whiten, A., & Flynn, E. (2010). The transmission and evolution of experimental microcultures in groups of young children. *Developmental Psychology, 46* (6), 1694–1709.

Whiten, A., McGuigan, N., Marshall-Pescini, S., & Hopper, L. (2009). Emulation, imitation, over-imitation, and the scope of culture for child and chimpanzee. *Philosophical Transactions of the Royal Society of London, Series B: Biological Sciences, 364*, 2417–2428.

Williamson, R. A., Meltzoff, A. N., & Markman, E. M. (2008). Prior experiences and perceived efficacy influence 3-year-olds' imitation. *Developmental Psychology, 44* (1), 275–285.

Williamson, R. A., Jaswal, V. K., & Meltzoff, A. N. (2010). Learning the rules: observation and imitation of a sorting strategy by 36-month-old children. *Developmental Psychology, 46*(1), 57–65.

Wood, J. N., & Hauser, M. D. (2008). Action comprehension in non-human primates: motor simulation or inferential reasoning? *Trends in Cognitive Sciences, 12*, 461–465.

17 What Can Imitation Do for Cooperation?

Cecilia Heyes

Does imitation play a significant role in human ultra-cooperation? Is our ability to copy body movements an important part of the matrix of cognitive skills that deliver collective action and information sharing on an unprecedented scale? For at least a century, psychologists and biologists have answered these questions with a firm "yes." This chapter also says "yes," but questions the traditional picture of how and why imitation supports cooperation. I'll argue that imitation is not a "module" or cognitive adaptation for cooperation, that it contributes to collective action and information sharing in a wider variety of ways than has been typically assumed, and that its celebrated contribution—to the cultural inheritance of technological skills—may be only a very small part of what imitation does for cooperation.

Two types of imitation have been identified, which I'll call "simple" and "complex."[1] It is widely assumed that simple imitation and complex imitation are cognitive adaptations—that they are based on distinct cognitive processes, which evolved independently via gene-based selection to fulfill different social functions. Simple imitation is thought to function as "social glue"—to enhance an individual's sense of belonging to a social group—in a way that could facilitate collective action. Complex imitation is thought to be an adaptation for a certain kind of information sharing—the cultural inheritance of technological skills.

Section 1 reviews recent research showing that simple and complex imitation are based on the same core cognitive mechanisms, and section 2 considers whether these and other imitation-related mechanisms constitute cognitive adaptations. Section 3 looks at recent studies of simple imitation suggesting that a virtuous circle of unconscious imitation and prosocial attitudes keeps the members of human social groups in a constant state of readiness for collective action and information sharing. Section 4 argues that complex imitation contributes to human cooperation primarily by promoting the cultural inheritance of communicative-gestural rather than instrumental-technological skills, and that it is particularly effective when it is "dumb," that is, not guided by rational mechanisms.

More broadly, this chapter discusses both the proximal psychological mechanisms and the evolutionary consequences of imitation. I argue that the proximal mechanisms are more ancient (primarily sections 1 and 2), and that the evolutionary consequences are more diverse (sections 3 and 4), than has typically been assumed. These two themes are related, but in a specific way. With one exception (section 4.2), I am *not* suggesting that imitation has particular evolutionary effects *because* it depends on ancient and relatively simple psychological mechanisms. In most cases, complex, recently evolved cognitive adaptations could in principle have the same evolutionary consequences. They could, but the experimental evidence indicates that they don't—that imitation does not, in fact, depend on cognitive adaptations. The point I want to make is that simple, old psychological mechanisms are enough; imitation need not be based on complex cognitive adaptations in order to support human ultra-cooperation.

1 Cognitive Mechanisms of Imitation

1.1 Simple and Complex Imitation

Simple imitation is also known as "mimicry" (Tomasello, 1996), "automatic imitation" (Heyes et al., 2005), "priming," and "response facilitation" (Byrne & Russon, 1998). It occurs when an observer copies body movements that are already part of his behavioral repertoire. For example, when two people are in conversation, it is common for each to copy the other's incidental gestures, such as ear-touching and foot-wagging (Chartrand & Bargh, 1999). Simple imitation is currently a focus of study in social cognitive psychology and cognitive neuroscience, where most people assume that the mechanisms underlying this kind of imitation consist of simple connections between event representations (Chartrand & Van Baaren, 2009).

Complex imitation is also known as "imitation learning" (Tomasello, 1996), "true imitation" (Zentall, 2006), "observational learning" (Carroll & Bandura, 1982) and "program-level imitation" (Byrne & Russon, 1998). It occurs when an observer copies a "novel" sequence of body movements, a sequence she had not performed before observing the model. For example, in one of many experimental demonstrations of complex imitation, adults observed and then reproduced a novel sequence of semaphore-like movements of the hand and arm (Carroll & Bandura, 1982). Complex imitation is a focus of study in developmental and comparative psychology, where it is assumed to involve a variety of complex psychological mechanisms. These include symbolic coding, program extraction, perspective-taking, and intention-reading.

The assumption that complex imitation depends on complex psychological processes has been motivated by task analysis rather than empirical data. To imitate observed body movements, my cognitive system has to translate visual information from the model into motor output that looks the same as the model's behavior from a third-person perspective. In cases where the first-person and third-person views of an action are very different (e.g., facial and whole-body movements), it is far from obvious how the cognitive system solves this "correspondence problem"—how it works out which of my potential actions corresponds to the one I saw you perform (Heyes & Bird, 2007). "Symbolic coding," "program extraction," "perspective-taking," and "intention-reading" are all broad-brush candidate solutions to the correspondence problem. They are broad-brush in that no one has specified in any detail what these processes involve. They are candidate solutions because each is thought to involve abstract, flexible representations of action, and it is plausible that representations of this kind could solve the correspondence problem by an up-and-down route. In other words, the cognitive system could solve the correspondence problem by taking a relatively low-level visual representation of an observed action, recoding it "up" into an abstract representation—a symbolic code, program, perspective or intention—and then recoding it "down" into a motor program.

The difficulty of the correspondence problem varies with the difference between first-person and third-person views of an action sequence, a difference that is maximal for facial gestures and minimal for vocalizations. The challenge posed by the correspondence problem does not vary with sequence novelty. The cognitive system has to solve the correspondence problem for simple as well as complex imitation—to enable copying of familiar as well as novel body movements. Curiously, it is seldom acknowledged that simple imitation poses the correspondence problem. As a result, when researchers suggest that "mimicry" is mediated by a simple, evolved cognitive mechanism (e.g., Tomasello, 1996), it is not clear whether they have overlooked the correspondence problem or are suggesting that the mechanism in question solves this problem in an unspecified but simple way.[2]

1.2 Associative Sequence Learning

Recent research in experimental psychology and cognitive neuroscience has provided evidence in support of a new "associative sequence learning" (ASL) model of imitation (see figure 17.1; Heyes & Ray, 2000; Catmur, Walsh & Heyes, 2009). This model suggests that the correspondence problem is solved in the same way for both simple and complex imitation, and that the solution does not involve up-and-down recoding. According to ASL, the correspondence problem is solved for any given action by a direct connection between a visual and a motor representation of that action. These connections, or "matching vertical associations," are forged in the course of development by the same domain- and species-general processes of associative learning that produce Pavlovian and instrumental conditioning in the laboratory (see Schultz & Dickinson, 2000, for a review of associative learning). These processes strengthen excitatory connections between pairs of event representations when the occurrence of the two events is correlated, that is, when they occur relatively close together in time (contiguity) and one event is predictive of the other (contingency).

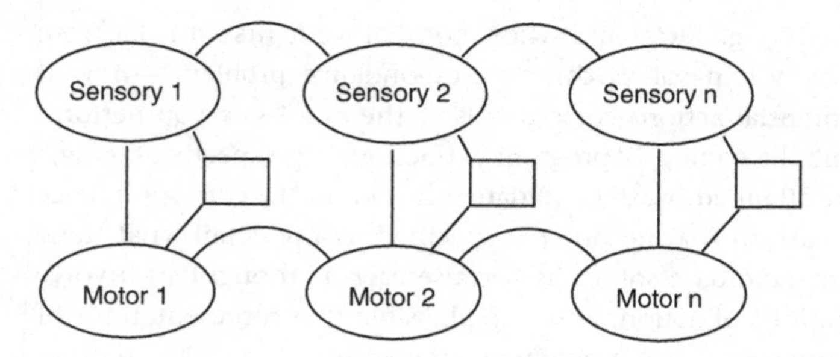

Figure 17.1
The associative sequence learning (ASL) model of imitation. The upper row of ovals depicts sensory (visual), and the lower row motor, representations of successive units in a sequence of body movements. The straight vertical lines, connecting sensory and motor representations of the same action unit, are the matching vertical associations (MVAs) that solve the correspondence problem for both simple and complex imitation. The curved lines represent the horizontal learning processes involved in complex imitation. These encode the sequence of visual events (upper row) and, via the matching vertical associations, the sequence of motor events (lower row), during action observation. The boxes represent acquired equivalence cues, which are not discussed in this article.

In terms of their internal structure, the processes of associative learning could just as easily produce nonmatching as matching vertical associations. If the sight of one action, X, is correlated with the performance of a different action, Y, associative learning will strength the connection between a visual representation of X and a motor representation of Y, supporting counterimitative rather than imitative behavior. According to ASL, matching vertical associations predominate because certain features of the human developmental environment ensure that we more often experience correlations between observation and execution of the same action than of different actions (Heyes, 2005). For example, experience of the former kind comes from direct self-observation (e.g., looking at your own hands in motion), mirror self-observation (using reflective surfaces), being imitated by others (especially facial imitation of infants by adults), synchronous activities of the kind involved in dance, sports, and military training, and indirectly via the use of action words (Ray & Heyes, 2010). The ASL model suggests that it is the relative paucity of these kinds of experience in the lives of nonhuman animals that explains why their imitation repertoires are more limited than those of humans.

ASL assumes that in cases of complex imitation, when an observer copies a novel sequence of actions, the operation of matching vertical associations is guided by processes that encode the serial order of visual stimuli. These "horizontal" processes learn what the novel action sequence "looks like." The representation they construct would be sufficient for subsequent recognition of the sequence, and to distinguish it from sequences containing the same components in a different order. However, for *imitation* of a novel action—to turn vision into matching action—the visual sequence representation formed by horizontal processes must activate, in the appropriate order, a matching vertical association for each element of the sequence.

Unlike several of the processes invoked by up-and-down accounts of complex imitation, horizontal mechanisms are not dedicated to the processing of body movements or even of social stimuli. The ASL model assumes that the same horizontal mechanisms encode the serial order of inanimate stimuli. It does not deny that processes like "intention-reading" are sometimes involved in overt imitative performance. In adult humans, the activation of motor representations via MVAs makes imitation possible, not obligatory, and intention-reading will sometimes determine whether overt imitative behavior is inhibited or allowed to occur. What the ASL model implies is that, while processes like intention-reading sometimes modulate both imitative and non-imitative behavior, they don't play a distinctive, necessary, or integral role in imitation. Specifically, they don't help to solve the correspondence problem.

Unlike previous accounts of the cognitive mechanisms mediating imitation, ASL has been explicitly tested against alternative models. These experiments have examined both simple and complex imitation, using behavioral and neurophysiological measures, and probed the model's claims about both vertical and horizontal processes. Supporting the idea that matching vertical associations are forged by associative learning, these studies have shown that novel sensorimotor experience can enhance (Press, Gillmeister & Heyes, 2007), abolish (Heyes et al., 2005), and even reverse (Catmur, Walsh & Heyes, 2007; Catmur et al., 2008) simple imitative behavior. It has been widely reported that humans typically show "automatic imitation" of various hand and foot movements (Heyes, 2011): In tasks that require us to ignore the sight of these movements, we nonetheless respond faster and more accurately when the required action matches an observed body movement. Hand opening is faster when observing hand opening than when observing hand closing, foot lifting is faster when observing foot lifting than hand lifting, and so on. These imitative effects appear to be relatively impervious to the actor's intentions, but they can be changed by retraining. For example, without explicit training, passive observation of index finger movement activates muscles that move the index finger more than muscles that move the little finger. However, after training in which people were required to respond to index finger movements with little finger movements, and vice versa, this pattern was reversed. Observation of index finger movement activated little finger muscles more than index finger muscles, implying that associative learning had converted automatic imitation into automatic counterimitation (Catmur et al., 2008; Catmur, Mars, Rushworth & Heyes, 2011).

Similarly, experiments examining complex imitation have provided evidence that it involves the same kind of sequence learning processes as non-imitative tasks; that these processes do not depend on intention-reading (Leighton et al., 2008); and that they do not show the flexibility one would expect if complex imitation were achieved by an up-and-down route (Bird & Heyes, 2005). For example, when people are required to imitate a sequence of movements involving the selection of a pen and its placement in one of two containers, they show exactly the same pattern of errors as when they are instructed to perform the same movements by flashing geometric shapes. Error patterns are indicative of underlying cognitive processes. Therefore, these results indicate that the same sequence encoding mechanisms are recruited in imitative and nonimitative tasks, and by stimuli that do and do not support the attribution of intentions (Leighton, Bird & Heyes, 2010). Regarding flexibility, Bird & Heyes (2005) found that people who had observed a complex sequence of key-pressing movements could reproduce that sequence only when they used exactly the same digits as the model. For example, when the model used her fingers, they could imitate the sequence with their fingers but not with their thumbs.

2 Cognitive Adaptations for Imitation

If the ASL model is broadly correct, the core mechanisms of imitation—the processes that solve the correspondence problem—are not cognitive adaptations. They were not favored by natural selection because individuals with those mechanisms outreproduced others by virtue of being better able to imitate. Matching vertical links are a product of associative learning. Associative learning is almost certainly an adaptation, but for the detection of causal relationships between events, not for imitation specifically. The horizontal sequence processing mechanisms may also be an adaptation, and of a more specific kind. Very few species show sequence-processing capabilities even remotely comparable to those of humans (Pearce, 2008). However, the evidence that these horizontal processes are domain-general—that they operate in the same way on sequences of animate and inanimate stimuli—suggests that they are not an adaptation for imitation.

Viewed through the lens of the ASL model, the human capacity for imitation is "epistemically engineered" (Sterelny, 2003) or socially constructed. Consider the contexts in which we receive correlated experience of seeing and doing the same action, the experience that creates matching vertical associations: direct self-observation, mirror self-observation, being imitated by others, and synchronous activities of the kind involved in sports, dance, and military training (Ray & Heyes, 2011). They are all contexts created or tightly regulated by cultural artifacts and practices. Direct self-observation is the least "social" context, but it is still subject to powerful cultural constraints. In contemporary Western cultures, infants spend a large proportion of their waking hours gazing at their own hands in motion (White, Catel & Held, 1964), an activity that would rarely be possible in cultures, past and present, where infant swaddling is the norm.

Although it now seems unlikely that the core mechanisms of imitation are cognitive adaptations, it is still possible that both cultural and genetic evolutionary processes have played a part in making us "Homo imitans" (Meltzoff, 1988), a species that can imitate an extraordinary range of actions with remarkable facility. It is possible that some of the social practices that foster the development of imitation have been favored by cultural group selection. For example, groups that trained their novices to dance and fight via synchronous drills may have been more successful than other groups, in part because these drills made the novices better imitators. In addition, gene-based individual selection could have favored the evolution of mechanisms—in adults and infants—that speed up learning of matching vertical associations. Candidate mechanisms of this kind include an attentional bias toward hands-in-motion (del Giudice, Manera & Keysers, 2009), which could promote learning by self-observation, and tendencies to produce and attend to "natural pedagogical" cues (Csibra & Gergely, 2009). These tendencies could make imitation of infants by adults a more potent source of matching vertical associations. On the production side, natural pedagogical cues include "motherese," the high-contrast intonation we tend to use when addressing infants and children, and on the receptive side, they include attentional biases toward both motherese and direct gaze.

Two features of these response tendencies and attentional biases are worth noting. First, they could be mediated by relatively simple perceptual and motoric behavior-control mechanisms; a tendency to orient toward a particular class of stimulus does not require anything as complex as intention-reading. Second, even if we call these low-level mechanisms "cognitive" and assume that they are biological adaptations, it would make them cognitive adaptations, but not cognitive adaptations *for imitation*. The hands-in-motion, motherese, and direct gaze attentional biases may speed up learning to imitate, but it's unlikely that this was a significant factor in their evolution. The hands-in-motion bias is likely to be an adaptation for precise visuomotor control of action, and natural pedagogical cues appear, as their name suggests, to be a generic method of ensuring that infants attend to, and therefore learn from, adults.

It has recently been suggested that humans have "a special kind of motivation for imitation" (Tennie, Call & Tomasello, 2009, 2412; see also Carpenter, 2006). It appears that, unlike chimpanzees, human children do not merely imitate in order to acquire more effective methods of solving instrumental problems. Experiments demonstrating "overimitation" or "overcopying" (e.g., Lyons, Young & Keil, 2007)—the imitation of incidental details of a model's technique—suggest that human children are "socially motivated." Above and beyond any instrumental benefits, children imitate because they just want to act in the same way as others.

This is a plausible and interesting suggestion. It has been assumed for too long that differences between chimpanzees and children in imitation tasks are due to ability rather than motivation (Shea, 2009), and it is not difficult to imagine how a desire to be like others could enhance the fidelity of cultural inheritance of behavior. However, it is far from clear that overimitation represents a motivational adaptation rather than a product of enculturation. From early infancy, children are selectively rewarded by adults for imitation. For example, Pawlby (1977) found that, whenever infants performed an action shortly after it had been performed by their mother, the mother responded with smiles and a general tone of encouragement, and that "a sense of special achievement was conveyed to the infant." Confirming the effects of this kind of training to imitate, Waxler and Yarrow (1975) found in a free-play session that infants who were rewarded more frequently for imitation exhibited imitation more often and across a broader range of behaviors. Thus, social motivation to imitate may well be human specific and an important part of the story about human cultural inheritance, but we shouldn't assume that it is an adaptation.

In sections 1 and 2, I've argued that neither the core nor the peripheral mechanisms of imitation are dedicated cognitive adaptations—they did not evolve via gene-based selection specifically to make imitation possible. Recent evidence suggests that the core mechanisms—those that solve the correspondence problem—are products of cultural epistemic engineering, and that the peripheral mechanisms—that assist the engineering process—are nonspecific and/or also cultural in origin. If imitation mechanisms are not cognitive adaptations, they could not be cognitive adaptations that have "evolved for" cooperative functions—social bonding in the case of simple imitation, and cultural inheritance in the case of complex imitation. However, as I hope to show in sections 3 and 4, this does not in any way undermine the view that imitation promotes human cooperation.

3 Cooperative Effects of Simple Imitation

Anecdotal reports and folk wisdom have long suggested that people inadvertently copy each other's gestures and mannerisms, and that this tendency somehow facilitates their interaction. In the last ten years, research on this kind of simple imitation—known as the chameleon effect (Chartrand & Bargh, 1999), or nonconscious mimicry (van Baaren et al., 2009)—has made enormous progress. Carefully controlled experiments in seminaturalistic settings have confirmed that the effect is pervasive, and postinteraction interviews provide no evidence that imitatees are aware of being copied, or that imitators are aware of, or intend to, imitate. Yet more important, recent experiments have shown that simple imitation is causally related to "prosocial" or cooperative attitudes and behavior (Chartrand & van Baaren 2009; van Baaren et al., 2009).

The causal relationships go in both directions: Being imitated makes people more cooperative, and when people are thinking and feeling in a cooperative way, they are more likely to imitate others. When a person has been imitated by an interaction partner, he or she likes the partner more (Chartrand & Bargh, 1999), judges the partner to be more persuasive (van Swol, 2003), and reports that the interaction was smoother and more enjoyable (Tanner et al., 2008) than when not being imitated. In negotiation exercises, dyads in which one person was asked to imitate the other secured better outcomes, both for themselves and for the group (Maddux, Mullen & Galinsky, 2008). Further evidence that the cooperative effects of being imitated do not only benefit the imitator comes from a study showing that, when they have been imitated, people are more willing to help others with simple tasks and donate more money to charity (van Baaren et al., 2004).

Other studies have examined the effect of cooperative attitudes on imitation. These show that people imitate a person they like more than a person they don't like, imitate members of an in-group more than members of an out-group (Likowski et al., 2008; Stel et al., 2010), and imitate more after they have been primed with words such as "we" and "ours" than after priming with words such as "me" and "mine" (Leighton, Bird, Orsini & Heyes, 2010). They also indicate that people imitate more when they are feeling ostracized or socially excluded, and under these circumstances they imitate members of the group that has excluded them more than members of other groups (Lakin, Chartrand & Arkin, 2008). This implies that, even when people are apparently unaware of giving or receiving simple imitation, their behavior is sensitive to the fact that imitation tends to elicit cooperative attitudes from the imitatee.

These bidirectional causal relationships raise the possibility that, when the members of a social group are in face-to-face contact, they are constantly maintaining one another in a cooperative frame of mind—in a state of readiness for collective action and information sharing—via a virtuous circle of simple imitation and prosocial attitudes. The idea that this virtuous circle functions to maintain, rather than to change, cooperative groups is consistent with research examining imitation across group boundaries. This shows not only that the members of other ethnic (Heider & Skowronski, 2008) and religious (Yabar et al., 2006) groups are imitated less than the members of one's own group, but also that being imitated by a member of a different ethnic group makes the imitatee like that imitator less, rather than more (Likowski et al., in press).

Very little is known about the psychological mechanisms supporting the virtuous circle. Even if we assume that matching vertical associations are the mechanism that makes imitation possible, we still need to understand how being imitated produces cooperative attitudes, and how cooperative attitudes engender imitation. Because the virtuous circle seems to be so well tuned to maintaining cooperative relationships, so fit for its purpose, it's tempting to assume that these mechanisms are dedicated, and perhaps sophisticated, cognitive adaptations. Future research may support this assumption, but that is far from inevitable. The virtuous circle could be underwritten by two relatively simple, nonspecific psychological gadgets: contingency detection and disinhibition.

Contingency detection, the capacity to detect when the movements of another person covary with one's own, could be the basis for the imitation to cooperation relationship. In other words, the imitatee's cognitive system need not know that the imitator's movements are topographically similar, only that movements of the imitatee's body predict those of the imitator's body. There is evidence that infants can detect contingencies between their own actions and those of external objects from an early age, and that detecting these contingencies is associated with positive emotion (Gergely & Watson, 1999). If the contingency-detection mechanism that generates these emotional reactions is the same one that mediates associative learning, then there is reason to believe that it is both simple and phylogenetically ancient (Rolls, 2000).

Disinhibition, or the release of inhibitory control, could underwrite the relationship between cooperative attitudes and imitative behavior. This suggestion is consistent with evidence that patients with lesions of the prefrontal cortex, an area involved in the inhibition of overlearned or automatic response tendencies, show compulsively imitative behavior (De Renzi, Cavalleri & Facchini, 1996). These inhibitory processes are complex and more highly developed in humans than in any other species. However, the possibility I am raising is that cooperative attitudes promote imitation not by using these complex inhibitory processes, but by switching them off. When we like someone, or perceive that person to be "like me," there is a release of inhibition that allows the activation of motor representations via matching vertical associations to produce overt imitative behavior. Thus, when we are cooperatively motivated, imitative tendencies, which are normally suppressed, are allowed to "get out" and influence observable behavior.

To find out whether basic mechanisms such as contingency detection and disinhibition are driving the virtuous circle, it will be necessary to test whether contingent, nonimitative behavior has the same prosocial effects as imitation, and whether cooperative attitudes "release" not only imitative behavior but also other overlearned reactions to social and inanimate stimuli. If experiments of this kind confirm that the virtuous circle depends on simple, nonspecific mechanisms, it would help to explain the early coevolution of human cooperation. As Sterelny (2003) has pointed out, language and cultural inheritance of information are not only instruments of cooperation but also mighty cooperative achievements. For a complete account of the origins of human cooperation, we need to find the precursors of these cooperative feats, the processes that got human cooperation off the ground. If it is based on simple psychological mechanisms, the virtuous circle linking simple imitation with cooperative attitudes is a strong candidate for this groundbreaking role.

4 Cooperative Effects of Complex Imitation

Discussion of the cooperative effects of complex imitation has been dominated by the possibility that complex imitation plays a crucial role in a certain kind of information sharing—the cultural inheritance of instrumental or technological skills. Much contemporary psychological research on this issue compares the behavior of children and chimpanzees in tasks where they are allowed to observe an expert performing a novel instrumental action (e.g., manipulating objects to obtain a reward) and then attempt to obtain the reward themselves. Focal questions are whether children are more likely than chimpanzees to imitate than to "emulate" (defined below) the expert's behavior, and, if so, whether this difference provides evidence that the capacity to imitate is a key psychological requirement for the cultural inheritance of instrumental skills (e.g., Tennie, Call & Tomasello, 2009; Whiten et al., 2009).

4.1 Imitation and Emulation

The distinction between imitation and emulation (Tomasello, 1996) is, I think, a very important one, but it is sometimes drawn in a confusing way. Performance of an instrumental task involves two kinds of transformation: a sequence of changes in the spatial properties of both the actor's body parts (body movements) and the objects on which he is acting (object movements). At its valuable root, the imitation–emulation distinction draws attention to *what* is copied by the observer of instrumental task performance: the body movements (imitation) or the object movements (emulation) or both. So far, so good. But this *what* distinction is sometimes confounded with a *how much* distinction: between the copying of a sequence and the copying of an endpoint.

When the *what* and *how much* distinctions are pulled apart, we have four possibilities: endpoint imitation—copying the endpoint of a sequence of body movements; endpoint emulation—copying the endpoint of a sequence of object movements; sequence imitation—copying a sequence of body movements; and sequence emulation—copying a sequence of object movements.

When the *what* and *how much* distinctions are confounded, "imitation" refers to copying a sequence of body movements, including the endpoint, while "emulation" refers to copying the endpoint, only, of a sequence of object movements (end/object). Thus, imitation is cast as "process copying" and emulation as "product copying" (Tennie, Call & Tomasello, 2009), and it seems that imitation is more likely than emulation to result in high-fidelity transmission of behavior—transmission of the kind needed for cumulative cultural evolution. If an observer copies only the endpoint of a model's action (the terminal configuration of the fingers, or of the parts of an instrument, or both), and there is more than one sequence that can culminate in this endpoint (the order and dynamics of the body part and/or object movements), it is almost inevitable that the *sequence* will be transmitted with lower fidelity. The fact that sequence copying is likely to lead to higher-fidelity transmission of sequences than endpoint copying is important in its own right. However, it does not imply that body movement copying is associated with higher-fidelity transmission than object movement copying.

Once the confusion between *what* and *how much* is penetrated, it becomes clear that the comparison that matters with respect to the cultural inheritance of skills is between sequence imitation and sequence emulation, and it is my hunch that, for most instrumental skills, sequence imitation will result in *lower-* fidelity copying than sequence emulation. A novice watching an expert potter, flint knapper, basket weaver, or computer programmer would do better to attend to, and copy, the sequence of object movements effected by these skills than to focus on the expert's body movements. A combination of both sequence imitation and sequence emulation is likely to be associated with the highest transmission fidelity, but if for some reason—local, developmental, or evolutionary—it has to be one or the other, my money is on sequence emulation coming out on top. Perhaps there are exceptions, but it seems that in the case of instrumental skills, actions on objects, the cultural wisdom lies in the object transformations rather than the body movements. If this is correct, copying the sequence of object movements (sequence emulation) will either ensure that the sequence of body movements is also inherited, or it won't matter, from the perspective of cultural evolution, how the object movements are effected by the actor's body.

To test the hunch that sequence emulation is more important than sequence imitation for the cultural inheritance of instrumental skills, it would be helpful to use eye-tracking to find out what novices are looking at most when they are learning a technological skill by observation. Are they, for example, studying more closely the movements of the flint knapper's fingers, hands, and arms, implying sequence imitation, or the angle and velocity at which one stone makes impact on the other, implying sequence emulation?

If my hunch is correct, imitation is much less important for the cultural inheritance of instrumental–technological skills than has previously been thought. However, once the *what* and *how much* distinctions have been dissociated, it becomes clear that imitation—sequence imitation—is likely to be indispensable for the cultural transmission of the other major category of skills—"communicative" or, more broadly, "gestural" skills. Many potent, culture-specific signs and rituals do not involve objects. They consist exclusively of conventional (i.e., instrumentally arbitrary) sequences of body movements, and therefore cannot be learned by sequence emulation.

These gestural skills are seldom considered by psychologists interested in cultural inheritance, but their importance in defining groups and promoting cooperation is recognized in anthropology and the humanities (Corbeill, 2004). They include the sequences of body movements that enable group members to communicate without words, and thereby to coordinate their activities when there are no words (e.g., when the message is ineffable, and before language coevolved), and when words are dangerous (e.g., when a group is stalking prey). They also include the sequences of body movements, such as those involved in ritualistic dancing, that enable group members to bond—to achieve the states of trust and commitment required for cooperative action—through the expression of common religious beliefs, and the sharing of heightened states of arousal. In addition to providing proximal support for cooperative action—by communicating information, promoting trust and commitment, and indicating who is and who is not part of the cooperative group—these gestural skills may also contribute to the kind of within-group homogeneity and between-group diversity necessary for the evolution of cooperation via cultural group selection (Boyd & Richerson, 2005).

4.2 Smart (Complex) and Dumb Imitation

I've called the imitation of novel actions "complex imitation" because it's traditionally assumed that the imitation of novel actions involves more complex psychological mechanisms than the imitation of familiar actions, and, in some measure, I agree with this tradition. The ASL model outlined in section 1.2 suggests that, although the same core mechanisms solve the correspondence problem for simple and complex imitation, something extra—horizontal, or serial order, processing—is needed to explain complex imitation.

Horizontal processing is complex in the sense of being computationally demanding, but in other respects it is dumb. Encoding the serial order of visual events, including observed body movements, is demanding on working memory and executive processes. However, horizontal processing can occur without consciousness or an intention to learn, and doesn't help the novice to make rational choices about what they should and should not imitate. The ASL model implies that rational decisions of this kind are an optional extra, not an intrinsic part, of complex imitation; rational decision making and imitation have distinct coevolutionary roots, and in adults as well as children and nonhuman animals, complex imitation often occurs without awareness or deliberation (Bird & Heyes, 2005).

In this respect, the ASL model is very different from theories suggesting that complex imitation necessarily involves the observer representing what the model intended to achieve (Tomasello, 1996), or to teach the observer (Csibra & Gergely, 2009). These theories suggest that complex imitation is smart in the selfishly rational sense: that an observer won't imitate an action unless she either understands what the outcome will be, and desires that outcome, or has reason to believe that copying the model's action represents the most efficient means for her to achieve a current goal. Shea (2009) has pointed out that smart imitation—imitation regulated by selfishly rational calculation—is likely to have negative effects on the fidelity of cultural inheritance, and therefore on the "evolvability" of cultural skills and practices. Studies reporting smart imitation imply that novices pick and choose which observed behaviors to copy, using the model as a source of elements and ideas, but reinventing rather than faithfully adopting the expert's method of getting the job done (e.g., Gergely, Bekkering & Kiraly, 2002). In contrast, dumb imitation—imitation that does not depend on rational calculation—has a better chance of allowing culturally accumulated wisdom to be propagated across generations of learners. I'd like to add two things to Shea's analysis. The first is a footnote, and the second a more substantive hypothesis about the importance of dumb imitation with respect to cooperation.

The footnote is to point out that dumb imitation is not unselective imitation; it won't result in novices copying every incidental detail of a model's behavior. On any given occasion when an expert demonstrates a skilled action, she may fumble or sneeze, and, at a low-level of analysis, the trajectories of her body movements will be slightly different from those she uses on other occasions. However, to learn any significant skill by imitation, a novice must watch many repetitions of the skilled action, performed by a single model or by a number of different models. Across these demonstrations the important elements of the action will persist while the incidentals will vary. Therefore, any horizontal sequence learning process sensitive to the frequency of elements and element transitions—that is, any sequence learning process worthy of the name—will be selective; it will encode the core, recurrent features of the sequence, filtering out accidents, idiosyncracies, and random variation.

The hypothesis is that, unlike smart imitation, dumb imitation can produce group markers—badges of group membership—that would be extremely difficult to fake. The members of a social group often have subtle behavioral characteristics in common; they walk with a certain gait, fiddle with their hair in a certain way, or use facial expressions that are minutely different from those of other groups. These group-specific characteristics can modulate cooperative behavior—by eliciting trust and other prosocial attitudes—but they are poor targets for smart imitation. In many cases, the distinctive features of a group's behavior can't be consciously detected, and therefore can't be copied (i.e., faked) by out-group members via a process that depends on conscious, rational calculation. It's possible that, in other cases, careful scrutiny permits the detection of group-specific behavior, but deliberate imitation of this behavior would yield an inauthentic caricature of the original.[3] However, these behavioral badges *can* be acquired via dumb imitation. The horizontal processes described by the ASL model allow a novel behavior to be learned by observation, and to be imitated without awareness or deliberation. Instead, they depend on the frequency with which the novel behavior is observed. In the case of group-specific postures, gestures, and mannerisms, frequency of observation is likely to covary with time spent in the company of group members—a fair indicator of genuine group membership.

5 Summary

For many years, those interested in the evolutionary origins and consequences of imitation have focused on the imitation of novel actions and its potential contribution to the cultural inheritance of technological skills. Drawing on recent research in experimental psychology, social cognitive psychology, and cognitive neuroscience, I have tried to adjust this focus in several ways. First, I have suggested that the imitation of novel actions (complex imitation) is continuous, in terms of its origins and underlying mechanisms, with the imitation of familiar actions (simple imitation), and therefore a full assessment of the cooperative functions of imitation must encompass them both. Second, I have reviewed research on simple imitation indicating that the members of human social groups are constantly copying each other's gestures and mannerisms. This "unconscious mimicry" both promotes, and is promoted by, prosocial attitudes, creating a virtuous circle that maintains group members in a state of readiness for cooperative action. Finally, I have argued that complex imitation is unlikely to play a major role in the cultural inheritance of technological skills, but that it is crucial for the cultural inheritance of gestural skills—sequences of body movements that promote cooperation via communication and social bonding, and, potentially, by creating the conditions necessary for cultural group selection. Provided that it does not depend on conscious, deliberative processes, complex imitation may also allow subtle, group-specific features of body movement to function as highly reliable badges of cooperative group membership.

Throughout the discussion, I have presented arguments and evidence suggesting that, although imitation has many adaptive effects, it is not "an adaptation" (*sensu* Williams, 1966). The core mechanisms of imitation—those that translate visual input from a model into matching motor output—are constructed by associative learning, a domain-general cognitive process, from sensorimotor experience provided by the sociocultural environment. Much of this experience depends on artifacts (e.g., mirrors) and practices (dance training) that are themselves cooperative achievements. Therefore, the human capacity to imitate both engenders, and is engendered by, cooperation.

Notes

1. "Imitation" refers to copying by an observer of a feature of the body movement of a model. "Copying" implies a specific causal relationship between observation of a feature of a model's body movement, *fm*, and execution by the observer of a body movement with the same feature, *fo*. This excludes not only cases in which *fm* and *fo* co-occur by chance, but also other examples of "social learning" in which *fo* is caused by observation of a property of the model other than *fm*, or in which the effect of observing *fm* is not specific to the production of *fo* (Heyes, 2001).

2. Meltzoff and Moore are exceptional in having pointed out very clearly that simple imitation of facial expressions poses the correspondence problem. Like those who discuss the correspondence problem in the context of complex imitation, they assume that it is solved by an up-and-down route, involving "supramodal coding" (Meltzoff & Moore, 1997).

3. Thanks to Ben Fraser for this suggestion.

References

Bird, G., & Heyes, C. M. (2005). Effector-dependent learning by observation of a finger movement sequence. *Journal of Experimental Psychology: Human Perception and Performance, 31*, 262–275.

Boyd, R., & Richerson, P. (2005). *Culture and the evolutionary process*. Chicago: University of Chicago Press.

Byrne, R. W., & Russon, A. E. (1998). Imitation: A hierarchical approach. *Behavioral and Brain Sciences, 21*, 667–684.

Carpenter, M. (2006). Instrumental, social, and shared goals and intentions in imitation. In S. J. Rogers & J. H. G. Williams (Eds.), *Imitation and the social mind: Autism and typical development* (pp. 48–70). New York: Guilford Press.

Carroll, W. R., & Bandura, A. (1982). The role of visual monitoring in observational learning of action patterns. *Journal of Motor Behavior, 14*, 153–167.

Catmur, C., Gillmeister, H., Bird, G., Liepelt, R., Brass, M., & Heyes, C. (2008). Through the looking glass: Counter-mirror activation following incompatible sensorimotor learning. *European Journal of Neuroscience, 28*, 1208–1215.

Catmur, C., Mars, R., Rushworth, M., & Heyes, C. M. (2011). Making mirrors: Premotor cortex stimulation enhances mirror and counter-mirror motor facilitation. *Journal of Cognitive Neuroscience, 23*, 2352–2362.

Catmur, C., Walsh, V., & Heyes, C. (2007). Sensorimotor learning configures the human mirror system. *Current Biology, 17*, 1527–1531.

Catmur, C., Walsh, V. & Heyes, C. (2009). Associative sequence learning: The role of experience in the development of imitation and the mirror system. *Philosophical Transactions of the Royal Society, Series B: Biological Sciences, 364*, 2369–2380.

Chartrand, T. L., & Bargh, J. A. (1999). The chameleon effect: The perception-behavior link and social interaction. *Journal of Personality and Social Psychology, 76*, 93–910.

Chartrand, T. L., & van Baaren, R. (2009). Human mimicry. *Advances in Experimental Social Psychology, 41*, 219–274.

Corbeill, A. (2004). *Nature embodied: Gesture in Ancient Rome*. Princeton, NJ: Princeton University Press.

Csibra, G., & Gergely, G. (2009). Natural pedagogy. *Trends in Cognitive Sciences, 13*, 148–153.

del Giudice, M., Manera, V., & Keysers, C. (2009). Programmed to learn? The ontogeny of mirror neurons. *Developmental Science, 12*, 350–363.

De Renzi, E., Cavalleri, F., & Facchini, S. (1996). Imitation and utilisation behaviour. *Journal of Neurology, Neurosurgery, and Psychiatry, 61*, 396–400.

Gergely, G., Bekkering, H., & Kiraly, I. (2002). Rational imitation in preverbal infants. *Nature, 415*, 755.

Gergely, G., & Watson, J. S. (1999). Early socio-emotional development: Contingency perception and the social-biofeedback model. In P. Rochat (Ed.), *Early social cognition: Understanding others in the first months of life* (pp. 101–136). Mahwah, NJ: Lawrence Erlbaum.

Heider, J. D. & Skowronski, J. J. (2008). Ethnicity-based similarity and the chameleon effect. (Submitted.)

Heyes, C. M. (2001). Causes and consequences of imitation. *Trends in Cognitive Sciences, 5*, 253–260.

Heyes, C. M. (2005). Imitation by association. In S. Hurley & N. Chater (Eds.), *Perspectives on imitation: From mirror neurons to memes* (Vol. 1, pp. 157–176). Cambridge, MA: MIT Press.

Heyes, C. M. (2011). Automatic imitation. *Psychological Bulletin, 137*, 463–483.

Heyes, C. M., & Bird, G. (2007). Mirroring, association and the correspondence problem. In P. Haggard, Y. Rossetti, & M. Kawato (Eds.), *Sensorimotor foundations of higher cognition, attention & performance XX*. Oxford: Oxford University Press.

Heyes, C., Bird, G., Johnson, H., & Haggard, P. (2005). Experience modulates automatic imitation. *Cognitive Brain Research, 22*, 233–240.

Heyes, C. M., & Ray, E. D. (2000). What is the significance of imitation in animals? *Advances in the Study of Behavior, 29*, 215–245.

Lakin, J. L., Chartrand, T. L., & Arkin, R. M. (2008). I am too just like you: The effects of ostracism on nonconscious mimicry. *Psychological Science, 14*, 334–339.

Leighton, J., Bird, G., Charman, T., & Heyes, C. M. (2008). Weak imitative performance is not due to a functional mirroring impairment in adults with autism spectrum disorders. *Neuropsychologia, 46*, 1041–1049.

Leighton, J., Bird, G., Orsini, C., & Heyes, C. M. (2010). Social attitudes modulate automatic imitation. *Journal of Experimental Social Psychology, 46*, 905–910.

Leighton, J., Bird, G., & Heyes, C. M. (2010). "Goals" are not an integral component of imitation. *Cognition, 114*, 423–435.

Likowski, K. U., Muhlberger, A., Seibt, B., Pauli, P., & Weyers, P. (2008). Modulation of facial mimicry by attitudes. *Journal of Experimental Social Psychology, 44*, 1065–1072.

Likowski, K. U., Schubert, T. W., Fleischmann, B., Landgraf, J. & Volk, A. (In press). Positive effects of mimicry are limited to the ingroup.

Lyons, D. E., Young, A. G., & Keil, F. C. (2007). The hidden structure of overimitation. *Proceedings of the National Academy of Sciences of the United States of America, 104*, 19751–19756.

Maddux, W. W., Mullen, E., & Galinsky, A. (2008). Chameleons bake bigger pies: Strategic behavioural mimicry facilitates integrative negotiations outcomes. *Journal of Experimental Social Psychology, 44*, 461–468.

Meltzoff, A. N. (1988). The human infant as homo imitans. In T. Zentall & B. Galef (Eds.), *Social learning: Psychological and biological perspectives*. Hillsdale, NJ: LEA.

Meltzoff, A. N., & Moore, M. K. (1997). Explaining facial imitation: a theoretical model. *Early Development & Parenting, 6*, 179–192.

Pawlby, S. J. (1977). Imitative interaction. In H. Schaffer (Ed.), *Studies in mother-infant interaction* (pp. 203–224). New York: Academic Press.

Pearce, J. M. 2008. *Animal learning and cognition: An introduction* (3rd Ed.). London: Psychology Press, Taylor & Francis Group.

Press, C., Gillmeister, H., & Heyes, C. (2007). Sensorimotor experience enhances automatic imitation of robotic action. *Proceedings of the Royal Society, Series B: Biological Sciences, 274*, 2509–2514.

Ray, E. D., & Heyes, C. M. (2011). Imitation in infancy: The wealth of the stimulus. *Developmental Science, 14*, 92–105.

Rolls, E. T. (2000). Precis of The brain and emotion. *Behavioral and Brain Sciences, 23*, 177–234.

Schultz, W., & Dickinson, A. (2000). Neuronal coding of prediction errors. *Annual Review of Neuroscience, 23*, 473–500.

Shea, N. (2009). Imitation as an inheritance system. *Philosophical Transactions of the Royal Society, Series B: Biological Sciences, 364*, 2429–2443.

Stel, M., Blascovich, J., McCall, C., Mastop, J., Van Baaren, R. B., & Vonk, R. (2010). Mimicking disliked others: Effects of a priori liking on the mimicry-liking link. *European Journal of Social Psychology, 40*, 867–880.

Sterelny, K. (2003). *Thought in a hostile world*. London: Blackwell.

Tanner, R., Ferraro, R., Chartrand, T. L., Bettman, J., & Van Baaren, R. (2008). Of chameleons and consumption: The impact of mimicry on choice and preferences. *Journal of Consumer Research, 34*, 754–766.

Tennie, C., Call, J. & Tomasello, M. (2009). Ratcheting up the ratchet: On the evolution of cumulative culture. *Philosophical Transactions of the Royal Society, Series B: Biological Sciences, 364*, 2405–2415.

Tomasello, M. (1996). Do apes ape? In C. M. Heyes & B. G. Galef (Eds.), *Social learning in animals: The roots of culture* (pp. 319–346). New York: Academic Press.

van Baaren, R. B., Janssen, L., Chartrand, T. L. & Dijksterhuis, A. (2009). Where is the love? The social aspects of mimicry. *Philosophical Transactions of the Royal Society, Series B: Biological Sciences, 364*, 2381–2389.

van Baaren, R. B., Holland, R. W., Kawakami, K., & van Knippenberg, A. (2004). Mimicry and prosocial behavior. *Psychological Science, 15*, 71–74.

van Swol, L. M. (2003). The effects of nonverbal mirroring on perceived persuasiveness, agreement with an imitator, and reciprocity in group discussion. *Communication Research, 30*, 461–480.

Waxler, C. Z., & Yarrow, M. R. (1975). An observational study of maternal models. *Developmental Psychology, 11*, 485–494.

White, B. L., Catel, P., & Held, R. (1964). Observations on the development of visually guided reaching. *Child Development, 35*, 349–364.

Whiten, A., McGuigan, N., Marshall-Pescini, S. & Hopper, L. M. (2009). Emulation, imitation, over-imitation, and the scope of culture in child and chimpanzee. *Philosophical Transactions of the Royal Society, Series B: Biological Sciences, 364*, 2417–2428.

Williams, G. C. (1966). *Adaptation and natural selection*. Princeton, NJ: Princeton University Press.

Yabar, Y., Johnston, L., Miles, L., & Peace, V. (2006). Implicit behavioral mimicry: Investigating the impact of group membership. *Journal of Nonverbal Behavior, 30*, 97–113.

Zentall, T. R. (2006). Imitation: Definitions, evidence, and mechanisms. *Animal Cognition, 9*, 1435–1448.

18 The Role of Learning in Punishment, Prosociality, and Human Uniqueness

Fiery Cushman

1 Introduction

At your local natural history museum rows of tiny dripping noses press on display cases, peering at the impalas, the grizzlies, and the komodo dragons. Like the glass that separates stuffed noses from stuffed animals, something separates humans from other animals—something substantial, but hard to see. Some combination of accumulated changes must explain language, science, culture, art, and civilization—in short, why humans build the museums and other animals inhabit them.

This essay focuses on just one element of that change: the uniquely rich, complex, and successful range of human prosocial behaviors. Even more narrowly, it focuses on the role of punishment in enforcing prosociality. In its approach, however, it aims for a broader insight: to illustrate the important relationship between abstract evolutionary models of behavior and the specific psychological mechanisms that actually produce behavior. This natural union improves evolutionary models, clarifies the structure of psychological mechanisms, and helps to reveal the foundations of human uniqueness.

Evolutionary theorists posit a simple relationship between punishment and prosocial[1] behavior (e.g., Boyd & Richerson, 1992; Clutton-Brock & Parker, 1995). In a population where some individuals punish antisocial behavior, it pays to be prosocial selectively with the punishers. And, in a population where some individuals behave prosocially only when threatened with punishment, it pays to punish antisociality. Put in concrete terms: I ought to stop stealing from you if you hit me in retaliation, and therefore the immediate costs of retaliation might be worth the long-term benefit of securing your property. This codependent relationship between punishment and prosocial behavior is well understood.

Far less understood are the psychological mechanisms that actually produce prosocial and punitive behaviors. At first glance, one might assume that psychology does not matter to the larger evolutionary question. Can't we understand and model the abstract relationships between evolutionary strategies without troubling ourselves with their psychological implementation? By analogy, formal models of economics do not concern themselves with the molecular structure of coins or bills.

This is a seductive perspective. It certainly simplifies the task of modeling the evolution of social behavior to ignore the underlying mechanisms. Ultimately, however, it is flawed. The functional design of punishment and prosociality depends critically on psychological details, in much the same way that formal economic models cannot ignore the peculiar irrationalities of human actors (Tversky & Kahneman, 1981). Psychological details explain when, how, and whom organisms decide to punish. They explain why punitive strategies (and especially reactive aggression) are more often observed than rewarding strategies among nonhuman animals. And they explain the pervasive, complex, and flexible nature of human social behavior.

Of course, the benefits of integration run the opposite way as well. Our understanding of the psychological mechanisms supporting punitive and prosocial behavior is enriched by considering their functional design.

This essay takes up both challenges: first, to demonstrate how the functional design of punishment and prosociality mirror psychological constraints, and second to demonstrate how the psychological mechanisms underlying punishment and prosociality are illuminated by considering their functional design. I argue that punishment is a specialized behavioral adaptation that exploits the ability of social partners to learn. Implicit in this argument is a distinction between specialized behavioral adaptations that function in a fixed manner in limited contexts, and general mechanisms of learning and behavioral choice that function flexibly across diverse contexts (Fodor, 1983; Hirschfeld & Gelman, 1994; Shiffrin & Schneider, 1977; Spelke, 2000). Many organisms can flexibly learn to avoid behaviors that have negative consequences: eating toxic foods, approaching open flames, jumping in cold water, and so on. This general capacity for learning presents an opportunity for social exploitation. Specifically, organisms can use punishment to teach social partners to act prosocially by exploiting their general learning capacity. As we will see, the evolutionary dynamics of this relationship between punishment and prosociality make it likely that punishment will operate via a specialized behavioral adaptation that is relatively fixed and limited, while prosociality will be supported in part by general, flexible mechanisms of learning and behavioral choice.

The general ability to learn associations between behavior and consequence is highly constrained in most organisms, however. For instance, the consequence must be rapid and salient in order for learning to occur (Mackintosh, 1975; Renner, 1964; Rescorla & Wagner, 1965; Schwartz & Reisberg, 1991). Thus, psychological constraints on learning will influence the functional design of punishment. In the abstract, any sort of punishment could motivate any sort of prosocial behavior. In reality, punishment must conform itself to the circumscribed ability of organisms to learn.

There is one species with a substantially expanded learning abilities: humans. It is no accident, therefore, that we also exhibit a uniquely flexible and productive prosocial behavior. To the extent that punishment (and also reciprocity and reward) depends on general learning mechanisms to motivate prosocial action, humans' uniquely powerful capacities in learning, reasoning, and behavioral choice stand to vastly expand their range of prosocial behavior beyond that of nonhuman animals.

I conclude by considering an irony of the human situation: Our punitive instincts (and possibility our instincts for reciprocity and reward) may not have "caught up" with our new capacities for learning, reasoning, and deciding. In some respects, the functional design of human punishment may still be adapted to the substantially more limited minds of our nonhuman ancestors.

2 Specialized versus General Mechanisms in Psychology

Psychological detail is starkly absent in most evolutionary models of punishment and prosociality. In these models a population of agents interacts and reproduces, leading to evolutionary change over successive generations. The modeler specifies a set of behavioral strategies that the agents can employ. For instance, possible strategies include "always behave prosocially," "punish antisocial behavior with 80 percent probability," and "cease antisocial behavior if it has been punished twice." These strategies are typically specified in very abstract terms, which is appropriate to the task of creating formal models that can be generalized across diverse cases. But these abstract strategies could be implemented at a psychological level in many different ways. Below, I describe a very coarse distinction between two classes of psychological mechanism: general mechanisms of associative learning and behavioral choice versus specialized behavioral adaptations.

Consider a rat that pushes a lever in order to avoid an electric shock. Decades of psychological research suggests that the rat's learned behavior is guided by something like a simplified calculation of expected value (reviewed in Daw & Doya, 2006). That is, the rat learns to associate certain behavioral choices with their likely future consequences in terms of subjective value, conditioned on some set of perceptual inputs. It then selects behaviors as a function of the value associated with each. The rat continuously updates these associations as it experiences punishments and rewards following its behavior. Critically, it has broad flexibility[2] in the kinds of associations that it can form. This allows it to adaptively adjust its behavior, guiding it toward optimal patterns of choice.

At the other extreme, a behavioral strategy can depend on an innate, rigid psychological response, what is called a "fixed action pattern" in behavioral biology and ethology. A classic example of a fixed action pattern is the motor routine by which a goose retrieves an egg that rolls out of its nest (Lorenz & Tinbergen, 1938). This behavior does not seem to be learned and regulated by general, flexible cognitive processes employing associative learning or value maximization. Rather, geese appear to have an innate mechanism that recognizes the perceptual input of an egg rolling out of the nest and automatically triggers a highly specific motor routine for retrieval.

Interpreted literally, most formal models of the evolution of social behavior use behavioral strategies like the fixed action pattern of the goose. They are innate, fixed over the course of the lifetime, and do not involve associative learning or the computation of expected value (Boyd & Richerson, 1992; Clutton-Brock & Parker, 1995; Maynard Smith, 1982; Nowak, 2006). A handful of studies, however, model social behavior using general, flexible learning mechanisms something like those used by the rat (Gutnisky & Zanutto, 2004; Macy & Flache, 2002; Nowak & Sigmund, 1993). These models demonstrate that, in principle, it is possible to achieve prosociality without the biological evolution of a domain-specific prosocial strategy.

So, which is the more accurate model of social behavior: the goose, or the rat? That question motivates much of the remainder of this essay. Before charging into the fray, it will help to arm ourselves with two general observations.

First, general learning mechanisms are "free" from an adaptive perspective. Basic mechanisms supporting associative learning and reward-maximizing choice—the essence of operant conditioning—exist in fruit flies, zebra fish, pigeons, rats, sophomores, and virtually animal in between. Thus, if we can explain some behavior in terms of general learning, then it will usually not be necessary to postulate any further adaptation. For instance, imagine that we observe a dog retrieve a newspaper. This behavior might be the product of general learning, or it might be the product of a specialized adaptation. All else being equal, if the behavior can be explained in terms of general learning processes, this hypothesis is more likely than the alternative hypothesis that the dog has a specialized adaptation for newspaper retrieval. The general learning hypothesis comes for free, whereas the specialized adaptation hypothesis requires some additional evolutionary event. This is an argument from parsimony.

Second, general learning mechanisms show characteristic constraints. First, and most obviously, general learning mechanisms require experience. Rats aren't born knowing when to push a lever; this behavior is only acquired given sufficient experience. Moreover, successful association between stimuli, behavioral choice, and punishment or reward requires special conditions. For instance, learned associations typically require (1) salient events that occur within (2) a relatively short period of time (Mackintosh, 1975; Renner, 1964; Rescorla & Wagner, 1965; Schwartz & Reisberg, 1991). These constraints may explain why geese do not rely on general learning processes to acquire egg-retrieval behavior. The relevant experiences are probably infrequent, and each negative experience is very costly to reproductive success. The feedback (one less chick born than egg laid) is probably not very salient to goose, and it comes only after a long temporal delay.

In summary, general learning processes are adaptively "free," but mechanistically constrained. By contrast, specific behavioral adaptations require new adaptive events, but they can move beyond the constraints of general learning processes. With these considerations in mind we can assess, first, whether punishment and prosociality are more likely to be supported by specific behavioral adaptations versus general learning processes and, second, how these psychological details affect the functional design of punishment.

3 The Evolutionary Dynamics of Punishment and Prosociality

The next two sections argue from complementary perspectives that we should expect organisms to punish like geese (using specialized adaptations) and behave prosocially like rats (using general mechanisms of learning and choice). Some technical detail is required, but I will begin by sketching the argument from evolutionary dynamics in broad strokes.

First, consider punishment. If punishment depends on a specialized mechanism, it can be inflexible: We might punish certain situations no matter how we estimate the costs or benefits. By contrast, if punishment depends on general learning mechanisms, it will be flexible: We will only punish situations when the anticipated benefits outweigh the anticipated costs. It turns out that there is a great benefit to inflexible punishment. Think of an unruly toddler who throws a tantrum whenever his parents attempt discipline: If parents flexibly decide whether to punish (like the rat), then they may conclude that the costs are greater than the benefits and give up. In this case, the toddler can strategically persist in misbehavior until his parents learn that punishment is hopeless—so much the worse for the parents. But if the parents are inflexibly committed to punishment, the toddler cannot profit from persistent tantrums. Rather, his best strategy is to behave. Hence, it pays to punish inflexibly, like the goose.

Behavioral flexibility is favored for prosociality, however. Imagine that the toddler's parents wisely adopt a strategy of inflexible punishment for misbehavior, but that his grandmother is a doting pushover. If the toddler is an inflexible devil, he benefits with grandma but pays the costs of punishment with his parents. If the toddler is an inflexible angel, he benefits with his parents but misses the valuable opportunity to exploit grandma's dotage. The optimal path for this toddler is to flexibly adopt prosociality depending on the costs and benefits: Behave around the parents, misbehave with grandma. In short, it pays to adopt prosociality flexibly, like a rat. Now, it is possible to imagine a specialized adaptation that facilitates this contingent choice, finely tuned to distinguishing the enforcers from the pushovers and selecting innately specified appropriate behavioral responses to each. General learning mechanisms provide this behavior strategy "for free," however, using past experience to learn when to be naughty and when to be nice.

With this rough argument on the table, I'll now turn to a more detailed consideration of the evolutionary dynamic between punishment and prosociality and the psychological mechanisms that we might expect to support each. It will help to give labels to the strategies we have considered. Imagine an interaction in which an agent (A) can act either prosocially or antisocially, and in response a partner (P) can either punish or not punish. Here are the six strategies that we must consider:

Fixed Prosociality A behaves prosocially toward P
Fixed Antisociality A behaves antisocially toward P
Contingent Prosociality A behaves prosocially toward P only if P punishes antisocial behavior
Fixed Punishment P always responds to A's antisocial behavior by punishing A
Fixed Nonpunishment P never responds to A's antisocial behavior by punishing A
Contingent Punishment P responds to A's antisocial behavior by punishing A only if this tends to decrease A's antisocial behavior

I will assume that punishment is more costly than nonpunishment and prosociality is more costly than antisociality (setting aside the future benefits of social partners' behavior).

A "ratlike" learner that maximizes expected utility based on past experiences will adopt contingent prosociality (only pay the costs of prosociality if antisociality carries the greater cost of punishment) and contingent punishment (only pay the costs of punishment if it succeeds in inducing prosocial behavior). Meanwhile, it is possible to imagine "gooselike" fixed action patterns that are either fixed or contingent in their operation. As noted in the last section, explaining contingent behavior by "adaptively free" general learning mechanisms is more parsimonious than invoking a specialized behavioral adaptation, all else being equal. Thus, I will be treating contingent strategies as products of general learning and choice mechanisms, and fixed strategies as products of specialized adaptations.

The following sections ask whether the coevolutionary dynamic between prosociality and punishment can emerge from various combinations of the strategies listed above. In particular, they ask whether a population of *antisocial nonpunishers* can be successfully invaded to yield a population of *prosocial punishers*. That is, can prosociality and punishment jointly arise where neither existed before?

3.1 Fixed Prosociality and Fixed Punishment

A population of antisocial nonpunishers is unlikely to be invaded by either fixed prosociality or fixed punishment. The combination of these strategies does not provide a reliable path toward prosociality enforced by punishment.

To begin with, fixed prosociality is a clear loser in a population that never punishes antisocial behavior—this strategy pays the costs of generosity with no contingent benefit. On the other hand, fixed prosociality would be favored over fixed antisociality in a population of fixed punishers. In this population, fixed prosociality avoids the costs of punishment for antisocial behavior.[3] So, could a population of fixed punishers emerge and maintain stability? No: Fixed punishment is disfavored whenever agents adopt prosociality or antisociality as fixed strategies. The fixed punishment strategy pays an extra cost (of punishment) whenever it encounters an antisocial agent, but this extra cost does not yield any contingent future benefit. The fixed antisocial agent is just as likely to adopt antisociality in future interactions, whether or not P adopts a punitive strategy. Likewise, fixed prosocial agents are just as likely to adopt prosocial behavior in future interactions, again whether or not P adopts a punitive strategy. Thus, the costs associated with punishment yield no selective benefit to punishers when either fixed prosociality or fixed antisociality dominates.

One helpful lens through which to view this interaction is the tragedy of the commons. Over time, a sufficient number of punishers can make fixed prosociality stable, but they pay a cost to do so. Unfortunately, their costly efforts are exploitable by nonpunishers, who avoid the costs of punishment but equally reap the benefits of prosociality.

3.2 Contingent Prosociality and Fixed Punishment

One solution to this tragedy of the commons is clear: Punishment will be favored if it yields *exclusive* benefits to the punisher. This condition is met when social partners engage in contingent prosociality. That is, if agents apply the rule, "Only act prosocially when it is enforced by punishment," and if they are sufficiently adept at discriminating between social partners, then they will end up adopting prosociality only in interactions with individuals who punish. So, is it possible for contingent prosociality and fixed punishment to invade a population of antisocial nonpunishers?

Figure 18.1 charts the evolutionarily dynamics between these two strategies. First, consider a social environment in which neither strategy is employed (upper left corner). Clearly, punishment will not be favored in this environment; it pays the costs of punishing antisocial acts without any contingent benefits. However, the strategy of contingent prosociality is neutral in this environment. An individual that adopts contingent prosociality will never experience punishment for its antisocial actions, and therefore will consistently behave in an antisocial manner (assumed to be fitness maximizing). If the strategy of contingent prosociality attained sufficient frequency in a population, the strategy of contingent punishment can then also be favored (lower left corner). This presents a plausible path from a state in which there is neither punishment nor contingent prosociality to a state in which there is both punishment and contingent prosociality (from the upper left to lower right of figure 18.1), which

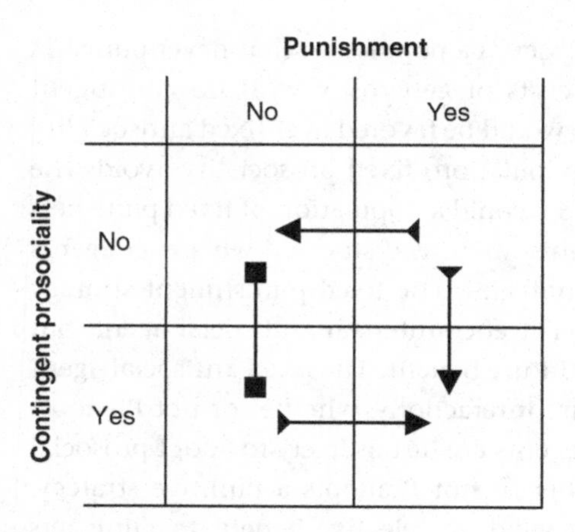

Figure 18.1

is evolutionarily stable. The key first step is the emergence of a strategy that responds to the punishment of antisocial behavior by switching to prosocial behavior.

As described above, it is possible that contingent prosociality could be supported either by general learning mechanisms or by a specialized adaptation. Which is more likely? Again, we return to argument by parsimony: General learning processes are adaptively free (in the sense that they don't require any new adaptive event)—they are sitting there, waiting to be exploited. By contrast, a specialized adaptation for contingent prosociality is relatively unlikely, especially because there is no selective pressure favoring such an adaptation prior to the emergence of punishment.

This point is best appreciated in concrete terms. Imagine a population of rats that have no specialized adaptations for punishment or contingent prosociality, but do have standard, general learning mechanisms. A punitive strategy that induces prosociality by exploiting rats' general learning mechanisms can immediately invade this population. By contrast, a punitive strategy that induces prosociality by exploiting a specialized adaptation for contingent prosociality must wait until such a specialized adaptation emerges. As described above, the emergence of such a specialized adaptation is not disfavored in fitness terms; it is a neutral change. But, the emergence of such a specialized innate behavior is unlikely given the absence of any selective pressure that favors it.

Once punishment has invaded a population, it becomes relatively easier for specialized adaptations for contingent prosociality to emerge. Consider again a hypothetical population of rats. Suppose that a punitive strategy has invaded this population by exploiting the general learning processes available in the population. But recall that general learning processes are characterized by certain constraints: The rat must experience sufficient punishment, this punishment must be timely and salient, and so on.

Within this social environment of punishment, there is a selective pressure for the emergence of specialized mechanisms that detect punishment and respond contingently with prosocial behavior more quickly and reliably than generalized learning processes (Cushman & Macendoe, 2009). In this case, general learning mechanisms establish an initial behavioral repertoire that facilitates the subsequent emergence of specialized adaptations. The ability of general learning mechanisms to pave the way toward specialized adaptations is a well-studied evolutionary phenomenon known as the "Baldwin effect." Thus, while I have argued in this section that punitive strategies are most likely to have emerged by exploiting contingent prosociality as a property of general learning processes, the present dynamic of punishment and prosociality may be supported by more specialized psychological mechanisms of contingent prosociality, or by some mix of specialized and general mechanisms. I return to this point in section 6.

3.3 Contingent Punishment

So far I have argued that contingent prosociality is necessary for *fixed* punitive strategies to be successful and, moreover, that contingent prosociality is relatively more likely to be supported by general learning mechanisms. I now turn to the same question regarding punishment. Can *contingent* punishment coevolve with prosociality?

On its face, contingent punishment looks superior to fixed punishment. Whereas fixed punishment pays the cost of punishing individuals who never respond with contingent prosociality (fixed antisocial actors), contingent punishment avoids these costs. Simply put, contingent punishers learn not to bother punishing where it can't help, and focus the costs of punishment solely where it maximizes benefits: changing the behavior of contingent prosocialists.

Despite these apparent advantages, however, contingent punishment does not provide a reliable path toward prosociality. The difficulty is that, faced with contingent punishment, both fixed and contingent prosociality are disfavored strategies. Rather, individuals do best by adopting fixed antisocial behavior. After all, a purely antisocial actor can largely avoid punishment in an environment dominated by contingent punishment by teaching social partners, "don't bother punishing me." It thereby reaps the rewards of antisociality, while avoiding the costs of punishment. Thus, the strategy of contingent punishment will tend to suppress prosociality, rather than to promote it.

More formally, this argument has two facets. First, in a population dominated by fixed antisocial behavior, the emergence of contingent punishment will not promote prosocial behavior. This follows directly from the logic of the previous paragraph. Neither fixed nor contingent prosociality outperforms fixed antisociality when played against contingent punishment.

Second, in a population that has achieved the fixed punishment–contingent prosociality dynamic described in the previous section, contingent punishment may be unable to invade. Specifically, contingent punishment is at best neutral, and possibly inferior, compared with fixed punishment. If contingent punishers must learn to adopt punishment, then they suffer the costs of this learning process (lost opportunities for prosociality obtained via punishment) compared with fixed punishers, who adopt the optimal punitive strategy immediately. At best, if contingent punishers have a strong initial bias toward punishment and a capacity to *un*learn punishment if it is unsuccessful, then they fare no worse (but no better) than fixed punishers when playing against contingent prosocialists. Critically, the only circumstance in which contingent punishment outperforms fixed punishment is when playing against a *fixed antisocial* partner. However, a population dominated by fixed punishers presents an extremely unforgiving environment for fixed antisocial players. Thus, fixed punishment sustains a social environment unfavorable to invasion by contingent punishment.

In summary, a stable dynamic between prosocial and punitive behavior requires inflexible punishment. This requirement of inflexibility among certain behavioral strategies is well recognized (Frank, 1988). In principle, a sufficiently sophisticated cognitive mechanism capable of general strategic reasoning could recognize this inflexibility requirement and adopt fixed prosociality. Humans sometimes do this: the doctrine of "mutual assured destruction" is an example. But the kind of general learning mechanisms possessed by most animals are unlikely to support this kind of abstract strategic reasoning. Rather, these learning mechanisms will tend to support punitive behavior only when it demonstrably promotes prosocial behavior: that is, flexibly. Consequently, the punitive behaviors of nonhumans are more likely to be the product of a specialized adaptation resembling "fixed punishment" than the product of general learning processes.

As we have seen, the evolutionary dynamics of codependent punishment and prosociality suggest that punishment more likely emerged as a specialized adaptation, whereas contingent prosociality was more likely initially supported by general learning processes. To put it another way, punishment is a mechanism that exploits general learning processes; it gets social partners to adopt prosocial behavior roughly by operant conditioning. This sets up a clear prediction about the functional design of punishment. Punishment should be designed to match the constraints of general learning processes, obtaining the maximum response from social partners at the minimum cost. Thus, to understand the functional design of punishment we will need to understand the psychology of learning. Section 5 reviews experimental evidence that supports this functional match, and section 7 considers its relevance to human behavior in particular.

First, however, section 4 illustrates how certain structural aspects of many social interactions also favor specialized adaptation or general learning processes for punishment and prosociality. This discussion depends not on considerations of evolutionary dynamics and arguments from parsimony, but rather on the other conceptual tool we established in section 2: the psychological constraints characteristic of general learning processes.

4 The Structure of Social Interactions and the Constraints of Learning

As noted above, general learning processes typically require that the reinforcement of actions occurs in a relatively quick and salient manner (Mackintosh, 1975; Renner, 1964; Rescorla & Wagner, 1965; Schwartz & Reisberg, 1991). For instance, if you want to train your dog to pick up the newspaper, it makes more sense to reward her with a biscuit each time she fetches than with a new collar for Christmas. General learning mechanisms are simply not sufficient to associate a paper fetched in May with a new neck accessory in December.

Turning to the relationship between punishment and prosociality, certain kinds of social interactions provide the opportunity for quick, salient reinforcement, while others do not. This provides an additional basis on which to predict whether, and when, punishment and prosociality are likely to be supported by general learning processes versus specialized adaptations. By analogy, if dogs readily adopt some behavior incentivized by immediate biscuits, it is plausible that their behavior depends on general learning mechanisms. But, if dogs adopt some behavior incentivized by collars at Christmas, it is unlikely that that behavior was learned by general mechanisms—rather, it indicates a specialized adaptation.

Consider again the social interaction in which an agent (A) harms a social partner (P). Possible harmful actions by A include aggression, resource theft, territorial violation, or sexual contact. Then, P punishes A by physical aggression. As a consequence, A does not perform this harmful action in future encounters. As above, we are faced with two questions. First, did A adopt prosociality because of a general learning mechanism or a specialized adaptation? Second, did P punish because of a general learning mechanism or a specialized adaptation? I will approach these questions in the following way: Is it plausible that each of these behaviors could be a product of general learning mechanisms, given the constraint that those mechanisms require quick and salient feedback?

In order for A to adopt prosociality via general learning processes, P's punishment should follow A's harmful action very quickly. If we assume that P is present when the harmful action occurs, there is no obstacle to rapid punishment: P can initiate physical aggression toward A immediately. On the other hand, relatively more delayed forms of punishment will be disfavored. For instance, if P punishes A by destroying a food resource of A's several days hence, A is relatively less likely to associate this punishment with her prior harmful act.

Similarly, in order for P to adopt punishment via general learning processes, A's desisting from future harm must follow P's punishment very quickly. In some cases this criterion will be easy to meet. For instance, if A is encroaching on P's territory and P's punishment drives A away without resistance,[4] this positive reinforcement of P's punishment occurs immediately. However, in other cases this criterion will be hard to meet. For instance, if A consumes some resource of P's and P punishes A, the positive reinforcement of P's punishment only occurs at some point in the future when A next has the opportunity to steal resources from P, but instead desists. In this case, the basic temporal structure of the social interaction—the fact that opportunities for A to steal from P arise only occasionally—makes it relatively easier for A to adopt prosociality via general learning processes (because P's punishment can be immediate and takes the form of a relatively salient aggressive action) but relatively harder for P to adopt punishment via general learning processes (because A's prosociality must be delayed and takes the form of a relatively nonsalient "omission" of harm).

To summarize, antisocial acts can be punished immediately, facilitating learning. But the value of punishment is harder to learn, because the behavioral changes it promotes follow at a longer temporal delay. Thus, general mechanisms of learning and behavioral choice may be sufficient to support prosocial action, whereas specialized psychological mechanisms may be required to support punishment.

Similar considerations may help to resolve a puzzle concerning punishment. A provocative series of studies shows that it is preferable to promote prosocial behavior in social partners by withholding aid from antisocial actors, rather than actively punishing antisocial actors (Dreber et al., 2008; Rand, Dreber, et al., 2009; Rand, Ohtsuki & Nowak, 2009). These studies model dyadic social interactions in which each actor has three choices of behavior. She can cooperate with her partner (e.g., share food), which is costly to herself and beneficial to her partner. She can defect against her partner (e.g., withhold food sharing), which is costless to herself and yields no benefit to her partner.[5] Or, she can punish her partner (e.g., by physical attack), which is costly to herself and even more costly to her partner. In brief, it turns out that both individual- and group-level fitness is maximized when players respond to defection with reciprocal defection. Both individual- and group-level fitness is lower when players respond to defection with punishment. This finding is puzzling because it predicts that we should not observe costly punishment as a response to antisocial behavior—rather, we should observe reciprocal antisociality. Yet, costly punishment is a common response to antisocial behaviors both in experimental and natural settings (Clutton-Brock & Parker, 1995; Fehr & Gachter, 2002; Henrich et al., 2005). Why?

This puzzle may be partially explained by the considerations introduced in this section: the structure of social interactions and the constraints of general learning mechanisms. When A defects against P, P may have an immediate opportunity to respond with costly punishment of A—for instance, by physical aggression. By contrast, the opportunity for P to respond with defection against A is necessarily delayed until a circumstance arises in which P has an opportunity to cooperate with A. The temporal delay imposed by the structure of the social interaction may prevent reciprocal defection from effectively exploiting general learning processes to promote future prosociality.

Of course, the same point applies if we reframe "reciprocal defection" as "reciprocal altruism"—the identical strategy framed in terms of prosociality-for-prosociality, rather than antisociality-for-antisociality. Acts of reciprocal prosociality must await appropriate opportunities to be performed. If you share food with me, for instance, I may not be able to reward this prosocial action until a situation arises where I am the one with surplus food. Thus, while mathematics favors reciprocal altruism, there are difficult psychological obstacles to implementing it via general mechanisms of learning and choice. This may explain the key role for trust in human cooperation (Knack & Keefer, 1997; McNamara, et al., 2008; Mui, Mohtashemi & Halberstadt, 2002; Silk, 2003; Zak & Knack, 2001). Possibly, trust functions as a specific behavioral adaptation that facilitates reciprocal exchanges of goods without requiring a general psychological mechanism to successfully associate prosocial acts with subsequent rewards.

The magnitude of the cognitive constraints on learned social behavior is difficult to overstate. For instance, experimental work in pigeons shows that even a short delay between choice and reinforcement can have severe consequences for cooperation in a prisoner's dilemma (Baker & Rachlin, 2002). In this study, pigeons played an iterated prisoner's dilemma game against a computer that adopted the strategy Tit-for-Tat, cooperating one trial after the pigeon cooperated, and defecting one trial after the pigeon defected. The delay between trials was varied. When there was no enforced delay between trials—that is, when pigeons experienced defection from the computer immediately following their own act of defection—cooperation rates averaged 64 percent. But when reciprocated defection was delayed by just 18 seconds, cooperation rates dropped by a quarter, to 48 percent. This finding illustrates just how severe the constraint of rapid and salient response to antisocial behavior will be for punishment (or reward, defection, etc.) to promote social behavior by exploiting general processes of associative learning and behavioral choice in nonhuman animals (see also Stephens, McLinn & Stevens, 2002).

In summary, the standard temporal structure of social interactions often allows punishment to follow rapidly after antisocial acts, but prevents contingent prosociality from following rapidly after punishment. This property makes possible a learned association between performing antisocial behaviors and receiving a punitive response, but makes more difficult a learned association between responding punitively and obtaining future benefits via prosociality. This affords an additional basis on which to predict that prosociality depends on ratlike learning mechanisms, while punishment depends on a gooselike specialized adaptation. Moreover, the same temporal constraints will often make punishment a more effective "teaching strategy" than reward. Physical aggression can often be employed for swift and salient reinforcement, whereas reward must often be delayed until an appropriate opportunity or resource is available.

5 The Functional Design of Punishment

Human punishment furnishes more than its share of puzzles. Why do we execute prisoners who are soon to die of natural causes? Why do we punish a malicious shooter who hits his target more than one who misses? Why do we excuse people for past crimes after a period of several years? Why is it illegal to push a child in a pond to drown, but perfectly legal not to throw a life preserver toward a drowning child?

Framed as moral, philosophical, and legal puzzles, these questions have tickled and tortured scholarly minds for centuries. But they can also be framed as psychological puzzles, and in this capacity the arguments developed above offer insight. Two clear predictions follow from the claim that punishment is a specialized adaptation that exploits general learning processes in order to promote prosocial behavior among social partners. First, punishment should operate in a relatively inflexible manner, that is, more like the fixed action pattern of the goose than the learning behavior of the rat. Second, punitive behavior should be functionally designed in ways that reflect the particular constraints of general learning processes. These predictions allow us to understand puzzles of punitive behavior in terms of functional design.

5.1 Retribution

Philosophical and legal scholarship identifies several possible motives for costly punishment. These include deterrence (establishing a policy that discourages future harmful behavior), incapacitation (directly preventing future harmful behavior, e.g., by imprisonment or death), and retribution (harming morally responsible harm-doers for reasons of "just desert"). Notice that incapacitation and deterrence treat punishment as instrumentally valuable: It is a useful behavior because it maximizes the welfare of possible future victims. This kind motivational structure is compatible with punishment as a product of general mechanisms of learning and behavioral choice, which also operate roughly by maximizing expected value. By contrast, retribution accords punishment itself primary value—retributive punishment occurs not because it is expected to bring secondary benefits, but rather because it is considered to be a necessary or deserved response. This motivational structure is more compatible with punishment as specialized behavioral response.

Several lines of psychological research suggest a basic process of assigning blame and punishment (Cushman, 2008; Darley & Shultz, 1990; Heider, 1958; Shaver, 1985; Shultz, Schleifer & Altman, 1981; Weiner, 1995), and in its details it is fundamentally retributive. When a harm occurs, we begin by seeking out individuals who are causally responsible. We then assess the harm-doers' mental states at the time of their actions, determining whether they had a culpable mental state such as intent to harm or foresight. Finally, we assign punishment to the causally responsible parties in proportion both to the degree of the harm and the degree of their culpable mental state. On its face, this basic model of punishment fits best with a retributive motive for punishment, as opposed to deterrence or incapacitation. It does not contain any explicit calculation of the probability of future transgression, as would be predicted if deterrence were the primary psychological motivation underlying punishment. Rather, it treats punishment itself as an object of primary value, as would be predicted if retribution were the primary psychological motivation underlying punishment.

Psychological studies have directly contrasted the predictions of incapacitation or deterrence as motivations for punishment against the predictions of a retributive motivation for punishment, consistently favoring the latter[6] (Carlsmith, 2006; Carlsmith, Darley & Robinson, 2002; Darley, Carlsmith & Robinson, 2000). Contrary to the predictions of an incapacitation or deterrence motivation, judgments of deserved punishment are not strongly modified by the probability that a perpetrator will reoffend or the probability that future offenses will go undetected, two factors that should increase the amount of punishment assigned.

Additionally, several studies of actual punitive behavior in structured economic exchanges show that people punish harmful acts even in one-shot, anonymous interactions (Fehr & Gachter, 2002; Henrich et al., 2005). This is a situation in which the punisher clearly has no personal stake in deterrence of future harms. It has been argued that the adaptive function of punishment in one-shot interactions is to deter future harms perpetrated against third parties (Fehr & Gachter, 2002; Gintis et al., 2003). However, there is some evidence that people are more likely to punish in a one-shot interaction if they have been harmed themselves than if a third party was the victim (Carpenter & Matthews, 2004). This casts doubt on the view that punishment is primarily motivated by a concern with future harms against third parties. Rather, the structure of punishment in one-shot interactions appears to be an inflexible, retributive response: You harmed me, so I harm you. Subjects' self-reported motivations match this conclusion: In one study of third-party punishment, 14 percent of subjects said that they punished third parties in order to reduce the incidence of future harms (deterrence), 56 percent said they punished third parties in order to "get back" at those who acted antisocially (retribution), and 30 percent said they were motivated by both factors (Carpenter & Matthews, 2004).

Of course, retributive motivations might reliably produce deterrent or incapacitative effects. In fact, I have taken pains to argue that the best way to understand the functional value of punishment is precisely in terms of deterrence—that is, eliciting contingent prosociality in future interactions. But the likely effects of punishment, and its adaptive function, need not constitute the psychological motivations that underlie it. Punishment may be adaptive for deterrent reasons at an "ultimate" adaptive level, and yet be instantiated by retributive mechanisms at a "proximate" psychological level. By analogy, the consumption of sugars and fats has future energetic benefits, and presumably the adaptive function of that consumption is to obtain the nutritive effects. The principal psychological motivation underlying the consumption of sugars and fats appears to be their taste, however, and not a learned association between consumption and future energetic states. Having an innate taste for sugar or fat circumvents the problem of learning by brute association which properties of potential foodstuffs are correlated with which future energetic states. Similarly, having an innate taste for punishment would circumvent the problem of learning associatively how to elicit future prosociality from social partners. Also, it would meet the inflexibility requirement introduced in section 3: the requirement that punishers cannot be "taught out of punishment" by intransigent antisocial actors.

In summary, the standard psychological processes underlying individual punishment of harmful actions are best characterized by a retributive motivation, and not by reasoning about the long-term benefits of punishment. Retributive motives are typically triggered when a person performs an action that causes harm, and subsequent punishment depends both on the severity of the harmful outcome and also on the degree to which that harmful outcome was intended. Retributive behaviors are surprisingly inflexible, operating even in contexts where interactions are one-shot and anonymous. This psychological model of retribution matches the inflexibility requirement discussed above: Punitive behavior will tend to be maintained even against social partners who fail to adopt prosociality. As we have seen, the apparent irrationality of this inflexible strategy actually has important consequences for the maintenance of codependent punishment and prosociality.

5.2 Punishing Accidents

People often judge that some punishment is deserved for unintentionally harmful behaviors—that is, for accidents. Our sensitivity to accidental outcomes appears to be substantially greater for punitive judgment than for judgments of moral permissibility or wrongness (Cushman, 2008). This is surprising: Why is it that we tend to punish accidents to a greater degree than we actually consider them wrongful? Could the punishment of accidental outcomes reflect the function of exploiting general learning mechanisms to promote prosocial behavior?

Outcome-sensitive punishment has been observed in several vignette studies (Cushman, 2008; Darley et al., 2000), and it is also widespread in the laws concerning negligent behavior.[7] More recently, it has been demonstrated in actual behavior using a probabilistic economic game (Cushman et al., 2009). In this "trembling hand" game, one player allocated money between herself and her partner either selfishly (everything for herself), fairly (an even split), or generously (everything for her partner). But, her allocation was subject to a probabilistic roll of a die—for instance, attempting to be selfish had a 4/6 chance of a selfish allocation, a 1/6 chance of a fair allocation, and a 1/6 chance of a generous allocation. Thus, the allocator could have selfish intent matched with a generous outcome (or vice versa). Finally, her partner was given the chance to respond by decreasing or increasing the allocator's payoff (i.e., punishing or rewarding the allocator). In doing so, the partner could respond to the allocator's intent, or the actual outcome of the allocation (even if unintended), or both. On average, responders punished both stingy intent and accidentally stingy outcomes. If anything, they weighted accidental outcomes slightly more than intent.

It is surprising that people's judgments of deserved punishment are strongly influenced by accidental outcomes, because their judgments of "moral wrongness" are not (Cushman, 2008). For instance, consider two drunk drivers, one who runs into a tree and another who runs into a person. People will tend to say that both behaved equally wrongly, but that the murder deserves substantially more punishment. This finding contradicts the commonsense assumption that we punish actions simply in proportion to their wrongfulness. To be sure, the moral status of an intention and an act play an important role in judgments of deserved punishment—but accidental outcomes count for a lot, too. The fact that judgments of moral wrongness and punishment differ in this way may explain why the punishment of accidents has been a point of particular concern in law and philosophy (Hall, 1947; Hart & Honore, 1959; McLaughlin, 1925; Nagel, 1979; Williams, 1981).

So, why do we punish accidents? Possibly because people learn from the punishment of accidents. To borrow a helpful term from education theory, accidents are "teachable moments." A may invade P's territory, eat P's food, consort with P's mate, and so on, with no knowledge of P's claims. These transgressions are, in some sense, unintentional—the acts of walking, eating, and mating are intentional, but their transgressive nature is unforeseen. Still, by punishing, P has the opportunity to teach A the boundaries of his territory, his property, and his relationships. Consider an even more unintentional harm: A loses his grip of a log and drops it on P's foot. The harm is wholly unintentional, but punishment may teach A to exert greater care in future circumstances. It also teaches A what matters ("my foot") and how much it matters ("as much as the whap you're about to get"). In these cases, A's unintentional harm provides an opportunity for P to teach a valuable lesson.

Of course, P's punishment will accomplish little if the type of behavior produced by A could not be successfully controlled even in future interactions. Indeed, experimental evidence suggests that the punishment of accidents is restricted to behaviors that could, in principle, be controlled (Alicke, 2000; Cushman, et al., 2009). This criterion is paralleled in Anglo-American law, as well (Kadish, Schulhofer, & Steiker, 2007). For instance, suppose a driver's brakes fail and he hits a pedestrian. Clearly the harm to the pedestrian is not intentional. However, the driver will tend to be exposed to greater liability if he failed to have his brakes maintained properly, and lesser liability if the flaw in the breaks was inherent to their original manufacture. In the former case, the brake failure was controllable by the driver; in the latter case, it was not. The factor of controllability also plays a key role in punitive behavior in the trembling hand game (Cushman, et al., 2009). When the allocator has some probabilistic control over the allocation by choosing one of three dice, she is punished for accidental outcomes, as described above. But, when the allocator has no probabilistic control over the allocation—when the allocator is forced to roll a single die where stingy, fair, or generous outcome is equally likely—she is punished less, if at all, for accidental outcomes.[8] Again, a focus on controllability makes sense from the functional perspective of modifying social partners' future behavior. When A's behavior is controllable, P's punishment can effectively modify A's future behavior ("I know you didn't mean to be stingy this time, but I'm going to show you what will happen if you don't watch out"). When A's behavior is not controllable, P's punishment cannot modify A's future behavior, and so there is no value to teaching a lesson.

So, accidents may be punished because they are "teachable moments." But to work, they must teach a lesson that the transgressor is able to learn. If I punish you for swinging a hammer at a nail and hitting my thumb, will you learn not to hit my thumb (the lesson I am hoping for)? Or will you learn instead not to aim for nails (a lesson I have no reason to teach)? This depends on the structure of the learning mechanisms themselves. The key factor is whether the learning mechanism associates reward or punishment with the intended action, or instead with the outcome actually produced.

This is a fundamental distinction in reinforcement learning (Daw & Shohamy, 2008; Sutton & Barto, 1999). *Model-free* mechanisms of learning associate experienced reinforcement with the action selected (i.e., the agents intent). Thus, if a ball is thrown toward the plate and hits the batter, a model-free mechanism reduces the value associated with "throwing the ball at the plate." By contrast, *model-based* mechanisms of learning associate experienced reinforcement with the outcome produced. These mechanisms will instead reduce the value of "hitting a batter," selecting future actions based on some model that relates possible actions (like throwing a ball) to outcomes (like hitting a batter). As the name suggests, model-based learning mechanisms require a working causal model of the world, one that relates actions to outcomes. For this reason, they are relatively harder to implement than model-free learning mechanisms. But, they will tend to learn from the punishment of accidental outcomes more effectively.

This raises a key empirical issue: In social contexts, do people learn from punishment in a model-free way ("don't perform that action again!"), or in a model-based way ("don't cause that outcome again!")? This question has been assessed using a modified version of the trembling hand game implemented in a two-player game of darts (Cushman & Costa, in preparation). One player throws darts at a multicolored board, and her shots will win or lose money for a second player. But the thrower doesn't know which colors will win money for her partner, and which colors will lose money. Her partner has the opportunity to teach the thrower, however, by rewarding or punishing the thrower after each throw. Critically, the thrower has to call her shots, and she receives a bonus from the experimenter every time she hits the color she calls. Thus, she has a clear incentive to be honest about what she is aiming for. The responder therefore knows both the thrower's intended target and, of course, what she actually hits.

Suppose that the thrower aims for a high-value target, but hits a low-value target. Should the responder reward the thrower (to encourage aiming at the high-value target) or punish the thrower (to discourage hitting the low-value target)? This question was assessed by using a confederate in the place of the responder who adopted either an intent-based policy of reward and punishment or an outcome-based policy of reward and punishment. The results showed that the thrower learned the relative value of the targets significantly better when the responder adopted an outcome-based punitive strategy, compared to an intent-based punitive strategy. Moreover, an outcome-based simulation of the thrower's learning process matched throwers' observed behavior substantially better than a intent-based simulation.

Thus, evidence suggests that the actual structure of human learning processes in social situations makes the punishment of accidental outcomes advantageous. The experimental results presented here are only an initial foray into the complicated psychology underlying human learning in social contexts. Minimally, however, they suggest that when we exploit accidental outcomes as "teachable moments," our social partners are receptive pupils. Punishing accidents may contradict our moral attitude that "it's the thought that counts." But, as a matter of functional design, it is an effective way to leverage learning mechanisms in order to maximize future prosocial behavior.

5.3 Salience and the Action–Omission Distinction

Some of the events, objects, and properties we encounter tend to "pop out" and capture our attention (Parkhurst, Law & Niebur, 2002), and these salient stimuli are the easiest to learn about (Mackintosh, 1975; Rescorla & Wagner, 1965; Schwartz & Reisberg, 1991). Thus, for punishment to successfully discourage antisocial behavior via general learning processes, both the punishment and the antisocial act should be salient to the learner. The learner needs to notice that he is being punished and to infer what prompted that punishment. This constraint may explain the preference for punishing harmful actions versus harmful omissions.

Consider a puppy with two antisocial traits: He wets the carpet (an action) and fails to pick up the newspaper (an omission). Imagine trying to induce prosocial behavior by punishing the puppy. Every time he wets the carpet you punish him, and every morning when he fails to bring in the newspaper you punish him. Intuitively, you might guess that the puppy will learn to stop wetting the carpet, but will never learn to fetch the newspaper. One way of putting this is that "wetting the carpet" is a more salient event to the dog than "not picking up the newspaper." Even if the puppy knew that he was being punished for not doing something, how would he know what that something is? After all, at any moment there is an infinity of actions we are *not* performing.

Both experimental evidence (Hineline & Rachlin, 1969; Hoffman & Fleshler, 1959) and formal modeling (Randløv & Alstrøm, 1998) bear out this intuition. Animals tend to respond to punishment by performing innate, stereotypic avoidance behaviors. It has been suggested that most "novel" behaviors successfully conditioned by contingent nonpunishment are in fact close variants of innate avoidance behaviors, and that truly novel behaviors have only been obtained via punishment with extreme effort on the part of the experimenter (Bolles, 1970). On the other hand, schedules of contingent reward are more successful at conditioning novel behaviors. The way that they succeed, by employing processes of "shaping" and "chaining," is also revealing. In shaping, the experimenter begins by rewarding a behavior already in the animal's repertoire and then restricts reinforcement to variants of the behavior ever closer to the desired action. In chaining, the experimenter rewards the performance of several individual behaviors when performed sequentially. These techniques presumably work because they target the performance of salient actions for immediate reinforcement, narrowing the space of potential hypotheses that an organism must consider when associating its behavioral choices with pleasant or aversive outcomes.

Notably, the action–omission distinction is widely reflected in the law and intuitive moral judgment. Although a legal "duty of care" does mandate prosocial action (i.e., punishes an antisocial omission) in certain specialized cases such as a parent's obligation to provide active support for a child, criminal law is overwhelmingly focused on the punishment of harmful actions, and not harmful omissions (Kadish et al., 2007). This distinction between actions and omissions is reflected in ordinary people's moral judgments as well (e.g., Cushman, Young & Hauser, 2006; Spranca, Minsk & Baron, 1991). Moreover, it appears that the action–omission distinction is particularly acute in judgments of deserved punishment, compared with judgments of moral wrongness (Cushman & Young, 2011).

To restate the present proposal in very general terms: Actions constitute a more salient target for learning than do omissions. Thus, to discourage antisocial actions, punishment (of the performance of the antisocial action) is preferred over reward (of the absence of antisocial action). By contrast, to encourage prosocial actions, reward (of the performance of the prosocial action) is preferred over punishment (of the absence of the antisocial action). Section 3 argued that punishment has a relative temporal advantage over reward, however: Your fists are always available for immediate punishment, whereas opportunities for reward may be fewer and farther between. Combining these proposals, basic learning constraints appear to make it easier for organisms to discourage each other's antisocial actions than to encourage each other's prosocial actions. Section 7.6 considers ways in which human cognition can move beyond these constraints.

5.4 Limited Capacity

When individuals have severe impairments in their general mechanisms of learning or behavioral control, punishment cannot effectively leverage those general mechanisms to promote prosocial behavior. Consequently, one might expect retributive motivations to be lessened for perpetrators incapable of learning or behavioral control. There is some suggestive evidence from legal codes, which commonly differentiate between perpetrators with full versus diminished mental capacity (Kadish, Schulhofer & Steiker, 2007). Psychological research also suggests that considerations of mental capacity affect people's judgments of deserved punishment (Darley, Carlsmith & Robinson, 2000; Fincham & Roberts, 1985; Robinson & Darley, 1995). For instance, Darley, Carlsmith, and Robinson (2000) found that subjects assigned substantially less punishment to a person who murdered as a consequence of hallucinations resulting from an inoperable brain tumor, compared to a person who murdered out of jealous rage. Subjects indicated that the tumor patient should be detained in a mental health facility, clearly evincing a nonretributive sensitivity to preventing future harm, but they did not report a motivation to see him punished with prison time.

In another study (Robinson & Darley, 1995), only 16 percent of subjects assigned punishment (versus civil commitment) in a case where a schizophrenic man killed an innocent bystander under the belief that the bystander was about to attack him. Darley and Robinson compare this case to another involving "mistaken identity" by a sane perpetrator. In this case, a shop owner is robbed and chases the burglar. He misidentifies another man as the burglar during the chase, gets in a fight with that man, and kills him. In this case, 100 percent of subjects assigned punishment. Why does the schizophrenic who mistakes an innocent as an assailant receive no punishment (but rather civil commitment), while the sane man who mistakes an innocent as an assailant receives substantial punishment? A simple analysis of the mental state of the perpetrator at the time of the crime is not sufficient to account for this discrepancy—both individuals murdered in a fight due to mistaken identity. A critical factor may be the greater capacity of sane versus schizophrenic individuals to learn from the experience of punishment and successfully regulate future behavior on the basis of that learning. Future research should explore, first, the extent to which an evaluation of diminished capacity is central to the human retributive instincts and, second, the specific categories of impairment that trigger such an assessment.

5.5 Immediacy

General learning processes form associations between behavior and reinforcement more efficiently when the delay between the two is short. Consequently, punishment should tend to follow as quickly as possible after the commission of a harmful act, and, given a long enough delay, the motivation for punishment should be extinguished. Notably, many legal systems impose a statute of limitations on the prosecution of a criminal act. For instance, in my home state of Massachusetts, standard criminal misdemeanors carry a statute of limitations of six years. However, this period is lengthened for some felonies, and there is no statute of limitations on murder.

It is unclear whether the statute of limitations reflects an underlying feature of the psychology of punishment in ordinary people. Additionally, it should be noted that the statute of limitations on most criminal offenses is on the scale of years, whereas temporal constraints on general learning processes in nonhuman animals often apply on the scale of minutes (Renner, 1964). A key direction for future research is to test whether something analogous to a "statute of limitations" is a fundamental feature of human retributive psychology, and whether it plausibly reflects the temporal constraints imposed by domain-general learning processes in nonhuman animals.

5.6 Conclusions

I have argued that several features of human punishment—retributive motives, the punishment of accidents, the preference for actions over omissions, the limited capacity excuse, and the statute of limitations—may reflect its functional design. Specifically, each of these features can be sensibly interpreted as elements of a specialized mechanism that uses punishment to induce prosocial behavior among social partners by exploiting their general mechanisms of learning and behavioral control. Like any attempt to understand complex behavior in adaptive terms, this proposal is speculative. Moreover, a general argument for functional design cannot, by itself, distinguish between the influences of biological adaptation, cultural adaptation, or human reasoning. Nevertheless, it shows how otherwise puzzling and disparate features of human social behavior can begin to cohere into a more sensible and unified schema by considering a simple question: What is this behavior designed to do?

6 Specialized Mechanisms of Prosocial Behavior

If punishment matches the constraints of learned prosocial behavior, then it certainly must be the case that prosocial behavior is learned. But is it? Or, alternatively, is prosocial behavior accomplished by specialized behavioral mechanisms more like the goose's egg retrieval than the rat's lever-press? Several lines of evidence are suggestive of innate mechanisms supporting prosociality in humans. However, there remains substantial scope for learned prosociality to have shaped the functional design of punishment.

The case for innate prosociality begins with adaptive considerations. The widespread existence of punishment and reward in human societies imposes a selective pressure to rapidly adopt prosocial behavior when it is enforced. As noted in section

3, the Baldwin effect describes a tendency for general learning mechanisms to pave the way for specific adaptations. Along these lines, some form of domain-specific innate preparation to adopt prosocial behavior might be favored over pure reliance on general mechanisms of learning and behavioral choice. An initial attempt to test this claim in an agent-based simulation model shows that a moderate bias toward prosocial behavior is favored in an evolving population where the punishment of antisocial behavior dominates (Cushman & Macendoe, 2009).

Empirical evidence is more compelling than adaptive theory, and here too an innate preparation for prosocial behavior is suggested. Just as economists talk about a "taste for retribution," researchers have identified tastes for generosity, fairness, and cooperation (Gintis et al., 2005). Across diverse experimental paradigms, humans choose behaviors that provide benefits for others at a cost to themselves, without the motivation of reciprocal reward or punishment (Batson & Shaw, 1991; Henrich et al., 2005). Moreover, prosocial behavior appears to emerge developmentally early. Human infants and some apes spontaneously engage in prosocial behavior, for instance, by picking up a pen that a stranger has dropped out of reach and returning it to him (Warneken & Tomasello, 2006).

At the same time, there is also compelling evidence that prosocial behavior has a substantial learned component. To begin with, there is substantial cross-cultural variation in prosocial behavior as measured by standard behavioral-economic paradigms (Henrich et al., 2005; Herrmann, Thöni & Gächter, 2008). Individuals determine their levels of prosocial behavior in part by assessing the behavior of peers (Frey & Meier, 2004). Prosocality is also acquired developmentally through experiences that direct attention to others' feelings and activate empathy (Hoffman, 2000). There is also evidence specifically for a role of punishment in learned prosociality. Research using economic games also shows that levels of prosocial behavior are sensitive to rates of punishment (Fehr & Gächter, 2002). Additional evidence comes from studies of psychopaths, who unquestionably lack the taste for generosity and fairness. Notably, psychopaths do not show an exclusive deficit in prosocial emotions, but rather a more general deficit in processing negative feedback (i.e., punishment) and integrating it into future behavioral plans (Blair, 2003).

Two broad conclusions are warranted. First, there are both innate and learned contributions to human prosociality. Second, human adults maintain flexibility in their prosocial behavior, adjusting levels of prosociality according to cultural norms and past experience of punishment and reward. So, where does this leave the argument that punishment should be adapted to match the constraints of general learning processes? One possible consideration is the origin of punishment: Although innate preparations for prosocial behavior currently exist, section 3 argued that punitive strategies probably originated by exploiting general learning mechanisms. Perhaps the current structure of punitive instincts still reflects the original functional design. On the other hand, perhaps not. This argument has little appeal because it depends on unverifiable speculation about evolutionary stasis.

A second argument depends not on distant origins of punishment, but rather on its present scope. Let's begin with the strong hypothesis there is a fully formed, innate "taste for generosity": People are born valuing prosocial behaviors and devaluing antisocial behaviors. For instance, imagine that sharing food feels intrinsically good, and stealing food feels intrinsically bad. Insofar as these innate mechanisms cause people to share and not to steal, punishment will be unnecessary. But, in some circumstances, the importance of food may outweigh the intrinsic disutility of theft for an agent. In simple terms, hunger will sometimes hurt more than shame. When an agent engages in theft despite its intrinsic disutility, punishment then plays a critical role by assigning an additional source of disutility to theft: the disutility of the punishment. In the future, the agent must weigh its hunger not only against its own intrinsic guilt, but also the prospect of extrinsic retaliation. For punishment to be effective in this manner, of course, the agent must associate the performance of antisocial actions with future punishment. The critical point is that punishment is required only when the intrinsic (possibly innate) value of prosociality is insufficiently motivating. Thus, however much prosocial behaviors are valued via innate mechanisms, punishment may still be required to exploit general processes of learning and behavioral control to obtain prosocial behavior. It is precisely when altruism fails that punishment must work.[9]

Adopting a somewhat weaker hypothesis, prosociality might depend on specialized learning mechanisms rather than an innate valuation of prosociality. Thus, for instance, the human mind may be finely tuned to detect punishment that follows antisocial action, even when delayed or unobvious. Analogous mechanisms certainly exist outside the social domain; for instance, rats are innately prepared to reliably associate between novel tastes and subsequent illness (Garcia, Ervin & Koelling, 1966; Garcia & Koelling, 1996). Within the social domain, animals use specialized behavioral routines to communicate violent threats without paying the costs of engaging in actual violent behavior (Maynard Smith & Price, 1973). In contexts where organisms deploy specialized learning mechanisms, the functional design of punishment should reflect the specific constraints of those mechanisms, rather than the constraints of general mechanisms of learning and behavioral choice.

Yet, once again, where innate preparations end, general learning mechanisms must suffice. Prosocial behaviors that fall outside of the scope of innate preparation must be supported by general mechanisms of learning and choice. And, when specialized learning mechanisms fail to sufficiently motivate prosocial behaviors, general mechanisms of learning can still be leveraged to provide additional motivation. These might be termed arguments from "scope": However large the scope of innate preparation for prosociality, it can be expanded via dependence on general mechanisms of learning and choice.

The argument from scope is particularly important when considering human social behavior, which exhibits broad flexibility across diverse contexts that could not have been anticipated on the timescale of biological evolution. I have argued at length that "ratlike" general learning mechanisms are highly constrained, and that aspects of human retributive instincts match those constraints. In the following section, however, I will argue that humans' general learning mechanisms are vastly less constrained, and this explains much about the unique complexity and successes of human social behavior.

7 Cognition and Social Behavior in Humans

Humans possess profoundly more sophisticated cognitive abilities than nonhuman animals, and prosociality is far more widespread and flexible in humans than in nonhuman animals. There is broad agreement that this no coincidence. On the one hand, it has often been argued that the existing demands of a complex social life may have provided a key selective pressure toward the development of more powerful general cognitive abilities (e.g., Byrne & Whiten, 1989; Trivers, 1971). On the other hand, it has been argued that the prior development of powerful cognitive abilities allowed complex social systems to emerge (Stevens, Cushman & Hauser, 2005), including especially the cognitive mechanisms that support cultural transmission (Richerson, Boyd & Henrich, 2003). To claim a single direction of causation probably misstates a fundamentally coevolutionary relationship; in any event, complex social life demands powerful cognition. The relationship between punishment and prosociality that we have considered above—and, critically, the distinction between specialized cognitive mechanisms versus general learning and reasoning mechanisms—helps to illuminate why. As we will see, general processes of learning and reasoning are critically important to the richness of human social life, first, because they expand our capacity to learn from punishment and reward and, second, because they expand our capacity to identify acts warranting punishment and reward.

Consider an illustrative example. As I have argued above, the general learning ability of dogs dooms the strategy of training your dog to fetch the newspaper by withholding doggie treats each Christmas if he fails to do so. The temporal delay between the dog's behavior and the reinforcement, combined with the minimal salience to the dog of "not fetching the newspaper" and "not receiving a doggie treat on Christmas," make it highly unlikely that the dog will form the necessary learned association. However, a similar strategy might be much more effective in training your son to fetch the newspaper. You can explain what you want him to do, and the consequences of failure. He can rapidly comprehend this connection, has an available conceptual structure that relates prosocial action to reciprocal holiday rewards, and might be sufficiently motivated by that distant reward to modify his present behavior. Your son's general capacity to (1) acquire information via language, (2) rapidly integrate new information into rich conceptual models of the world, (3) and use that conceptual knowledge to guide behavioral planning and choice allows him to respond to social punishments and rewards far more flexibly than your dog.

7.1 Language

Language has a transformational impact on human learning. Without language, knowledge about the world will typically only be obtained via direct observation of or interaction with the relevant phenomenon; with language, the experience of one individual can ultimately support knowledge among others (Tomasello, 1999). Thus, for instance, I know a great deal about Rwanda, retirement, and ribosomes despite very little direct interaction with each. More particularly, language plays a key role in acquiring conceptual abstractions. It allows us to generalize from the three stooges, the three kings, and the three tenors to the conceptual abstraction "three," embedded within broader concepts of counting and numerosity (Carey, 2004). We can generalize from moving objects to "velocity," from unsavory characters to "psychopath," from missed chances to "opportunity cost."

Conceptual abstraction may be possible without language, but linguistic symbols learned from social partners create a cognitive placeholder in the mind. They focus attention on fruitful generalizations and allowing the learner to gradually fill the empty conceptual structure with rich, productive content. Like the grain of sand that starts a pearl, linguistic symbols provide a nucleus around which concepts can grow. Moreover, formal models show that the kinds of conceptual abstractions supported by language cascade downward to support learning at lower, more concrete levels, as well (Goodman, Ullman & Tenenbaum, 2009). Thus, for instance, learning conceptual abstractions such as "belief" or "cause" can support the acquisition of knowledge about particular beliefs and particular causal relationships.

These consequences of language greatly enhance the potential of punishment (and reward) to elicit prosocial behavior via general learning processes. First, language allows threats of punishment and promises of reward to be communicated in advance of the relevant behavior. Absent language, threats and promises can only be inferred by the experience or observation of past instances of punishment and reward for sufficiently similar behaviors. Second, language allows a behavior to be readily associated with punishment or reward at a long temporal delay. Absent language, behavior will typically only be associated with rewards and punishments when they follow immediately. Third, language allows for the rapid communication of novel and complex behavioral expectations that do not already exist in the behavioral repertoire of social partners. That is, language provides a rapid solution to the problem of "shaping" new behaviors, described in section 5.3. I have already used the example of "bring me the newspaper" as a demand that is easy to communicate by language and relatively harder to communicate without language. Demands like "bring a ten-percent tithe," "bring the murderer dead or alive," and "bring these tools back by October" fall into the same category; and precisely these kinds of demands are central to the complexity of human social life.

7.2 Conceptual Models

Of course, the power of language to communicate is constrained by the power of language-users to comprehend. Here, again, humans have fundamentally different mental resources available compared with nonhuman animals, certainly in magnitude and possibly in kind. Human thought is supported by rich mental models that make use of conceptual abstractions and can be productively combined (Carey, 2009; Fodor, 1975; Roberts & Mazmanian, 1988; Sloman, 1996; Thompson & Oden, 2000). Three particular conceptual competencies are likely to have a large impact on prosocial behavior. The first is our understanding of others' mental states—their perceptions, sensations, goals, and beliefs—which allows us to rapidly and reliably infer what social partners want from us (Saxe, Carey & Kanwisher, 2004; Tomasello et al., 2005; Warneken & Tomasello, 2006). The second is our ability to construct complex causal theories relating spatially and temporally distant events (Carey, 2009; Murphy & Medin, 1985; Waldmann & Holyoak, 1992), which allows us to predict the likely consequences of our behavioral choices on social partners' welfare. The third is our ability to construct appropriate analogies between situations, and to infer abstract principles on the basis of those analogical constructions (Gentner, Holyoak & Kokinov, 2001). Combining these three competencies, humans have the capacity to infer from specific experiences of punitive behavior ("When I steal apples from Billy, he punches me") to a general model of punitive behavior ("When my behavior interferes with others' goals, they punish me").[10] Conversely, we have the ability to appreciate how a linguistically communicated rule stated in abstract terms ("Do onto others as you would have others do onto you") applies to particular circumstances ("Don't steal apples from Billy").

Each of these three aspects of humans' conceptual knowledge allows us to infer the appropriate course of action without direct experience of past punishment for a particular behavior. A conceptual abstraction such as "Help others achieve their goals—eventually they will do the same for you" depends on mental state inference, the association of temporally distant events, and abstraction across diverse cases. Critically, it can effectively guide behavior in novel, unfamiliar circumstances. Without such a conceptual abstraction, an organism must wait for specific feedback for each category of action it can perform in order to learn optimal patterns of social behavior.

7.3 Planning and Choice

Finally, humans have a greatly enhanced capacity to use complex conceptual knowledge to guide behavioral planning and choice. Humans are much more flexible in means-end reasoning than nonhuman animals, accomplishing large goals by constructing a hierarchy of smaller subgoals through planning (Badre & D'Esposito, 2009; Conway & Christiansen, 2010; Koechlin & Jubault, 2006). This expands the range of prosocial actions that one person can undertake on behalf of another—not just sharing food, but sharing a plough, so to speak.

Additionally, humans have a much greater ability to inhibit impulsive or habitual responses in order to maximize value in the distant future (Rachlin, Raineri & Cross, 1991). In economic terms, humans have a very shallow rate of temporal discounting: a dollar tomorrow is deemed nearly as valuable as a dollar today. Nonhuman animals discount future rewards several orders of magnitudes more steeply, often devaluing rewards by more than half within a single minute (Mazur, 1987; Richards et al., 1997; Stevens, Hallinan & Hauser, 2005). Temporal discounting has important consequences on the stability of prosocial behavior (Stephens, McLinn & Stevens, 2002). Even if your dog could understand that fetching the newspaper in May is linked to rewards at Christmas, it would probably not experience those distant rewards as sufficiently motivating; by contrast, your son is likely to weight the prospect of future reward much more heavily. Among humans, the punishments and rewards for social behavior typically occur at a long delay. Our ability to experience the motivational force of delayed reinforcement, and to incorporate it into complex behavioral plans, is critical to the functioning of human social life.

7.4 Cognition and the Punisher

So far, I have focused on the way that powerful cognitive mechanisms expand the ability of humans to learn from punishment. At the same time, they can also expand the circumstances in which humans choose to punish. The retributive impulse to punish people who cause you harm is limited by the capacity to identify the relevant causal relationship. The learning and reasoning mechanisms possessed by nonhuman animals will typically limit inferences of causal responsibility for harm to direct observation. In humans, the assignment of causal responsibility for harm can extend across miles, years, and long causal chains. For example, consider many Americans' urge for retribution against Osama bin Laden for the September 11 attacks. Thus, the simple rule "punish those who harm you" inherits the tremendously sophisticated ability to assign causal responsibility for harm, affording a powerful motivation to kill a man half a world away whose role in causing the harm was decisive, to be sure, but also very indirect. So, just as uniquely human cognitive abilities afford much greater ability to learn from punishment, they afford much greater ability to assign blame. Our "taste for retribution" can be a specialized behavioral adaptation and still be profoundly enhanced by very general improvements in cognitive ability.

7.5 Domain Generality and the Diversity of Human Social Behavior

Each of the three capacities I have considered—linguistic communication, conceptual abstraction, and controlled behavioral choice—functions in diverse domains of human thought and behavior. There is no sense in which they are limited to the specific problem of supporting punitive or prosocial behavior. It may be that social demands provided a key selective pressure favoring the emergence of these domain-general capacities (Byrne & Whiten, 1989). If so, they could be regarded as adaptations to social life. But they are clearly not specialized mechanisms in the sense of the fixed action pattern of the goose; rather, their power lies specifically in their flexibility, like the general learning processes of the rat (yet much more powerful still).

Past discussions of cognitive contributions to human prosocial behavior have not emphasized the importance of flexible, domain-general mechanisms as opposed to narrowly deployed specialized mechanisms. It has been asserted that the specific problems of punishment, reward, and prosociality require cognitive capacities like individual recognition (Trivers, 1971), memory (McElreath et al., 2003; Stevens et al., 2005; Trivers, 1971), quantitative representations of value (McElreath et al., 2003; Silk, 2003; Stevens et al., 2005), the capacity to track reputation (McElreath et al., 2003; Mohtashemi & Mui, 2003; Nowak, 2006), and motivations for reciprocity (Axelrod, 1984; Fehr & Henrich, 2003; Trivers, 1971), retributive punishment (Fehr & Gachter, 2002), and the valuation future rewards (Stephens, McLinn & Stevens, 2002; Stevens et al., 2005). Surely, these capacities are critical. Still, what seems to make humans unique is that they can be flexibly engaged across arbitrarily diverse circumstances.

To see why domain generality is so important in the human case, consider a foil: the specialized cognitive mechanisms that support food caching in the scrub jay (Clayton & Dickinson, 1998; Emery & Clayton, 2001). These birds cache thousands of food items in different locations and retrieve them months later, solving a challenging memory task. They retrieve the resources according to their relative nutritional content, and are sensitive to variable rates of decay between food types, solving a challenging valuation problem (Clayton & Dickinson, 1998). They control the impulse to consume the resources immediately and instead cache them for use months in the future, solving a challenging intertemporal choice problem. They even exhibit sensitivity to the perceptual access of other birds to their hidden caches (Emery & Clayton, 2001), solving a challenging behavioral prediction problem. The problems that food-caching birds solve are much like the problems inherent to social interaction. But, these birds' psychological solutions appear to be specialized to the task of food caching—more like the goose, less like the rat. As far as we know, these specialized mechanisms don't have much of an impact at all on prosocial behavior. In fact, they don't even seem to affect these birds' foraging behaviors beyond the narrow domain of food caching. At the risk of stating the obvious, scrub jays have not learned to plant seeds in order to harvest them later, or to arrange food in ways that promote the growth of nutritionally valuable insects, or (returning to the social domain) to trade food between each other. Yet, humans do all of these things. It is precisely because our reasoning and learning mechanisms are not specialized—because of their extraordinary flexibility—that they are so transformational.

7.6 Reward and Prosociality

The same powerful and flexible cognitive mechanisms that help people learn from punishment can also help them learn from rewards. As I argued above, the task of using rewards to elicit prosociality via general learning processes is particularly difficult because of the basic temporal structure of social interactions. Fists and teeth are always available for immediate punishment; by contrast, opportunities to aid another, or to share resources, are not always available for immediate reward. This may explain why punishment is relatively common among nonhuman animals (Clutton-Brock & Parker, 1995) whereas evidence for reciprocal reward has been stubbornly difficult to establish (Hammerstein, 2003). But human capacities for linguistic communication, conceptual abstraction, and controlled behavioral choice go a long way toward alleviating the constraints of general learning processes. We are able to learn about delayed rewards, comprehend how they are contingent on our own behavior, and motivate our behavior according to them.[11]

In section 5.3 I argued that the constraint of salience favors rewarding and punishing actions rather than omissions. This suggests a match between the punishment of harmful actions (versus the reward of omitting harm) and the reward of helpful actions (versus the punishment of omitting help). Combining this observation with the argument that punishment has key advantages over reward as a "teaching" device given cognitive constraints on learning in nonhuman animals, we can predict that nonhuman animals will generally avoid harmful actions toward each other (learned from punishment), but will not generally seek out helpful actions toward unrelated others (because of the difficulty of learning from reward). The human ability to leverage language, abstract thought, and long-term planning and choice to make strategies of contingent reward a viable strategy thus stands to expand the general boundaries of prosocial behavior from "do not harm" to "provide help"—from a social world of libertarian individualism to a world of collective action. This argument depends on a number of steps and warrants a healthy skepticism. Nevertheless, its implications are substantial. What people will accomplish by aiming to help each other vastly exceeds what they will accomplish by simply aiming not to harm each other. If this captures a rough distinction between human and nonhuman social behavior, then the unique flourishing of human social life can readily be understood.

7.7 Conclusion

I have reviewed three related aspects of human cognition that are radically different from their nonhuman counterparts: the capacity for linguistic communication, the capacity for reasoning about complex causal relations involving conceptual abstractions, and the capacity for controlled behavioral planning and choice. Collectively, these capacities greatly expand the capacity of humans to learn contingencies between the social consequences of their behavior and the contingent rewards and punishments of social partners. They also greatly expand the contexts in which humans can comprehend the impact of others' behavior on their own well-being, potentially expanding the range of circumstances that invoke retributive (or rewarding) motivations. In short, complex and powerful cognition can explain the complex and productive social life of humans. But, the relevant cognitive capacities are not specific to the social domain, much less to morality alone. To the contrary, the very feature of human cognition that explains its transformative role in social life is its domain generality.

8 Conclusion: The Irony of Punishment

There is an apparent tension in my argument. On the one hand, human punishment apparently matches some constraints of "general learning processes" possessed by our nonhuman ancestors. Put simply, our punitive instincts treat people like rats. On the other hand, the stunning complexity of human social behavior derives from new and powerful domain-general processes of learning and reasoning. That is, the foundation of human prosociality and cooperation is our ability to learn very differently, and much more effectively, than rats. Here is the tension: If it is so important to human social life that we learn much better than other animals, why would aspects of our punitive instincts be designed as if we learned just like other animals? Shouldn't punishment be tailored to the constraints—and the possibilities—of human learning?

There is a tension here, but not a logical contradiction. It is possible that some aspects of the psychology of punishment are relics of an earlier social environment, better suited to the learning mechanisms of our pets than our peers. Perhaps they work well enough in modern human life, just not quite optimally. Our taste for sugar and fat is often discussed in similar terms: unquestionably adaptive for our ancestors, but tuned suboptimally to our present circumstances. At least two aspects of punishment might be similar.

One is the punishment of accidental outcomes. I reviewed several different studies demonstrating that judgments of deserved punishment are strongly affected by the degree of harm caused. Let me add one more favorite example to the mix: the legal penalties associated with drunk driving. In Massachusetts, if a drunk driver falls asleep at the wheel, hits a tree, and gets picked up by the police, he can expect a fine of several hundred dollars. He might also be forced to enroll in an outpatient treatment program, or even have his license suspended for several months. But if he falls asleep at the wheel, hits a pedestrian, and kills her, he will receive between two and a half and fifteen years in prison. These are radically different punishments for the same behavior.

I argued that we punish harmful outcomes, even when accidental, because it is the most effective way to teach social partners what we want them to avoid and how much we want them to avoid it. In an experiment, dart-throwers learned the value of several targets better when rewarded and punished on the basis of outcomes, and worse when rewarded and punished on the basis of intent. This darts game was designed to reflect a time in our evolutionary history when the only way to communicate the value of others' behavior was to reward and punish. But, of course, humans do have language, and the ability to infer what social partners value much better than our nearest primate relatives. Every drunk driver knows the value of a pedestrian life, though he may disregard it. We do not need to punish drunk drivers who kill in order to teach them that we value of others' lives; we need to punish them to disincentivize future drunk driving by themselves and others. That disincentive could operate as strongly if we punished all acts of drunk driving moderately, rather than punishing cases that cause no harm minimally and cases that cause death maximally.

To be sure, there are instances where intentions simply cannot be known and outcomes are the most reliable proxy. But, just as surely, there are cases where our knowledge of intent is quite reliable, and yet we still have a retributive impulse to grade punishment according to the degree of harm caused. Understanding the origins of that retributive impulse may help us to decide whether, on reflection, to endorse it. If I am right that its origins trace to a social environment very unlike today's, then a skeptical eye is warranted.

A second aspect of punishment that may be poorly adapted to our present situation is, quite simply, punishment itself. Recall that recent experimental and theoretical results illustrate striking benefits at the individual and group level when prosocial behavior is enforced via reciprocated prosociality, rather than the threat of punishment (Dreber et al., 2008; Rand, Dreber, et al., 2009; Rand, Ohtsuki, et al., 2009). In these studies, the enforcement of prosociality by punishment tended to devolve into cycles of costly retribution, while reciprocal prosociality tends to evolve into cycles of productive cooperation. Moreover, developmental research suggests that focusing children's empathy on the suffering of victims is a far more successful method of promoting prosocial behavior than punishing their transgressions (Hoffman, 2000).

Yet still, we punish—why? I have suggested that punishment was adaptively preferable for much of our evolutionary history because it afforded more immediate and salient responses to antisocial acts than the withholding of reward. This property was critical when social partners used general learning processes to adopt prosociality in the face of punishment, and when those general learning processes were highly constrained. But humans have the capacity to communicate and comprehend the contingency between behavior and delayed reward, along with the capacity to be appropriately motivated by its prospect. Consequently, in some circumstances, punishment itself may be an outmoded and very costly impulse, compared with the possibilities of reciprocation and reward.[12]

As we have seen, there is an instructive comparison between our "taste for punishment" and our taste for sugar and fat. In both cases, these motivations circumvent the problem of learning a more general associative relationship. We do not need to learn which foods are associated with future energetic states in order to be motivated to consume; likewise, we do not need to learn that the punishment of antisocial behavior can promote prosociality in order to be motivated to act retributively. Moreover, the relative inflexibility of an innate retributive motivation avoids the unsustainable evolutionary dynamic of punishing only those who can, or do, learn from punishment.

But our taste for sugar and fat comes with a definite cost in the modern world, where high-calorie foods are more widely available than our evolved tastes anticipated. Much the same is true of our taste for retribution. Cognitive abilities unique to humans enable strategies of punishment and reward to support vastly more complex and productive forms of prosocial behavior than in nonhuman animals. Still, in some respects, the structure of our retributive taste—and perhaps even the taste itself—is adapted to those very constraints on learning that the human mind brilliantly exceeds.

Notes

1. In this essay, I define prosocial behavior as a behavior that has worse fitness consequences than some alternative (the antisocial choice) for the agent, while having better fitness consequences than the antisocial alternative for some social partner. Grooming, sharing food, and alarm calling could all be prosocial behaviors, on this definition. Choosing not to steal food from a social partner, or not to encroach on his territory, could also be prosocial behaviors. In accounting for the costs and benefits of a prosocial behavior, I am specifically excluding the contingent response of a social partner (e.g., punishment or reciprocation). Thus, grooming with the expectation of reciprocation, or respecting another's property under the threat of punishment, both count as prosocial behaviors.

2. Broad, but certainly not unlimited (Garcia & Koelling, 1996).

3. Note, however, that contingent prosociality performs better still. A fixed prosocial strategy avoids punishment by punishing social partners, but misses the opportunity to exploit nonpunitive partners by behaving antisocially selectively toward them. By contrast, a contingently prosocial strategy avoids punishment by punishers, while selectively exploiting nonpunishers.

4. Of course, if A retaliates against P's punishment, then the most likely learned association for P in the short term is: Punishing A is costly and harmful.

5. Of course, to respond to defection with defection might be regarded as punishment of a sort—call it "passive" punishment (failing to provide a benefit), and contrast it with "active" punishment (imposing a cost). When using the term "punishment" in this chapter, it is the active sort

that I have in mind. Part of the argument of this section is that active punishment can more salient and rapid, and therefore more learnable, than passive punishment (i.e., defection).

6. These studies target the punitive judgments of ordinary, nonexpert respondents to psychological surveys. A separate but potentially related issue is the structure of the actual legal system, which in some instances is better described by retributive motives and at other times by deterrence or incapacitation motives.

7. In the Anglo-American tradition, when a person performs a negligent act, she assumes liability for the consequences of that negligent behavior. But, if no harm occurs, there is no liability. As the American judge Benjamin Cardozo wrote, "proof of negligence in the air, so to speak, will not do" (*Palsgraf v. Long Island Railroad Co.*, 1928). If a harm is caused, then typically the extent of liability is increased in proportion to the degree of harm caused.

8. When the allocator has no intentional control over the allocation, the responder's behavior is still, on average, affected by the allocation amount. However, this pattern of behavior may be best understood as a product of inequity aversion (Fehr & Schmidt, 1999) rather than retributive punishment.

9. Similarly, there may be variation between individuals in levels of intrinsic prosociality, in which case punishment would play a key role in those individuals who are only minimally motivated by intrinsic prosocial concerns.

10. For an initial attempt to model the role of individual- versus group-level inferences about social behavior in the context of punishment and prosociality, see Cushman & Macendoe (2009).

11. Money also serves to eliminate the temporal delay between action and reward, replacing trust for token. There is no obvious analogue in the domain of punishment—i.e., no system of symbolic, immediate exchanges of sanctions.

12. On the other hand, there are surely cases where punishment is required. Consider a thief: What kind of incentive are your rewards to him? There is nothing you can give him that he can't steal. To stop his behavior may simply require punishment.

References

Alicke, M. (2000). Culpable control and the psychology of blame. *Psychological Bulletin, 126*(4), 556–574.

Axelrod, R. (1984). *The evolution of cooperation*. New York: Basic Books.

Badre, D., & D'Esposito, M. (2009). Is the rostro-caudal axis of the frontal lobe hierarchical? *Nature Reviews: Neuroscience, 10*, 659–669.

Baker, F., & Rachlin, H. (2002). Self-control by pigeons in the prisoner's dilemma. *Psychonomic Bulletin & Review, 9*(3), 482–488.

Batson, C. D., & Shaw, L. L. (1991). Evidence for altruism: Toward a pluralism of prosocial motives. *Psychological Inquiry, 2*(2), 107–122.

Blair, J. (2003). Neurobiological basis of psychopathy. *British Journal of Psychiatry, 182*, 5–7.

Bolles, R. (1970). Species-specific defense reactions and avoidance learning. *Psychological Review, 77*(1), 32–48.

Boyd, R., & Richerson, P. J. (1992). Punishment allows the evolution of cooperation (or anything else) in sizeable groups. *Ethology and Sociobiology, 113*, 171–195.

Byrne, R., & Whiten, A. (1989). *Machiavellian intelligence*. Oxford: Oxford University Press.

Carey, S. (2004). Bootstrapping and the origin of concepts. *Daedalus, 133*(1), 59–68.

Carey, S. (2009). *The origins of concepts*. Cambridge, MA: MIT Press.

Carlsmith, K. (2006). The roles of retribution and utility in determining punishment. *Journal of Experimental Social Psychology, 42*(4), 437–451.

Carlsmith, K., Darley, J., & Robinson, P. (2002). Why do we punish? Deterrence and just deserts as motives for punishment. *Journal of Personality and Social Psychology, 83*(2), 284–299.

Carpenter, J. P., & Matthews, P. H. (2004). *Social reciprocity*. Institute for the Study of Labor.

Clayton, N. S., & Dickinson, A. (1998). Episodic-like memory during cache recovery by scrub jays. *Nature, 395*, 272–274.

Clutton-Brock, T. H., & Parker, G. A. (1995). Punishment in animal societies. *Nature, 373*, 209–216.

Conway, C. M., & Christiansen, M. H. (2010). Sequential learning in non-human primates. *Trends in Cognitive Sciences, 5*(12), 539–546.

Cushman, F. A. (2008). Crime and punishment: Distinguishing the roles of causal and intentional analyses in moral judgment. *Cognition, 108*(2), 353–380.

Cushman, F. A., & Costa, J. (In preparation). Why do we punish accidents? An experimental investigation.

Cushman, F. A., Dreber, A., Wang, Y., & Costa, J. (2009). Accidental outcomes guide punishment in a "trembling hand" game. *PLOS One 4*(8), e6699. doi:6610.1371/journal.pone.0006699.

Cushman, F. A., & Macendoe, O. (2009). The coevolution of punishment and prosociality among learning agents. In N. Taatgen & H. van Rijn (Eds.), *Proceedings of the Thirty-First Annual Conference of the Cognitive Science Society*. Austin, TX: Cognitive Science Society.

Cushman, F. A., Young, L., & Hauser, M. D. (2006). The role of conscious reasoning and intuitions in moral judgment: Testing three principles of harm. *Psychological Science, 17*(12), 1082–1089.

Cushman, F. A., & Young, L. (2011). Patterns of moral judgment derive from non-moral psychological representations. *Cognitive Science, 35*, 1052–1075.

Darley, J., Carlsmith, K., & Robinson, P. (2000). Incapacitation and just deserts as motives for punishment. *Law and Human Behavior, 24*(6), 659–683.

Darley, J. M., & Shultz, T. R. (1990). Moral rules—their content and acquisition. *Annual Review of Psychology, 41*, 525–556.

Daw, N., & Doya, K. (2006). The computational neurobiology of learning and reward. *Current Opinion in Neurobiology, 16*(2), 199–204.

Daw, N., & Shohamy, D. (2008). The cognitive neuroscience of motivation and learning. *Social Cognition, 26*(5), 593–620.

Dreber, A., Rand, D., Fudenberg, D., & Nowak, M. (2008). Winners don't punish. *Nature, 452*(7185), 348.

Emery, N. J., & Clayton, N. S. (2001). Effects of experience and social context on prospective caching strategies by scrub jays. *Nature, 414*, 443–446.

Fehr, E., & Gächter, S. (2002). Altruistic punishment in humans. *Nature, 415*, 137–140.

Fehr, E., & Henrich, J. (2003). Is strong reciprocity a maladaptation? On the evolutionary foundations of human altruism. In P. Hammerstein (Ed.), *The genetic and cultural evolution of cooperation*. Cambridge, MA: MIT Press.

Fehr, E., & Schmidt, K. (1999). A theory of fairness, competition, and cooperation. *Quarterly Journal of Economics, 114*(3), 817–868.

Fincham, F. D., & Roberts, C. (1985). Intervening causation and the mitigation of responsibility for harm doing. *Journal of Experimental Social Psychology, 21*(2), 178–194.

Fodor, J. A. (1975). *The language of thought*. New York: Crowell.

Fodor, J. A. (1983). *The modularity of mind*. Cambridge, MA: MIT Press.

Frank, R. H. (1988). *Passion within reason: The strategic role of the emotions*. New York: Norton.

Frey, B. S., & Meier, B. (2004). Social comparisons and pro-social behavior: Testing "conditional cooperation" in a field experiment. *American Economic Review, 94*(5), 1717–1722.

Garcia, J., Ervin, F., & Koelling, R. (1966). Learning with prolonged delay of reinforcement. *Psychonomic Science, 5*(3), 121–122.

Garcia, J., & Koelling, R. (1996). Relation of cue to consequence in avoidance learning. In L. D. Houck & L. C. Drickamer (Eds.), *Foundations of animal behavior: Classic papers with commentaries* (p. 374). Chicago: University of Chicago Press.

Gentner, D., Holyoak, K. J., & Kokinov, B. N. (Eds.). (2001). *The analogical mind*. Cambridge, MA: MIT Press.

Gintis, H., Bowles, S., Boyd, R., & Fehr, E. (2003). Explaining altruistic behavior in humans. *Evolution and Human Behavior, 24*, 153–172.

Gintis, H., Bowles, S., Boyd, R., & Fehr, E. (2005). Moral sentiments and material interests: Origins, evidence, and consequences. In H. Gintis, S. Bowles, R. Boyd, & E. Fehr (Eds.), *Moral sentiments and material interests*. Cambridge, MA: MIT Press.

Goodman, N., Ullman, T., & Tenenbaum, J. (2009). Learning a theory of causality. In N. Taatgen & H. van Rijn (Eds.), *Proceedings of the Thirty-First Annual Conference of the Cognitive Science Society.* Austin, TX: Cognitive Science Society.

Gutnisky, D., & Zanutto, B. (2004). Cooperation in the iterated prisoner's dilemma is learned by operant conditioning mechanisms. *Artificial Life, 10*(4), 433–461.

Haley, K., & Fessler, D. (2005). Nobody's watching? Subtle cues affect generosity in an anonymous economic game. *Evolution and Human Behavior, 26*, 245–256.

Hall, J. (1947). *General principles of criminal law*. Indianapolis: Bobbs-Merrill.

Hammerstein, P. (2003). Why is reciprocity so rare in social animals? A protestant appeal. In P. Hammerstein (Ed.), *Genetic and cultural evolution of cooperation* (pp. 83–94). Cambridge, MA: MIT Press.

Hart, H. L. A., & Honore, T. (1959). *Causation in the law*. Oxford: Clarendon Press.

Heider, F. (1958). *The psychology of interpersonal relations*. New York: Wiley.

Henrich, J., Boyd, R., Bowles, S., Camerer, C., Fehr, E., Gintis, H., et al. (2005). "Economic Man" in cross-cultural perspective: Behavioral experiments in 15 small scale societies. *Behavioural Brain Research, 28*, 795–855.

Herrmann, B., Thöni, C., & Gächter, S. (2008). Antisocial punishment across societies. *Science, 319*(5868), 1362.

Hineline, P. N., & Rachlin, H. (1969). Escape and avoidance of shock by pigeons pecking a key. *Journal of the Experimental Analysis of Behavior, 12*, 533–538.

Hirschfeld, L. A., & Gelman, S. A. (1994). *Mapping the mind: Domain specificity in cognition and culture*. Cambridge: Cambridge University Press.

Hoffman, H., & Fleshler, M. (1959). Aversive control with the pigeon. *Journal of the Experimental Analysis of Behavior, 2*(3), 213.

Hoffman, M. (2000). *Empathy and moral development*. New York: Cambridge University Press.

Kadish, S. H., Schulhofer, S. J., & Steiker, C. S. (2007). *Criminal law and its processes* (8th Ed.). New York: Aspen.

Knack, S., & Keefer, P. (1997). Does social capital have an economic payoff? A cross-country investigation. *Quarterly Journal of Economics, 112*(4), 1251–1288.

Koechlin, E., & Jubault, T. (2006). Broca's area and the hierarchical organization of human behavior. *Neuron, 50*(6), 963–974.

Lorenz, K., & Tinbergen, N. (1938). Taxis und Instinkthandlung in der Eirollbewegung der Graugans/Directed and instinctive behavior in the egg rolling movements of the gray goose. *Zeitschrift für Tierpsychologie, 2*, 1–29.

Mackintosh, N. (1975). A theory of attention: Variations in the associability of stimuli with reinforcement. *Psychological Review, 82*(4), 276–298.

Macy, M., & Flache, A. (2002). Learning dynamics in social dilemmas. *Proceedings of the National Academy of Sciences of the United States of America*, *99*(Suppl. 3), 7229.

Marr, D. (1982). *Vision*. New York: Freeman.

Maynard Smith, J. (1982). *Evolution and the theory of games*. Cambridge: Cambridge University Press.

Maynard Smith, J., & Price, G. (1973). The logic of animal conflict. *Nature*, *246*(5427), 15–18.

Mazur, J. E. (1987). An adjusting procedure for studying delayed reinforcement. In M. L. Commons, J. E. Mazur, J. A. Nevin & H. Rachlin (Eds.), *The effect of delay and of intervening events on reinforcement value: Quantitative analysis of behavior* (Vol. 5, pp. 55–73). Hillsdale, NJ: Lawrence Erlbaum.

McElreath, R., Clutton-Brock, T., Fehr, E., Fessler, D., Hagen, E., Hammerstein, P., et al. (2003). Group report: The role of cognition and emotion in cooperation. In P. Hammerstein (Ed.), *Genetic and cultural evolution of cooperation* (pp. 125–152). Cambridge, MA: MIT Press.

McLaughlin, J. A. (1925). Proximate cause. *Harvard Law Review*, *39*(2), 149–199.

McNamara, J. M., Stephens, P. A., Dall, S. R. X., & Houston, A. (2008). Evolution of trust and trustworthiness: Social awareness favours personality differences. *Proceedings of the Royal Society, Series B: Biological Sciences*, *276*(1657), 605–613.

Mohtashemi, M., & Mui, L. (2003). Evolution of indirect reciprocity by social information: the role of trust and reputation in evolution of altruism. *Journal of Theoretical Biology*, *223*(4), 523–531.

Mui, L., Mohtashemi, M., & Halberstadt, A. (2002). A computational model of trust and reputation. In J. F. Nunamaker & R. H. Sprague, Jr. (Eds.), *Proceedings of the 35th Hawaii International Conference on Systems Sciences*. Los Alamitos, CA: IEEE Computer Society Press.

Murphy, G., & Medin, D. (1985). The role of theories in conceptual coherence. *Psychological Review*, *92*(3), 289–316.

Nagel, T. (1979). *Mortal questions*. Cambridge: Cambridge University Press.

Nowak, M. A. (2006). Five rules for the evolution of cooperation. *Science*, *314*, 1560–1563.

Nowak, M. A., & Sigmund, K. (1993). A strategy of win-stay, lose-shift that outperforms tit-for-tat in a prisoner's dilemma game. *Nature*, *364*, 56–58.

Palsgraf v. Long Island Railroad Co., 162 N.E. 99 (New York Court of Appeals 1928).

Parkhurst, D., Law, K., & Niebur, E. (2002). Modeling the role of salience in the allocation of overt visual attention. *Vision Research*, *42*(1), 107–123.

Rachlin, H., Raineri, A., & Cross, D. (1991). Subjective probability and delay. *Journal of the Experimental Analysis of Behavior*, *55*, 233–244.

Rand, D., Dreber, A., Ellingsen, T., Fudenberg, D., & Nowak, M. (2009). Positive interactions promote public cooperation. *Science*, *325*(5945), 1272.

Rand, D., Ohtsuki, H., & Nowak, M. (2009). Direct reciprocity with costly punishment: Generous tit-for-tat prevails. *Journal of Theoretical Biology, 256*(1), 45–57.

Randløv, J., & Alstrøm, P. (1998). Learning to drive a bicycle using reinforcement learning and shaping. In J. W. Shavlik (Ed.), *Proceedings of the Fifteenth International Conference on Machine Learning*. San Francisco: Morgan Kaufmann.

Renner, K. (1964). Delay of reinforcement: A historical review. *Psychological Bulletin, 61*(5), 341–361.

Rescorla, R., & Wagner, A. (1965). A theory of Pavlovian conditioning: Variations in the effectiveness of reinforcement and nonreinforcement. In A. H. Black & W. F. Prokasey (Eds.), *Classical conditioning II: Current research and theory* (pp. 64–99). New York: Appleton-Century-Crofts.

Richards, J. B., Mitchell, S. H., de Wit, H., & Seiden, L. S. (1997). Rate Currencies and the foraging starling: The fallacy of the averages revisited. *Behavioral Ecology, 7*, 341–352.

Richerson, P., Boyd, R., & Henrich, J. (2003). Cultural evolution of human cooperation. In P. Hammerstein (Ed.), *Genetic and cultural evolution of cooperation* (pp. 357–388). Cambridge, MA: MIT Press.

Roberts, W. A., & Mazmanian, D. S. (1988). Concept learning at different levels of abstraction by pigeons, monkeys, and people. *Journal of Experimental Psychology: Animal Behavior Processes, 14*, 247–260.

Robinson, P. H., & Darley, J. M. (1995). *Justice, liability, and blame*. Boulder: Westview Press.

Saxe, R., Carey, S., & Kanwisher, N. (2004). Understanding other minds: Linking developmental psychology and functional neuroimaging. *Annual Review of Psychology, 55*, 1–27.

Schwartz, B., & Reisberg, D. (1991). *Learning and memory*. New York: W. W. Norton.

Shaver, K. G. (1985). *The attribution of blame: Causality, responsibility, and blameworthiness*. New York: Springer.

Shiffrin, R. M., & Schneider, W. (1977). Controlled and automatic human information processing: II. Perceptual learning, automatic attending, and a general theory. *Psychological Review, 84*(2), 127–190.

Shultz, T. R., Schleifer, M., & Altman, I. (1981). Judgments of causation, responsibility, and punishment in cases of harm-doing. *Canadian Journal of Behavioural Science, 13*(3), 238–253.

Silk, J. B. (2003). Cooperation without counting: The puzzle of friendship. In P. Hammerstein (Ed.), *Genetic and cultural evolution of cooperation* (pp. 37–54). Cambridge, MA: MIT Press.

Sloman, S. (1996). The empirical case for two systems of reasoning. *Psychological Bulletin, 119*, 3–22.

Spelke, E. (2000). Core knowledge. *American Psychologist, 55*, 1233–1243.

Spranca, M., Minsk, E., & Baron, J. (1991). Omission and commission in judgment and choice. *Journal of Experimental Social Psychology, 27*(1), 76–105.

Stephens, D. W., McLinn, C. M., & Stevens, J. R. (2002). Discounting and reciprocity in an iterated prisoner's dilemma. *Science, 298*, 2216–2218.

Stevens, J., Hallinan, E., & Hauser, M. (2005). The ecology and evolution of patience in two New World monkeys. *Biology Letters, 1*(2), 223.

Stevens, J. R., Cushman, F. A., & Hauser, M. D. (2005). Evolving the psychological mechanisms for cooperation. *Annual Review of Ecology Evolution and Systematics, 36*, 499–518.

Sutton, R., & Barto, A. (1999). *Reinforcement learning*. Cambridge, MA: MIT Press.

Thompson, R., & Oden, D. (2000). Categorical perception and conceptual judgments by nonhuman primates: The paleological monkey and the analogical ape. *Cognitive Science, 24*(3), 363–396.

Tomasello, M. (1999). *The cultural origins of human cognition*. Cambridge, MA: Harvard University Press.

Tomasello, M., Carpenter, M., Call, J., Behne, T., & Moll, H. (2005). Understanding and sharing intentions: The origins of cultural cognition. *Behavioral and Brain Sciences, 28*, 675–735.

Trivers, R. L. (1971). The evolution of reciprocal altruism. *Quarterly Review of Biology, 46*, 35–57.

Tversky, A., & Kahneman, D. (1981). The framing of decisions and the psychology of choice. *Science, 211*, 453–463.

Waldmann, M., & Holyoak, K. (1992). Predictive and diagnostic learning within causal models: Asymmetries in cue competition. *Journal of Experimental Psychology: General, 121*(2), 222–236.

Warneken, F., & Tomasello, M. (2006). Altruistic helping in human infants and young chimpanzees. *Science, 311*(5765), 1301.

Weiner, B. (1995). *Judgments of responsibility: A foundation for a theory of social conduct*. New York: Guilford Press.

Williams, B. (1981). *Moral luck*. Cambridge: Cambridge University Press.

Zak, P., & Knack, S. (2001). Trust and growth. *Economic Journal, 111*, 295–321.

19 Our Pigheaded Core: How We Became Smarter to Be Influenced by Other People

Hugo Mercier

(Other) people's gullibility is a common source of complaint in the political world. Republicans lament that Democrats naively trust the "liberal media." Democrats wonder how Republicans can be so credulous as to believe Fox News. In this kind of attack, gullibility is often equated with lack of sophistication; the subtext is "How can they be so stupid?" ("Sophisticated" is a common antonym of "gullible"). Indeed, there seems to be a widespread intuition that the best way to influence people is to stop them from thinking. Politicians, newscasters, and educators are wont to dumb down their messages; ad men try to distract us so that their slogans will remain unexamined; interrogators try to break suspects' ability to reason through continuous questioning or sleep deprivation (or worse). Yet it is possible to argue that this intuition is profoundly misguided and that, overall, the best way to influence people is to tap into their most sophisticated psychological mechanisms, especially the complex calibration of trust and reasoning. In support of this claim, I will first offer an evolutionary argument and then proceed to "peel off" the mechanisms we use to evaluate information to reveal older mechanisms that are harder to influence: our pigheaded core.

The Joys and Dangers of Communication

All the important decisions we make in our life are profoundly affected by communicated information: whom we befriend, who is our partner(s), where we live, what career we follow. It would be difficult to deny that our ability to communicate and all the things it enables, from collective action to cultural transmission, have played a major role—probably *the* major role—in the evolutionary success of our species. Underlying the importance of communication is its amazing efficiency: On a first approximation, language is very close to telepathy. However, great dangers accompany such a powerful means of influence. Given the uncanny ability of other individuals to create representations in our minds, and the power these representations can have in shaping our decisions, there is an ever present possibility of abuse. People can manipulate each other; senders can lie, deceive, and more generally take advantage of listeners. Even if one were to discount the possibility of purely Machiavellian motives, a simple lack of consideration of other people's interests would lead to widespread trouble. (One could truthfully state that "I want you to want to give me $10,000"—it would still be a bad idea to comply.)

Dawkins and Krebs were among the first to draw attention to the evolutionary problems raised by communication (Dawkins & Krebs, 1978; Krebs & Dawkins, 1984). For communication to be evolutionarily stable, it has to benefit both senders and receivers. If senders do not benefit from communication, they stop sending (they become "mute"); if it is the receivers who lack benefits, then they stop receiving (they become "deaf"). That communication often breaks down in zero sum games, even in an intensely communicative species as ours, is immediately apparent from the behavior of, for instance, poker players: The "poker face" is an extreme example of what happens when there is no incentive whatsoever to communicate.

Once senders can influence receivers, they will often have incentives to take advantage of them. Alarm calls could be used to distance others from a valuable resource, food calls could be used to lure others to a given place, and so on. This means that in a great many cases the interests of the receivers—and, therefore, the stability of communication—have to be secured by some additional mechanism. Many such devices have been discovered across the animal world. For instance, if a signal is costly it can be used as a reliable indicator of the capacity and motivation of the sender to pay these costs (Zahavi & Zahavi, 1997). The incentives of senders and receivers can also be more easily aligned when their genetic interests overlap—although examples such as conflicts between mothers and their fetuses dramatically attest that this is far from being a foolproof solution (Haig, 1993). The stability of human communication cannot easily be explained through these means: We routinely communicate with nongenetically related individuals and the bulk of our communication is cheap talk (for economists, Farrell & Rabin, 1996) or low-cost signaling (for biologists, Maynard Smith, 1994). Another solution is to take advantage of the specific characteristics of a communication domain (Sterelny, 2012). The structure of the communicative environment—such as redundancies between different sources—can provide a cheap way to evaluate the validity of a piece of information. Some contents can also be easily verified—demonstrations of skills, for instance, are hard to fake. When communication only bears on a limited domain, dedicated mechanisms relying on the specifics of the situation can evolve. In humans, this could be the case for some forms of emotional contagion, for instance (see note 4 below). But human communication differs from everything else that exists in the animal kingdom because we can talk about practically anything. This means that other mechanisms, which do not rely on the specifics of a given domain, are likely to have evolved in order to maintain the stability of this very general form of communication. Receivers can *filter* communicated information; they can use a variety of means to determine which communicated information they should pay heed to.

Trust calibration may be the most obvious of these mechanisms: When we are told something, we weight it differently depending on its source—a trusted doctor or a quack, a friend or a stranger, and so on. But, given the huge influence of communication on our lives, trust calibration can only be one of the many devices we use to evaluate communicated information. Some of these mechanisms deal mostly with the "compliance" level: Should I act on the basis of communicated information? Others are dedicated to epistemic vigilance: maximizing the chances that the beliefs we accept are true (Mascaro & Sperber, 2009; Sperber et al., 2010).

If we assume that, like other complex adaptations (Pinker & Bloom, 1990), human communication—consisting mostly of, but not restricted to, language—evolved relatively gradually since our last common ancestors with the chimpanzees, becoming increasingly efficient and influential, it follows that the filtering mechanisms must also have evolved gradually, in line with communication's ever growing importance in our ancestors' lives. There are two ways for such filtering mechanisms to evolve, each of which may have played a role at different points of our evolution. Filtering mechanisms can evolve in order to *reject more and more information*. Let's imagine that at some stage of evolution individuals are not very discriminating and that as a result they *accept too much information*. Since receivers are not very discriminating, this might leave a window for senders to evolve toward more skillful manipulation. Receivers are then under pressure to evolve better filtering mechanisms—otherwise, they stop benefiting from communication. This can be dubbed the "Machiavellian" view. But filtering mechanisms could also have evolved in order to *accept more and more information*. Here we start from a stage in which receivers *reject too much information*. If, then, communication becomes more important, receivers are under pressure to evolve better filtering mechanisms so that they can accept information they were previously rejecting.[1,2]

Which of these two processes played the major role in the evolution of human filtering mechanisms? Intuitively, it seems that erring on the side of caution should be the most effective strategy. However, making a strong a priori argument to that effect would require estimates of the costs and benefits of rejecting too much true information and of accepting too much false information that may not be within our reach. Moreover, we cannot examine directly the psychological mechanisms of our ancestors. At best, we can hope to make informed inferences regarding their communicative system, but that would not get us very far in the present endeavor. However, we may be able to find remnants of earlier filtering mechanisms in modern humans. Much in the same way as linguists study "fossils" in modern languages to make inferences regarding previous stages of language evolution (e.g., Jackendoff, 1999), it might be possible to observe older filtering mechanisms still at play in modern humans. Accordingly, the present chapter reviews evidence related to these older filtering mechanisms and tries to show that they tend to err on the side of caution—that they reject too much information rather than not enough—and that new mechanisms evolved to make us accept more information.

For the fossil approach to work, however, these older mechanisms need to have been preserved. More specifically, the new mechanisms would have become fine-grained regulators of these older mechanisms. This entails a specific, highly modular view of the mind. Another perspective could also be envisioned, one in which the older mechanisms are "cannibalized": They are modified and put to new uses. Here again, it might be possible to make an a priori argument in favor of the former view. According to this former view, the main role played by the new mechanisms would be to override the negative verdicts of the older mechanisms when they deem a piece of communicated information to be, in fact, beneficial. With such a design, if, for some reason, the new mechanisms are unable to function properly, then the older mechanisms will take over and protect the individual from the costs of accepting too much information. General principles of design—which apply from cells to complex artifacts—would seem to favor this solution, as it is more modular, more robust, and it relies more on regulation (Carlson & Doyle, 2002; Kitano, 2004; Wagner, 2005). However, as is the case for the previous argument, these a priori reasons are unlikely to sway most readers. So I will rely on the evidence reviewed above to try and tip the scales in favor of the hypothesis that older filtering mechanisms are still present but that they have been increasingly regulated by layers of more recent filtering mechanisms.

In what follows, I try to show (i) *that older, less sophisticated mechanisms filtering communicated information are still present in humans* and (ii) *that these mechanisms are overly cautious and reject more information than more recent mechanisms*. Layers of filtering mechanisms will be peeled off to reveal our "pigheaded core." To do so, I have to find situations, experimental or otherwise, in which the workings of the more recent layers are perturbed, and the action of the older mechanism exposed, starting with the most extreme case: that of subliminally presented information.

The Core: Competition between Goals

Humans are endowed with a wealth of dedicated mechanisms that filter incoming communicated information (Sperber et al., 2010). What is left if we try to remove all these dedicated mechanisms? What is left is a very simple mechanism that, even if it is not its main purpose, can already protect us against communicated information—even subliminally communicated information: competition between goals.

In cognitively complex species, many goals or plans are competing at any given point for control of our motor system (Sperber, 2005). Stimuli perceived in the environment compete with each other as well as with previous plans in such a manner that there nearly never is a mandatory behavioral reaction to any type of input. This picture is controversial, but a review of the literature is out of question here, so I will only offer an example rendered convincing by the counterintuitiveness of its conclusion.

Neurobiologists have devoted a lot of attention to the tail-flip escape response of the crayfish, to such a point that it has become "one of the best-understood neural circuits in the animal kingdom" (Edwards, Heitler & Krasne, 1999, p. 153, to which this section is heavily indebted). At first sight, this might seem to be a poster child for the reflex; a single neuron commands a single behavior that is ecologically crucial: a flight response. But careful research has shown that the action of this "reflex" is in fact modulated by many factors. If escape is blocked by physical constraints, then it is not triggered, probably to prevent the animal from hurting itself. When the crayfish is feeding, the perception of threat has to be more serious or more persistent to trigger escape: This makes sense given that there is a higher cost associated with the departure from the food resources (Krasne & Lee, 1988). Likewise, if the crustacean is already involved in an incompatible behavior—such as backward walking or the defense posture—then the tail-flip escape response is strongly inhibited (Beall, Langley & Edwards, 1990). Finally, and even more impressively, the social status of the crayfish heavily influences its escape response: If the individual is a subordinate, it moderates its behavior and switches to "flexible, nonreflex ('voluntary') types of escape" (Krasne, Shamsian & Kulkarni, 1997, p. 709). Dominants, however, seem to ignore the presence of subordinates and go about their behavior unperturbed (in a manner paralleled by human dominants; see, e.g., Fiske, 1993). All of these things are performed routinely by crayfish. While this is far from a demonstration, it shows at least that any assumption of "reflexiveness" should be very seriously tested before it is accepted and that, in the meantime, it seems reasonable to assume that the vast majority—probably all—of our behaviors are the result of some kind of modulation, including competition among different goals.

Mechanisms designed to regulate motor control in such a way are very old and have clearly not evolved specifically to deal with problems related to filtering communicated information. Still, their action should play an important role in this domain because goals generated or influenced by communicated information—as any other goal—have to win this competition to cause any kind of behavioral effect. According to the present argument, when it comes to communicated information, this mechanism should be expected to have a very stringent baseline. Our behavior should only be minimally influenced by communicated information when we are deprived of any means of evaluation. But it is not easy to test for such a baseline because in the vast majority of cases we have at least one means to evaluate the information, namely, its source. Even when the source is someone we know very little about, or is anonymous, we can always venture educated guesses about what kind of individual she is and how much trust we can grant her. There is one case, however, in which we seem to be deprived of any means to evaluate a piece of communicated information: subliminal influence.

In subliminal influence people are exposed to stimuli that lie completely outside of their awareness. The most famous cases are words flashed on a screen too quickly for people to consciously perceive them. Since the 1950s, subliminal influence has been a recurrent source of fears ("I'm being constantly manipulated!") and hopes ("The tapes that I listen while I sleep will make me smart and self-confident"). However, what people thought were conclusive results proved to be the fraudulent invention of a disgruntled ad man (Weir, 1984) and the subsequent forty years of research failed to detect any sign of subliminal influence (see, e.g., Greenwald et al., 1991; Moore, 1982; Pratkanis & Aronson, 1992). More recently, some results have started to surface showing reliable effects of subliminal stimuli. The big difference introduced by these new experiments is that the stimuli are coherent with the previous goals of the participants. For instance, thirsty participants flashed with subliminal words related to thirst drink more water than participants flashed with neutral words. The exact same stimuli, however, have no effect whatsoever on nonthirsty participants (Strahan, Spencer & Zanna, 2002; see also Berridge & Winkiehnan, 2003; Dijksterhuis & Bargh, 2001). As Bargh puts it: "The main reason for the recent success is that researchers are taking the consumer's (experimental participant's) current goals and needs into account" (Bargh, 2002, pp. 282–283). So, while subliminal influence may make you drink a little more water if you are thirsty (thanks to fifty years of intensive research), good old influence can make most of us inflict enough electric shocks to risk killing a fellow human being (Milgram, 1974). Given the power many people attributed to subliminal influence, this outcome is rather ironic and confirms the idea that the older filtering mechanisms are not very responsive to communicated information.

Ostensive and Nonostensive Communication

We have seen that when filtering mechanisms are stripped to their bare minimum—when stimuli are perceived subliminally—they make it very hard for people to be influenced: Only small changes in behaviors that were already planned are allowed. This is coherent with the hypothesis that core mechanisms are still present today that protect us by severely diminishing the influence of communicated information. However, there are many layers between this core and the most recent filtering mechanisms. To keep peeling off the layers of filtering mechanisms, it is useful to introduce the distinction between ostensive and nonostensive communication.

Here we will define an ostensive stimulus as one that aims at attracting an audience's attention (Sperber & Wilson, 1995). The bulk of human communication is ostensive-inferential: Senders ostensively provide evidence that will help the receiver infer the sender's meaning (Sperber & Wilson, 1995). However, other channels of communication are nonostensive. For instance, if something frightens me, my feelings are likely to be reflected in a facial expression. This facial expression is communicative, but it is not ostensive: I do not intend to attract your attention to the fact that my face is harboring this expression. By contrast, I could mime the expression of fear in an ostensive manner—I could start by making eye contact. In this case, I would be trying to ostensively communicate something to you—for instance, that I think that something frightening is going to happen in the movie we are watching.

There are several differences between ostensive and nonostensive communication, but the most relevant to the present endeavor is that ostensive communication provides receivers with an additional layer of protection against communicated information. The reason is that when a receiver infers a sender's meaning, the output of this inference is in a metarepresentational format: "sender means P," which, in the case of a standard assertion, entails "sender intends me to believe P." It makes a lot of sense, from a filtering point of view, for the "P" not to be automatically disembedded from its metarepresentational context (see Sperber, 1997). While it is so embedded, it is mostly harmless: Thinking that you intend me to believe P when I doubt P or even have reasons to believe that it is a lie is not only innocuous, but can be highly informative (I will trust you less in the future; I can try to understand your motive for lying to me).

Even though the lack of automatic disembedding makes sense from the present point of view, it is not generally accepted. In particular, one line of experiment purports to show that our mental systems begin by automatically accepting any communicated information before examining it and, potentially, rejecting it (Gilbert, Krull & Malone, 1990; Gilbert, Tafarodi & Malone, 1993; see also Recanati, 1997, for a theoretical argument). In one of these experiments, participants were told they would have to learn Hopi words. They were then presented with sentences such as "A monishna is a star," followed shortly by TRUE or FALSE, indicating the veracity of the preceding sentence. In some cases, however, participants were distracted during their processing of the TRUE or FALSE indication. Later, they had to complete a recognition task: The same statements about Hopi words were presented, and the participants had to determine whether they were true or false. The authors predicted that if disembedding is not automatic, then interruptions of the TRUE indication would tend to produce more recognition as being false. On the other hand, if disembedding is automatic, then the opposite pattern should emerge: Interrupted FALSE indications should lead to remembering the statement as being true. Supporting the author's hypothesis, the latter pattern of results was observed, leading them to conclude that "you can't not believe everything you read" (Gilbert, Tafarodi & Malone, 1993). But, while these results seem to contradict the logic of good design for filtering mechanisms, several caveats are in order.

More recent experiments have highlighted the many shortcomings in the materials used by Gilbert and his colleagues. First, the participants have no previous knowledge that could help them evaluate the sentence. When this is not the case—particularly when the sentences contradict some previous beliefs of the participants—the effect disappears completely and performance is near ceiling (Richter, Schroeder & Wöhrmann, 2009). Second, the statements used in the original experiments become completely irrelevant if false (knowing that "A monishna is not a star" is quite useless), which is not true of much of the information we encounter. For instance, knowing that "Patrick is a good father" is false is very informative. So while participants who discover that "A monishna is a star" is false have no motivation to remember this information, participants discovering that "Patrick is a good father" is false should be motivated and able to remember that Patrick is not a good father. This is precisely what happens: The original effect all but disappears if the sentences used are informative when false (Hasson, Simmons & Todorov, 2005). Finally, in the original experiments the source of the information is both trustworthy and quite irrelevant. There is no a priori reason to doubt a computer spouting out translations of Hopi words, and there is very little interest in learning about the truthfulness of this computer. This is exceedingly rare outside the psychology laboratory. Most of the information we get is from people about whom we have a great deal of information (we even track the reputation of journalists, or at least of newspapers or news networks), and whose lies or mistakes are highly relevant (can I trust this friend, this colleague, this newspaper?). Accordingly, when the source is specified and is relevant (the best friend of the participant), then the effect, once again, disappears (Bergstrom & Boyer, submitted).

We can conclude from these experiments that outside the psychology laboratory, disembedding is never—or nearly never—automatic.[3] By contrast, nonostensive communication need not benefit from the protection offered by the metarepresentational format of ostensive communication. For instance, if I become afraid when I see a frightened individual, the format of the mental representation elicited in the receiver is not of the form "sender intends me to believe that there is something frightening," but could be of the form "there is something frightening." By getting rid of a layer of protection—the metarepresentational embedding—nonostensive communication could be abused much more easily by senders. Following the logic of the present argument, it should yield less influence than ostensive communication.[4] Nonostensive communication, however, should still have much more influence than subliminal stimuli because of one major difference: The source is identifiable. Knowing the source provides at least two means to ensure that communicated information can be safely accepted. One is that it is possible to punish a misleading source, provided we find out we have been misled and we can remember who was responsible. Such a punishment can vary widely in form, from direct physical harm to gossip or even simply making it less likely to believe the person in the future. Even though the latter is not intuitively construed as punishment, it can make communication harder for the source, which can certainly exert a certain cost. Whatever form it takes, punishment increases the costs for senders of communicating misleading information, making them less likely to do so. But the main advantage offered by the knowledge of a source of information is that we can use our previous knowledge about the source to evaluate whether we should accept what she communicates: For instance, has she been reliable in the past?

These considerations lead to the following predictions: (i) Nonostensive communication should have more influence than subliminally presented stimuli but (ii) less than ostensive communication, and (iii) people should rely both on compatibility with their previous goals and on knowledge about the source to evaluate nonostensive communication. In what follows, the focus will be on emotional signals (such as facial displays of anger, fear, etc.) as examples of nonostensive communication.

Point (i) is relatively trivial. Various emotional displays can exert an influence through communication: Blushing can make us decide to pursue further a romantic interest, seeing his mother's angry face can make a child start preparing an excuse for those missing cookies, and so on. Nothing even remotely as effective has been demonstrated with subliminally presented faces: Such stimuli may elicit an increase in amygdala activation (Morris, Öhman & Dolan, 1998; Whalen et al., 1998) and some minor facial movements (Dimberg, Thunberg & Elmehed, 2000), but not more. This is related to the fact that subliminally presented conditioned stimuli elicit much lower responses than supraliminal conditioned stimuli in general (Olsson & Phelps, 2004).

Point (ii) may be less obviously true, but the examination of a few cases should be sufficient to make the point. You are watching someone on TV. He is smiling and his voice reflects his happy mood. If you have no reason to dislike him, this can certainly be quite enjoyable: Moods can be transmitted through such signals (e.g., Neumann & Strack, 2000). Now it's November 4, 2008, and the man is Barack Obama. He's delivering his victory speech. For millions of people, this moment is not simply enjoyable, it is fraught with an emotional intensity that will forever be stamped in their mind. Had Obama's smile been meant to convey that he had won the election, this simple gesture, now ostensive, could have provoked infinitely more emotion than that procured through mere contagion. Cultural productions reflect the power of ostensive communication: We read uplifting stories instead of listening to the prosody of someone in a good mood and we watch horror movies instead of looking at scared faces. It would be hard to deny that our most intense emotions have come either from events in our own lives or from stories told by people we hold dear, rather than through watching facial displays or other such nonostensive signals.

Finally, point (iii) is probably the most contentious. Contrary to all the talk about "automaticity" in nonostensive signals such as facial displays of emotions (which is often understood as mandatoriness in the sense of Fodor, 1983), the present theory predicts that the reaction of individuals faced with such signals should be heavily modulated by their source. While previous goals should still play a role in the behavioral outcome, potential conflicts between these goals and the communicated information can now be overcome if the source is deemed to be reliable enough. Unfortunately, although there is a wealth of information about how adults (Petty & Wegener, 1998) and children (Clément, in press; Harris, 2007; Mascaro & Sperber, 2009) use the source of ostensive communication in their evaluation of it, there is much less research on source effects for nonostensive communication. Still, the little there is points toward a strong and reliable effect of the source (for a more extensive review, see Dezecache, Mercier and Scott-Phillips, submitted). Empathic blushing—due to vicarious embarrassment—is observed when someone is looking at oneself (on tape) or a friend doing something embarrassing; but the effect is much reduced if it is a stranger performing the same action (Shearn et al., 1999). We "automatically" imitate the facial expression of anger (or other negative emotions) of people belonging to positively valued groups, but not of people from other groups (Bourgeois & Hess, 2007; Mondillon et al., 2007). Responses to facial displays of pleasure or pain are often empathic, but they can become *counter*empathic (smiling in reaction to an expression of pain, and vice versa) if we expect to compete with the other person (Lanzetta & Englis, 1989). If someone has been unfair to us, for some of us (the males) it is the reward, and not the empathy, system that will be activated when that individual is hurt (Singer et al., 2006).[5] Evaluation of the source can also be more fine-grained than a simple good or bad: Political attitudes toward leaders influence reactions to their emotional displays (McHugo, Lanzetta & Bush, 1991).

That emotional contagion is modulated by its source is also clearly apparent in the way it contributes to spreading emotions. If emotional contagion were truly automatic, emotions should spread like wildfire (as implicitly argued by, e.g., Christakis & Fowler, 2009). And sometimes this is exactly what seems to be happening in instances of "mass hysteria": whole schools erupt with inexhaustible laughter (e.g., Ebrahim, 1968) and factories are swept with strange emotional symptoms (e.g., Stahl & Lebedun, 1974). While these observations lend superficial support to the automatic view of emotional contagion, a deeper examination mostly highlights its limitations. First of all, these episodes, remarkable as they may be, are still very rare: Their scarcity is as much to be explained as their existence, and it is hard to explain from a purely automatic point of view. Second, echoing the experimental findings reviewed above, the spread of the emotions is heavily modulated by the source of the displays. The emotions and the behaviors they generate do not spread to people who might be brought in to investigate the case, or to other strangers. Instead, they spread among individuals who know and are close to each other: pupils in a classroom, workers in the same plant. Finally, these apparently strange behaviors always seem to serve a goal of the individuals, even if this goal is unconscious. For instance, "mass hysteria" in work settings typically arises when the workers are "suddenly exposed to what they perceive to be an imminent threat" (Evans & Bartholomew, 2009, p. 364). It is thus not surprising to find a very good correlation between the severity of the symptoms and the level of job dissatisfaction (Stahl & Lebedun, 1974). Likewise, mass hysteria in schools typically targets "a socially cohesive group . . . exposed to a stressful stimuli" (Evans & Bartholomew, 2009, p. 384). So, far from being a testimony to the automaticity of emotional contagion, such cases of "mass hysteria" show the transmission of emotions to be modulated by the predicted factors: source and compatibility with previous goals.

Emotion contagion is far from being the only kind of nonostensive communication in humans. It is a convincing example, however, because it is generally thought to be automatic and therefore powerful. According to the present argument, it is precisely *the opposite* that is the case: The limited influence nonostensive communication has is made possible by its nonautomaticity (more precisely, by the fact that it is modulated by the source). Even less automatic mechanisms, such as ostensive communication, yield even more influence.

Are Smart People Gullible?

I started this review by peeling off as many layers of filtering mechanisms as possible. When people have to rely on the most primitive of filtering mechanisms, communicated information barely has any influence on them. When people have more cues to the reliability of communicated information—such as its source in the case of nonostensive communication—filtering mechanisms allow for more ample influence. This influence is still weaker, however, than that observed when people have the luxury of examining communicated information while it is embedded in a metarepresentational context before deciding whether they should accept it. This last type of communication—ostensive communication—is by far the most important in humans. It is therefore to be expected that several filtering mechanisms should be dedicated precisely to ostensively communicated information. In the last step in the process, I look at reasoning as one of the latest addition to the list of filtering mechanisms and try to see what happens when we peel it off. In line with the hypothesis laid out at the outset, if this layer of filtering mechanism is removed, then people should accept less information. But to show that, I must first make the case that reasoning is indeed such a filtering mechanism.

The content of communicated information ("Is it compatible with my plans and beliefs?"), as well as its source ("Is she reliable?"), is routinely used in the process of evaluation. Yet these mechanisms still lead to the rejection of some potentially beneficial information. Sometimes we are wrong; our beliefs are mistaken, our plans suboptimal. When this is the case, we would be better off being influenced by others. Trust can help solve this problem by allowing people to override basic compatibility checking when information comes from a reliable source. But trust is far from perfect. For instance, it can be very long in the making, but it is easily lost (Slovic, 1993). In many cases, people would benefit from accepting a piece of information even when trust alone would lead to its rejection. A solution to this dilemma is for the sender to provide reasons supporting the information she wishes to communicate, reasons that can then be examined by the receiver. For instance, Mary might not believe Peter when he tells her that John is a womanizer: She does not know Peter that well, and she previously had a rather positive opinion of John. Without the ability to provide reasons, the situation would stall and the evil John might woo Mary. But if Peter is able to find reasons, he could support his statement with additional information such as: "He slept with Rita and then never called back" (a piece of information Mary can check with her friend) or "Remember the way he flirted with Sarah even though his girlfriend Britney was there?" (an event from which Mary might have drawn no inference). Mary can then evaluate these arguments and, if they are deemed sufficient, change her mind about John.

The ability to find good arguments, however, does not come for free (Mercier, 2010). Even for a skilled language user, the ability to make someone else *understand* our statements is not the same as the ability to make someone *accept* these statements. In the example above, Peter can easily lead Mary to understand that he wants her to believe that John is a womanizer. Given that the skills required to perform this task have done their duty perfectly (Mary understood what Peter meant), they would be of no use in the next step—acceptance. For this, a new cognitive device (or a set of devices) is required. Given that the task of this device is to find and evaluate reasons, we can call it "reasoning" (Mercier & Sperber, 2009).[6]

This view of reasoning as a specific mental mechanism is very much in line with a host of empirical work in psychology that goes under the umbrella of "dual process theories" (Evans, 2008; Kahneman, 2003). According to these theories, the mind can usefully be divided into two broad categories of processes: intuitions and reasoning.[7] A theoretically grounded way to frame this distinction is to express it in terms of intuitive and reflective inferences (Mercier & Sperber, 2009). The vast majority (in humans) or the whole (in other animals) of inferences are intuitive: They are performed without attending to the reasons why they are performed. For instance, we very quickly form impressions of the people we meet (e.g., Fiske, Cuddy & Glick, 2007). These impressions are barely ever based on a conscious, explicit assessment of reasons. Instead they are influenced by factors that we would be hard put to introspect (the shape of the face) or that we would even disown if we knew about their effect (the color of the skin). On the other hand, we are sometimes willing, and able, to revise these initial assessments on the grounds of reasons we consciously ponder. For instance, we might decide that our assessment of an individual was biased by some factor—gender, ethnicity, clothing—and try to revise it accordingly. Faced with visitors from a foreign culture, we often have to remind ourselves that "strange" behaviors that might intuitively lead us to a negative assessment are to be evaluated in light of different norms.

Following the Cartesian tradition, standard dual process theories tend to ascribe to reasoning a mostly individual function: By allowing us to correct the errors of our intuitions, reasoning should lead us toward better decisions and epistemic improvement (Evans & Over, 1996; Kahneman, 2003; Stanovich, 2004). This is where the view presented here parts way with classical models. Instead of seeing reasoning as a prop of individual cognition, the argumentative theory suggests that reasoning evolved for a profoundly social purpose: finding reasons to convince other people and evaluate these reasons so as to be convinced only when we should be (Mercier & Sperber, 2011).[8] In this perspective, reasoning is one of the most recent—if not the most recent—layer of filtering mechanism to have been added to our cognitive makeup. In line with the argument offered at the beginning of this chapter, it should therefore allow people to be *more* influenced by communication. This prediction clashes with the Cartesian view of the good reasoner as the über-skeptic, rejecting naïve beliefs through constant doubt and careful examination. In what follows I offer a defense of the present view in four points: (i) Despite the confirmation bias, people can be swayed by good arguments, even when their conclusions conflict with some previous beliefs; (ii) reasoning does a better job than other mechanisms at transmitting counterintuitive beliefs; (iii) trying to stop people from reasoning does not lead to good results in terms of influence; and (iv) reasoning more leads to more diverse beliefs, often including more false beliefs.

(i) According to the argumentative theory, reasoning is designed in part to produce arguments to convince other people. Reasoning should find reasons that support one's point of view or rebut the interlocutor's, and not the opposite. A confirmation bias is therefore to be expected. And, indeed, people exhibit a strong and robust confirmation bias (Nickerson, 1998). Although such a bias is predicted by the hypothesis, it could be problematic for the present argument. The prevalence of the confirmation bias could lead to the conclusion that people are bad at evaluating arguments. Indeed, many experiments have shown that people are biased in their evaluation of arguments, being more prone to discover flaws or find counterarguments when the conclusion disagrees with their own opinion (see Mercier & Sperber, 2011). In some cases, their prior beliefs completely skew the way they treat arguments, to the point that being presented with contradictory evidence can even strengthen these prior beliefs (Batson, 1975; Burris, Harmon-Jones & Tarpley, 1997; Lord, Ross & Lepper, 1979; Tormala & Petty, 2002). One could think that, because of the confirmation bias, reasoning leads people to be *less* influenced by communication, not more—they become more pigheaded once again, contrary to the earlier prediction. It is therefore important to emphasize that this is, in fact, not the case.

Two factors prevent us from drawing strong negative conclusions from the studies mentioned above. First, in most cases, despite a biased assessment, the arguments—to the extent that they are strong—still change the participants' attitudes (Petty & Wegener, 1998). Experiments reporting polarization of prior beliefs following counterarguments depend on a very specific set of circumstances, rarely obtained (Kuhn & Lao, 1996; Miller et al., 1993; Pomerantz, Chaiken & Tordesillas, 1995). Second, these experiments never put participants in the context of a dialogue. Instead, the participants are faced with a (generally) written argument, and its evaluation is soon succeeded by a phase in which the participants produce counterarguments (when they do not agree with the conclusion). But there is no one to refute these counterarguments or to propose new arguments for the other side. In such an artificial context, the confirmation bias present in the production of arguments can proceed unimpeded and make the evaluation appear much more biased than it actually is. If the participants were dealing with an interlocutor able to defend the opposing point of view, they would not have such leisure to be biased. Accordingly, the confirmation bias can be much attenuated in group settings (Kuhn, Shaw & Felton, 1997). Moreover, in many cases members who hold a minority view in a debate are able to prevail—for instance when they have understood a reasoning problem (Laughlin & Ellis, 1986; Moshman & Geil, 1998). For this to be the case, all the other group members must have been convinced to change their views, in spite of any confirmation bias they might have: Even though they certainly try, at first, to defend their views, they run short of argument at some point and come to accept the correct answer.[9] We can conclude that the confirmation bias does not stop people from changing their minds when they are confronted with good arguments.

(ii) In a more historical perspective, it is possible to take the existence of science as a testimony to the power of reasoning to change people's minds, even to the point where they come to accept counterintuitive beliefs. If one might claim that most of us acquire our scientific knowledge (outside our area of expertise, for scientists) through trust in authority, such was clearly not the case when this knowledge started to emerge. The history of science is a long, uphill battle in which people become convinced of otherwise weird things by good arguments and evidence. Indeed, most of the scientific knowledge we have now is deeply counterintuitive: Continents move, time changes with speed, life emerged out of inert chemicals, the universe has more than three dimensions, and so on (see Cromer, 1993). We should therefore expect people to exhibit a strong confirmation bias in such cases. While this has no doubt been the case—scientists are far from being exempt from this bias (Mahoney, 1977; Nickerson, 1998)—it has not stopped them from, eventually, accepting the arguments and the conclusions they support, counterintuitive as these conclusions may be. Other groups that mostly rely on reasoning to transmit their beliefs have arrived at very counterintuitive constructs. Mathematicians have created objects that only a few people can represent, logicians have constructed conditionals that yield paradoxical conclusions, philosophers have developed postmodern relativism, and economists can sustain a strange faith in the free market. By comparison, beliefs that spread through other means tend to be much more intuitive. It can be argued that religious beliefs, for instance, spread mostly through trust, whether it is in one's parents, one's friends, or in a more official representative of a given religion (see, e.g., Stark & Bainbridge, 1980). Accordingly, religious beliefs can be counterintuitive, but not too much so (Boyer, 1996a,b). Even within religion, it might be that the most counterintuitive beliefs are the ones arrived at through painstaking theological argument, and that these beliefs require more reasoning to spread. The historical evidence could thus be interpreted as showing that reasoning is more effective than other means of communication in making people accept deeply counterintuitive beliefs.

(iii) Given the widespread idea that reasoning is generally used to resist influence, it should come as no surprise that people have tried to deprive individuals of their ability to reason in order to get them to accept some beliefs. Unfortunately, some of the people sharing this view had total control over other people and no scruples. During the war in Korea, 7,190 Americans were held captive by the Korean and the Chinese. Many of them were subject to intense "brainwashing," from long daily (or twice-daily) lectures on the benefits of communism and the pitfalls of capitalism to group discussions on the same topic (Jowett, 2006). Possibly in order to make the prisoners more receptive to the arguments, they were treated very harshly, deprived of sleep, and underfed: not the best conditions for reasoning. Yet only twenty-one soldiers defected and chose to go to China after the war—not exactly a success. Moreover, these prisoners later recanted, came back to the United States, and said that they had done this to put an end to their terrible treatment (Streatfeild, 2007). This null rate of success is to be compared with the two following numbers. First, 2,730 POWs died while in captivity. Killing people is apparently easier than changing their minds. This shows how hard it is to convince people without trust or good arguments. Second, 50,000 Chinese and Korean POWs decided to defect without the help of any brainwashing. This comparison demonstrates that an actual display (the United States offers a much superior material comfort) can be much more effective than communication.

People have also charged some new religious movements with brainwashing their recruits and using techniques to deprive them of their reasoning ability, from sleep deprivation to overworking. Yet these new religious movements encounter, on the whole, very little success: Most of their recruits choose to leave the cults/religions quite quickly and their growth rate has been close to zero (Anthony & Robbins, 2004; Streatfeild, 2007). By comparison, sects such as the early Christians or the Mormons have been much more successful using normal means of influence: gaining people's trust little by little, reaching people through their friends and families, and so on (Stark, 1996; Stark & Bainbridge, 1980). Techniques associated with brainwashing can also be used by interrogators to obtain compliance from suspects. Yet, here again, when the suspect is highly motivated and intelligent, the only chance is to talk to him and get to a position where a modicum of trust has been established and well-targeted arguments can be used (Streatfeild, 2007, p. 375; Alexander & Bruning, 2008). On the whole, trying to reduce people's capacity to reason has proven to be a very inefficient—indeed, often counterproductive—strategy.

Besides the evolutionary argument, there also is a good mechanistic reason why stopping people from reasoning will rarely make it easier to convince them. When people use reasoning to evaluate arguments, part of the process is a search for counterarguments. Conviction will then be stronger if the attempt is unsuccessful. But if someone is prevented from reasoning, she will not be able to engage in such a search for counterarguments. As a result, she is much less likely to be convinced by the argument. Given that arguments are typically offered in support of conclusions that would not otherwise be accepted, this means that the receiver will revert to her earlier evaluation and reject the conclusion. This is a very good design for a filtering mechanism: If it is disrupted, it reverts to a negative assessment; only its good functioning can allow more information to be accepted. Experimental results confirm this prediction: People who think more about good arguments tend to be more convinced by them, probably because they did not manage to come up with good enough counterarguments (if they had, they would not change their minds, however good the original argument might be; e.g., Cacioppo, Petty & Morris, 1983; Petty & Cacioppo, 1979), and, of particular interest to academics, a study showed that people who have had a coffee are more alert and change their minds more often in response to good arguments (Martin et al., 2005).

(iv) Both the classical view of reasoning and the argumentative theory agree that on the whole more or better reasoning should be conducive to more accurate beliefs. Where the two theories differ is in the way this result is achieved. In the classical view, one of the most important uses of reasoning is to critically evaluate our own beliefs as well as those of others. In this case, epistemic improvement is mostly achieved through a weeding out of unjustified beliefs, resulting in a higher share of justified beliefs. According to the argumentative theory, on the other hand, reasoning achieves epistemic improvement by allowing us to accept more justified beliefs, mostly stemming from communication. As a result, even if the predicted share of true beliefs were the same in both cases, the quantity of beliefs that the use of reasoning leads to would be different: fewer beliefs for the classical view, more according to the argumentative theory.[10] In turn, this difference in quantity must logically give rise to a difference in diversity. Thus, according to the argumentative theory, people who reason more, or better, should have more diverse beliefs. It might then be expected that people who reason more have *more* false beliefs than other people, but that they also have *even more* true beliefs than them.

A cursory examination of the beliefs of people with impeccable reasoning credentials seems to confirm this prediction. From Newton, who spent more time on biblical numerology than on physics, to Linus Pauling, who was convinced that vitamin C was the ultimate secret to a healthy life, brilliant intellectuals have had the habit of accepting weird beliefs. Some paranormal beliefs correlate positively with variables usually associated with more or better reasoning (such as cognitive ability or level of education) (Tobacyk, Miller & Jones, 1984). Mensa members seem to be particularly prone to belief in extrasensory perception (Shermer, 1997). Smart, educated people are often among the early adopters of novel beliefs (see Vyse, 1997, in the case of new age, Shermer, 1997, in the case of UFOs and alien abductions, and Wallace, 2009, in the case of rejection of vaccination). While such evidence falls short of being conclusive, it makes it hard to deny that good reasoning is very far from foolproof and that even the best reasoners have a tendency to accept weird beliefs that quite often turn out to be wrong.

The four points above converge toward the conclusion that the more people reason, the more they can be influenced: Despite the confirmation bias, people can change their minds when faced by good arguments, even to the point of accepting counterintuitive beliefs; people deprived of reasoning are very hard to influence, whereas people who rely a lot on reasoning often entertain a wider range of beliefs, including quite a few wrong ones.

Conclusion

This chapter set out to explore some of the filtering mechanisms we use when dealing with communicated information. It suggested that these mechanisms would have evolved in successive layers, allowing communicated information to be increasingly influential. It defended the following claims:

- When we have very limited means of evaluation, as in the case of subliminal influence, we revert to old mechanisms of competition between goals that leave communicated information with only small effects, and only when those effects are compatible with our previous goals.
- Because nonostensive communication can, and does, use the source of information to modulate its effects, it can yield more influence than subliminal influence. However, it does not enjoy the layer of protection offered by ostensive communication, and so it is much less influential.
- Reasoning is a recently evolved filtering mechanism. By allowing us to understand and evaluate arguments, reasoning makes it possible to communicate beliefs that would otherwise have very little chance of being accepted by receivers: It increases the amount of information efficiently transmitted.

From an evolutionary perspective, the results presented here should come as no surprise. We know that communication exerts vastly more influence in humans than in any other primate species. It is reasonable to assume that this influence has been ever growing during our evolution. Given that the effects of communication must be held in check by filtering mechanisms, it follows that these mechanisms should have evolved to allow for this increasing influence of communicated information.

Although they may not be surprising from such a point of view, these results still run against some widely shared intuitions: that less cognitively gifted people should be easier to influence, that preventing people from reasoning should make them more manipulable, and so on. These intuitions explain the success—in popular opinion, not in fact—of subliminal influence. Subliminal influence may have been harmless, but other manipulation attempts based on these intuitions took on a much more somber aspect, from sleep deprivation to sensory isolation. It is therefore quite comforting to think that such attempts have little chance of encountering regular success (as the CIA has found out after a great many experiments; see Streatfeild, 2007). On the contrary, the most efficient way to influence people is to win their trust and to use cogent arguments.

Acknowledgments

I wish to thank Brett Calcott, Ben Fraser, Richard Joyce, Dan Sperber, and Kim Sterelny for insightful comments on this chapter.

Notes

1. The increase in acceptance of communicated information is to be understood in relative terms: the ratio of information accepted to information rejected. However, given that the total amount of information should also increase (this is the point of the evolution of more sophisticated filtering mechanisms in the first place), it is possible that the absolute amount of information rejected will also increase.

2. Even though this chapter does not dwell on them, in tandem with these new filtering mechanisms, senders should evolve new mechanisms to help receivers filter information more efficiently, so that senders can exert more influence on receivers through an increase in accepted information. Moreover, given that senders are the initiators of communication, they are likely to be its greatest beneficiary, and so we should expect them to bear most of the costs of the mechanisms that will make communication more efficient.

3. One proviso should be mentioned: In the experiments conducted so far, the source has never been someone very close to the participant, someone who the participant could trust—nearly—completely, such as a close parent or, in some domains, a teacher. It could therefore be argued that in such cases disembedding is automatic. However, Bergstrom and Boyer (submitted) have found that disembedding is not automatic when the statements come from a close friend. Moreover, the reason this is the case might be that if a close friend is caught telling something false, this is much more relevant than if it is a stranger who is thus caught. So it would seem reasonable to speculate that such a result would also be obtained with, say, a family member.

4. This generalization can easily accept exceptions. In particular, some emotional contagion carries little cost if one is deceived, but important costs for incredulousness. Panic could be an example: The costs of panicking when one shouldn't are much lower than those of not panicking when one should.

5. Simply seeing someone experiencing an event that should cause pain is not typically part of communication. However, to the extent that empathy could easily be manipulated (by faking pain), it makes sense that individuals should be wary of empathizing with others not deemed to be reliable (and, indeed, mice [Langford et al., 2006] and monkeys [Masserman, Wechkin & Terris, 1964] will empathize only with selected individuals). See De Vignemont and Singer (2006) and Hein and Singer (2008) on the modulation of empathic responses.

6. This assumes that understanding and acceptance are two distinct processes, and acceptance is not taken for granted after understanding. This runs against some views of language comprehension that rely on "interpretive charity"—i.e., people need to assume most statements to be true to be able to understand them (Davidson, 1984). A forceful critique of these views from a perspective congenial to the one adopter here can be found in Sperber et al. (2010).

7. Whereas most authors use "reasoning" to describe the two levels, describing them as "system 1 and system 2" (Stanovich, 2004), "heuristic and analytic" (Evans, 2007), or "associative and rule-based" (Sloman, 1996), here we follow Kahneman (2003) and Mercier and Sperber (2009) in using "intuition" for the first type of mechanism and restricting "reasoning" to the second.

8. There is no space here to present, even briefly, the empirical evidence that supports the theory, but it is comprehensively presented in Mercier and Sperber (2011). Other evidence can be found in Mercier (2011a), which defends the idea that argumentation is a human universal, and Mercier (2011b), which shows that children are skilled arguers from very early on.

9. Lately, the capacity of deliberation to revise people's attitudes has also come under attack in political science. According to some critics of deliberative democracy, debates among groups of citizens are rather futile and only rarely succeed in changing minds (partly because of the confirmation bias) (Goodin & Niemeyer, 2003; Hibbing & Theiss-Morse, 2002; Sunstein, 2002). This conclusion, however, mostly stems from a misinterpretation of the empirical data (Mercier & Landemore, 2012) and problems in the way the effects of debates are measured (Mackie, 2006).

10. The strength of this conclusion rests on how the classical view construes reasoning. If it construes it as working mostly through self-criticism, then the difference between the predictions of the two views should be large. On the other hand, if reasoning, on the classical view, also relies significantly on the construction of new beliefs, then the gap between the predictions of the two views is smaller. Still, as long as there is a difference in the use of our critical abilities between the two views—if they are aimed mostly at ourselves for the classical view, and mostly at communicated information for the argumentative view—then this conclusion should hold.

References

Alexander, M., & Bruning, J. (2008). *How to break a terrorist: The U.S. interrogators who used brains, not brutality, to take down the deadliest man in Iraq*. New York: Free Press.

Anthony, D., & Robbins, T. (2004). Conversion and "brainwashing" in new religious movements. In J. R. Lewis (Ed.), *The Oxford handbook of new religious movements* (pp. 243–297). Oxford: Oxford University Press.

Bargh, J. A. (2002). Losing consciousness: Automatic influences on consumer judgment, behavior, and motivation. *Journal of Consumer Research, 29*(2), 280–285.

Batson, C. D. (1975). Rational processing or rationalization? The effect of discontinuing information on stated religious belief. *Journal of Personality and Social Psychology, 32*(1), 176–184.

Beall, S. P., Langley, D. J., & Edwards, D. H. (1990). Inhibition of escape tailflip in crayfish during backward walking and the defense posture. *Journal of Experimental Biology, 152*, 577.

Bergstrom, B., & Boyer, P. Submitted. Who mental systems believe: Effects of source on judgments of truth.

Berridge, K. C., & Winkiehnan, P. (2003). What is an unconscious emotion? (The case for unconscious "liking.") *Cognition and Emotion, 17*(2), 181–211.

Bourgeois, P., & Hess, U. (2007). The impact of social context on mimicry. *Biological Psychology, 77*(3), 343–352.

Boyer, P. (1996a). Cognitive limits to conceptual relativity: The limiting case of religious categories. In J. Gumperz & S. Levinson (Eds.), *Rethinking linguistic relativity* (pp. 203–231). Cambridge: Cambridge University Press.

Boyer, P. (1996b). What makes anthropomorphism natural: Intuitive ontology and cultural representations. *Journal of the Royal Anthropological Institute, 2*(1), 83–97.

Burris, C. T., Harmon-Jones, E., & Tarpley, W. R. (1997). By faith alone: Religious agitation and cognitive dissonance. *Basic and Applied Social Psychology, 19*(1), 17–31.

Cacioppo, J. T., Petty, R. E., & Morris, K. J. (1983). Effects of need for cognition on message evaluation, recall, and persuasion. *Journal of Personality and Social Psychology, 45*(4), 805–818.

Carlson, J. M., & Doyle, J. (2002). Complexity and robustness. *Proceedings of the National Academy of Sciences of the United States of America, 19*(99), 2538–2545.

Christakis, A. N., & Fowler, J. H. (2009). *Connected: The surprising power of our social networks and how they shape our lives*. Boston: Little, Brown.

Clément, F. (In press). To trust or not to trust? Children's social epistemology. *Review of Philosophy and Psychology*.

Cromer, A. (1993). *Uncommon science: The heretical nature of science*. New York: Oxford University Press.

Davidson, D. (1984). Radical interpretation. In D. Davidson (Ed.), *Inquiries into truth and interpretation* (pp. 125–140). Oxford: Clarendon Press.

Dawkins, R., & Krebs, J. R. (1978). Animal signals: Information or manipulation? In J. R. Krebs & N. B. Davies (Eds.), *Behavioural ecology: An evolutionary approach* (pp. 282–309). Oxford: Blackwell Scientific.

De Vignemont, F., & Singer, T. (2006). The empathic brain: How, when, and why? *Trends in Cognitive Sciences, 10*(10), 435–441.

Dezecache, G., Mercier, H., & Scott-Phillips, T. (Submitted). An evolutionary approach to emotional communication.

Dijksterhuis, A., & Bargh, J. A. (2001). The perception-behavior expressway. In M. P. Zanna (Ed.), *Advances in experimental social psychology* (Vol. 33, pp. 1–40). San Diego, CA: Academic Press.

Dimberg, U., Thunberg, M., & Elmehed, K. (2000). Unconscious facial reactions to emotional facial expressions. *Psychological Science, 11*, 86–89.

Ebrahim, G. J. (1968). Mass hysteria in school children: Notes on three outbreaks in East Africa. *Clinical Pediatrics, 7*(7), 437.

Edwards, D. H., Heitler, W. J., & Krasne, F. B. (1999). Fifty years of a command neuron: The neurobiology of escape behavior in the crayfish. *Trends in Neurosciences, 22*(4), 153–161.

Evans, H., & Bartholomew, R. (2009). *Outbreak! The encyclopedia of extraordinary social behavior*. New York: Anomalist Books.

Evans, J. S. B. T. (2007). *Hypothetical thinking: Dual processes in reasoning and judgment*. Hove: Psychology Press.

Evans, J. S. B. T. (2008). Dual-processing accounts of reasoning, judgment, and social cognition. *Annual Review of Psychology, 59*, 255–278.

Evans, J. S. B. T., & Over, D. E. (1996). *Rationality and reasoning*. Hove: Psychology Press.

Farrell, J., & Rabin, M. (1996). Cheap talk. *Journal of Economic Perspectives, 10*, 110–118.

Fiske, S. T. (1993). Controlling other people: The impact of power on stereotyping. *American Psychologist, 48*, 621–628.

Fiske, S. T., Cuddy, A. J. C., & Glick, P. (2007). Universal dimensions of social cognition: Warmth and competence. *Trends in Cognitive Sciences, 11*(2), 77–83.

Fodor, J. (1983). *The modularity of mind*. Cambridge, MA: MIT Press.

Gilbert, D. T., Krull, D. S., & Malone, P. S. (1990). Unbelieving the unbelievable: Some problems in the rejection of false information. *Journal of Personality and Social Psychology, 59*(4), 601–613.

Gilbert, D. T., Tafarodi, R. W., & Malone, P. S. (1993). You can't not believe everything you read. *Journal of Personality and Social Psychology, 65*(2), 221–233.

Goodin, R. E., & Niemeyer, S. J. (2003). When does deliberation begin? Internal reflection versus public discussion in deliberative democracy. *Political Studies, 51*(4), 627–649.

Greenwald, A. G., Spangenberg, E. R., Pratkanis, A. R., & Eskenazi, J. (1991). Double-blind tests of subliminal self-help audiotapes. *Psychological Science, 2*(2), 119–122.

Haig, D. (1993). Genetic conflicts in human pregnancy. *Quarterly Review of Biology, 68*, 495–532.

Harris, P. L. (2007). Trust. *Developmental Science, 10*, 135–138.

Hasson, U., Simmons, J. P., & Todorov, A. (2005). Believe it or not: On the possibility of suspending belief. *Psychological Science, 16*(7), 566–571.

Hein, G., & Singer, T. (2008). I feel how you feel but not always: The empathic brain and its modulation. *Current Opinion in Neurobiology, 18*(2), 153–158.

Hibbing, J. R., & Theiss-Morse, E. (2002). *Stealth democracy: Americans' beliefs about how government should work*. Cambridge: Cambridge University Press.

Jackendoff, R. (1999). Possible stages in the evolution of the language capacity. *Trends in Cognitive Sciences, 3*(7), 272–279.

Jowett, G. S. 2006. Brainwashing: The Korean POW controversy and the origins of a myth. In G. S. Jowett & V. O'Donnell (Eds.), *Readings in propaganda and persuasion: New and classic essays* (pp. 201–210). Thousand Oaks: Sage Publications.

Kahneman, D. (2003). A perspective on judgment and choice: Mapping bounded rationality. *American Psychologist, 58*(9), 697–720.

Kitano, H. (2004). Biological robustness. *Nature Reviews: Genetics, 5*(11), 826–837.

Krasne, F. B., & Lee, S. C. (1988). Response-dedicated trigger neurons as control points for behavioral actions: Selective inhibition of lateral giant command neurons during feeding in crayfish. *Journal of Neuroscience, 8*(10), 3703.

Krasne, F. B., Shamsian, A., & Kulkarni, R. (1997). Altered excitability of the crayfish lateral giant escape reflex during agonistic encounters. *Journal of Neuroscience, 17*(2), 709.

Krebs, J. R., & Dawkins, R. (1984). Animal signals: Mind-reading and manipulation? In J. R. Krebs & N. B. Davies (Eds.), *Behavioural ecology: An evolutionary approach* (pp. 390–402). Oxford: Blackwell Scientific.

Kuhn, D., & Lao, J. (1996). Effects of evidence on attitudes: Is polarization the norm? *Psychological Science, 7*, 115–120.

Kuhn, D., Shaw, V. F., & Felton, M. (1997). Effects of dyadic interaction on argumentative reasoning. *Cognition and Instruction, 15*, 287–315.

Langford, D. J., Crager, S. E., Shehzad, Z., Smith, S. B., Sotocinal, S. G., Levenstadt, J. S., et al. (2006). Social modulation of pain as evidence for empathy in mice. *Science, 312*(5782), 1967.

Lanzetta, J. T., & Englis, B. G. (1989). Expectations of cooperation and competition and their effects on observers' vicarious emotional responses. *Journal of Personality and Social Psychology, 56*(4), 543–554.

Laughlin, P. R., & Ellis, A. L. (1986). Demonstrability and social combination processes on mathematical intellective tasks. *Journal of Experimental Social Psychology, 22*, 177–189.

Lord, C. G., Ross, L., & Lepper, M. R. (1979). Biased assimilation and attitude polarization: The effects of prior theories on subsequently considered evidence. *Journal of Personality and Social Psychology, 37*(11), 2098–2109.

Mackie, G. (2006). Does democratic deliberation change minds? *Politics, Philosophy & Economics, 5*(3), 279.

Mahoney, M. J. (1977). Publication prejudices: An experimental study of confirmatory bias in the peer review system. *Cognitive Therapy and Research, 1*(2), 161–175.

Martin, P. Y., Laing, J., Martin, R., & Mitchell, M. (2005). Caffeine, cognition, and persuasion: Evidence for caffeine increasing the systematic processing of persuasive messages. *Journal of Applied Social Psychology, 35*(1), 160–161.

Mascaro, O., & Sperber, D. (2009). The moral, epistemic, and mindreading components of children's vigilance towards deception. *Cognition, 112*, 367–380.

Masserman, J. H., Wechkin, S., & Terris, W. (1964). "Altruistic" behavior in rhesus monkeys. *American Journal of Psychiatry, 121*(6), 584–585.

Maynard Smith, J. (1994). Must reliable signals always be costly? *Animal Behaviour, 47*, 1115–1120.

Maynard Smith, J., & Harper, D. (2003). *Animal signals*. Oxford: Oxford University Press.

McHugo, G. J., Lanzetta, J. T., & Bush, L. K. (1991). The effect of attitudes on emotional reactions to expressive displays of political leaders. *Journal of Nonverbal Behavior, 15*(1), 19–41.

Mercier, H. (2010). The social origins of folk epistemology. *Review of Philosophy and Psychology, 1*(4), 499–514.

Mercier, H. (2011a). On the universality of argumentative reasoning. *Journal of Cognition and Culture, 11*(1–2), 85–113.

Mercier, H. (2011b). Reasoning serves argumentation in children. *Cognitive Development, 26*(3), 177–191.

Mercier, H., & Landemore, H. (2012). Reasoning is for arguing: Understanding the successes and failures of deliberation. *Political Psychology, 33*(2), 243–258.

Mercier, H., & Sperber, D. (2009). Intuitive and reflective inferences. In J. S. B. T. Evans & K. Frankish (Eds.), *In two minds*. New York: Oxford University Press.

Mercier, H., & Sperber, D. (2011). Why do humans reason? Arguments for an argumentative theory. *Behavioral and Brain Sciences, 34*(2), 57–74.

Milgram, S. (1974). *Obedience to authority: An experimental view*. New York: Harper & Row.

Miller, A. G., Michoskey, J. W., Bane, C. M., & Dowd, T. G. (1993). The attitude polarization phenomenon: Role of response measure, attitude extremity, and behavioral consequences of reported attitude change. *Journal of Personality and Social Psychology, 64*(4), 561–574.

Mondillon, L., Niedenthal, P. M., Gil, S., & Droit-Volet, S. (2007). Imitation of in-group versus out-group members' facial expressions of anger: A test with a time perception task. *Social Neuroscience, 2*(3–4), 223.

Moore, T. E. (1982). Subliminal advertising: What you see is what you get. *Journal of Marketing, 46*(2), 38–47.

Morris, J. S., Öhman, A., & Dolan, R. J. (1998). Conscious and unconscious emotional learning in the human amygdala. *Nature, 393*(6684), 467–470.

Moshman, D., & Geil, M. (1998). Collaborative reasoning: Evidence for collective rationality. *Thinking & Reasoning, 4*(3), 231–248.

Neumann, R., & Strack, F. (2000). Mood contagion: The automatic transfer of mood between persons. *Journal of Personality and Social Psychology, 79*(2), 211–223.

Nickerson, R. S. (1998). Confirmation bias: A ubiquitous phenomena in many guises. *Review of General Psychology, 2*, 175–220.

Olsson, A., & Phelps, E. A. (2004). Learned fear of "unseen" faces after Pavlovian, observational, and instructed fear. *Psychological Science, 15*(12), 822.

Petty, R. E., & Cacioppo, J. T. (1979). Issue involvement can increase or decrease persuasion by enhancing message-relevant cognitive responses. *Journal of Personality and Social Psychology, 37*, 349–360.

Petty, R. E., & Wegener, D. T. (1998). Attitude change: Multiple roles for persuasion variables. In D. Gilbert, S. Fiske & G. Lindzey (Eds.), *The handbook of social psychology* (Vol. 1, pp. 323–390). Boston: McGraw-Hill.

Pinker, S., & Bloom, P. (1990). Natural language and natural selection. *Behavioral and Brain Sciences, 13*(4), 707–784.

Pomerantz, E. M., Chaiken, S., & Tordesillas, R. S. (1995). Attitude strength and resistance processes. *Journal of Personality and Social Psychology, 69*(3), 408–419.

Pratkanis, A. R., & Aronson, E. (1992). *Age of propaganda: The everyday use and abuse of persuasion*. New York: W. H. Freeman.

Recanati, F. (1997). Can we believe what we do not understand? *Mind & Language, 12*(1), 84–100.

Richter, T., Schroeder, S., & Wöhrmann, B. (2009). You don't have to believe everything you read: Background knowledge permits fast and efficient validation of information. *Journal of Personality and Social Psychology, 96*, 538–558.

Ruys, K. I., & Stapel, D. A. (2008). Emotion elicitor or emotion messenger? Subliminal priming reveals two faces of facial expressions. *Psychological Science, 19*(6), 593–600.

Scott-Phillips, T. C. (2008). Defining biological communication. *Journal of Evolutionary Biology, 21*(2), 387–395.

Shearn, D., Spellman, L., Straley, B., Meirick, J., & Stryker, K. (1999). Empathic blushing in friends and strangers. *Motivation and Emotion, 23*(4), 307–316.

Shermer, M. (1997). *Why people believe weird things*. New York: Henry Holt.

Singer, T., Seymour, B., O'Doherty, J. P., Stephan, K. E., Dolan, R. J., & Frith, C. D. (2006). Empathic neural responses are modulated by the perceived fairness of others. *Nature, 439*(7075), 466.

Sloman, S. A. (1996). The empirical case for two systems of reasoning. *Psychological Bulletin, 119*(1), 3–22.

Slovic, P. (1993). Perceived risk, trust, and democracy. *Risk Analysis, 13*(6), 675–682.

Sperber, D. (1997). Intuitive and reflective beliefs. *Mind & Language, 12*(1), 67–83.

Sperber, D. (2005). Modularity and relevance: How can a massively modular mind be flexible and context-sensitive? In P. Carruthers, S. Laurence & S. Stich (Eds.), *The innate mind: Structure and contents*. Oxford: Oxford University Press.

Sperber, D., Clément, F., Heintz, C., Mascaro, O., Mercier, H., Origgi, G., & Wilson, D. (2010). Epistemic vigilance. *Mind & Language, 25*(4), 359–393.

Sperber, D., & Wilson, D. (1995). *Relevance: Communication and cognition*. Oxford: Blackwell.

Stahl, S. M., & Lebedun, M. (1974). Mystery gas: An analysis of mass hysteria. *Journal of Health and Social Behavior, 15*(1), 44–50.

Stanovich, K. E. (2004). *The robot's rebellion*. Chicago: University of Chicago Press.

Stark, R. (1996). *The rise of Christianity: A sociologist reconsiders history*. Princeton, NJ: Princeton University Press.

Stark, R., & Bainbridge, W. S. (1980). Networks of faith: Interpersonal bonds and recruitment to cults and sects. *American Journal of Sociology, 85*(6), 1376–1395.

Sterelny, K. (2012). *The evolved apprentice: How evolution made humans unique*. Cambridge, MA: MIT Press.

Strahan, E. J., Spencer, S. J., & Zanna, M. P. (2002). Subliminal priming and persuasion: Striking while the iron is hot. *Journal of Experimental Social Psychology, 38*(6), 556–568.

Streatfeild, D. (2007). *Brainwash: The secret history of mind control*. New York: Thomas Dunne Books.

Sunstein, C. R. (2002). The law of group polarization. *Journal of Political Philosophy, 10*(2), 175–195.

Tobacyk, J., Miller, M. J., & Jones, G. (1984). Paranormal beliefs of high school students. *Psychological Reports, 55*, 255–261.

Tormala, Z. L., & Petty, R. E. (2002). What doesn't kill me makes me stronger: The effects of resisting persuasion on attitude certainty. *Journal of Personality and Social Psychology, 83*(6), 1298–1313.

Vyse, S. A. (1997). *Believing in magic: The psychology of superstition*. New York: Oxford University Press.

Wagner, A. (2005). Robustness, evolvability, and neutrality. *FEBS Letters, 579*(8), 1772–1778.

Wallace, A. (2009). An epidemic of fear: How panicked parents skipping shots endangers us all. *Wired*, October 19.

Weir, W. (1984). Another look at subliminal "facts." *Advertising Age*, October 15, 46.

Whalen, P. J., Rauch, S. L., Etcoff, N. L., McInerney, S. C., Lee, M. B., & Jenike, M. A. (1998). Masked presentations of emotional facial expressions modulate amygdala activity without explicit knowledge. *Journal of Neuroscience*, *18*(1), 411.

Zahavi, A., & Zahavi, A. (1997). *The handicap principle: A missing piece of Darwin's puzzle*. Oxford: Oxford University Press.

20 Altruistic Behaviors from a Developmental and Comparative Perspective

Felix Warneken

Humans are not oblivious to the needs of others. Indeed, people will act on the behalf of others, even in situations that are of no obvious immediate benefit to the actor and may involve some kind of cost to him or her. Philosophers and scientists have debated the origins of these altruistic tendencies for centuries. More recently, empirical methods in the behavioral sciences and insights from evolutionary theory have provided some answers and added new questions to this perennial debate about human altruism. Importantly, some of the most illuminating research has broadened the scope beyond the altruistic behaviors exhibited by adult humans, and put it into context by determining their origins, both in terms of the phylogenetic roots and concerning the development of altruistic behaviors in human ontogeny. A powerful method to accomplish this goal is to study human children and compare their behavior to our closest living evolutionary relatives, chimpanzees (*Pan troglodytes*) and bonobos (*Pan paniscus*). By studying young children, scientists can assess the psychological capacities with which humans are equipped early in life that prepare them to develop altruistic behaviors, and by examining their development, scientists elucidate the interplay between these biological predispositions and social learning. Furthermore, studies of chimpanzees and bonobos allow inferences concerning which aspects of human altruism may have already been present in the last common ancestor of apes and humans, versus those components of the human mind that are species-unique and emerged only in the human lineage. In this chapter, I will therefore explore some of the major new insights that have emerged from the behavioral sciences about the origins of human altruism in ontogeny and phylogeny.

Specifically, I will focus on the proximate mechanisms for altruism that underpin behaviors where an individual does something beneficial for another. In contrast to perspectives from evolutionary theory that focus on the fitness consequences of altruistic behaviors, the current framework examines the psychology of altruism: What motivational factors makes individuals want to help others, and what cognitive capacities are required to do so effectively? In this formulation, behaviors that I here consider as "psychologically altruistic" may not necessarily result in lifetime fitness costs at the ultimate level. On the other hand, altruism defined at the ultimate level—in terms of fitness costs and benefits—does not necessarily depend on complex psychological mechanisms, as "a mindless organism can be an evolutionary altruist" (Sober, 2002, p. 17). However, the behaviors of such a "mindless altruist" may be confined to a very restricted set of situations and cannot be modified when confronted with novel exigencies. Thus, I here want to emphasize that one reason to pay special attention to the proximate mechanisms is that humans engage in a variety of altruistically motivated behaviors that are likely unparalleled anywhere else in the animal kingdom—even if the fitness consequences of these behaviors do not fundamentally differ from those in other species. That is, humans appear to possess psychological processes to perform flexible, situation-specific acts of helping, sharing, and other types of altruistically motivated behaviors that make humans special—perhaps unique—at the proximate level. By studying these, we can learn more about the factors that enable and constrain certain types of altruistically motivated behaviors and explain to what extent different forms are human-unique or mirror behaviors that are found in other animals as well.

Consequently, here I propose that human altruistic behaviors are not based on a unitary psychological trait, but that humans engage in a variety of altruistic behaviors that likely recruit diverse psychological mechanisms. Specifically, recent research indicates that humans and other apes may display greater or lesser altruistic tendencies in different domains of activity. Warneken and Tomasello (2009a,b) proposed the following typology to classify different altruistic behaviors based on the commodity provided to the recipient: *comforting* others by providing emotional support, *sharing* valuable goods such as food, *informing* others by providing useful information, and *helping* others to achieve goals by acting for them. Traditionally, developmental psychology has focused on comforting (or empathic intervention), but in recent years, several new empirical insights have emerged concerning the latter three categories, often complemented by comparative studies with chimpanzees. Therefore, in the current chapter, I will focus on recent findings on sharing, informing, and helping, starting with developmental studies with children and moving on to comparative studies with chimpanzees. I will use these studies to develop the argument that human altruism is not due to socialization alone. Specifically, it has been proposed in many places that a long period of socialization practices such as external rewards by adults, explicit teaching, or the internalization of social norms through imitation are the main or only factor inculcating altruistic motivations into otherwise self-focused and selfish children (Bar-Tal, 1982; Dovidio & et al., 2006; Henrich et al., 2005). In contrast to this view, I will argue that humans have a biological predisposition to develop altruistic behaviors, and that socialization practices can build on this predisposition, which is apparent early in human ontogeny and is also expressed in certain forms of altruistic behaviors in chimpanzees. I will conclude this chapter with a proposal to investigate how factors that are important mediators of altruistic behaviors in adults, such as reciprocity and cultural norms, begin to play a role during the emergence of altruistic behavior in childhood.

Sharing

Sharing of resources can be regarded as perhaps the prototypical form of altruistic behavior. By definition, it involves an immediate cost, as one person is giving up a valuable resource that benefits another individual. Obviously, the effects on biological fitness are difficult to establish in most cases, and concrete resources can serve only as a proxy at best. Nevertheless, it can provide insight into the mechanims involved, that is, the motivation to sacrifice a concrete resource for the benefit of another individual. Correspondingly, sharing of resources is the most studied form of altruism among anthropologists investigating food-sharing in hunter-gatherers, behavioral economists using the dictator or the ultimatum game, or biologists observing food-sharing among animals. Despite the importance of this behavior, surprisingly few systematic studies have been conducted with young children, which invites the question: To what extent are young children able to detect that other individuals lack a resource and are willing to share with them—even if doing so involves a cost to themselves?

One of the few studies investigating this question is an experiment by Brownell and colleagues, which indicates that 2-year-old children begin to take into account another person's needs when they have the opportunity to share with others (Brownell, Svetlova & Nichols, 2009). Specifically, toddlers were confronted with an apparatus offering the choice to either pull one rope that would deliver a snack to themselves and a snack to an adult bystander (1/1 option) or to pull another rope that would bring a snack to themselves, but no snack to the bystander (1/0 option) (task adapted from Silk et al., 2005). Thus, the children could provide food to the bystander at no cost to themselves (in terms of either resources or effort). Interestingly, whereas 18-month-old children chose randomly, 25-month-old children more often chose the 1/1 option, benefiting both themselves and the bystander simultaneously. Notably, they did so only in a condition in which the bystander had verbalized her desire for the food, indicating that young children require explicit cues to note the other person's need in this context. This shows the importance of cues provided by the recipient to elicit sharing. This is also the conclusion from a study by Dunfield and colleagues, in which 18- and 24-month-old children performed costly sharing acts by giving some of their own food to an adult who expressed a desire by making a sad face and requesting sharing with a palm-up gesture (Dunfield et al., 2011). Taken together, these experiments show that young children are willing to share in some circumstances and that the absence of sharing might not always be due to a reluctance to give up a resource, but rather to the lack of social cognitive understanding about the other person's need.

Later on during development, children begin to engage in sharing with absent individuals (for an overview see Gummerum, Hanoch & Keller, 2008). For example, Moore (2009) shows that, even though the recipient was not present during the test, 4- to 5-year-old children more often chose equal rewards for both themselves and another child over a selfish option with a higher payoff for the themselves only—at least when the recipient was a friend. Toward school-age, costly sharing becomes more common, and also occurs in situations in which children have the choice to share with anonymous others. Blake and Rand (2010) point out that their own experiment and other studies using the dictator game (in which the subject divides up a resource between oneself and another individual) indicate that children's tendency to give at least something increases continuously over development between 3 and 9 years of age. Moreover, these experiments in which children allocate actual resources, as well as studies using hypothetical situations, converge on the finding that between 5 and 7 years of age, children most often share according to equality, even if the alternative would be to obtain a larger reward for themselves (Blake & Rand, 2010; Damon, 1977; Fehr, Bernhard & Rockenbach, 2008; Hook & Cook, 1979). Thus, from early on in ontogeny, children begin to share resources with others. Initially, it appears crucial that the recipient is copresent and expresses her desire. It seems that later during development, children begin to share with absent individuals, adhering to equality norms even if this requires self-sacrifice on the part of the actor.

This finding stands in contrast to several studies with chimpanzees, who in similar sharing situations do not seem to act on behalf of others even if the costs to themselves are minimal or nonexistent. Specifically, in Silk et al. (2005), chimpanzees chose randomly when confronted with the 1/1 and 1/0 option, and in Jensen et al. (2006), chimpanzees did not act altruistically even when doing so would only have required pulling a rope that moved an otherwise unobtainable board with a piece of food toward another conspecific (see Vonk et al., 2008, for variations on this food-provision paradigm leading to similar results). Thus, these studies indicate that chimpanzees are not particularly inclined to actively share resources with others even if this would come at no cost to themselves.

Chimpanzees appear to be quite limited in their tendency to share resources, even in mutualistic situations in which both individuals could potentially benefit. For example, studies by Melis and colleagues showed that in problem-solving tasks in which chimpanzees can only succeed if they together simultaneously pull in a board with food on it, they mostly failed when the food was clumped in the middle of the board, making it easy for the more dominant individual to monopolize. They only succeeded when the rewards were spread apart so that each individual was able to access their portion without potential interference from the other chimpanzee. Moreover, success was highly dependent on the degree of social tolerance between the partners: The only dyads who cooperated successfully on this mutualistic task were those who had in pretests co-fed peacefully over a food resource rather than trying to monopolize it (Melis, Hare & Tomasello, 2006a). Thus, social tolerance appears to be an important constraining factor for chimpanzee cooperation.

The importance of social tolerance over food is also highlighted by cross-species comparisons. Specifically, bonobos are known to be generally more socially tolerant over food, as indexed by their tendency to co-feed without individuals trying to monopolize a resource. As a consequence, when they were tested in the same board-pulling task as the chimpanzees, the location of the food as either clumped in the middle or spread apart did not matter. In contrast to chimpanzees, who mainly succeeded when the rewards were spread apart, bonobos were equally likely to succeed in the clumped and the dispersed version (Hare et al., 2007). Thus, probably because there was no expectation of competition over the spoils at the end, it was of less importance whether the rewards were monopolizable or not. These experiments demonstrate that chimpanzees are highly competitive over monopolizable resources and are thus less inclined to actively provision food for others or cooperative mutualistically with others.

Naturalistic observations have also shown that when sharing occurs in chimpanzees, it predominately consists in passive sharing, in which one individual lets the other individual take a resource, rather than transferring it to the other proactively (Boesch & Boesch, 1989). In particular, meat-sharing after cooperative hunts occur mainly in response to harassment (Gilby, 2006) in which the possessor lets another individual have part of the bait. This is probably not due to generosity, but because sharing parts of the carcass is less costly than trying to defend it, that is, risking getting in a fight or potentially losing the carcass altogether. Also, in mother–infant pairs, infant-initiated food transfers in which the infant attempts to take food and the mother tolerates it are more common than mother-initiated food transfers (Ueno & Matsuzawa, 2004). Moreover, the latter occurred exclusively with less desirable parts of the fruits such as seeds and husks.

Thus, striking differences between humans and chimpanzees emerge when it comes to resource sharing. Chimpanzees appear to be less inclined to actively provide food, especially when it involves giving up a valuable resource. They appear to have a strong tendency to monopolize valuable resources, which often precludes cooperative behaviors, including those that would result in mutualistic outcomes. Interestingly, this constraint appears to be mitigated in bonobos, who are better able to deal with situations that involve monopolizable resources. Human children are willing to share resources, even if it involves a cost, but at least early in development, sharing occurs mainly in situations in which recipients actively express their need; older children demonstrate proactive sharing behaviors, indicating that sharing not only requires the willingness to give up a resource, but also the social-cognitive capacity to understand other people's needs and desires.

Informing

Another important form of altruistically motivated behavior becomes apparent in the gestural communication of young children: Infants will point to things other people are searching for. These kinds of behaviors are obviously of low immediate cost (requiring neither great effort nor sacrifice of material resources), but they provide another example for an altruistic motivation to act on behalf of another person's goal or need. Specifically, a series of studies demonstrates that from the time that infants begin to point (at around one year of age), they already use this simple but efficient communicative means for altruistic purposes by providing helpful information to others. In one such study, a protagonist sat at a desk and used an object such as a hole-puncher on a stack of papers (Liszkowski et al., 2006). After she had left, a second "evil" experimenter took the hole-puncher as well as an irrelevant object and placed them on different platforms behind the desk out of the protagonist's view. Upon returning to the desk to punch more papers, the protagonist looked around in bewilderment, unable to locate the hole-puncher. Infants pointed more often to the hole-puncher than the distracter object, indicating that they were trying to help her in her search and were able to differentiate between the relevant and the irrelevant object. A follow-up study demonstrated that 1-year-old children can also differentiate between an object that disappeared unbeknownst to the protagonist and an object that she saw disappear (Liszkowski, Carpenter & Tomasello, 2008). Thus, prelinguistic infants use pointing not only imperatively to make other people do things for themselves, but also to provide information that other people need. This form of altruism is interesting because it requires the cognitive capacity, on some level, to differentiate between one's own state of knowledge about the world and that of another person's, and to provide the lacking piece of information to the ignorant other. Thus, in contrast to comforting (based on a determination and intervention toward the emotional states of others), and in contrast to sharing (based on the provision of a concrete resource), altruistic acts of informing others critically depend on a cognitive capacity to take into account and act on epistemic states.

This rather sophisticated cognitive capacity required for acts of informing might explain why, despite the low costs of the act itself, this form of altruistic behavior is not ubiquitous across the animal kingdom. More specifically, chimpanzees do not appear to have developed a communicative system that would allow for such acts. Chimpanzees do not point for each other (Leavens, 2004; Tomasello, 2006). Captive chimpanzees do point in interactions with humans, but it seems like they do so only imperatively, to entice them to do things for them such as bringing an object; there is no conclusive evidence that they point in order to provide information needed by another individual (Tomasello, 2008). More concretely, in a study by Bullinger et al. (2011), chimpanzees reliably pointed to a misplaced tool that they needed to extract a reward for themselves (a selfish "for-me" condition), but rarely did so when an experimenter needed the tool to retrieve a reward for herself (an altruistic "for-you" condition). In an experiment closely matching this scenario, 2-year-old children pointed equally often in both conditions, demonstrating that they use pointing gestures for selfish and altruistic purposes. The interpretation by Bullinger and colleagues is that in contrast to human children, chimpanzees do not engage in informative pointing with an altruistic motivation to help the other, and in fact do not even use pointing as a means to convey a piece of information the ignorant experimenter is lacking (letting the experimenter know where the tool was), but rather point in order to direct her behaviorally to the tool. Thus, they do not aim to change the epistemic states of others, but try to directly influence the other's behavior for their own purposes.

As a matter of fact, chimpanzees do not even seem to be able to comprehend pointing gestures when they would be useful to them. A large number of experiments demonstrates that chimpanzees fail to use gestural cues in object-choice paradigms in which an experimenter is using a pointing gesture, or, in a variation of this, placing a marker to indicate the location of food under one of two opaque containers (see Call & Tomasello, 2005, for an overview). One-year-old human children have no problem with this (Behne, Carpenter & Tomasello, 2005), but chimpanzees appear to be unable to interpret the helpful communicative gestures of others. Interestingly, when the same situation (in which a piece of food is in one of two opaque containers) is framed as a competitive one, chimpanzees are suddenly successful. Hare and Tomasello (2004) directly compared a competitive and a cooperative version of this object-choice task: When chimpanzees saw a competitor (human or chimpanzee) unsuccessfully reaching for one of two containers with a hand gesture very similar to pointing, they were able to infer that this one was the container with food in it, and chose accordingly when it was their turn. However, when a cooperative experimenter pointed to the correct container, they chose at random. Thus, chimpanzees were able to infer the location of the food based on the other's attempt to snatch the food but did not understand the communicative intention to inform them of the correct location.

Why can't chimpanzees comprehend communicative cues produced by a helpful experimenter? One possible explanation for why chimpanzees fail is that they do not seem to grasp the helpful communicative intent of others. Specifically, these gestures become meaningful only under the premise that the receiver views the pointing gesture as part of a joint collaborative activity, with the receiver being able to understand that the sender has the helpful communicative intention to guide their searching—that you are producing this gesture for me to help me find the object (for details, see Tomasello, 2006, 2008). At the same time, chimpanzees do understand on a practical level that other humans can be helpful (or at least useful). This is highlighted in examples such as imperative pointing (to have a human give them something) or imperative giving (like handing a container so that the human opens it [Tomasello, 2008]). Taken together, although the reasons are rather unclear, the experimental evidence suggests that chimpanzees appear to have a fundamental lack of understanding about gestures as devices that can be used to convey helpful information, both when they are in the role of the sender and in that of the receiver.

Instrumental Helping

Human altruistic behavior is expressed not only in the tendency to, on occasion, share material resources or information, but also in such mundane acts as picking up a dropped object, holding open a door for someone, or trying to fix something for others if they fail to succeed. To engage in these behaviors in a competent way, the helper has to be able to represent the other person's goal and to have the motivation to act on behalf of the other's goal. These behaviors—for which I use the term "instrumental helping"—differ from sharing and informing, as different commodities are involved: Rather than giving up a resource or providing a piece of information, the helper assists instrumentally by contributing to the other's goal fulfillment through his own action. These behaviors also differ from empathic intervention (such as comforting a person who is in distress) because, rather than being affected by the emotional state of others and acting in response to this state, acts of instrumental helping require a cognitive understanding of goal-directed action.

Do young children engage in acts of instrumental helping? With regard to the social-cognitive component of helping, it is a well-established finding that infants from around 12 to 18 months of age understand other people's behaviors in terms of their intentions. For instance, infants can differentiate accidental from purposeful actions (for an overview, see Tomasello et al., 2005). They can determine whether a similar environmental outcome was either the result of a purposeful act (when the outcome matches the goal) or an accident (when goal and outcome do not match). Similarly, they can infer the goals that another person was trying to achieve without actually witnessing the intended outcome (Meltzoff, 1995). Thus, infants already appear to possess the crucial social-cognitive component necessary for instrumental help.

With regard to the motivational component of helping, children as young as 12 months begin to show concern for others in distress and sometimes intervene by comforting them (for an overview, see Eisenberg, Fabes & Spinrad, 2006). However, these instances are all based on the infant's empathic responses to the emotional needs of another person. In contrast to these instances of empathic intervention (or what one could also call "emotional helping"), little was known about whether infants would also perform acts of instrumental helping: helping another person achieve an unfulfilled goal.

To explore this issue, Warneken and Tomasello (2006) presented 18-month-old infants with ten different situations in which an adult was having trouble achieving a goal. This range of tasks presented the children with a variety of difficulties in discerning the adult's goal. For instance, in one task, while using clothespins to hang towels on a clothesline, an experimenter accidentally dropped a clothespin on the floor and unsuccessfully reached for it. In this case, the child had the opportunity to help by picking up the clothespin and handing it to the experimenter. In another task, the experimenter was trying to put a stack of magazines into a cabinet but could not open the doors because his hands were full; the child could thus potentially help by opening the doors for him. The finding of this experiment was that children displayed spontaneous, unrewarded helping behaviors when another person is unable to achieve his goal (and performed these behaviors significantly less often in control conditions where no help is necessary). This helping was observed across a range of situations: Infants handed the experimenter out-of-reach objects; completed an action after his failed attempt at stacking books; opened the door of a cabinet for him; and brought about the experimenter's goal by different means: Rather than using the experimenter's wrong approach to try to squeeze his hand through a tiny hole in order to retrieve an object from a box, children lifted a flap on the side of the box and gave the object to him. In another study, even 14-month-old children acted helpfully, although only with cognitively less demanding tasks (such as a person reaching for an object), but failed to do so in situations with presumably more complex goals and more complex types of intervention (like the cabinet task, [Warneken & Tomasello, 2007]). This is also exemplified in a study by Svetlova and colleagues, in which children from 1.5 to 2.5 years were more likely to help with concrete goals (such as reaching for a dropped object) than with tasks that involved at least one more inferential step in how to help (such as bringing a blanket to a person who is shivering) (Svetlova, Nichols & Brownell, in press). In order to help appropriately, younger children in particular more often needed explicit communicative cues from the helpee in the latter kind of situation than in tasks with concrete action-goals. Taken together, these studies show that shortly after their first birthdays, human children begin to spontaneously help others, becoming more flexible in their ability to intervene in various types of situations over the second year of life, and requiring fewer direct communicative cues also in contexts in which the type of intervention does not follow straightforwardly from situational cues alone.

During the second year of life, children even become able to infer another's goal based on what the person does or does not know. Specifically, in one helping situation (Buttelmann, Carpenter & Tomasello, 2009), a protagonist put his toy into box A, which was then moved to box B either while the protagonist was absent (ignorant condition) or present (knowledge condition). In the ignorant condition, when the protagonist tried but failed to open box A, children did not help him to open box A (the previous location of the toy) but instead opened box B (where the toy actually was), indicating that they did not just blindly join into the erroneous action to open box A, but inferred that the protagonist was actually trying to get at the toy (which was now in a new place). Children in the knowledge condition, however, were more likely to open box A that the protagonist was trying to open, presumably reasoning that he was not trying to get at his toy (which he witnessed having moved to box B) but must have another goal in mind. Therefore, this study indicates that young children actually help other people with their goals (and not just complete a concrete action such as the failure to open something) and are able to infer goals from representations about the other person's state of knowledge or ignorance.

These young children help even when the costs of helping are slightly raised. As an example, when children are having fun playing with a novel toy (a box with buttons that light up and produce sounds), they continue to help even if it means they have to disengage from this activity (Warneken & Tomasello, 2008). As another example, children do not stop helping over repeated trials even if they have to surmount an array of obstacles, something that can be quite challenging for a toddler who just learned to walk (Warneken et al., 2007). Children are thus motivated to help another person even if it involves opportunity costs or effort to do so.

Children are thus willing to put some effort in helping—but do they expect to be rewarded in return? One experiment directly addressed this question of whether 18-month-old children are motivated by the other person's goal or by an immediate benefit for themselves by varying whether the helpee would offer a reward in return for their helping effort (Warneken et al., 2007). The results could not be clearer: Children helped by picking up objects for which the experimenter was unsuccessfully reaching, irrespective of whether they received a rewarded for their help. Rewarding was not necessary to elicit helping, nor did it increase the rate of helping. Thus, what determined children's helping was the other's unfulfilled goal, not an immediate benefit for themselves. Using a crucial distinction from motivational psychology, we may thus ask whether such acts of helping are intrinsically or extrinsically motivated: Do children help one another because the helpful act itself is inherently rewarding, or only because the helpful act is instrumental in bringing about separate outcomes such as material rewards? To investigate the effects of rewards on young children's helping more closely, Warneken and Tomasello (2008) took advantage of a curious feature of intrinsic motivation: It is a well-established phenomenon that intrinsic rewards can be undermined by salient extrinsic rewards (what has also been called the "overjustification effect") (Deci, 1971; Lepper, Greene & Nisbett, 1973). Warneken and Tomasello found that children who had received a material reward for helping during an initial test phase were subsequently less likely to engage in further helping than those children who had not received such a reward (see also Fabes et al., 1989, for school-aged children). This rather surprising finding provides even further evidence for the hypothesis that children's helping is driven by an intrinsic rather than an extrinsic motivation. Rewards are often not only superfluous, but can even have detrimental effects as they may undermine children's intrinsic altruistic motivation.

In sum, this series of studies demonstrates that the ontogenetic origins of altruistic helping are apparent in early childhood. Infants as young as 14 months of age display spontaneous, unrewarded helping behaviors when another person is unable to achieve his goal. Throughout the second year of life, children become increasingly flexible in their ability to read others' intentions and intervene in different kinds of situations. Human infants use their emerging mind-reading capabilities not only for their own ends, but also to help others. They are willing to help multiple times and continue to help when the costs for helping are raised. Further experiments confirm that infants are actually motivated by the other's goal and not by an immediate benefit for themselves, as external rewards are not necessary to elicit helping nor do they increase the rate of helping. On the contrary, children appear to have an initial inclination to help that maybe be diminished by extrinsic rewards.

Would chimpanzees perform such acts of instrumental helping? A series of experiments shows that under certain circumstances, chimpanzees do in fact act on behalf of others. In an initial study, Warneken and Tomasello (2006) adapted the helping scenarios originally used with human infants and found that human-raised chimpanzees helped their human caregiver even in the absence of an external reward. Specifically, they handed her objects for which she was unsuccessfully reaching (and they did not do so in the control condition, in which she was not reaching for them). These chimpanzees were thus able to determine the experimenter's goal and had the motivation to help her with the goal in the absence of a reward. However, in contrast to the out-of-reach tasks, these chimpanzees did not help reliably in the other types of tasks (opening a door for the other, using different means to open the novel box for the other, etc.). Perhaps importantly, this discrepancy in chimpanzees' willingness to help in different types of tasks parallels a finding with 14-month-old human infants, whose helping was also limited to tasks involving out-of-reach objects (Warneken & Tomasello, 2007). These findings suggest that the occurrence of helping may be influenced by the complexity of the goal structure and the type of intervention necessary. This once again shows the importance of taking into account the social-cognitive demands of certain tests of altruism. These human-raised chimpanzees were in principle willing to help, but showed this behavior only in contexts in which the other's goal was easy to identify.

It should be noted, though, that the subjects were human-raised chimpanzees who helped a caregiver with whom they maintained a close relationship. Chimpanzees with such a rearing history often possess superior social-cognitive skills and develop behaviors not found in individuals with less human contact (Bering, 2004; Call & Tomasello, 1996; Tomasello & Call, 2004; Tomasello & Carpenter, 2005). However, in a follow-up study, chimpanzees with a different rearing history also helped. Specifically, Warneken et al. (2007) tested a sample of semi–free ranging chimpanzees who were born in the wild and now live in a sanctuary in Uganda. These chimpanzees spend the day in the forest of an island where they come to a human shelter for feeding and sleeping. They thus have regular contact with humans, but have not been exposed to rearing practices comparable to those of human-raised chimpanzees in zoos (as the chimpanzees from the initial helping study had been). Most importantly, they were tested by a human who had not interacted with them before the experiment.

One major finding of this study was that, just like human infants tested in a similar situation, chimpanzees helped over consecutive trials by handing the out-of-reach object when the experimenter indicated that he was trying to get the object, and they did so irrespective of being rewarded. This indicates that the chimpanzees were motivated to help the experimenter with his unachieved goal, and not by the possibility of retrieving a material reward for themselves. Moreover, chimpanzees continued to help even when doing so involved not only picking up the object, but also required that they first climb into a raceway to retrieve the object for the other (Warneken et al., 2007, Experiment 2). Taken together, these two experiments show that also semi–free ranging chimpanzees perform acts of helping toward a human stranger, even when helping is made effortful and they receive no immediate benefit for themselves.

However, all of these instances of chimpanzee helping involved humans as recipients. The question remains: Will chimpanzees help other chimpanzees? As mentioned before, studies on sharing of resources among conspecifics came to a negative conclusion (Jensen et al., 2006; Silk et al., 2005). However, experiments with negative results all involved the active provision of food, which, owing to the tendency of chimpanzees to compete over monopolizable resources, might preclude altruistic behaviors. Moreover, in these experiments, the need for help might not have been obvious to the subjects, as the recipients often remained passive during the trials, and did not have an opportunity to actively try to access the desired object. To this end, Warneken et al. (2007) created a situation in which the need for help would potentially be more salient, by having a chimpanzee actively struggle with a problem. Namely, one chimpanzee (the recipient) was faced with the problem that a door leading to a room containing a piece of food was fixed with a chain that he could not unlock. Only if the other chimpanzee (the subject) released this chain from another room could the recipient enter. Results showed that chimpanzees helped in the majority of cases by releasing the chain. They did so significantly more often than in control conditions, in which releasing the chain would either not help the recipient or no recipient was present. This shows that subjects were attentive to the recipient's goal: They were more likely to release the chain if the recipient was unsuccessfully trying to enter through that door. Thus, this experiment shows that when the problem is made more salient, chimpanzees use a novel skill to help conspecifics gain access to food in a novel situation. In a related study, Yamamoto and colleagues showed that cues by the recipient are critical in eliciting altruistic behaviors. In their study, chimpanzees passed a tool to another individual who needed it to retrieve food, but did so almost exclusively in situations in which they previously observed the conspecific attempting to reach for the tool (Yamamoto, Humle & Tanaka, 2009). Chimpanzees virtually never handed over the tool proactively in the absence of such a cue. The importance of cues was also found in a study by Melis and colleagues (2010), in which the subject could release a bag containing a reward to allow it to slide down a chute toward a recipient. Chimpanzees performed this behavior more often when the recipient was actively trying to access the reward (by pulling a rope, which in some conditions was attached to the bag) or communicate with the subject, than when the recipient remained passive. In addition, Melis and colleagues manipulated whether the reward was a piece of food or a token (that the recipient could later exchange for food), testing the hypothesis that the presence and necessity to actively provision food might impede helping. However, this study revealed that the tendency to help was not diminished with food rewards (as compared to tokens). Thus, the main factor predicting helping was again the activity of the recipient.

These findings elucidate both the ontogenetic and phylogenetic origins of human altruism. Human children possess social-cognitive capacities that enable them to express these tendencies in a variety of ways, as highlighted by the early emergence of helping, informing, and sharing. These results challenge the view that altruism is imposed by the social environment (Bar-Tal, 1982; Dovidio et al., 2006; Henrich et al., 2005). Infants show altruistic tendencies at an age when socialization could not yet have had a major impact on their development. Also, the internalization of norms or value systems is inapplicable to 1-year-old children. Moreover, we can assume that these infants had had few opportunities to help and receive reinforcement for helping. And even if it is the case that they have had some exposure to helping and reinforcement, they also help in novel situations with unfamiliar adults. Thus, these early achievements are unlikely to be due to socialization practices alone. A more plausible explanation for the various results from these studies is that altruistic acts in young children reflect a natural predisposition to develop these altruistic behaviors. Socialization, in turn, can then build on these early, preexisting tendencies.

These behaviors are not completely absent in chimpanzees, providing further evidence for the notion that the ontogenetic emergence of these behaviors is based on a biological predisposition. Moreover, it indicates that the altruistic tendencies seen in humans have deep phylogenetic roots, potentially dating back at least to the last common ancestor of humans and chimpanzees. Specifically, chimpanzees also on occasion act on behalf of others, specifically in instrumental helping situations. Moreover, the fact that they show some flexibility in helping (toward different recipients who pursued different goals) indicates that chimpanzees can utilize their social-cognitive skills in reading others' goals for altruistic purposes. However, chimpanzee altruism appears to be much more restricted and more fragile than what we see already in young humans. This appears to be due in part to limitations on their social cognition, as they do not seem to engage in the communicative acts that characterize human intentional communication and neither comprehend nor produce gestures aimed at helping other individuals. Moreover, chimpanzees appear to have a strong tendency to monopolize food, which often predominates and in particular precludes instances of sharing. Thus, rather than asking whether altruistic tendencies are present or absent in chimpanzees, research should move forward and explore the specific circumstances under which chimpanzees do or do not act altruistically.

Mechanisms to Sustain and Facilitate Altruism

The studies reported above highlight that young children, and, to some extent, non-human apes, possess the necessary cognitive and motivational capacities to engage in some forms of altruistically motivated behavior. But what are the biological consequences of these types of psychological processes? Evolutionary theory suggests that this type of unconstrained altruism is not an evolutionary stable strategy, because individuals that provide help unconditionally will accrue high fitness costs. Thus, biology suggests that individuals must also be selective, biasing their altruistic efforts only toward certain social partners under certain circumstances. That is, altruistic psychological mechanisms that detect the needs of others and act on them must be complemented by other mechanisms that prevent the altruist from being exploited by others (e.g., Cosmides & Tooby, 1992). Thus, a mature altruist has to possess both the psychological mechanisms to engage in acts of altruism and the psychological mechanisms to decide whom to help and when, in order to be viable at the ultimate level. Multiple theories have proposed how altruistic motivations can be biologically beneficial at the level of lifetime reproductive success. For example, individuals might direct altruistic acts toward kin (Hamilton, 1964) or toward those who reciprocate over those who reap the benefits without reciprocating in the future (Trivers, 1971). In this section, I thus explore the development of these altruistic "safeguards."

Importantly, these safety measures need not be in place ontogenetically at the same time when children begin to engage in altruistic behavior. For instance, the ability to detect cheaters who profit from helpful acts but do not reciprocate in the future is potentially of less relevance early in ontogeny when children are mainly surrounded by family members and protected by their parents. The ability to distinguish between altruists and cheaters probably becomes important only later in life, as interaction with strangers increases. Thus, some proposals suggest that children start out as rather indiscriminate altruists who become more selective as they grow older (Hay, 1994; Warneken & Tomasello, 2009a). Specifically, children's emerging social-cognitive understanding and new experiences should lead to more selective helping, informing, and sharing. Here I will therefore address whether (or at what point in development) such selectivity characterizes the social behaviors of human children—and to what extent this selectivity characterizes the social behaviors of chimpanzees. Many models from evolutionary theory and empirical findings from social psychology and behavioral economics with adults have focused on direct and indirect reciprocity as important mechanisms that support cooperative behaviors. Direct reciprocity refers to cases in which commodities are exchanged between two individuals over repeated encounters (by, for example, taking turns in helping each other over time). In cases of indirect reciprocity, two individuals might not directly interact again, but the interaction is registered by third parties, and thus altruistic acts (or lack thereof) can affect one's reputation and thus future opportunities to interact positively with others (Nowak, 2006). As such, I will here focus on when during ontogeny these factors begin to influence altruistically motivated behaviors in children.

Even though direct reciprocity as originally proposed by Trivers (1971) was conceived as a simple way to sustain cooperation in a way that was mutually beneficial in the long term, more recent examinations of reciprocal strategies such as Tit-for-Tat have suggested that they may actually be quite cognitively demanding. For example, as the temporal delay between paying a cost to help others and ultimately recouping it in terms of benefits received, psychological capacities including delay of gratification (in which individuals forgo an immediate benefit in favor of a higher future benefit), identifying those individuals with whom one will interact again in the future, and the ability to detect cheaters become increasingly important (Stevens & Hauser, 2004). Moreover, reciprocal behaviors can have diverse psychological bases, ranging from symmetrical relations where two individuals associate at high rates and thus show apparent reciprocity, to more contingent reciprocity where individuals account for the costs and benefits accrued in their relationship (Brosnan & de Waal, 2002). When do children show these capacities and start to engage in broadly reciprocal behaviors? To what extent are these behaviors based on mental accounting of others' actions and their consequences?

In developmental psychology, reciprocity has been mainly studied in the domain of sharing and distributive justice. When asked about their explicit judgments, preschoolers at 4 to 5 years of age show no understanding of reciprocity (Berndt, 1979) and do not refer to reciprocity in their justifications about how a resource should be divided before 6 to 7 years of age (Damon, 1975; De Cooke, 1992). Using a simpler paradigm, Olson and Spelke (2008) demonstrated that when 3.5-year-olds are asked how a puppet character should divide up a resource between two other characters, they advise giving more to the character who had shared with the puppet previously versus to the one who hadn't shared. Thus, at this age children appear to possess some basic notion of reciprocity in the sense of two individuals treating each other alike. However, it is unclear how these judgments of others' actions translate into young children's *own* altruistic behaviors, such as sharing their own resources or putting effort into helping. In fact, there is only one correlational study on sharing in 3-year-olds indicating that they perhaps engage in reciprocal altruism. Specifically, when one child had lots of toys and the other one was toy deprived, only those "rich" children who shared with a "poor" child after prompts from their mother in the initial phase of the experiment received toys back when roles of "rich" and "poor" subject were switched in the second phase (Levitt et al., 1985). Concerning instrumental helping, one experimental study shows that 21-month-old infants will selectively help a person who was willing but unable to share a toy with them over a person who was unwilling to do so (Dunfield & Kuhlmeier, 2010). Specifically, even though during the exposure phase, the outcome of either person's action was the same (the child didn't receive the toy), children were attentive to the "nice" person's intention who was trying but failing to give them a toy (which accidentally rolled away) in contrast to the "mean" person's intention (who offered the toy, but then withdrew it in a teasing fashion). During the subsequent test phase, both persons reached for a single object, and results showed that children preferentially gave it to the previously "nice" unable person over the "mean" unwilling person. Thus, children kept track of how each person had behaved toward them before and adjusted their helping accordingly, thus demonstrating that they possess a critical capacity required for contingent reciprocal altruism by keeping track of who had had helpful intentions in the past. Nevertheless, it is still unclear whether children engage in calculated reciprocity by performing an altruistic act in the present in anticipation of reciprocation in the future (Brosnan & de Waal, 2002; de Waal & Luttrell, 1988).

Regarding indirect reciprocity, recent research has examined how infants and children encode third-party interactions, as well as observing how dyadic interaction influences how the children direct their subsequent altruistic behaviors. For example, infants in the first year of life differentiate between "helpers" and "hinderers," namely agents displayed as geometric shapes who either assist another agent to climb up an incline or push it down (Kuhlmeier, Wynn & Bloom, 2003), and show a preference for the helper by touching the helper over the hinderer in a forced-choice paradigm (Hamlin, Wynn & Bloom, 2007). At least by 3 years of age, children use such discriminatory abilities in their own altruistic behaviors by avoiding the provision of help to mean people: After they have observed how one "mean" adult harmed another individual, they are less likely to subsequently help the mean adult over a helpful or neutral person. This includes situations in which the mean person tried but failed to harm another individual, demonstrating that children actually pay attention to the intentions rather than just an actual harmful or nonharmful outcome (Vaish, Carpenter & Tomasello, 2010). Therefore, this small number of studies on direct and indirect reciprocity indicates that between 2 and 3 years of age, children begin to selectively help those who had good intentions toward them, and also selectively help those based on their intentions toward others in third-party interactions.

Finally, another critical component of indirect reciprocity is not just how individuals use third-party information to decide with whom to interact and provide benefits, but also how individuals manage their own reputation so they can receive assistance from others. That is, how is children's behavior influenced by concerns about how their display of altruistic behavior (or lack thereof) might affect their own future interactions with others? To the best of my knowledge, there is no study specifically addressing this issue with regard to altruistic behaviors, but related findings come from studies of children's self-presentation behaviors and dispositional praise. What becomes apparent from these studies is that we cannot expect young children to strategically manage their reputation as it appears in the eye of the beholder. Self-presentation behavior—in which individuals try to shape others' evaluations of their public self (also called "impression management")—does not appear to be used (Aloise-Young, 1993) or understood (Banerjee, 2002) by children before around 8 years of age. Moreover, only after around 8 years of age is children's prosocial behavior affected by dispositional praise. Specifically, only the prosocial behaviors of these fairly old children are influenced positively when adults provide internal attributions for the prosocial act, highlighting the child's altruistic personality by stating that the child is "a nice and helpful person" (e.g., Grusec & Redler, 1980). This finding indicates that changes in children's self-image mediate prosocial behavior only after they have gained an understanding of personality traits as stable entities (Eisenberg, Fabes & Spinrad, 2006). Thus, future studies must address when during their development children's altruistic behaviors begin to be mediated by reputational concerns.

It is still a matter of debate whether chimpanzees have the capacity to engage in reciprocal altruism (Hauser, McAuliffe & Blake, 2009; Mitani, 2006; Stevens & Hauser, 2004; Schino & Aureli, 2010; for social animals more generally, see Hammerstein, 2003, and Clutton-Brock, 2009). Correlational studies indicate that individuals will engage in reciprocal relationships (e.g., de Waal, 1997; Koyama, Caws & Aureli, 2006; Muller & Mitani, 2005), although it remains contested whether this can be explained by contingent reciprocity with a causal relationship between giving and receiving favors. One experimental study indicated that there might be a weak tendency for contingency-based reciprocation (Melis, Hare & Tomasello, 2008). Using the door-opening paradigm from Warneken et al. (2007) described above, chimpanzees tended to help a previously helpful individual more than an unhelpful one. This indicates that contingency-based reciprocation within a short time-interval can slightly alter the basic tendency to help conspecifics, but it accounts for only a small proportion of the variance (as in correlational studies, de Waal, 1997; Koyama, Caws & Aureli, 2006). However, again in a food-provision context without active solicitation from a recipient similar to those used by Silk et al. (2005) and Jensen et al. (2007), chimpanzees do not appear to take advantage of the opportunity to engage in contingent reciprocity (Vonk et al., 2008; Brosnan et al., 2009; Yamamoto & Tanaka, 2010). It is not clear why contingent reciprocity does not appear to emerge as a consistent strategy in chimpanzees (at least when tested in short-term interactions in the laboratory). Chimpanzees possess the critical components necessary for contingent reciprocity, including the ability to delay gratification (Rosati et al., 2007), detect whether someone was unwilling or unable to give food to them (Call et al., 2004), help others (see above), retaliate against uncooperative individuals (Jensen, Call & Tomasello, 2007) and select partners for mutualistic cooperation (Muller & Mitani, 2005, for observational studies in the wild; Melis, Hare & Tomasello, 2006b, for experimental evidence). It is possible that contingent reciprocity operates over longer time frames than those employed in experiments (Melis, Hare & Tomasello, 2006b; Schino & Aureli, 2009). It is also possible that in most situations that could initiate bouts of reciprocal exchange, immediate selfish interests predominate: The myopic tendency to accrue resources that interferes with sharing more generally also jeopardizes the long-term benefits that could be gained from reciprocity. Thus, although there is evidence that altruistic and other cooperative interactions are characterized by reciprocity, it remains unclear what mechanisms account for these patterns. In particular, there is no strong experimental evidence that chimpanzees engage in bouts of contingency-based reciprocation.

Beyond the issue of direct reciprocity, do chimpanzees cooperate with others based on their reputation? Do they take into account how their (un)cooperative behavior affects their own reputation in the eye of the beholder? To the best of my knowledge, there is currently no published experiment that speaks to these issues. However, some experimental evidence indicates that chimpanzees learn about the cooperative behaviors of others (broadly construed) through third-party observation. Specifically, in Subiaul et al. (2008), after being trained over several sessions to differentiate between a "nice" and a "mean" experimenter who either gave them food or refused to do so in a direct interaction, chimpanzees observed how a novel pair of "mean" and "nice" experimenters treated another chimpanzee in a feeding situation. When the subject now had the opportunity to beg from these two new individuals, they showed a preference for the previously "nice" person, indicating that they had learned from observing other people's behaviors. Russell, Call, and Dunbar (2008) came to a similar conclusion based on an experiment in which chimpanzees first monitored an interaction among three humans, in which a "nice" person gave food to a human beggar and a "mean" person slapped the beggar's hand when he reached for the food. This was observed in four incidents. Results showed that chimpanzees then approached the previously "nice" over the "mean" experimenter when both had food in front of them, presumably because they thought that they would have more luck with him. Taken together, these experiments suggest that chimpanzees learn from observing third-party interactions, and that they apply this information to feeding situations in which they are in the position of the recipient. However, it is not totally clear from these results that the chimpanzees learn that the experimenter is a social agent with "nice" or "mean" intentions or just a more or less reliable food dispenser. Last but not least, even if one grants chimpanzees the ability to attribute these social qualities to the two agents and use it for their own choice with whom to interact, these results do not address the question of whether chimpanzees will alter their *own* cooperative behavior after observing others' cooperation, let alone whether they are influenced by the knowledge that others are watching them. Insight into the relationship between reputation formation and cooperative behaviors remains an important focus for future studies.

Conclusion

I have attempted to provide a comprehensive overview of the variety of human altruistic behaviors as they emerge in human ontogeny and phylogeny, focusing on the development of young children and comparative studies with chimpanzees and bonobos. Recent studies indicate that humans and other apes share some of the core capacities required for altruistic behaviors, especially in the form of instrumental helping. The expression of altruism depends to some extent on the domain of activity, such as helping others, sharing resources or sharing information, which suggests that different proximate mechanisms might be at play and that such mechanisms are not necessarily all present in humans and other apes. The basic motivations and cognitive capacities for altruistic behavior emerge early in human ontogeny, perhaps reflecting a biological predisposition that we share with our closest evolutionary relatives. Future research should investigate the transformation of these early forms of human altruism into the mature form seen in adults, while paying special attention to the critical proximate mechanisms that serve to maintain altruistic behaviors evolutionarily. Humans may have created unique social mechanisms to preserve and foster the basic altruistic tendencies found in young children and other apes, resulting in altruistic behaviors not found outside the human species.

Acknowledgments

Thanks to Alexandra Rosati, Brett Calcott, Richard Joyce, Kim Sterelny, and Ben Fraser for helpful comments on a previous version of the manuscript.

References

Aloise-Young, P. A. (1993). The development of self-presentation: Self-promotion in 6- to 10-year-old children. *Social Cognition, 11*(2), 201–222.

Banerjee, R. (2002). Children's understanding of self-presentational behavior: Links with mental-state reasoning and the attribution of embarrassment. *Merrill-Palmer Quarterly, 48*(4), 378–404.

Bar-Tal, D. (1982). Sequential development of helping behavior: A cognitive-learning approach. *Developmental Review, 2*, 101–124.

Bering, J. (2004). A critical review of the "enculturation hypothesis": The effects of human rearing on great ape social cognition. *Animal Cognition, 7*(4), 201–212.

Behne, T., Carpenter, M., & Tomasello, M. (2005). One-year-olds comprehend the communicative intentions behind gestures in a hiding game. *Developmental Science, 8*, 492–499.

Berndt, T. J. (1979). Lack of acceptance of reciprocity norms in preschool children. *Developmental Psychology, 15*(6), 662–663.

Blake, P. R., & Rand, D. G. (2010). Currency value moderates equity preference among young children. *Evolution and Human Behavior, 31*, 210–218.

Boesch, C., & Boesch, H. (1989). Hunting behavior of wild chimpanzees in the Tai National Park. *American Journal of Physical Anthropology, 78*, 547–573.

Brosnan, S., & De Waal, F. (2002). Regulation of vocal output by chimpanzees finding food in the presence or absence of an audience. *Evolution of Communication, 4*(2), 211–224.

Brosnan, S. F., Silk, J. B., Henrich, J., Mareno, M. C., Lambeth, S. P., & Schapiro, S. J. (2009). Chimpanzees (*Pan troglodytes*) do not develop contingent reciprocity in an experimental task. *Animal Cognition, 12*(4), 587–597.

Brownell, C., Svetlova, M., & Nichols, S. (2009). To share or not to share: When do toddlers respond to another's needs? *Infancy, 14*(1), 117–130.

Bullinger, A. F., Zimmerman, F., Kaminski, J., & Tomasello, M. (2011). Different social motives in the gestural communication of chimpanzees and human children. *Developmental Science, 14*(1), 58–68.

Buttelmann, D., Carpenter, M., & Tomasello, M. (2009). Eighteen-month-old infants show false belief understanding in an active helping paradigm. *Cognition, 112*, 337–342.

Call, J., Hare, B., Carpenter, M., & Tomasello, M. (2004). "Unwilling" versus "unable": Chimpanzees' understanding of human intentional action. *Developmental Science, 7*(4), 488–498.

Call, J., & Tomasello, M. (1996). The effect of humans on the cognitive development of apes. In A. E. Russon & K. A. Bard (Eds.), *Reaching into thought: The minds of the great apes* (pp. 371–403). Cambridge: Cambridge University Press.

Call, J., & Tomasello, M. (2005). What do chimpanzees know about seeing revisited: An explanation of the third kind. In N. Eilan, C. Hoerl, T. McCormack & J. Roessler (Eds.), *Joint attention: Communication and other minds* (pp. 45–64). Oxford: Oxford University Press.

Clutton-Brock, T. (2009). Cooperation between non-kin in animal societies. *Nature, 462*(5), 51–57.

Cosmides, L., & Tooby, J. (1992). Cognitive adaptations for social exchange. In J. H. Barkow, L. Cosmides & J. Tooby (Eds.), *Evolutionary psychology and the generation of culture* (pp. 163–228). Oxford: Oxford University Press.

Damon, W. (1975). Early conceptions of positive justice as related to the development of logical operations. *Child Development, 46*, 301–312.

Damon, W. (1977). *The social world of the child.* London: Jossey-Bass.

Deci, E. L. (1971). Effects of externally mediated rewards on intrinsic motivation. *Journal of Personality and Social Psychology, 18*(1), 105–115.

De Cooke, P. A. (1992). Children's understanding of indebtedness as a feature of reciprocal help exchanges between peers. *Developmental Psychology, 28*(5), 948–954.

de Waal, F. B. M. (1997). The chimpanzee's service economy: Food for grooming. *Evolution and Human Behavior, 18*, 375–386.

de Waal, F. B. M., & Luttrell, L. M. (1988). Mechanisms of social reciprocity in three primate species: Symmetrical relationship characteristics or cognition? *Ethology and Sociobiology, 9*, 101–118.

Dovidio, J. F., Piliavin, J. A., Schroeder, D. A., & Penner, L. A. (2006). *The social psychology of prosocial behavior.* Mahwah, NJ: Lawrence Erlbaum.

Dunfield, K. A., & Kuhlmeier, V. A. (2010). Intention mediated selective helping in human infants. *Psychological Science, 21*, 523–527.

Dunfield, K. A., Kuhlmeier, V. A., O'Connell, L. J., & Kelley, E. A. (2011). Examining the diversity of prosocial behavior: Helping, sharing, and comforting in infancy. *Infancy, 16*(3), 227–247.

Eisenberg, N., Fabes, R. A., & Spinrad, T. (2006). Prosocial development. In N. Eisenberg (Ed.), *Handbook of child psychology* (Vol. 3): *Social, emotional, and personality development* (6th Ed., pp. 646–718). Hoboken, NJ: John Wiley.

Fabes, R. A., Fultz, J., Eisenberg, N., May-Plumlee, T. & Christopher, F. S., (1989). Effects of rewards on children's prosocial motivation: A socialization study. *Developmental Psychology, 25*(4), 509–515.

Fehr, E., Bernhard, H., & Rockenbach, B. (2008). Egalitarianism in young children. *Nature, 454*, 1079–1083.

Gilby, I. C. (2006). Meat sharing among the Gombe chimpanzees: Harassment and reciprocal exchange. *Animal Behaviour, 71*, 953–963.

Grusec, J., & Redler, E. (1980). Attribution, reinforcement, and altruism: A developmental analysis. *Developmental Psychology, 16*(5), 525–534.

Gummerum, M., Hanoch, Y., & Keller, M. (2008). When child development meets economic game theory: An interdisciplinary approach to investigating social development. *Human Development, 51*, 235–261.

Hamlin, K. J., Wynn, K., & Bloom, P. (2007). Social evaluation by preverbal infants. *Nature, 450*, 557–560.

Hamilton W.D. (1964). The genetical evolution of social behaviour I and II. *Journal of Theoretical Biology, 7*, 1–52.

Hammerstein, P. (2003). Why is reciprocity so rare in social animals? A protestant appeal. In P. Hammerstein (Ed.), *Genetic and cultural evolution of cooperation* (pp. 83–94). Cambridge, MA: MIT Press.

Hare, B., Melis, A. P., Woods, V., Hastings, S., & Wrangham, R. (2007). Tolerance allows bonobos to outperform chimpanzees in a cooperative task. *Current Biology, 17*, 619–623.

Hare, B. A., & Tomasello, M. (2004). Chimpanzees are more skillful in competitive than in cooperative cognitive tasks. *Animal Behaviour, 68*(3), 571–581.

Hauser, M., McAuliffe, K., & Blake, P. R. (2009). Evolving the ingredients for reciprocity and spite. *Philosophical Transactions of the Royal Society of London, 364*, 3255–3266.

Hay, D. F. (1994). Prosocial development. *Journal of Child Psychology and Psychiatry, and Allied Disciplines, 35*(1), 29–71.

Henrich, J., Boyd, R., Bowles, S., Camerer, C. F., Fehr, E., Gintis, H., et al. (2005). "Economic man" in cross-cultural perspective: Behavioral experiments in 15 small-scale societies. *Behavioral and Brain Sciences, 28*, 795–855.

Hook, J. G., & Cook, T. D. (1979). Equity theory and the cognitive ability of children. *Psychological Bulletin, 86*(3), 429–445.

Jensen, K., Call, J., & Tomasello, M. (2007). Chimpanzees are vengeful but not spiteful. *Proceedings of the National Academy of Sciences of the United States of America, 104*(32), 13046–13050.

Jensen, K., Hare, B., Call, J., & Tomasello, M. (2006). What's in it for me? Self-regard precludes altruism and spite in chimpanzees. *Proceedings of the Royal Society of London, Series B: Biological Sciences, 273*, 1013–1021.

Koyama, N. F., Caws, C., & Aureli, F. (2006). Interchange of grooming and agonistic support in chimpanzees. *International Journal of Primatology, 72*(5), 1293–1309.

Kuhlmeier, V., Wynn, K., & Bloom, P. (2003). Attribution of dispositional states by 12-month-olds. *Psychological Science, 14*(5), 402–408.

Leavens, D. A. (2004). Manual deixis in apes and humans. *Interaction Studies: Social Behaviour and Communication in Biological and Artificial Systems, 5*(3), 387–408.

Lepper, M. R., Greene, D., & Nisbett, R. E. (1973). Undermining children's intrinsic interest with extrinsic reward: A test of the "overjustification" hypothesis. *Journal of Personality and Social Psychology, 28*(1), 129–137.

Levitt, M. J., Weber, R. A., Clark, M. C., & McDonnell, P. (1985). Reciprocity of exchange in toddler sharing behavior. *Developmental Psychology, 21*(1), 122–123.

Liszkowski, U., Carpenter, M., Striano, T., & Tomasello, M. (2006). Twelve- and 18-month-olds point to provide information for others. *Journal of Cognition and Development, 7*(2), 173–187.

Liszkowski, U., Carpenter, M., & Tomasello, M. (2008). Twelve-month-olds communicate helpfully and appropriately for knowledgeable and ignorant partners. *Cognition, 108*(3), 732–739.

Melis, A. P., Hare, B., & Tomasello, M. (2006a). Engineering cooperation in chimpanzees. *Animal Behaviour, 72*, 275–286.

Melis, A. P., Hare, B., & Tomasello, M. (2006b). Chimpanzees recruit the best collaborators. *Science, 311*, 1297–1300.

Melis, A. P., Hare, B., & Tomasello, M. (2008). Do chimpanzees reciprocate received favours? *Animal Behaviour, 76*(3), 951–962.

Melis, A. P., Warneken, F., Jensen, K., Schneider, A.-C., Call, J., & Tomasello, M. (2010). Chimpanzees help conspecifics obtain food and non-food items. *Proceedings of the Royal Society of London, Series B: Biological Sciences, 278*(1710), 1405–1413.

Meltzoff, A. N. (1995). Understanding the intentions of others: Re-enactment of intended acts by 18-month-old children. *Developmental Psychology, 31*(5), 1–16.

Mitani, J. C. (2006). Reciprocal exchange in chimpanzees and other primates. In P. Kappeler & C. Van Schaik (Eds.), *Cooperation in primates: Mechanisms and evolution* (pp. 101–113). Heidelberg: Springer.

Moore, C. (2009). Fairness in children's resource allocation depends on the recipient. *Psychological Science, 20*(8), 944–948.

Muller, M., & Mitani, J. C. (2005). Conflict and cooperation in wild chimpanzees. In P. J. B. Slater, J. Rosenblatt, C. Snowdon, T. Roper & M. Naguib (Eds.), *Advances in the study of behavior* (pp. 275–331). New York: Elsevier.

Nowak, M. A. (2006). Five rules for the evolution of cooperation. *Science, 314*(5805), 1560–1563.

Olson, K. R., & Spelke, E. S. (2008). Foundations of cooperation in preschool children. *Cognition, 108*, 222–231.

Rosati, A. G., Stevens, J. R., Hare, B., & Hauser, M. D. (2007). The evolutionary origins of human patience: Temporal preferences in chimpanzees, bonobos, and human adults. *Current Biology, 17*, 1663–1668.

Russell, Y. I., Call, J., & Dunbar, R. I. M. (2008). Image scoring in great apes. *Behavioural Processes, 78*, 108–111.

Schino, G., & Aureli, F. (2009). Reciprocal altruism in primates: Partner choice, cognition and emotions. *Advances in the Study of Behavior, 39*, 45–69.

Schino, G., & Aureli, F. (2010). Primate reciprocity and its cognitive requirements. *Evolutionary Anthropology, 19*, 130–135.

Silk, J., Brosnan, S., Vonk, J., Henrich, J., Povinelli, D., Richardson, A. S., et al. (2005). Chimpanzees are indifferent to the welfare of unrelated group members. *Nature, 437*, 1357–1359.

Sober, E. (2002). The ABCs of altruism. In S. G. Post, L. G. Underwood, J. Schloss & W. B. Hurlbut (Eds.), *Altruism and altruistic love* (pp. 17–28). Oxford: Oxford University Press.

Stevens, J., & Hauser, M. (2004). Why be nice? Psychological constraints on the evolution of cooperation. *Trends in Cognitive Sciences, 8*(2), 60–66.

Subiaul, F., Vonk, J., Okamoto-Barth, S., & Barth, J. (2008). Do chimpanzees learn reputation by observation? Evidene from direct and indirect experience with generous and selfish strangers. *Animal Cognition, 11*(4), 611–623.

Svetlova, M., Nichols, S., & Brownell, C. (In press). Toddlers' prosocial behavior: From instrumental to empathic to altruistic helping. *Child Development.*

Tomasello, M. (2006). Why don't apes point? In S. Levinson & N. Enfield (Eds.), *Roots of human sociality* (pp. 506–524). Oxford: Berg.

Tomasello, M. (2008). *Origins of human communication*. Cambridge, MA: MIT Press.

Tomasello, M. (2009). *Why we cooperate*. Cambridge, MA: MIT Press.

Tomasello, M., & Call, J. (2004). The role of humans in the cognitive development of apes revisited. *Animal Cognition, 7*, 213–215.

Tomasello, M., & Carpenter, M. (2005). The emergence of social cognition in three young chimpanzees. *Monographs of the Society for Research in Child Development, 70*(1), 1–152.

Tomasello, M., Carpenter, M., Call, J., Behne, T., & Moll, H. (2005). Understanding and sharing intentions: The ontogeny and phylogeny of cultural cognition. *Behavioral & Brain Sciences, 28*(5), 675–691.

Trivers, R. (1971). The evolution of reciprocal altruism. *Quarterly Review of Biology, 46*(1), 35–57.

Ueno, A., & Matsuzawa, T. (2004). Food transfer between chimpanzee mothers and their infants. *Primates, 45*(4), 231–239.

Vaish, A., Carpenter, M., & Tomasello, M. (2010). Young children selectively avoid helping people with harmful intentions. *Child Development, 81*(6), 1661–1669.

Vonk, J., Brosnan, S. F., Silk, J. B., Henrich, J., Richardson, A. S., Lambeth, S. P., Schapiro, S. J., & Povinelli, D. J. (2008). Chimpanzees do not take advantage of very low cost opportunities to deliver food to unrelated group members. *Animal Behaviour, 75*, 1757–1770.

Warneken, F., Hare, B., Melis, A. P., Hanus, D., & Tomasello, M. (2007). Spontaneous altruism by chimpanzees and young children. *PLoS Biology, 5*(7), 1414–1420.

Warneken, F., & Tomasello, M. (2006). Altruistic helping in human infants and young chimpanzees. *Science, 311*, 1301–1303.

Warneken, F., & Tomasello, M. (2007). Helping and cooperation at 14 months of age. *Infancy, 11*(3), 271–294.

Warneken, F., & Tomasello, M. (2008). Extrinsic rewards undermine altruistic tendencies in 20-month-olds. *Developmental Psychology, 44*(6), 1785–1788.

Warneken, F., & Tomasello, M. (2009a). The roots of human altruism. *British Journal of Psychology, 100*, 455–471.

Warneken, F., & Tomasello, M. (2009b). Varieties of altruism in children and chimpanzees. *Trends in Cognitive Sciences, 13*(9), 397–482.

Yamamoto, S., Humle, T., & Tanaka, M. (2009). Chimpanzees help each other upon request. *PLoS ONE, 4*(10), e7416. doi:10.1371/journal.pone.0007416.

Yamamoto, S., & Tanaka, M. (2010). The influence of kin relationship and reciprocal context on chimpanzees' other-regarding preferences. *Animal Behaviour, 79*, 595–602.

21 Culture-Gene Coevolution, Large-Scale Cooperation, and the Shaping of Human Social Psychology

Maciek Chudek, Wanying Zhao, and Joseph Henrich

Standard evolutionary approaches to understanding human sociality, and in particular to understanding cooperation and altruism, have yielded a wide range of important insights. Noteworthy examples include explorations of how natural selection has shaped our psychological capacities for recognizing and helping kin, and differentially investing in offspring (e.g., Daly & Wilson, 1999; DeBruine, 2002). However, these canonical evolutionary approaches to cooperation—based on kinship, reciprocity, reputation, and signaling—fail both empirically and theoretically when they aim to explain *larger-scale human cooperation*—that is, cooperation and exchange among hundreds or thousands of unrelated, ephemeral interactants. Empirically, these approaches do not take sufficient note of the peculiarities of human sociality, cooperation, and exchange (e.g., Burnham & Johnson, 2005), which a proper theory should address. Theoretically, much work from this perspective does not pay sufficiently close attention to what the mathematical evolutionary models of cooperation do, *and do not*, predict about patterns of cooperation (e.g., Price, Tooby & Cosmides, 2002).

In this chapter, we aim to highlight these deficiencies and provide an alternative approach to human social psychology and cooperation rooted in culture-gene coevolutionary (CGC) theory. We begin by presenting five challenges to explaining larger-scale human cooperation and sociality. Next we explain why these challenges cannot be met by the canonical approaches, including by extensions of the canonical approaches that assume psychological mismatches between modern and ancestral social environments. Finally, with these challenges in mind, we summarize a large body of theory and evidence that applies the logic of natural selection to developing hypotheses about our psychological capacities for cultural learning. We consider how *cultural evolutionary processes*, unleashed by these newly evolved capacities, interact with genetic evolution to shape our evolved social psychology. Building directly on the formal evolutionary modeling in this arena, we hypothesize that humans may possess a *norm-psychology* that includes default settings, expectations, memory biases, and inferential mechanisms that facilitate and influence the acquisition of culturally transmitted rules and motivations. We explain how, over centuries and millennia, competition among social groups has driven cultural evolution to increasingly favor the social norms that harness and extend aspects of our evolved psychology to foster and sustain large-scale cooperation in modern societies. We close by assessing how well both our approach and mismatch hypotheses meet the five challenges.

The Five Challenges of Human Cooperation

Many species both live in social groups and cooperate, often to substantial degrees and in large groups (e.g., bees, ants, and wasps). However, larger-scale cooperation in humans has a set of interesting patterns that are not found in other species and jointly need to be explained. The five challenges are (Henrich & Henrich, 2007):

(1) *Species differences* Why is the scale and intensity of human cooperation (at least for some societies) so different from that found among other primates or similar mammals? Any theory that purports to explain larger-scale human cooperation should also be tested against other, similar species where we do not see high levels of cooperation. As we discuss below, many primate species live in small-scale stable social groups with lots of repeated interaction among many relatives, but do not cooperate like humans. All-purpose explanations that resort to features like "language" or "intelligence" only make the problem worse (Lachmann & Bergstrom, 2004; Henrich & Henrich, 2007).
(2) *Scale and intensity differences* Why does the scale and intensity of cooperation and sociality vary so dramatically among human societies, from societies entirely lacking collective action beyond the extended family (e.g., among the Matsigenka; see Johnson, 2003) to societies that routinely cooperate on the order of thousands or even millions of individuals, as in modern nation-states? That is, explaining human cooperation requires explaining why some societies do *not* cooperate much at all, despite economic and ecological pressures for greater cooperation. Whatever one's explanation, it must also account for why we do not see this level of variation in larger-scale cooperation in other primates.
(3) *Domain differences* Why do the domains of larger-scale cooperation vary so much from society to society? Comparative ethnography makes it clear that different social groups inhabiting the same ecology cooperate in different domains. Some cooperate only in warfare and fishing, while others, just downstream, cooperate only in house-building and communal rituals (Henrich et al., 2003; Henrich & Henrich, 2007). Why don't we observe this in other species?
(4) *Rapid intensification* How can we account for the increasing scale and intensity of human cooperation over the last 12,000 years? Theories of human cooperation need to explain this "scaling-up" on time-scales of only millennia as some human populations moved from small-scale communities with kin-based social organizations to vast cities and states with complex social, political, and economic institutions, organizations, and divisions of labor.

(5) *Noncooperative and maladaptive sociality* Why do the sanctioning and other incentive mechanisms that support cooperation, such as those based on punishment, reputation, and signaling, also enforce social behavior that is unrelated to cooperation, such as ritual practices, food taboos, and clothing? Why do these same mechanisms sometimes even sustain maladaptive practices, such as the consumption of dead relatives (Durham, 1991), penile subincision, female foot-binding, or clitoral infibulations (Mackie, 1996)?

Many theoretical efforts to explain larger-scale human cooperation aim only to explain its scale and intensity. However, as our challenges indicate, the puzzle of larger-scale human cooperation is both more difficult and more interesting than is generally recognized. In a later section we argue for the centrality of both cultural evolution and the emergence of social norms to addressing the five challenges, but first we address the canonical approaches to human cooperation.

The Mismatch Hypotheses

Mainstream evolutionary approaches to psychology build primarily from the assumption that humans are well adapted to an ancestral, small-scale, nomadic lifestyle and that much contemporary behavior can be understood as the (sometimes maladaptive) action of these ancestral adaptations in modern environments (e.g., Tooby & Cosmides, 1989; Buss, 1999). Consequently, many evolutionarily oriented researchers are inclined to believe that the high levels of human cooperation, fairness, and trust observed in contemporary societies, as well as those observed in findings from behavioral experiments among undergraduates, arise from a misapplication of cognitive heuristics evolved according to the logic of kinship and reciprocity (direct and indirect) for survival in ancestral small-scale societies. The argument is that these misfires lead people's intuitions to overestimate their relatedness to interactants, the likelihood of future interactions, and/or the possibility of reputation damage, even in genuinely anonymous, one-shot situations. These mechanisms "misfire," the story goes, because they are adapted to cues present in ancestral societies and cannot fully recalibrate to the novel cues of either modern society or behavioral experiments. While many find such *mismatch hypotheses* intuitively plausible, they have typically been stated only in abbreviated form (Johnson, Stopka & Knights, 2003; Dawkins, 2006; Nowak, Page & Sigmund, 2000; Levitt & Dubner, 2009; Trivers, 2004; Delton et al., 2010). They have not, to our knowledge, been systematically presented or defended, with two possible exceptions (Burnham & Johnson, 2005; Hagen & Hammerstein, 2006).

We want to emphasize that we are not taking issue with mismatch hypotheses in general, only with those mismatch hypotheses related to large-scale cooperation. Many mismatch hypotheses seem perfectly plausible to us, for example, those that explain that modern diets are so high in fat and sugar because these cues reliably led to healthier, more energy-rich diets in food-scarce ancestral environments. However, some mismatch hypotheses fitting the data does not imply that all mismatch hypotheses are correct. Hereafter, our references to mismatch hypotheses are concerned specifically with those derived from applying evolutionary conceptions (or misconceptions; see below) of kinship, reciprocity, signaling, and reputations to explaining large-scale cooperation, without considering the impacts of cultural evolution, or the interaction between genes and culture over a broad stretch of human evolutionary history.[1]

This section proceeds in three steps. First, we clarify several common misconceptions regarding our criticisms of these mismatch hypotheses. Second, we delineate several theoretical problems for mismatch hypotheses. In short, mathematical models of cooperation reveal basic conceptual problems for the intuitions expressed by mismatch proponents. The difficulty is severe enough that we doubt that these ideas deserve to get off the ground as cogent evolutionary hypotheses—despite their superficial plausibility. Third, putting aside theoretical problems, we compare the claims about human ancestral social life required by mismatch theories to the available empirical evidence about our evolutionary history. While the data are thin, and the inferences necessarily substantial, there is little empirical reason to suspect that ancestral human social lives were characterized by a lack of fitness-relevant interaction with (1) nonkin, (2) short-term interaction partners, or (3) situations without reputational repercussions. Nothing we know from studies of primates, paleoanthropology, or small-scale societies indicates that *these three features* of modern social environments were absent or unimportant in our evolutionary history, or that cognitive mechanisms adapted to ancestral conditions should have trouble calibrating to them. Finally, we review recent experimental findings from both human and nonhuman primates that challenge mismatch hypotheses, while remaining consistent with our CGC approach. We close by examining how well both our approach and mismatch hypotheses do against the five challenges of human cooperation.

Clarification and Theoretical Problems

Before we lay out our theoretical concerns with this approach, let's clarify a common misconception. Though culture-gene coevolution (CGC) also takes as its point of departure that humans possess evolved psychologies associated with kinship, status competition, and reciprocity, the available mathematical evolutionary theory is not consistent with the claim that *these mechanisms alone* can explain the scale of human cooperation, let alone the other four challenges presented above. We argue that (1) cultural evolution has harnessed and extended aspects of our *ancient social psychology* (e.g., kinship, status) to expand the sphere of human cooperation and that (2) this process may be sufficiently old and important that genetic evolution has responded, with the result that we possess a coevolved *tribal psychology* (Richerson & Boyd, 1998), including a norm-psychology, built on top of those ancient social instincts. Thus, the debate at hand is not a question of "evolved psychologies" versus "culture." It is a question of "only ancient evolved psychologies" (mismatch hypotheses) versus "ancient evolved psychologies plus cultural evolution plus gene-culture coevolved psychologies" (CGC hypotheses: Henrich & Henrich, 2007; Richerson & Boyd, 1998). These are alternative competing evolutionary accounts.

Kinship

Sociality mismatch hypotheses come in three distinct flavors—kinship, reciprocity and reputation; all are purely verbal extensions of formal modeling results. Kinship-mismatch builds on work by W. D. Hamilton (1964). Hamilton and his successors (West, Griffin & Gardner, 2007), with the aid of a few simplifying assumptions, translate formal models of complex evolutionary dynamics into a simpler, tractable insight that explains when kinship will promote cooperation: whenever $rb > c$, where b and c represent the benefits and costs of cooperation. Hamilton's r is a key quantity that, *given some simplifying assumptions*, is well approximated by the probability that two individuals' genes are identical by common descent. This value, sometimes called *relatedness*, can be readily calculated: For full siblings, parents, and children, $r=.5$; for half-siblings, $r=.25$; for cousins, $r=.125$. Testing the prediction that kinship promotes cooperation produces a very good fit—many species really do seem to recognize and preferentially help their kin, or at least avoid hurting them, including humans (Daly & Wilson, 1982).

Given just this, the kinship-mismatch hypothesis seems plausible: Ancestral humans lived in much smaller groups that probably consisted of numerous kin. If fitness interactions with nonkin were rare or inconsequential enough, couldn't our genome have evolved a reliable, cognitive-load-sparing heuristic of always assuming some minimum (nonzero) relatedness to interaction partners? Today those same cognitive heuristics may still drive us to cooperate with strangers, even when we could profit by defecting (i.e., not cooperating, either exploiting or not helping someone).

This may sound good, but it is deeply flawed. Even in communities of relatives, evolutionary theory predicts that individuals should try to accurately estimate how related they are to any particular interactant and moderate their degree of cooperation accordingly. Still, a mismatch theorist might argue that what's mismatched is not our ability to gauge relatedness, but our capacity to calibrate to interactants for whom $r = 0$; we heuristically always assume some relatedness. This reveals a deeper problem: claiming ancestral humans interacted in isolated kin groups violates one of the assumptions used to derive the simplified form of Hamilton's rule: *that competition is global*, not local. If we assume individuals interact only locally (not globally), Hamilton's rule becomes $\frac{r-\bar{r}}{1-\bar{r}}b > c$, where $\bar{r}$ is the average relatedness of the interacting, evolutionarily competing individuals (McElreath & Boyd, 2007). In such societies, individuals should selectively cooperate only with *closer kin* than average (i.e., where $r - \bar{r} > 0$). Such individuals would not cooperate with relatives who were less related than average and would in fact be *less* cooperative toward them than if interaction (and competition) were global and included strangers (Boyd, 1982; Gardner & West, 2004).

If anything, kin selection predicts that animals like humans will *almost never* cooperate unconditionally with *all* the members of their local group; they will reserve cooperation for those they are *more* related to. Even if ancestral humans interacted exclusively with kin, they would not have evolved default heuristics for cooperating with just anyone, least of all strangers, and—in fact—we might predict spite toward distant relatives and strangers.

Direct Reciprocity

The reciprocity-mismatch hypothesis extends important formal results about the ability of reciprocity to sustain cooperation (Trivers, 1971; Axelrod & Hamilton, 1981; Boyd & Lorberbaum, 1987). With some simplifying assumptions, these too can boil down to a rule: cooperation can be sustained when $\omega b > c$, where ω is the probability of an individual interacting again with the same partner, and b and c are, again, the partner's benefits and actor's costs respectively (Henrich & Henrich, 2007). When this is true, dyadic partners can enjoy the benefits of continued cooperation, while unremitting defectors whose past behaviors reveal their ill intent soon find their partners turning against them.

Reciprocity-mismatch theories claim that our psychology mistakenly promotes cooperation in modern one-shot interactions because it is calibrated to ancestral environments where such interactions never, or almost never, occurred. The simplest, and unfortunately rather common, version of this argument focuses on the rarity of true one-shot interactions. Reciprocity theory shows that viable reciprocators cooperate only when the probability of future interactions (ω) with their partner exceed the cost-benefit ratio of a decision. For any interactant, there is a cost-benefit ratio at which one should defect, and adaptive reciprocators need to be sensitive to it. "One-shot" interactions are a limiting case whose ω might be, if anything, easier to estimate; but regardless, natural selection will have favored individuals who estimate ω as accurately as is feasible. There is no reason to suspect that the cost-benefit ratios of many interactions were not regularly close to zero. Reciprocators who cannot assess ω near zero create a niche for selfish, itinerant strangers. Whether they can assess precisely $\omega = 0$ is completely irrelevant.

A subtler version of reciprocity mismatch admits that human minds estimate ω, but since interactions with a very low probability of continuing rarely existed, natural selection calibrated a non-zero floor on our estimates, perhaps to lessen our cognitive load. Thus sometimes we continue cooperating even when the actual likelihood of continuing interaction drops below $\frac{c}{b}$. One reason to be skeptical of this line of argument is aging and death. In every environment humans have ever faced, the probability of an adult dying has increased with age. Survivorship curves for modern foragers (Gurven & Kaplan, 2007) approach zero as individuals enter their seventies, meaning that *ω approaches zero*. This effect is important for both young-old reciprocity dyads and old-old reciprocity partners. Two old reciprocators face an even stricter limit, since either might die before there is a chance to reciprocate *ω goes to zero even faster* (survival probability squared). Aging alone, exacerbated by disease and injury, generates ample need to calibrate for low-frequency interactions whose probably of continuing approaches zero. Unless one thinks aging and survival curves were dramatically different in our ancestral past (they didn't decline to zero with age), then successful reciprocators faced low ω values quite frequently, and reciprocators who could not respond adaptively to them (by not cooperating) would have been outcompeted by those who could.

Mismatch fans often point to early work by Axelrod (1984) indicating that successful strategies are generally "nice," meaning they cooperate on the first round of an interaction. This finding, however, is based on an assumption of independent pairing. Releasing this assumption, which was made purely for analytical convenience, by modeling the emergence of more realistic networks of reciprocators yields the opposite conclusion: Successful strategies are only "nice" until they establish a core set of relationships, which is constrained by many factors; then they begin defecting on everyone else immediately (Hruschka & Henrich, 2006). The logic of direct reciprocity does not predict *broad* cooperation or prosociality with new partners (repeated or not) under more plausible assumptions. These mismatch intuitions are built on an artifact of early reciprocity models.

Even if we ignore these concerns, modeling work also indicates that direct reciprocity cannot sustain cooperation as groups get larger (Boyd & Richerson, 1988; Joshi, 1987). That is, if we take the available evolutionary theory seriously, we should predict that individuals will be keenly sensitive to the size of the cooperative group, defecting as soon as it rises above a handful of individuals, and that most reciprocity-based cooperation should be limited to dyads (Henrich & Henrich, 2007). This means that direct reciprocity cannot sustain *larger-scale cooperation* even if we accept the typical image of ancestral human societies as small groups.[2]

In short, the available models of reciprocal altruism suggest that selection will favor individuals who attend to costs, benefits, group size, and the likelihood of future interactions, as well as to the costs of monitoring and sustaining their current relationships and to their current strategic ecology. A crucial and underappreciated finding is that no pure strategy is stable, even in the simplest models, so reciprocal altruists must consistently adjust their strategies. This is perhaps why reciprocity, despite all the hype, has rarely been observed in nonhumans (Clutton-Brock, 2009). In humans, adaptive cultural learning can hone reciprocity strategies to the details of local environments (Henrich & Henrich, 2007), which can explain why we observe reciprocity in humans. Overall, we think there is little doubt that human behavior and psychology are influenced by direct reciprocity, but theoretically speaking, there is no reason to suspect that this mechanism alone can sustain larger-scale cooperation—even if we assume a mismatch in social environments.

Indirect Reciprocity

Indirect reciprocity extends direct reciprocity by showing that cooperators can thrive if they withhold benefits from anyone who has a *reputation* for defecting on others, irrespective of whether they'll interact with them again personally (Ohtsuki & Iwasa, 2004; Leimar & Hammerstein, 2001). Perhaps, the mismatch argument runs, cooperation in ancestral environments was sustained by reputation, and groups were small enough that reputations were well known and accurate. Since reputations ensured that negative consequences usually followed noncooperative behavior, we evolved a cognitive heuristic for cooperation that misfires in anonymous modern interactions where reputation damage is unlikely.

This version of the mismatch hypothesis faces two central theoretical problems. First, as with direct reciprocity and for the same reasons, indirect reciprocity works best for interactions involving pairs of individuals—not larger-scale cooperation (say, interactions of thirty people building a house, or raiding a nearby village). As the number of interactants increases, the chances for stable cooperation rapidly collapse because the poor reputations of one or two individuals stop group members from cooperating, penalizing the whole cooperative group (Henrich & Henrich, 2007).

Second, indirect reciprocity can only sustain cooperation when high quality (both accurate and complete) reputational information is available (Taylor & Nowak, 2007). There are two forms of reputation in indirect reciprocity. The simpler version proposes that individuals directly observe interactions and judge the interactants (i.e., give them reputations) based on their behavior. This is only plausible for very small groups, and only for some limited types of interactions. On this account, when groups become large enough that many interactions aren't witnessed directly by most individuals (and even foraging bands are this large), well-adapted indirect reciprocators should stop cooperating. The more complex form of reputation involves using language to spread reputational information. Unfortunately, invoking this form simply introduces another cooperative dilemma, arguably even more pernicious than the one it solves (Lachmann & Bergstrom, 2004). You cannot just "assume language" to get cooperation. Communication, the sharing of valuable, abstract information, is itself a challenging cooperative dilemma, which cannot be solved by simply assuming language. Only by the addition of an ad hoc hypothesis that independently explains the origins of language could indirect-reciprocity mismatch theories escape this vicious explanatory regress. As far as we know, no such account has been offered in this context.

Some models extend indirect reciprocity via reputation to stabilize larger-scale cooperative dilemmas (Panchanathan & Boyd, 2004). This modeling work, mentioned above, shows that these mechanisms produce multiple stable equilibria, including noncooperative equilibria. Unless mismatch theorists invoke genetic group selection or (like CGC theorists) cultural group selection, there's no theoretical reason why indirect reciprocity should produce *cooperative* equilibria involving larger groups that spread across the species.

Social Life in Ancestral Societies

Let's now consider how mismatch hypotheses stack up against the empirical evidence, particularly the claims they make about ancestral environments. Kinship-mismatch requires not only that we lived in closely related groups (substantial coefficients of relatedness), but that those groups had sufficiently few nonkin members that any selective pressures to distinguish nonkin would have been trivial. The reciprocity (direct and indirect) versions of the mismatch hypothesis require that (1) low-frequency interactions were fitness *irrelevant*, such that insufficient selection pressure existed for the evolution of an ability to calibrate one's expectations of future interactions to be near zero; and that (2) interactions with individuals without reputations were similarly fitness irrelevant, such that humans have evolved an implicit assumption that costly reputational consequences always exist.

There are three kinds of evidence that anthropologists typically use to reconstruct the nature of ancestral environments: data from (1) nonhuman primates, from (2) ethnographically known foraging populations, and from (3) paleoanthropology (the stones and bones of ancestral societies).

Nonhuman primates like chimpanzees live in small-scale foraging societies, just as human ancestors did. How related are individuals in these groups? This question has been answered by long-term field observation and molecular genetics. The average relatedness between pairs of individuals in three different chimpanzee societies in Taï National Park in West Africa range from 0.012 to 0.15 for males and –0.003[3] to 0.045 for females (Vigilant et al., 2001). Over 92 percent of all the dyads in these foraging groups have r values less than 0.2. For an East African chimp community, values were even lower: males at $\bar{r} = -0.22$ and females at $\bar{r} = -0.082$. A recent study of chimps in an unusually large community in Kibale National Park, Uganda (Langergraber, Mitani & Vigilant, 2007) found the same pattern; average relatedness was about 0.02. This leaves little doubt that chimpanzees absolutely need to be able to calibrate to an r of zero, even when interacting with members of their own sex in their own group.

What about the reciprocity versions of the mismatch hypothesis: Do chimpanzees have infrequent interactions, perhaps with strangers, that have fitness relevant consequences? Absolutely. When chimpanzees from one group encounter chimpanzees from another group, violence is generally only avoided if the two groups are evenly matched. If a solitary male chimpanzee encounters a group of five stranger-chimpanzees, he'd best flee for his life, as the strangers are likely to try to kill him. During one 3.5-year study period in Kibale National Park in Uganda, researchers observed 95 boundary patrols by male coalitions; twelve resulted in intergroup violence and five in fatalities, along with four further fatal intergroup encounters in the months immediately following the study (Watts et al., 2006). Such coalitional, intergroup violence has also been observed in wild populations in Gombe National Park, Tanzania (Wilson, Wallauer & Pusey, 2004) and Budongo forest, Uganda (Newton-Fisher, 1999). Both strangers and nonrelatives are a fitness-relevant, even potent, component of the small-scale societies of our closest relative, undermining mismatch hypotheses' assumption that fitness-relevant interactions with non-kin and strangers are novel to modern environments.

This evidence is consistent with interactions between unfamiliar chimpanzees at research facilities; chimpanzees go ape at the mere sight of a stranger-chimp, and experimental interactions cannot be performed with stranger-chimpanzees. Thus, despite living in more kin-based groups involving mostly repeat interactions, selection appears to have favored in chimpanzees a keen psychological ability to distinguish both strangers and nonkin.

Another source of data on ancestral human societies comes from ethnographically and historically known foraging populations. How closely related are the smallest-scale foraging bands? Not very. In the most detailed study of precontact foraging band composition, data from 58 Ache (Paraguay) bands involving 980 adults and 20,614 dyadic relationships reveal an average degree of relatedness of 0.05. More importantly, about a quarter of a band were entirely unrelated, not even linked through marriage ties. Data from the Gebusi, a very small-scale population of forager-horticulturalists from New Guinea, shows that 28 percent of community coresidents are nonrelatives (Knauft, 1985). Less detailed data from 31 other foraging societies are consistent with the Ache picture (Hill et al., n.d.). These patterns are not surprising, given the importance of exogamy, band fluidity, and bilocality among hunter-gathers (Marlowe, 2003, 2004), and mean that unrelated group members form a substantial component of social life in ancestral human environments. We should be keenly able to distinguish relatives from nonrelatives.

So the kinship version of mismatch hypothesis is a nonstarter. But what about the reciprocity and reputation versions: Are there fitness-relevant interactions with low-frequency interactants and individuals for whom reputational information is unavailable? Are there interactions with strangers or individuals unlikely to be seen again?

Fehr and Henrich (2003) have summarized the ample evidence for these kinds of interactions. Here are three highlights:

(1) Life histories gathered from Australia foragers—living on a continent of hunter-gatherers—who had not seen a white man until their sixties show frequent journeys across vast territories, often on "walkabout," in which these foragers would encounter and briefly live with all manner of peoples they had never seen before and would never encounter again.
(2) The earliest European explorers found that both Australian and Fuegian (South American) foragers had special rituals designed for bringing strangers into camp—implying that encounters with strangers were sufficiently frequent and the strangers themselves sufficiently different to drive the emergence of special rituals.
(3) Among Kalahari foragers, periods of drought that hit approximately once in a person's lifetime brought distant bands together at shared waterholes. Suddenly people needed to interact with many strangers with whom they were unlikely to interact again in their lifetimes.

Based on ethnographic and ethnohistorical observations, interactions with strangers and ephemeral interactants are neither uncommon nor fitness irrelevant in human foraging societies. This is true even if one puts aside the massive evidence for intergroup conflicts, raids, rapes, ambushes, and theft by strangers and anonymous others (Keeley, 1997; Bowles, 2006).[4] As noted previously, the mere fact that people get old, contract deadly diseases, and suffer mortal wounds means that there is no shortage of selection pressures to recognize slim chances for future interactions.

One basis for the reputational mismatch hypothesis is the view that ancestral societies were so closely knit that no actions could be anonymous or secret and thus that secret decisions in behavior experiments are a novel feature of modern environments. Without even considering intergroup interactions, the ethnography of small-scale societies is full of clandestine actions within groups, in particular theft and adultery.[5] To illustrate this, table 21.1 shows two populations of foragers and two of forager-horticulturalists that have indigenous words for "secret" in their languages. There are no (indigenous) words for "computer" or "phone" in these languages, but they do have word(s) for "secret," which suggests that secrets are just as much a feature of small-scale societies as modern ones. People should have ample cognitive machinery for understanding that their actions can remain unknown to others in their group.

Overall, the ethnographic and ethnohistorical record does not provide support for the notion that ancestral foraging societies lacked fitness-relevant interactions with nonkin, low-frequency or short-term interactants, strangers, or situations with low probabilities of reputational damage. Instead, it suggests that people should be keenly able to distinguish kin from nonkin, strangers from frequent partners, and opportunities for action without reputational consequences.

Table 21.1
Words for "secret" in four small scale societies[a]

Group	Location	Word(s)
Hadza	Africa, Tanzania	chiki chikia
Ju (!Kung)	Africa, Kalahari	dcaa = to steal love secretly; have an affair. kaahn = secretly (adverb) g=om = to keep one's mouth shut (keep a secret)
Tsimane	Amazonia, Bolivia	jumve, camin
Machiguenga	Amazonia, Peru	Maireni

a. Thanks to Frank Marlowe, Glenn Sheppard, Mike Gurven, and Polly Weissner for this information, and to Rob Boyd for suggesting this approach.

We have presented a series of theoretical problems for mismatch hypotheses, showing that such hypotheses are actually inconsistent with the available theoretical work, and provided empirical evidence from both nonhuman primates and human foragers that challenge the assumptions required by mismatch hypotheses about the nature of ancestral environments. We next lay out a culture-gene coevolutionary approach to the puzzle of cooperation, which we believe provides a superior framework capable of meeting the five challenges.

Building a Theory for a Cultural Species

We begin by recognizing that humans are different from other species in a critical way. We are a cultural species: Compared to all other species, humans are heavily reliant on large assemblies of socially learned know-how that have accumulated over generations. Even the simplest hunter-gather populations are entirely dependent on these reservoirs of cultural information (including skills, motivations, preferences, and practices) related to such domains as finding water, processing food, making fire, cooking, gathering, medicinal plant use, tool manufacture, tracking, and animal behavior. Also culturally transmitted are extensive bodies of social rules, beliefs, and values about rituals, meat sharing, water ownership, community defense, marriage, and kinship relations. Of course, the existence of all this culturally transmitted stuff doesn't diminish the importance of a rich, evolved cognitive architecture—to the contrary, this reliance on transmitted information creates selection pressures that might not otherwise exist, for acquiring, storing, and organizing cultural knowledge about artifacts, foods, animals, plants, and human groups. Nevertheless, it does suggest that any complete account of human behavior—even for understanding the survival of Paleolithic foraging bands—requires a rich theory of cultural transmission and cultural evolution (Henrich & McElreath, 2003; Henrich, 2008).

How do we apply evolutionary theory to a cultural species? Our first step is to deploy the logic of natural selection, aided by formal evolutionary modeling, to generate hypotheses about the *learning strategies* by which individuals—be they toddlers or song birds—can best acquire new behaviors, including forms or elements of social behavior (Boyd & Richerson, 1985). Formal modeling permits researchers to systematically explore what kinds of strategies are favored by natural selection, and under what conditions (e.g., McElreath & Strimling, 2008). In these approaches, alongside direct experience (which, of course, is still important), learning strategies are divided into those that influence the adoption of a novel *cultural variant* (a belief, behavior, or anything else learned from others) based on (1) its *content*, what it actually is or says, and (2) its *context*, the number or type of people expressing the variant, and the manner and situations in which they do. These learning strategies evolve because they allow cultural learners to most effectively extract useful information from their social milieu and adapt to their local social and physical environments (Henrich & McElreath, 2007). Here are some examples.

Model-biased transmission is a context-based cultural learning strategy in which individuals use cues to figure out who, among their potential models (the people around them), is likely to possess adaptive information, and then to preferentially learn from these models. Theory suggests, and a wide range of empirical findings have confirmed, that both children and adults preferentially pay attention to and learn from models who are more skillful, competent, successful, and prestigious. Evidence also indicates that learners use cues of ethnic markers (dialect, dress, etc.), sex, and age (Efferson, Lalive & Fehr, 2008; Henrich & Gil-White, 2001), which further allow learners to selectively acquire the ideas, beliefs, practices, and preferences that are most likely to be useful to them later in life (McElreath, Boyd & Richerson, 2003). These learning strategies emerge early in childhood (Birch, Vauthier & Bloom, 2008; Nurmsoo & Robinson, 2009; Kinzler, Dupoux & Spelke, 2007; Birch & Bloom, 2002; Jaswal, 2004; Chudek et al., n.d.) and appear to influence many forms of cultural transmission, including social behavior, altruism (Henrich & Henrich, 2007, ch. 2), opinions, economic decisions, food preferences, strategies, beliefs, technological adoptions, and dialect (Mesoudi, 2008).

Conformist-biased transmission (Henrich & Boyd, 1998; Boyd & Richerson, 1985, ch. 7) is another context-based strategy, where learners integrate observations from multiple models.[6] By using strategies like "copy the majority" or "average what the prestigious individuals are doing," learners can piggyback on the learning costs already paid by many other individuals and reduce transmission errors (misperceptions of any single model's cultural variants), extracting cultural knowledge that is more adaptive, on average, than anything they could acquire from a single model or figure out on their own (Henrich & Boyd, 2002). Some empirical work supports these formal predictions (Efferson, McElreath et al., 2008; McElreath et al., 2005; Kohler, VanBuskirk & Ruscavage-Barz, 2004; Carpenter, 2004; Coultas, 2004).

Alongside *context* biases, evolutionary approaches to cultural learning suggest a rich set of hypotheses about how learners should evaluate the *content* of cultural variants. The general insight is that learners should pay particular attention to and remember variants likely to contain adaptive information. Specifically, those judged, *ceteris paribus*, to be more (1) fitness relevant, (2) actionable, and (3) plausible (i.e., compatible with evolved intuitions or existing cultural beliefs; see Henrich, 2009). Such content usually sparks strong emotional responses or greater attention, increasing its likelihood of being stored in memory and recalled later, and may motivate or potentiate adaptive responses, a topic we turn to in our discussion of the internalization of social norms. Empirical work shows content biases related to meat (Fessler, 2003), gossip and social interaction (Mesoudi, Whiten & Dunbar, 2006; Chudek et al., in prep.), disgust (Heath, Bell & Sternberg, 2001), dangerous animals (Barrett, 2007), inductive folkbiological generalizations (Henrich & Henrich, 2010), and social norms (O'Gorman, Wilson & Miller, 2008).

Grounded both theoretically and empirically, these reliably developing features of human cognition generate, as a by-product, cumulative cultural evolution. Cumulative cultural evolution creates *de novo* a whole new set of selection pressures, which open up evolutionary vistas not accessible to noncultural species. One of those new vistas involves large-scale cooperation and ultra-sociality (Richerson & Boyd, 1998).

Many Cultural Evolutionary Roads Lead to Larger-Scale Cooperation and Social Norms

How do such innate learning strategies help explain larger-scale cooperation? Our next step is to take theoretically and empirically established findings about human cultural learning and ask what happens when people use these learning strategies to ontogenetically adapt their social behavior to interactions with others. Specifically, what happens in larger-scale cooperative interactions when people use these individually adaptive learning strategies?

Cultural evolutionary game theory has now repeatedly demonstrated several different ways in which cultural learning can sustain larger-scale cooperation. These approaches all work through some combination of the peculiarities of cultural transmission (vis-à-vis genetic inheritance) combined with incentive mechanisms related to punishment, rewarding, signaling, and reputation (Boyd & Richerson, 1992; Gintis, Smith & Bowles, 2001; Henrich, 2009; Panchanathan & Boyd, 2004; Axelrod, 1986; Henrich & Boyd, 2001; Kendal, Feldman & Aoki, 2006). These models demonstrate how culturally transmitted patterns of behavior can, when common, make *immediately* prosocial choices (i.e., those that benefit others) more beneficial on average for individuals than *immediately* selfish ones (i.e., those that maximize personal gain at the time of the action). Below, we explain why these same mechanisms are unlikely to sustain cooperation by genetic inheritance alone.

All of these cultural evolutionary models give rise to stable behavioral regularities in social groups, or equilibria, in which deviations do not spread via adaptive learning. That is, stable states that could sensibly be called *social norms* arise as an emergent property of these systems. These stable states can sustain individually costly behaviors and thus can include highly cooperative norms. However, the same mechanism can also sustain any costly behavior, even if it does not benefit anyone. In fact, such mechanisms can sustain practices that hurt the group overall. That is, these evolutionary systems all have *multiple stable equilibria*. Stable states typically include those with high levels of cooperation, those with high levels of defection, and many others in which individuals pay costs to perform actions that do not help anyone (e.g., food taboos), and may even hurt individuals and the group (e.g., female infibulations).

Below, we discuss the question of *equilibrium selection*, how cultural evolutionary processes can "select" among this multiplicity of potential stable equilibria/norms. Some of these processes, cultural group selection in particular, can favor the spread of group beneficial norms, including those that stabilize cooperation. Others may be important as well. Happily, this theoretical situation is looking more and more like the world, with its diverse variety of norms and institutions, recorded by anthropologists and sociologists over the last 150 years. We expect shared, culturally transmitted, behavioral regularities stabilized by either punishment or other incentives (like reputational damage or improvement) that can, but need not be, prosocial or group-beneficial and may even be maladaptive for groups. These theoretical features are crucial, and will help us explain the differences in the scale and intensity of human cooperation across societies (Challenge 2), the variation in domains of cooperation (Challenge 3), and the presence of some neutral or maladaptive (Edgerton, 1992) social norms (Challenge 5).

Equilibrium Selection

Three broad theoretical approaches confront the problem of *equilibrium selection* (Henrich, 2006). The first, and perhaps the most intuitive, is that rational, forward-looking individuals will recognize the long-term payoffs available at stable cooperative equilibria (i.e., queuing rather than pushing), assume others are similarly sensible, and choose the prosocial state (Harsanyi & Selton, 1988). Though this may be important in some special cases, there are three reasons why it is unlikely to be the main driver of equilibrium selection. First, groups and individuals are usually quite bad at foreseeing the outcome of complex, probabilistic processes (Tversky, Kahneman & Slovic, 2005)—even with the aid of recent mathematical and computer technologies, this can be very hard and even theoretically impossible in sufficiently complex cases. Second, group decisions are often heavily influenced by leaders and coalitions whose interests diverge from the overall group. And third, as one looks across the globe, the world is still full of non-prosocial and even downright antisocial institutions that hurt the group as a whole (Edgerton, 1992).

The second mechanism is stochasticity (Young, 1998). Over long periods of time, rare accumulations of chance events shock the distribution of behaviors within a society from one equilibrium to another (Kendal, Feldman & Aoki, 2006). However, since these transits from one stable equilibrium to another are not equally likely, societies will spend more time at some stable states than others. This means that in the long run, some norms will be more common among societies than other norms, on the basis of stochastic fluctuations alone. Different evolved learning strategies can make these random, stochastic shifts more likely (e.g., a common evolved bias to learn from particular individuals) or less likely (e.g., an evolved disposition to conform). Stochastic movements can drive interacting societies to different equilibria, providing the raw material for the third mechanism: *cultural group selection*.

At equilibrium the individuals within a group have reached a certain harmony; however, groups at different equilibria also interact. This competition among groups with different stable norms will favor the spread of norms that best facilitate success and longevity in competition with other groups (Boyd & Richerson, 1990; Henrich, 2004). Competition among groups can take the form of warfare (with assimilation or extinction), demographic production, or more subtle forms in which individuals learn by observing more successful individuals from groups at more group-beneficial equilibria. This can lead to a differential flow of decisions, strategies, and even preferences from higher to lower payoff groups (Boyd and Richerson 2002a), or to differential migration from high payoff groups to lower payoff groups (Boyd & Richerson, 2009).[7] For group selection to spread cooperation, the dynamics operating between groups must outweigh the forces favoring selfishness within groups. This *can* occur when behavioral change within groups has reached an internal equilibrium,[8] which for cultural evolution may result from cognitive adaptations related to cultural learning such as biases for credibility enhancing displays (Henrich 2009) or dispositions to conform (Henrich & Boyd, 2001) or sanction–norm psychology (discussed below; see also Henrich & Henrich, 2007).

Building on this foundation of formal modeling, several lines of empirical evidence now support cultural group selection, including data from laboratory studies, ethnography, archaeology, and history. In the laboratory, Gurerk et al. (2006) have shown how individuals migrate from lower-payoff institutions to higher-payoff ones, and adopt the local norms of that group (Henrich, 2006). Ethnographically and ethnohistorically, Atran et al. (1999) have shown how conservation-oriented ecological beliefs spread from locally prestigious Itza Maya to Ladinos in Guatemala, and how highland Q'eqchi' Maya, with tightly bound cooperative institutions and commercially oriented economic production, are spreading at the expense of both Itza and Ladinos. Soltis et al. (1995), using quantitative data gleaned from New Guinea ethnographies, have shown that even the slowest form of cultural group selection (conquest) can occur in 500- to 1,000-year timescales. Kelly (1985) has demonstrated how differences in culturally acquired beliefs about bride-price fueled the Nuer expansion over the Dinka, and how different social institutions, underpinned by cultural beliefs about segmentary lineages, provided the decisive competitive advantage. Sahlins (1961) has argued that cultural beliefs in segmentary lineages facilitated both the Nuer and Tiv expansions. Using archaeological data, anthropologists are increasingly arguing for the importance of cultural group selection in prehistory (Flannery & Marcus, 2000; Spencer & Redmond, 2001), including competition among foragers (Young & Bettinger, 1992; Bettinger & Baumhoff, 1982). At the global level, Diamond (1997) has made a cultural group selection case for the European expansion after 1500 AD, as well as for the Bantu and Austronesian expansions.

There is little doubt that in the real world, equilibrium selection likely mixes all the mechanisms we have discussed above, along with some we have not considered. While competition between groups (cultural group selection) spreads cooperative norms that can sustain internal harmony, conflict among interest groups within a society can shift them toward less harmonious outcomes, as predatory elites and other self-interested subgroups establish equilibria/norms that disproportionately benefit themselves (Henrich & Boyd, 2008). Ideologically motivated groups with coercive power may even sustain equilibria maladaptive extremes, such as perhaps the construction of the immense idol heads on Easter Island, or the huge pyramid tombs of Ancient Egypt.

In closing this section, we briefly address three interrelated questions about this approach. To begin, why do these processes require cognitive adaptations for cumulative cultural evolution, and why are they unlikely to work by genetic evolution alone? There are three reasons. First, some of the mechanisms discussed rely on forms of transmission (e.g., conformist transmission) that do not exist for genes. Second, for those that do not rely on such mechanisms, genetic evolution might be able to produce stable equilibria, but equilibrium shifting in genetic evolutionary processes is extremely slow—too slow given the time available for human evolution, even when driven by competition among groups. Finally and most importantly: Assuming away these theoretical problems still does not lead to a world that meets the five challenges. In particular, it cannot explain variation in domain, scale, and intensity, rapid intensification on historical timescales, or our distinct patterns of maladaptively uncooperative behavior. Even if cooperation could evolve genetically via one of these mechanisms, it would not produce a world that fits with what we observe—it would not address the five challenges.

Next question: How did culturally transmitted cooperation become common in the very first group? Cultural group selection can spread cooperation once it is established in one group, but how can it spread when it is initially rare in all groups? As just noted, cultural transmission is noisier than genetic transmission and, for a variety of reasons, more subject to stochastic "peak shifting." For example, the evolved learning strategies reviewed above can help explain how punishment norms might initially spread: At first they are championed by prestigious individuals (model bias) and then preferentially learned because they're held by the majority (conformist bias), or they become linked with other content-biased ideas (e.g., successful religions typically have potent, moralizing deities; Henrich, 2009; Atran & Henrich, 2010).

Finally, we have assumed these equilibria are stable against most shocks, and that there is no gradual movement away from cooperation. But, what if these assumptions do not hold? What if all that these mechanisms of cultural transmission, punishment, reputation, and signaling do is slow down an inevitable decay to full defection within groups? Some modeling work suggests that even this is not a devastating problem, if competition among cultural groups is strong enough. Boyd and his collaborators (Boyd et al., 2003; Boyd, Richerson & Henrich, forthcoming) have shown that adding costly punishment to a model with intergroup competition and extinction can still dynamically sustain high levels of cooperation and punishment. This occurs even though *within a single isolated group* both cooperative and punishing strategies will decline to zero in the long run. Pursuing a similar strategy, Bowles and Choi have combined models involving warfare and altruism with empirical data on conflict and mortality rates from foraging populations. They show that, especially if one assumes that inheritance is cultural, actual rates of intergroup competition are sufficient to spread and sustain altruistic behavior (Choi & Bowles, 2007; Bowles, 2006). More generally, comparisons of genetic variation with cultural variation show that there is much more cultural variation among humans groups than genetic variation, indicating that we should expect cultural group selection to be substantially more important in humans than genetic group selection (Bell, Richerson & McElreath, 2009).

So, to summarize our culture-gene coevolutionary (CGC) account so far: The learning strategies that benefit individual members of a cultural species, when employed by many individuals, lead to stable group-wide patterns of behavior, that is, norms. Groups with more cooperative norms outcompete others by the various mechanisms of cultural group selection, spreading their cooperative norms.

Culture-Gene Coevolution and Norm Psychology

Recent investigations have documented many examples of the human genome responding and adapting to our cultural environment (for a review, see Laland, Odling-Smee & Myles, 2010). If norms are a ready by-product of a species heavily reliant on cultural learning, then social norms have likely been a ubiquitous and enduring part of our social environment. How might genetic evolution have adapted to their presence? When norms create stable equilibria, individuals who deviate from the normative behavior do worse on average than those who abide by them. Though the content of the norms themselves may change frequently, and even evolve to be more cooperative via cultural group selection, the need to figure out the local norms and follow them (at least sometimes) has remained a stable feature of our social environments. It's plausible that these conditions selected for a *norm psychology:* a suite of genetically evolved cognitive mechanisms for rapidly perceiving local norms and internalizing them.[9]

What would a norm psychology be like? First, it would need to include a capacity to (consciously or otherwise) recognize one's group's norms—perhaps by observing the frequency of behaviors or witnessing others' responses to deviations. It would include a capacity for cognitively representing observed norms, so they could influence later behavior. Since adherence to stable norms, by the definition of *stable*, is fitness enhancing, it should also include a default disposition to adopt the norms one witnesses, especially the most common ones.

Perhaps the most powerful way that culturally transmitted behavioral norms can stabilize cooperation is by dynamically coordinating punishment systems (e.g., Boyd, Gintis & Bowles, 2010; Sigmund et al., 2010; Panchanathan & Boyd, 2004). As such, an adaptive norm psychology should be sensitive to which norm-violations invoke retaliation from other individuals. What's more, most punishment systems that can stabilize cooperation include mechanisms for (sometimes indirectly) punishing non-punishers, which suggests selection for a norm psychology that also includes a disposition to punish norm-violators. This might, for instance, be brought about by an evolved capacity to feel emotions like "outrage."

Lastly, if norm-adherence really is robustly advantageous across most of the ecologies a cultural species faces, evolved norm psychology may well lead adaptive individuals to emotionally internalize norms—they almost always adhere to them because it just *feels right*—rather than just making them available to error-prone and effortful rational deliberation. Of course, such internalization would also entail costly norm adherence in situations where it's *not* the individually best choice. It's plausible that in societies where behavioral deviations are subject to coordinated punishment, these costs would be dwarfed by the costs of mistaken deviation.

Even purely genetic models show that reputation and punishment can make prosocial behavior individually advantageous in the long run, but what's peculiar is that many humans act prosocially even when anonymity ensures that these mechanisms don't apply (e.g., returning a stranger's wallet, anonymous charitable giving). Our CGC model of norm psychology helps explain why this is so. Cultural learning (a) makes norms—socially learned behavioral equilibria—abound and (b) magnifies the frequency of peak-shifting between equilibria, generating variation between societies while promoting homogeneity within them. Cultural group selection favors societies with cooperation-enhancing, often punishment-based norms. In such societies, individuals who recognize and emotionally internalize norms thrive. Eventually societies exist where most individuals genuinely *feel* that it's right to cooperate with others, even when a fully rational economic agent would know better. Norm psychology can turn *rationally* cooperative societies (where defection usually doesn't pay) into *highly* cooperative ones (where people cooperate even when defection does pay).

Predictions and Empirical Support for Norm Psychology

The above approach makes a number of empirical predictions about our evolved psychology. Here, we only have space to briefly present each prediction and sketch the available evidence.

Humans should possess cognitive mechanisms that scaffold the acquisition of norms. That is, individuals intuitively assume that social rules exist, that these rules influence people's behavior, that others ought to follow these rules, and that failure to adhere to these rules can have negative consequences. People assume such rules exist even when they do not know yet what the specific rules are. Recent work with younger children shows that (1) children spontaneously infer the existence of social rules, in one trial learning, (2) react negatively to deviations by others to a rule learned in one trial, and (3) spontaneously sanction norm violators (Rakoczy, Wameken & Tomasello, 2008).

Because cultural group selection has selectively favored mostly prosocial norms over tens of thousands of years, our cognitive intuitions should include prosocial default content and related learning biases. Recent work by Hamlin and colleagues indicates that 3-month-old preverbal infants are able to evaluate actions as either helpful or hurtful to another's goal. By 6 months, and possibly earlier, they use this information as a basis for their own desire to interact with these helping or hindering individuals (Hamlin, Wynn & Bloom, 2007; Hamlin, Wynn & Bloom, 2010).

Humans should readily acquire costly social norms via cultural learning. An extensive body of research demonstrates that children acquire context-specific prosocial norms by observing others perform actions consistent with such norms. Children also spontaneously enforce prosocial norms on other children, suggesting that merely observing a costly prosocial act being modeled induces normative inferences. The effects of such inferences persist in retests weeks or months after initial learning (Rushton & Campbell, 1977; Rushton, 1975; Mischel & Liebert, 1966; reviewed in Henrich & Henrich, 2007; Bryan, 1971; Bryan, Redfield & Mader, 1971).

Both adhering to social norms and punishing norm violators should be intrinsically or internally rewarding. The evidence supporting this comes from economic games (Camerer, 2003). Converging lines of evidence now indicate these games tap into social norms for dealing with ephemeral interactants (those with whom the actor lacks any relationship, based on kinship, status, reciprocity, etc.) in situations involving money (Henrich et al., 2010). First, findings in these experiments vary dramatically across human populations in ways that correlate with the relevant norms (Henrich et al., 2005; Herrmann, Thöni & Gächter, 2008). Second, in the United States and

Europe the costly prosocial behavior measured in such experiments develops slowly over the life course and does not hit its adult plateau until a person's mid-twenties (Sutter & Kocher, 2007; Carter & Irons, 1991; Henrich, 2008; Harbaugh, Krause & Liday, 2002). Third, results can be influenced by reframing the game in ways that tap into other social norms (Pillutla & Chen, 1999; Ross & Ward, 1996; Heyman & Ariely, 2004), and the same framing effects operate differently in different places (Hayashi et al., 1999). Fourth, chimpanzees—lacking the relevant norms—do not show any prosociality when placed in parallel experiments (Jensen, Call & Tomasello, 2007; Silk et al., 2005; Jensen et al., 2006; Vonk et al., 2008). Adult chimpanzees, unlike any human population ever studied, including 4-year-olds, actually behave according to game-theoretic predictions that assume pure self-interest.

Using these behavioral games, work in neuroeconomics shows that behaving according to acquired social norms—by cooperating, contributing to public goods or charity, and punishing in locally prescribed ways—activates the brain's rewards or reward anticipation circuits in the same manner as does obtaining a direct cash payment (Rilling et al., 2004; Fehr & Camerer, 2007; Tabibnia, Satpute & Lieberman, 2008; Sanfey et al., 2003; de Quervain et al., 2004). These studies variously show that: complying with norms "feels good" to brains in the same way that personally getting money does; and punishing by really hurting defectors (physically or monetarily) activates these reward circuits more than punishing symbolically. Activations of the brain's reward circuitry in these experiments predict behavioral outcomes.

Once internalized, norm adherence should be automatic such that norm violations require a cognitive override. Violating norms (not adhering to norms), such as breaking promises or inflicting harm on an individual for utilitarian reasons, requires deliberately overriding more automatic responses by brain regions responsible for cognitive control and abstract reasoning (Baumgartner et al., 2009; Greene et al., 2004).

Now let's see how our CGC theory and mismatch hypotheses stack up in meeting the five challenges of human cooperation.

Meeting the Five Challenges

(1) *Species differences* Why is the scale and intensity of human cooperation so different from that found among other primates? CGC hypothesizes a long-term interaction between genetic and cultural evolution, initiated when a species becomes sufficiently dependent on cultural learning. Thus, the theory applies only to highly cultural species, which rules out all primates and most mammals. In contrast, *mismatch hypotheses* purport to explain larger-scale human cooperation using evolutionary models that are equally applicable to other primate species (among many other species). Without an independent account of the evolutionary origins of the cognitive capacities that change the way these mechanisms apply to humans, such as CGC provides, there is

little room in this approach to even account for differences in cooperation between human and chimpanzee foragers.

(2) *Scale and intensity differences* Why do the scale and intensity of cooperation and sociality vary so dramatically among human societies? CGC proposes that larger-scale cooperation depends on local, culturally evolved norms and the institutions and divisions of labor they support. Different intensities of competition on different continents or in different regions, and in different ecologies, have favored the proliferation of norms that vary in the nature and degree of prosociality they sustain. Mismatch hypotheses have not addressed this issue. One approach might be to invoke the jukebox analogy (Tooby & Cosmides, 1992), and suggest that different environmental cues evoke different degrees of cooperation based on local affordances. Of course, we know that similar groups have revealed quite different degrees of cooperation even while inhabiting identical environments (Henrich et al., 2003).

(3) *Domain differences* Why do the domains of cooperation vary so much from society to society? For CGC, this is a prediction that follows directly from the notion that larger-scale cooperation is linked directly to local norms. Mismatch theorists have not tackled this one. They might again deploy the jukebox analogy, emphasizing varying costs and benefits across different domains.

(4) *Rapid intensification* How can we account for the increasing scale and intensity of human cooperation over the last 12,000 years? CGC explains this with cultural group selection. Climatic changes at the beginning of the Holocene created conditions favorable to agriculture and larger societies (Richerson, Boyd & Bettinger, 2001; Henrich et al., 2010). Those societies with norms and institutions best suited to sustaining ever expanding cooperative populations engaged in mutually beneficial transactions and spread at the expense of other groups. Mismatch theorists have not addressed this issue.

(5) *Noncooperative and maladaptive sociality* Why do the sanctioning mechanisms that enforce cooperation also sustain social behavior that is unrelated to cooperation, such as ritual practices, food taboos, and clothing choice? For CGC, this prediction emerged unexpectedly from mathematical models of larger-scale cooperation—punishment, reputation, and signaling can all sustain any costly action, not merely cooperation. Mismatch theorists might try to deny this claim, and instead propose distinct explanations for each noncooperative phenomenon.

Notes

1. Prior work has referred to these mismatch hypotheses descriptively (not snarkily) as "big mistake hypotheses" (Fehr & Henrich, 2003; Boyd & Richerson, 2002b; Henrich &, Henrich 2007). We adopt "mismatch" here in deference to terminology used by proponents of this point of view.

2. Large-scale cooperation does not mean lots of cooperation in different dyads. Large-scale cooperation involves large groups cooperating, in, for example, joint projects (armies, recycling, paying taxes, giving blood, etc.), not just the concurrent existence of many cooperating dyads.

3. Note that relatedness measures the genetic similarity between two individuals *relative to the population average*. Negative relatedness values imply that two individuals are less similar to each other than they are, on average, to a randomly selected member of their species (Grafen, 1985).

4. As far as evolution is concerned, sparing someone when you could profit by exploiting them is just as much a cooperative act as providing them with costly aid; thus these are all examples of cooperation-relevant interactions with strangers.

5. Much work in evolutionary psychology is based on the assumption that extra-pair copulations (adultery) and copulations with different males in relatively short periods of time were not only possible, but a serious factor in the lives of our foraging ancestors (Buss, 2007). If true, then secrets and anonymity (like about paternity) were indeed possible, and individuals could avoid reputational damage and revenge while engaging in clandestine activities. We think this fits well with ethnography, but not with the assumptions of some mismatch hypotheses. That is, mismatch assumptions about ancestral environments conflict with much work by those studying mating psychology.

6. A lively debate persists on the evolutionary foundations of conformist transmission (e.g., Nakahashi, 2007; Wakano & Aoki, 2007; Guzman, Rodriguez-Sickert & Rowthorn, 2007).

7. Recent theoretical work by Lehman et al. (Lehman & Feldman, 2008; Lehman, Feldman & Foster, 2008; Lehman et al., 2007) has sought to challenge some of these theoretical conclusions. Their models, however, make quite different assumptions (which are buried deep in the mathematics) about (1) the strength of cultural learning relative to the forces of mixing among Group and (2) the presence of multiple stable equilibria. Their efforts are not critiques but rather alternative hypotheses based on empirically difficult-to-support assumptions about human cultural transmission and social norms (Boyd, Richerson & Henrich, forthcoming).

8. Note that this does not imply that groups are internally homogeneous; equilibria can be polymorphic: stable mixes of different behaviors, where an increase in the frequency of any behavioral strategy makes the alternatives more advantageous. For example, consider a market in which everyone is selling peanut butter. In such an environment, a jelly seller will do well until there are too many jelly sellers, and not enough peanut butter sellers.

9. Is this "strong reciprocity"? The term "strong reciprocity" emerged in the late 1990s as a label for a set of social motivations that were implied by a large body of empirical findings mostly emerging from experimental economics laboratories. Some researchers tried to develop evolutionary theories to explain the origins of these motivations (Gintis, 2000; Fehr, Fischbacher & Gächter, 2002). It's important to differentiate the empirical findings (and their implied proximate motivations) from efforts at evolutionary explanations (which are not "strong reciprocity"). Our approach begins with culture-gene coevolution theory and derives predictions. These predictions regarding norm psychology can account for the patterns associated with strong reciprocity, as

empirical phenomena, but also go well beyond them by seating such patterns in the context of both cultural and genetic evolution.

References

Atran, S., & Henrich, J. (2010). The evolution of religion: How cognitive by-products, adaptive learning heuristics, ritual displays, and group competition generate deep commitments to prosocial religions. *Biological Theory*, *5*(1), 1–13.

Atran, S., Medin, D. L., Ross, N., Lynch, E., Coley, J., Ek, E. U., & Vapnarsky, V. (1999). Folkecology and commons management in the Maya Lowlands. *Proceedings of the National Academy of Sciences of the United States of America*, *96*, 7598–7603.

Axelrod, R. (1984). *The evolution of cooperation*. New York: Basic Books.

Axelrod, R. (1986). An evolutionary approach to norms. *American Political Science Review*, *80*(4), 1095–1111.

Axelrod, R., & Hamilton, W. D. (1981). The evolution of cooperation. *Science*, *211*(1), 390–396.

Barrett, H. Clark, & Broesch, J. (In press). Prepared social learning about dangerous animals in children. *Evolution and Human Behavior*. Available online April 2012.

Baumgartner, T., Fischbacher, U., Feierabend, A., Lutz, K., & Fehr, E. (2009). The neural circuitry of a broken promise. *Neuron*, *64*(5), 756–770.

Bell, A. V., Richerson, P. J., & McElreath, R. (2009). Culture rather than genes provides greater scope for the evolution of large-scale human prosociality. *Proceedings of the National Academy of Sciences of the United States of America*, *106*(42), 17671–17674.

Bettinger, R. L., & Baumhoff, M. A. (1982). The numic spread: Great Basin cultures in competition. *American Antiquity*, *47*(3), 485–503.

Birch, S. A. J., & Bloom, P. (2002). Preschoolers are sensitive to the speaker's knowledge when learning proper names. *Child Development*, *73*(2), 434–444.

Birch, S. A. J., Vauthier, S. A., & Bloom, P. (2008). Three- and four-year-olds spontaneously use others' past performance to guide their learning. *Cognition: International Journal of Cognitive Science*, *107*(3), 1018–1034.

Bowles, S. (2006). Group competition, reproductive leveling, and the evolution of human altruism. *Science*, *314*(5805), 1569–1572.

Boyd, R. (1982). Density-dependent mortality and the evolution of social interactions. *Animal Behaviour*, *30*(4), 972–982.

Boyd, R., Gintis, H., & Bowles, S. (2010). Coordinated punishment of defectors sustains cooperation and can proliferate when rare. *Science*, *328*(5978), 617.

Boyd, R., Gintis, H., Bowles, S., & Richerson, P. J. (2003). The evolution of altruistic punishment. *Proceedings of the National Academy of Sciences of the United States of America, 100*(6), 3531–3535.

Boyd, R., and Lorberbaum, J. P. (1987). No pure strategy is evolutionarily stable in the repeated prisoner's dilemma game. *Nature, 327*, 58–59.

Boyd, R., & Richerson, P. J. 1985. *Culture and the evolutionary process*. Chicago: University of Chicago Press.

Boyd, R., & Richerson, P. J. (1988). The evolution of reciprocity in sizable groups. *Journal of Theoretical Biology, 132*, 337–356.

Boyd, R., & Richerson, P. J. (1990). Group selection among alternative evolutionarily stable strategies. *Journal of Theoretical Biology, 145*, 331–342.

Boyd, R., & Richerson, P. J. (1992). Punishment allows the evolution of cooperation (or anything else) in sizable groups. *Ethology and Sociobiology, 13*(3), 171–195.

Boyd, R., & Richerson, P. J. (2002a). Group beneficial norms can spread rapidly in a structured population. *Journal of Theoretical Biology, 215*, 287–296.

Boyd, R., & Richerson, P. J. (2002b). Solving the puzzle of human cooperation. In S. Levinson (Ed.), *Evolution and culture*. Cambridge, MA: MIT Press.

Boyd, R., & Richerson, P. J. (2009). Voting with your feet: Payoff biased migration and the evolution of group beneficial behavior. *Journal of Theoretical Biology, 257*(2), 331–339.

Boyd, R., Richerson, P. J., & Henrich, J. (Forthcoming). Rapid cultural adaptation can facilitate the evolution of large-scale cooperation. *Behavioral Ecology and Sociobiology*.

Bryan, J. H. (1971). Model affect and children's imitative altruism. *Child Development, 42*(6), 2061–2065.

Bryan, J. H., Redfield, J., & Mader, S. (1971). Words and deeds about altruism and the subsequent reinforcement power of the model. *Child Development, 42*(5), 1501–1508.

Burnham, T. C., & Johnson, D. D. (2005). The biological and evolutionary logic of human cooperation. *Analyse & Kritik, 27*, 113–135.

Buss, D. (1999). *Evolutionary psychology: The new science of the mind*. Boston: Allyn & Bacon.

Buss, D. (2007). *Evolutionary psychology: The new science of the mind* (3rd Ed.). Boston: Allyn & Bacon.

Camerer, C. (2003). *Behavior game theory: Experiments in strategic interaction*. Princeton: Princeton University Press.

Carpenter, J. (2004). When in Rome: Conformity and the provision of public goods. *Journal of Socio-Economics, 4*(4), 395–408.

Carter, J. R., & Irons, M. D. (1991). Are economists different, and if so, why? *Journal of Economic Perspectives, 5*, 171–177.

Choi, J.-K., & Bowles, S. (2007). The coevolution of parochial altruism and war. *Science, 318*(5850), 636–640.

Chudek, Maciek, Sarah Heller, Susan Birch, and Joseph Henrich. (n.d.). Prestige-bias: Developmental evidence for biased cultural learning. Vancouver.

Chudek, M., Mesoudi, A., Collard, M., & Henrich, J. (In prep.). Social recall bias: An influence on the evolution of cultural information?

Clutton-Brock, T. (2009). Cooperation between non-kin in animal societies. *Nature, 462*(7269), 51–57.

Coultas, J. (2004). When in Rome . . . An evolutionary perspective on conformity. *Group Processes & Intergroup Relations, 7*(4), 317–331.

Daly, M., & Wilson, M. (1982). Homicide and kinship. *American Anthropologist, 84*, 372–378.

Daly, M., & Wilson, M. (1999). *The truth about Cinderella: A Darwinian view of parental love*. New Haven: Yale University Press.

Dawkins, R. (2006). *The God delusion*. Boston: Houghton Mifflin.

de Quervain, D. J., Fischbacher, U., Treyer, V., Schellhammer, M., Schnyder, U., Buck, A., et al. (2004). The neural basis of altruistic punishment. *Science, 305*, 1254–1258.

DeBruine, L. (2002). Facial resemblance enhances trust. *Proceedings of the Royal Society of London, Series B: Biological Sciences, 269*, 1307–1312.

Delton, A. W., Krasnow, M. M., Cosmides, L., & Tooby, J. (2010). Evolution of fairness: Rereading the data. *Science, 329*(5990), 389.

Diamond, J. M. (1997). *Guns, germs, and steel: The fates of human societies*. New York: W. W. Norton.

Durham, W. H. (1991). *Coevolution: Genes, culture, and human diversity*. Stanford: Stanford University Press.

Edgerton, R. B. (1992). *Sick societies: Challenging the myth of primitive harmony*. New York: Free Press.

Efferson, C., Lalive, R., & Fehr, E. (2008). The coevolution of cultural groups and ingroup favoritism. *Science, 321*(5897), 1844–1849.

Efferson, C., Lalive, R., Richerson, P. J., McElreath, R., & Lubell, M. (2008). Conformists and mavericks: The empirics of frequency-dependent cultural transmission. *Evolution and Human Behavior, 29*(1), 56–64.

Fehr, E., & Camerer, C. F. (2007). Social neuroeconomics: The neural circuitry of social preferences. *Trends in Cognitive Sciences, 11*(10), 419–427.

Fehr, E., & Henrich, J. (2003). Is strong reciprocity a maladaption? In P. Hammerstein (Ed.), *Genetic and cultural evolution of cooperation*. Cambridge, MA: MIT Press.

Fehr, E., Fischbacher, U., & Gächter, S. (2002). Strong reciprocity, human cooperation, and the enforcement of social norms. *Human Nature, 13,* 1–25.

Fessler, D. M. T. (2003). Meat is good to taboo: Dietary proscriptions as a product of the interaction of psychological mechanisms and social processes. *Journal of Cognition and Culture, 3*(1), 1–40.

Flannery, K. V., and Marcus, J. (2000). Formative Mexican chiefdoms and the myth of the "Mother Culture." *Journal of Anthropological Archaeology, 19*, 1–37.

Gardner, A., & West, S. A. (2004). Spite and the scale of competition. *Journal of Evolutionary Biology, 17*(6), 1195–1203.

Gintis, H. (2000). Strong reciprocity and human sociality. *Journal of Theoretical Biology, 206*(2), 169–179.

Gintis, H., Smith, E. A., & Bowles, S. (2001). Costly signaling and cooperation. *Journal of Theoretical Biology, 213*(1), 103–119.

Grafen, A. (1985). A geometric view of relatedness. *Oxford Surveys in Evolutionary Biology, 2,* 28–89.

Greene, J. D., Nystrom, L. E., Engell, A. D., Darley, J. M., & Cohen, J. D. (2004). The neural bases of cognitive conflict and control in moral judgment. *Neuron, 44*(2), 389–400.

Gurerk, O., Irlenbusch, B., & Rockenbach, B. (2006). The competitive advantage of sanctioning institutions. *Science, 312,* 108–111.

Gurven, M., & Kaplan, H. (2007). Longevity among hunter-gatherers: A cross-cultural examination. *Population and Development Review, 33*(2), 321–365.

Guzman, R. A., Rodriguez-Sickert, C., & Rowthorn, R. (2007). When in Rome, do as the Romans do: The coevolution of altruistic punishment, conformist learning, and cooperation. *Evolution and Human Behavior, 28*(2), 112–117.

Hagen, E. H., & Hammerstein, P. (2006). Game theory and human evolution: A critique of some recent interpretations of experimental games. *Theoretical Population Biology, 69*(3), 339–348.

Hamilton, W. D. (1964). The genetical evolution of social behavior. *Journal of Theoretical Biology, 7,* 1–52.

Hamlin, J. K., Wynn, K., & Bloom, P. (2010). Three-month-olds show a negativity bias in their social evaluations. *Developmental Science, 13,* 923–929.

Hamlin, J. K., Wynn, K., & Bloom, P. (2007). Social evaluation by preverbal infants. *Nature, 450*(7169), 557–559.

Harbaugh, W. T., Krause, K., and Liday, S. G. (2002). Bargaining by children. University of Oregon Working Paper no. 2002-4.

Harsanyi, J. C., & Selton, R. (1988). *A general theory of equilibrium selection in games*. Cambridge, MA: MIT Press.

Hayashi, N., Ostrom, E., Walker, J., & Yamagishi, T. (1999). Reciprocity, trust, and the sense of control—A cross-societal study. *Rationality and Society, 11*(1), 27–46.

Heath, C., Bell, C., & Sternberg, E. (2001). Emotional selection in memes: The case of urban legends. *Journal of Personality and Social Psychology, 81*(6), 1028–1041.

Henrich, J. (2004). Cultural group selection, coevolutionary processes, and large-scale cooperation. *Journal of Economic Behavior & Organization, 53*, 3–35.

Henrich, J. (2006). Cooperation, punishment, and the evolution of human institutions. *Science, 312*, 60–61.

Henrich, J. (2008). A cultural species. In M. Brown (Ed.), *Explaining culture scientifically*. Seattle: University of Washington Press.

Henrich, J. (2009). The evolution of costly displays, cooperation, and religion: Credibility enhancing displays and their implications for cultural evolution. *Evolution and Human Behavior, 30*(4), 244–260.

Henrich, J., Bowles, S., Smith, E. A., Young, H. P., Boyd, R., Sigmund, K., et al. (2003). The cultural and genetic evolution of human cooperation. In P. Hammerstein (Ed.), *Genetic and cultural evolution of cooperation*. Cambridge, MA: MIT Press.

Henrich, J., & Boyd, R. (1998). The evolution of conformist transmission and the emergence of between-group differences. *Evolution and Human Behavior, 19*, 215–242.

Henrich, J., & Boyd, R. (2001). Why people punish defectors: Weak conformist transmission can stabilize costly enforcement of norms in cooperative dilemmas. *Journal of Theoretical Biology, 208*, 79–89.

Henrich, J., & Boyd, R. (2002). On modeling cultural evolution: Why replicators are not necessary for cultural evolution. *Journal of Cognition and Culture, 2*(2), 87–112.

Henrich, J., & Boyd, R. (2008). Division of labor, economic specialization, and the evolution of social stratification. *Current Anthropology, 49*(4), 715–724.

Henrich, J., Boyd, R., Bowles, S., Camerer, C., Fehr, E., Gintis, H., McElreath, R., Alvard, M., Barr, A., Ensminger, J., Henrich, N. S., Hill, K., Gil-White, F., Gurven, M., Marlowe, F. W., Patton, J. Q., & Tracer, D. (2005). "Economic man" in cross-cultural perspective: Behavioral experiments in 15 small-scale societies. *Behavioral and Brain Sciences, 28*(6), 795.

Henrich, J., Ensminger, J., McElreath, R., Barr, A., Barrett, C., Bolyanatz, A., Cardenas, J. C., Gurven, M., Gwako, E., Henrich, N., Lesorogol, C., Marlowe, F., Tracer, D. P., & Ziker, J. (2010). Market, religion, community size, and the evolution of fairness and punishment. *Science, 327*, 1480–1484.

Henrich, J., & Gil-White, F. (2001). The evolution of prestige: Freely conferred deference as a mechanism for enhancing the benefits of cultural transmission. *Evolution and Human Behavior, 22*(3), 165–196.

Henrich, N., & Henrich, J. (2007). *Why humans cooperate: A cultural and evolutionary explanation.* Oxford: Oxford University Press.

Henrich, J., & Henrich, N. 2010. The evolution of cultural adaptations: Fijian taboos during pregnancy and lactation protect against marine toxins. *Proceedings of the Royal Society of London, Series B: Biological Sciences*, http://rspb.royalsocietypublishing.org/content/early/2010/07/26/rspb.2010.1191.short?rss=1.

Henrich, J., & McElreath, R. (2003). The evolution of cultural evolution. *Evolutionary Anthropology, 12*(3), 123–135.

Henrich, J., & McElreath, R. (2007). Dual inheritance theory: The evolution of human cultural capacities and cultural evolution. In R. Dunbar & L. Barrett (Eds.), *Oxford handbook of evolutionary psychology*. Oxford: Oxford University Press.

Herrmann, B., Thöni, C., & Gächter, S. (2008). Antisocial punishment across societies. *Science, 319*, 1362–1367.

Heyman, J., & Ariely, D. (2004). Effort for payment—a tale of two markets. *Psychological Science, 15*(11), 787–793.

Hill, K. R., & Walker, R., Bozicevic, M., Eder, J., Headland, T., Hewlett, B., Magdalena Hurtado, A., Marlowe, F. W., Wiessner, P., & Wood, B. (n.d.). Unique ancestral human social structure indicated by coresidence patterns in hunter-gatherer societies.

Hruschka, D., & Henrich, J. (2006). Friendship, cliquishness, and the emergence of cooperation. *Journal of Theoretical Biology, 239*(1), 1–15.

Jaswal, V. K. (2004). Don't believe everything you hear: Preschoolers' sensitivity to speaker intent in category induction. *Child Development, 75*(6), 1871–1885.

Jensen, K., Call, J., & Tomasello, M. (2007). Chimpanzees are rational maximizers in an ultimatum game. *Science, 318*(5847), 107–109.

Jensen, K., Hare, B., Call, J., & Tomasello, M. (2006). What's in it for me? Self-regard precludes altruism and spite in chimpanzees. *Proceedings of the Royal Society of London, Series B: Biological Sciences, 273*(1589), 1013–1021.

Johnson, A. (2003). *Families of the forest: Matsigenka Indians of the Peruvian Amazon.* Berkeley: University of California Press.

Johnson, D., Stopka, P., & Knights, S. (2003). The puzzle of human cooperation. *Nature, 421*, 911–912.

Joshi, N. V. (1987). Evolution of cooperation by reciprocation within structured demes. *Journal of Genetics, 66*(1), 69–84.

Keeley, L. (1997). *War before civilization*. Oxford: Oxford University Press.

Kelly, R. C. (1985). *The Nuer conquest.* Ann Arbor: University of Michigan Press.

Kendal, J., Feldman, M. W., & Aoki, K. (2006). Cultural coevolution of norm adoption and enforcement when punishers are rewarded or non-punishers are punished. *Theoretical Population Biology, 70*(1), 10–25.

Kinzler, K. D., Dupoux, E., & Spelke, E. S. (2007). The native language of social cognition. *Proceedings of the National Academy of Sciences of the United States of America, 104*(30), 12577–12580.

Knauft, B. M. (1985). *Good company and violence: Sorcery and social action in a Lowland New Guinea Society.* Berkeley: University of California Press.

Kohler, T. A., VanBuskirk, S., & Ruscavage-Barz, S. (2004). Vessels and villages: Evidence for conformist transmission in early village aggregations on the Pajarito Plateau, New Mexico. *Journal of Anthropological Archaeology, 23*(1), 100–118.

Lachmann, M., & Bergstrom, C. T. (2004). The disadvantage of combinatorial communication. *Proceedings of the Royal Society of London, Series B: Biological Sciences, 271*(1555), 2337–2343.

Laland, K. N., Odling-Smee, J., & Myles, S. (2010). How culture shaped the human genome: Bringing genetics and the human sciences together. *Nature Reviews: Genetics, 11*(2), 137–148.

Langergraber, K. E., Mitani, J. C., & Vigilant, L. (2007). The limited impact of kinship on cooperation in wild chimpanzees. *Proceedings of the National Academy of Sciences of the United States of America, 104*(19), 7786–7790.

Lehmann, L., & Feldman, M. W. (2008). The co-evolution of culturally inherited altruistic helping and cultural transmission under random group formation. *Theoretical Population Biology, 73*(4), 506–516.

Lehmann, L., Feldman, M. W., & Foster, K. R. (2008). Cultural transmission can inhibit the evolution of altruistic helping. *American Naturalist, 172*(1), 12–24.

Lehmann, L., Rousset, F., Roze, D., & Keller, L. (2007). Strong reciprocity or strong ferocity? A population genetic view of the evolution of altruistic punishment. *American Naturalist, 170*(1), 21–36.

Leimar, O., & Hammerstein, P. (2001). Evolution of cooperation through indirect reciprocity. *Proceedings of the Royal Society of London, Series B: Biological Sciences, 268*(1468), 745–753.

Levitt, S. D., & Dubner, S. J. (2009). *Super freakonomics: Global cooling, patriotic prostitutes, and why suicide bombers should buy life insurance*. Toronto: HarperCollins.

Mackie, G. (1996). Ending footbinding and infibulation: A convention account. *American Sociological Review, 61*(6), 999–1017.

Marlowe, F. W. (2003). The mating system of foragers in the standard cross-cultural sample. *Cross-Cultural Research, 37*(3), 282–306.

Marlowe, F. W. (2004). Marital residence among foragers. *Current Anthropology, 45*(2), 277–284.

McElreath, R., & Boyd, R. (2007). *Mathematical models of social evolution: A guide for the perplexed.* Chicago: University of Chicago Press.

McElreath, R., Boyd, R., & Richerson, P. J. (2003). Shared norms and the evolution of ethnic markers. *Current Anthropology, 44*(1), 122–129.

McElreath, R., Lubell, M., Richerson, P. J., Waring, T. M., Baum, W., Edsten, E., Efferson, C., & Paciotti, B. (2005). Applying evolutionary models to the laboratory study of social learning. *Evolution and Human Behavior, 26*(6), 483–508.

McElreath, R., & Strimling, P. (2008). When natural selection favors imitation of parents. *Current Anthropology, 49*(2), 307–316.

Mesoudi, A. (2008). An experimental simulation of the "copy-successful-individuals" cultural learning strategy: Adaptive landscapes, producer-scrounger dynamics, and informational access costs. *Evolution and Human Behavior, 29*(5), 350–363.

Mesoudi, A., Whiten, A., & Dunbar, R. (2006). A bias for social information in human cultural transmission. *British Journal of Psychology, 97*, 405–423.

Mischel, W., & Liebert, R. M. (1966). Effects of discrepancies between observed and imposed reward criteria on their acquisition and transmission. *Journal of Personality and Social Psychology, 3*, 45–53.

Nakahashi, W. (2007). The evolution of conformist transmission in social learning when the environment changes periodically. *Theoretical Population Biology, 72*(1), 52–66.

Newton-Fisher, N. E. (1999). Infant killers of Budongo. *Folia Primatologica, 70*(3), 167–169.

Nowak, M. A., Page, K. M., & Sigmund, K. (2000). Fairness versus reason in the ultimatum game. *Science, 289*(5485), 1773–1775.

Nurmsoo, E., & Robinson, E. J. (2009). Identifying unreliable informants: Do children excuse past inaccuracy? *Developmental Science, 12*(1), 41–47.

O'Gorman, R., Wilson, D. S., & Miller, R. R. (2008). An evolved cognitive bias for social norms. *Evolution and Human Behavior, 29*(2), 71–78.

Ohtsuki, H., & Iwasa, Y. (2004). How should we define goodness? Reputation dynamics in indirect reciprocity. *Journal of Theoretical Biology, 231*(1), 107–120.

Panchanathan, K., & Boyd, R. (2004). Indirect reciprocity can stabilize cooperation without the second-order free rider problem. *Nature, 432*(7016), 499–502.

Pillutla, M. M., & Chen, X. P. (1999). Social norms and cooperation in social dilemmas: The effects of context and feedback. *Organizational Behavior and Human Decision Processes, 78*(2), 81–103.

Price, M., Tooby, J., & Cosmides, L. (2002). Punitive sentiment as an anti-free rider psychological device. *Evolution and Human Behavior, 23*, 203–231.

Rakoczy, H., Wameken, F., & Tomasello, M. (2008). The sources of normativity: Young children's awareness of the normative structure of games. *Developmental Psychology, 44*(3), 875–881.

Richerson, P. J., Boyd, R., & Bettinger, R. L. (2001). Was agriculture impossible during the Pleistocene but mandatory during the Holocene? A climate change hypothesis. *American Antiquity, 66*(3), 387–411.

Richerson, P., & Boyd, R. (1998). The evolution of ultrasociality. In I. Eibl-Eibesfeldt & F. K. Salter (Eds.), *Indoctrinability, ideology, and warfare*. New York: Berghahn Books.

Rilling, J. K., Sanfey, A. G., Nystrom, L. E., Cohen, J. D., Gutman, D. A., Zeh, T. R., et al. (2004). Imaging the social brain with fMRI and interactive games. *International Journal of Neuropsychopharmacology, 7*, S477–S478.

Ross, L., & Ward, A. (1996). Naive realism: Implications for social conflict and misunderstanding. In T. Brown, E. Reed, & E. Turiel (Eds.), *Values and knowledge*. Hillsdale, NJ: Lawrence Erlbaum.

Rushton, J. P. (1975). Generosity in children: Immediate and long term effects of modeling, preaching, and moral judgement. *Journal of Personality and Social Psychology, 31*, 459–466.

Rushton, J. P., & Campbell, A. C. (1977). Modeling, vicarious reinforcement and extraversion on blood donating in adults: Immediate and long-term effects. *European Journal of Social Psychology, 7*(3), 297–306.

Sahlins, M. (1961). The segmentary lineage: An organization of predatory expansion. *American Anthropologist, 63*(2), 322–345.

Sanfey, A. G., Rilling, J. K., Aronson, J. A., Nystrom, L. E., & Cohen, J. D. (2003). The neural basis of economic decision-making in the ultimatum game. *Science, 300*, 1755–1758.

Sigmund, K., De Silva, H., Traulsen, A., & Hauert, C. (2010). Social learning promotes institutions for governing the commons. *Nature, 466*(7308), 861–863.

Silk, J. B., Brosnan, S. F., Vonk, J., Henrich, J., Povinelli, D. J., Richardson, A. S., et al. (2005). Chimpanzees are indifferent to the welfare of unrelated group members. *Nature, 437*, 1357–1359.

Soltis, J., Boyd, R., & Richerson, P. J. (1995). Can group-functional behaviors evolve by cultural group selection? An empirical test. *Current Anthropology, 36*(3), 473–494.

Spencer, C., & Redmond, E. (2001). Multilevel selection and political evolution in the Valley of Oaxaca. *Journal of Anthropological Archaeology, 20*, 195–229.

Sutter, M., & Kocher, M. (2007). Age and the development of trust and reciprocity. *Games and Economic Behavior, 59*, 364–382.

Tabibnia, G., Satpute, A. B., & Lieberman, M. D. (2008). The sunny side of fairness—preference for fairness activates reward circuitry (and disregarding unfairness activates self-control circuitry). *Psychological Science, 19*(4), 339–347.

Taylor, C., & Nowak, M. A. (2007). Transforming the dilemma. *Evolution: International Journal of Organic Evolution, 61*(10), 2281.

Tooby, J., & Cosmides, L. (1989). Evolutionary psychology and the generation of culture: Theoretical considerations. *Ethology and Sociobiology, 10*, 29–49.

Tooby, J., & Cosmides, L. (1992). The psychological foundations of culture. In J. Barkow, L. Cosmides, & J. Tooby (Eds.), *The adapted mind: Evolutionary psychology and the generation of culture.* New York: Oxford University Press.

Trivers, R. (2004). Behavioural evolution: Mutual benefits at all levels of life. *Science, 304*(5673), 964–965.

Trivers, R. L. (1971). The evolution of reciprocal altruism. *Quarterly Review of Biology, 46*, 34–57.

Tversky, A., Kahneman, D., & Slovic, P. (2005). Judgment under uncertainty: Heuristics and biases. In D. L. Hamilton (Ed.), *Social cognition: Key readings* (pp. 167–177). New York: Psychology Press.

Vigilant, L., Hofreiter, M., Siedel, H., & Boesch, C. (2001). Paternity and relatedness in wild chimpanzee communities. *Proceedings of the National Academy of Sciences of the United States of America, 98*(23), 12890–12895.

Vonk, J., Brosnan, S. F., Silk, J. B., Henrich, J., Richardson, A. S., Lambeth, S. P., Schapiro, S. J., & Povinelli, D. J. (2008). Chimpanzees do not take advantage of very low cost opportunities to deliver food to unrelated group members. *Animal Behaviour, 75*, 1757–1770.

Wakano, J. Y., & Aoki, K. (2007). Do social learning and conformist bias coevolve? Henrich and Boyd revisited. *Theoretical Population Biology, 72*(4), 504–512.

Watts, D. P., Muller, M., Amsler, S. J., Mbabazi, G., & Mitani, J. C. (2006). Lethal intergroup aggression by chimpanzees in Kibale National Park, Uganda. *American Journal of Primatology, 68*(2), 161.

West, S. A., Griffin, A. S., & Gardner, A. (2007). Evolutionary explanations for cooperation. *Current Biology, 17*(16), R661–R672.

Wilson, M. L., Wallauer, W. R., & Pusey, A. E. (2004). New cases of intergroup violence among chimpanzees in Gombe National Park, Tanzania. *International Journal of Primatology, 25*(3), 523–549.

Young, D., & Bettinger, R. L. (1992). The Numic spread: A computer simulation. *American Antiquity, 57*(1), 85–99.

Young, H. P. (1998). *Individual strategy and social structure: An evolutionary theory of institutions.* Princeton, NJ: Princeton University Press.

22 Suicide Bombers, Weddings, and Prison Tattoos: An Evolutionary Perspective on Subjective Commitment and Objective Commitment

Daniel M. T. Fessler and Katinka Quintelier

1 Introduction

Consider three hypothetical suicide bombers. The first seeks to die in a suicide attack because he believes that doing so is an effective means of achieving his pragmatic objectives, including obtaining access to sexual opportunities and ensuring the material and spiritual welfare of his family. The second seeks to die in a suicide attack because he is outraged at the treatment that he and others like him have received at the hands of their oppressors, feels that his honor has been tarnished, and therefore longs to visit vengeance upon his enemy. The third seeks to die in a suicide attack because, having already recorded a videotape detailing his plan, the social costs of backing out would be enormous. In common parlance, all three are said to be "committed" to terrorist acts. However, this broad term masks important differences among these cases. For example, the presence of alternative avenues may readily dissuade the first bomber; the second bomber is less easily discouraged, yet may nonetheless falter at the critical moment; the third bomber is more reliable still. In reality, terrorist organizations recognize these differences, and seek to create redundant motivations by employing all three facets in conjunction (see Moghadam, 2003). While practitioners thus appear to have hit upon successful recipes through trial and error or intuition, we believe that scholars studying such topics have not always been as successful, and hence that understanding commitment can be advanced by efforts to systematically define and contrast different types of commitment.

The central feature of our notion of commitment is that, at the time at which commitment is initiated, multiple courses of action present themselves to the actor; in committing, the actor is selecting one of these options to pursue, and, by definition, that choice is intended to endure despite the continued or anticipated attractiveness of alternatives. Importantly, markedly different mechanisms might generate the two features of "selecting among options" and "the durability of the choice." This heterogeneity is reflected in the remarks of Schelling, a seminal contributor to the modern study of commitment: "The ways to commit . . . are many. Legally, one files suit. Reputationally, one takes a public position. Physically, one gathers speed before taking an intersection. Emotionally, one becomes obsessed" (Schelling, 2001, p. 49). In this chapter, we propose that, as inspection of these instances suggests, commitment is an analytically diverse category that is useful inasmuch as it draws attention to some of the many possible factors that influence whether an individual's behavior will be consistent over time.

Reflecting the heterogeneity that we attribute to commitment, the literature on this topic is characterized by a profusion of terms and distinctions. This is not without problems. First, though often useful within a specific subfield of the commitment literature, many distinctions cannot be applied to commitment in the broader literature. For example, Frank (2001, p. 58) draws a fundamental distinction between commitment facilitated by emotions and that facilitated by contracts. Although this is a legitimate distinction, some important commitments involve neither exclusively internal motivations nor other parties, thus fitting into neither category.

The goals of this essay are first to disambiguate two major types of commitment, then to stress the importance of hitherto neglected forms of commitment, and finally to consider how disambiguating the processes at issue sheds light on the evolution of a variety of psychological and social phenomena. We begin by distinguishing between subjective commitment and objective commitment (section 2); we then introduce the notion of a commitment device (section 3). These concepts are all defined in relation to the committing individual—whether others play a role in committing, and what role they play, is irrelevant to these prior distinctions. In section 4 we then introduce various kinds of social interactions that can play a role in commitment, focusing on the costs and benefits that have an impact on third parties in this regard. In section 5 we turn to the communicative facets of commitment involving social interactions, arguing that signals stemming from objective commitments are more reliable than those deriving from subjective commitments. Many social commitments employ existing practices; hence, in section 6 we explore how culture shapes commitment, concluding with conjectures regarding the coevolution of culture and the psychology of commitment.

2 Subjective and Objective Commitment

By *subjective commitment* we mean an internal, psychological phenomenon wherein, either consciously or subconsciously, individuals appraise one course of action as intrinsically superior to other courses of action, leading them to pursue it. In contrast to the first suicide bomber in our opening example (who, by our criteria, would not be described as committed to a suicide attack), in cases of subjective commitment, the selected option is chosen not because it is deemed instrumentally superior, that is, more likely to achieve some objective separate from the course of action itself, but rather because this course of action is valued more highly in and of itself. As illustrated by the second suicide bomber in our opening example, plausible primary factors in subjective commitment are ongoing, recalled, and anticipated emotions, and internalized cultural norms and values relevant to the chosen act. Our notion bears resemblance to the concept of subjective commitment as frequently used in the literature, but draws a cleaner distinction than is sometimes made. For example, Nesse (2001, p. 16) uses the term "subjective commitment" to denote those commitments wherein "fulfillment depends on emotions and concerns about reputation." Because reputation is external to the individual, by our definition, commitments defined by reputational issues are not subjective. Granted, Nesse does stress *concerns about* reputation, but, in our terminology, this would only be relevant to the question of whether one is committed to upholding one's reputation. After all, individuals also have a subjective valuation of money, yet it would be nonsensical (or at least unproductive) to say that an actor who engages in an action because he is paid to do so is subjectively committed *to that course of action*—at most, he might be subjectively committed to earning money. By the same token, in our terminology, an actor who is himself indifferent between two options, but selects one over the other because he knows that others will praise him for doing so, is not subjectively committed to his choice. All this is not to say that reputation is not important for commitment. In section 4, we will argue that reputation can play a significant role in all forms of commitment; this, however, depends on the role of third parties in a focal actor's commitments, a topic that can be treated independently from the subjective or objective nature of the commitment.

As will prove important later, by our definition, in pure subjective commitments, no externally generated costs befall the actor should the selected course of action subsequently be abandoned in favor of another option. We stress the source of costs here because subjective commitment does entail costs, but these arise internally, primarily in the form of aversive emotional states, should the commitment be broken. In contrast, as we define it, *objective commitment* encompasses interactions with the external world that create a situation wherein the actor has narrowed the range of options, in that costs that are external to the actor will be incurred in the event of a subsequent change of course. The case of the third suicide bomber in our opening example illustrates the external nature of such costs: Once the videotape has been made, the bomber's own attitude toward the terrorist plan becomes far less relevant to the costs that he will suffer if he fails to carry it out—were he to back out, dissemination of the videotape would ensure the ostracism of him and his family. Note that for objective commitment to obtain, it is crucial that the actor takes steps, even if only by selecting one course of action over others, that change the costs of alternative courses of action. If such changes are the product of events not involving the actor's own actions, then the term "commitment" does not apply; for example, when robbed at gunpoint, the victim is coerced, not committed, to handing over his wallet.

While subjective and objective processes can both contribute to commitment to a given course of action, in their fullest manifestations, neither form of commitment requires the other form to ensure that commitment is successful, that is, that the selected course of action is followed to its conclusion. Strong subjective commitment requires no objective commitment because the actor's current ranking of the relative desirability of the various courses of actions is wholly predictive of the actor's future rankings—because the valuation of one option over others does not change over time, the course of action selected is pursued to its completion without fail. For example, our second hypothetical suicide bomber may be so firm in his belief in the moral rectitude of his cause, and so unwavering in his hatred of his foe, that neither concern for himself, empathy for his victims, nor obstacles in his way will lead him to abandon his plans. Conversely, strong objective commitment requires no subjective commitment because the alternative courses of action have been made prohibitively expensive (or, in some cases, eliminated entirely). For example, to protect themselves from security forces, some terrorist organizations make it difficult, or even fatal, to leave the organization once one has joined, thereby substantially obviating the need for unwavering ideological or emotional motives (see Crenshaw, 1987; Miller, 2006).

In all objective commitments, the act of choosing alters the cost-benefit ratios, or availability, of the various courses of action. The same is not true of many forms of subjective commitment, as, much of the time, there is no feedback from the course of action to the emotions and values that motivated the choice. However, a subclass of subjective commitments involves selecting a course of action that, by virtue of its inherent subsequent effects on the actor, generates secondary subjective motivations that reinforce the original hierarchy of preferences. For example, an individual who wishes to reduce his salt intake for health reasons will, if he adheres to a low-sodium diet, find that his appetite for salt eventually diminishes, making his new diet more palatable than his old diet. Likewise, a teenager who wishes to overcome the aversive aspects of cigarette smoking in order to appear fashionable will, if she smokes consistently, become chemically dependent on nicotine. Though dramatic, changes in bodily reactions to the actions at hand are not the only source of secondary subjective reinforcers; mere habit may offer mild forms thereof, if only because following practices that have become habitual requires less concentration than deviating from them. Nevertheless, as centuries of discussions in philosophy and theology attest, many of the most important manifestations of subjective commitment lack such secondary subjective reinforcers—when discussing sexual fidelity, courage in battle, and similar challenges, observers have long understood that it is precisely because the relative attractiveness of the options remains unaltered by the choice that strong forms of subjective commitment are often needed if subjective commitment alone is to determine whether the selected course of action is fulfilled. One partial solution to this challenge is to augment the initial choice with secondary choices, as discussed below.

3 Commitment Devices

In some circumstances, once a course of action has been selected from among the available options, provided that the option takes some time to be completed, it is possible to then additionally select a second course of action, unrelated to the first, that serves to increase the probability that the primary course of action will be pursued to its conclusion, that is, that the commitment will be successful. We define a *commitment device* as any action that is taken with the intention of increasing the probability that a commitment will be successful.[1] Although both subjective and objective commitment can exist, and can lead to successful commitment, without the use of commitment devices, both can also be bolstered by commitment devices.

Actors can seek to maintain subjective commitment by structuring their environment in a manner that reinforces or regenerates the initial motivation to commit. For example, after being demoted for drinking on the job, an alcoholic may be subjectively committed to giving up alcohol. However, anticipating that the shame and regret undergirding this commitment will fade over time, while the pain of alcohol withdrawal will increase, the alcoholic may seek to maintain his subjective commitment by posting a copy of his demotion letter on his liquor cabinet, thereby re-eliciting the subjective state that led to his decision. We suggest that, though often not described as such, commitment devices intended to bolster subjective commitment are quite common. Religious symbols and icons displayed in the home evoke feelings of piety and reinforce the choice to forgo temptation; photos of loved ones, lockets, and similar reminders are akin to minor emotional spark plugs, reinvigorating dedication to the goal of benefiting the depicted individuals (e.g., Gonzaga et al., 2008); and so on.

As is true with regard to subjective commitment, commitment devices are not an intrinsic part of objective commitment. Some choices are themselves costly or impossible to reverse, hence taking them objectively commits the actor to pursuing the chosen action. For example, as noted earlier, in some terrorist or criminal organizations, once one has joined the organization, subsequent options are limited to a choice between continued membership and death; in cases such as this, no additional actions need be taken to raise the cost of the alternatives in order to ensure that the selected course of action will be pursued to completion. Nevertheless, just as commitment devices can rekindle subjective commitment, so too can they bolster objective commitment. Moreover, unlike subjective commitment (in which commitment devices usually merely recreate or enhance the initial hierarchy of preferences by, for example, rekindling emotions that contributed to that hierarchy), objective commitment can be generated *de novo* when a commitment device creates costs or eliminates alternatives. For example, choosing to save money for retirement in itself involves no objective commitment, as the act of choosing does not generate tangible costs if the actor changes course at some later date. However, having chosen to save money for retirement, if one then places one's savings in a retirement account having substantial penalties for early withdrawal, one creates an objective commitment, as this commitment device alters the costs of premature spending.

4 Social Facets of Commitment: Investment

Neither subjective commitment nor objective commitment intrinsically requires interaction with other parties. For example, suicide can be pursued in isolation by simply stopping the consumption of food and drink (a process involving subjective commitment, as both the opportunity to change course and the costs of doing so continue to exist until the process is completed) or jumping off a precipice (a process involving objective commitment, as other options evaporate once the action is begun). However, although social interaction is not intrinsic to commitment, interesting ramifications develop when a social component occurs, as social interaction in the context of commitment raises two distinct classes of issues, namely investment and communication.

In political science, economics, and related fields, commitment is discussed primarily as a social phenomenon, principally in terms of its communicative facets. We will have much to say about this, but before turning to communication, we would like to address two arguably simpler facets of social aspects of commitment, namely, social interactions as commitment devices and social interactions as investments in other's commitments.

Because social interactions are themselves powerful elicitors of emotions, social interactions can bolster the emotional underpinnings of subjective commitment. Similarly, because we are prone to both imitate prestigious individuals and conform to the ideas held by a majority of the members of our group, others can have a strong influence on an actor's values. As a consequence of these and similar phenomena, social interactions can reinforce the hierarchy of preferences at the heart of subjective commitment. Together, these effects allow certain forms of social interaction to serve as commitment devices for subjective commitment. For example, our aforementioned alcoholic may join a self-help support group in which he is encouraged to revisit his shameful past failures, is provided with extensive input regarding the value of abstention, and is placed in relationships with successful role models. In this case, the members of the support group cooperate with the focal individual in order to help him fulfill his commitment. Resembling the manner in which objects can help sustain subjective commitment, the support group can operate as a subjective commitment device.

Social interactions can also be a powerful source of commitment devices in objective commitments, as interactions can be used to change the cost or availability of alternative courses of action. In some cases, others are directed to enact a commitment device on behalf of the actor, as when Odysseus instructed his crew to first tie him to the mast of their vessel and then ignore his subsequent orders until they had sailed beyond the point where he would be able to hear the Sirens' seductive songs. In other cases, others are themselves the source of some of the costs and benefits of various courses of action, such that they are part and parcel of the commitment device. For example, publicly swearing an oath or issuing a promise creates a context wherein failing to adhere to the selected course of action entails reputational costs that increase the incentive to adhere to the commitment.

Whenever others play a role in commitment devices, they pay some cost in doing so, even if, in the minimal case, it is only time and attention. This raises the question of why, absent nepotistic or cooperative motives that obtain beyond the given interaction, others are willing to pay such costs. In some cases, others are subject to larger contracts that encompass the given interaction—payroll officers are paid to manage payrolls, including deducting contributions to a retirement plan from the employees' wages; Odysseus's crew was obliged to follow his orders, including those concerning the Sirens; and so on. In other cases, despite the absence of a larger contract, other parties participate in a nonsymmetrical and nonreciprocal manner. For example, in the oath-swearing case, reputational factors are at stake precisely because observers benefit from knowing whether the focal actor is the sort of person who adheres to oaths, cares about others' opinions, and so on—information that is useful to the observers because they can then employ it in making decisions regarding possible relationships with the focal actor. In still other cases, participation is a nonsymmetrical reciprocal act; for example, in Alcoholics Anonymous, it is believed that an experienced sponsor benefits by mentoring a new member, as mentoring purportedly assists the sponsor in his own pursuit of sobriety. Most interesting, perhaps, are those cases in which participation is a symmetrical reciprocal act, that is, all parties are in pursuit of the same objective, and they each generate a commitment device for the other. For example, rotating savings and credit associations (ROSCAs) are a mechanism that, among other functions, provides reciprocal social commitment devices that generate objective commitment to save up lump sums. In a common variant of this institution, at scheduled intervals, each member contributes a fixed amount to a pool, and the pooled funds are given to a single member at each meeting; this cycle is repeated until all members have received the pool one time. Once an individual has joined a ROSCA group, she is objectively committed to adhering to the contribution schedule, as failing to do so will elicit the wrath of those pool recipients whose proceeds would thereby ultimately be less than their net contributions. Although the respective roles are sequential rather than simultaneous, these social commitment devices are relatively symmetrical, as, owing to simple self-interest, each member is motivated to perform the role of enforcer for each of the other members (Fessler, 2002). This symmetry means that the system can be self-sustaining, as it does not rely on any additional benefits to ensure that actors are willing to pay the costs of enacting commitment devices for one another.

In sum, the recognition that enacting commitment devices for others entails investment focuses the investigator's attention on the benefits that those who serve in such a capacity reap; the higher the costs, the greater the benefits must be if the commitment device is to be reliably enacted. Given this, reciprocal symmetrical arrangements such as ROSCAs in which each party performs the same function for the other will often be among the most stable such systems. However, as attentive readers may have noticed, costs and benefits do not need to coincide in time. This creates the possibility of defection: A focal actor may claim to commit to a certain action but then change her behavior as soon as third parties have invested. In the next section, we turn to the question of when signals of commitment are reliable.

5 Social Facets of Commitment: Communication and Reliability

The communicative aspect of commitment has long played a central role in scholarly treatments of the topic (e.g., Schelling, 1960). In both situations involving conflict (ibid.) and situations involving cooperation (Hirshleifer, 1987), the optimal course of action by one party is often contingent on the course of action selected by the other party. In such contexts, commitment is frequently conceptualized as a pledge that is communicated by one party with the intention of influencing the behavior of the other party (Schelling, 2001, p. 48; Hirshleifer, 1987; Nesse, 2001). A signaled commitment is a threat if the signaler pledges to do something at a cost to himself that inflicts a cost on the adversary. The intention is to change the adversary's behavior in a way that will be beneficial to the signaler. At the same time, the adversary's new behavior will also be more beneficial to the adversary in light of the possibility that the threat will be carried out. For example, during an armed robbery, the robber threatens to shoot the victim if he does not hand over his wallet. Shooting is bad for the victim but also bad for the robber, who faces harsher punishment if caught. Before the threat, handing over his wallet would not be in the victim's interest. When threatened, it becomes in both the victim's and the thief's interest that the victim do so. Conversely, a signaled commitment is a promise if the signaler pledges to do something at a cost to himself that will provide a benefit to the other party. The intention is again to change the other party's behavior in a way that will be more beneficial to the signaler. At the same time, the partner's new behavior will also be more beneficial to the partner in the event that the promise is kept. Promises thus differ from threats in that, in cases where the signal is veridical, a threat is only followed through if the adversary does not change behavior. As the examples suggest, threats occur primarily in cases of conflict, whereas promises are common in cases of cooperation.

Threats and promises are members of a larger class of signals describing purported future actions by the signaler that inform the recipient's optimal choice among the available courses of action. In deciding whether to act on such signals, recipients must, however, consider the signal's reliability, where "reliability" is defined in terms of the accuracy of the forecast that the signal provides concerning the signaler's future behavior. As we have seen, changing behavior contingent on a signal of commitment is beneficial only if the commitment is (or would otherwise be) followed through to completion. But how can the recipient be certain that the signal of commitment is reliable? In the robbery example, shooting is very costly to the robber, so he would do well to avoid it, while the victim, knowing this, would do well to keep his wallet if the threat is, in fact, an empty one. Among the factors that influence the reliability of signals of commitment, and hence their impact on recipients' behavior, the distinction between subjective and objective commitment plays a central role.

Subjective commitments can certainly be powerful determinants of behavior. Consider, for example, that the evolved psychological systems that motivate eating and drinking markedly increase the attractiveness of these behaviors as the period of deprivation progresses, yet subjectively committed hunger strikers nonetheless sometimes fulfill their threats by fasting to death in pursuit of political objectives. Nevertheless, despite such dramatic examples of the powerful and enduring nature of some subjective commitments, recipients of signals communicating courses of action that derive from subjective commitment face the problem that such signals can be unreliable for two reasons. First, if the focal actor's internal motivators change between the time of the signal and the time of the behavior of interest to the recipient, then the signal will inaccurately forecast the behavior. The fact that today someone is passionately in love, or fervently dedicated to a political cause, does not preclude the possibility that he or she will not be so tomorrow, or next year—people "fall out of" love, become disillusioned with causes, substitute new beliefs and values for old ones, and so on. Second, the focal actor may seek to deceive the recipient in order to manipulate the recipient's subsequent actions.

As prior investigators (e.g., Frank, 1988) have noted, signals associated with subjective commitment afford deceptive manipulation. By virtue of the fact that the determinants of the focal actor's commitment are internal, they cannot be directly observed by the recipient of the signal. As a consequence, it will often be relatively simple for the focal actor to send signals that falsely convey the nature or degree of the commitment—declarations of love or political dedication are unreliable because it is relatively easy to lie about one's emotions and values. In the abstract, there is no intrinsic connection between sentiments and statements; hence, recipients of statements of subjective commitment are often rightly skeptical of them. And yet, subjective commitment is an undeniably real phenomenon. Seeking to explain the ultimate functions of subjective commitment in part with regard to the question of signaling, investigators have attempted to grapple with the question of how, despite the above considerations, subjective commitment might nonetheless lead to signals of sufficient reliability as to provide a selective advantage to those with a capacity and propensity for subjective commitment.

Hirshleifer (1987) and Frank (1987, 1988) promulgated versions of what has become an influential theory of subjective commitment wherein many emotions are described as evolved mechanisms that not only generate subjective commitment, but, moreover, signal said to other parties by virtue of costly voluntary acts. Around the same time, Gauthier (1986) proposed a somewhat parallel account of internalized moral norms. Hirshleifer, Frank, and Gauthier all propose that witnessing acts that have some cost attached to them, particularly when numerous and distributed over a prolonged period, leads the observer to accurately infer the dispositional nature of the focal actor, that is, to discern that actor's chronologically stable hierarchy of preferences. In other words, through repeated actions, the actor establishes a reputation that accurately captures his enduring propensities. Such accounts hold that the honesty of the signal is maintained through budgetary constraints, cognitive constraints, or both.

Turning first to budgetary constraints, it is claimed that the larger the number of separate signals emitted, the greater the costs of employing a deceptive strategy. Habitually acting in a manner consistent with a given disposition is therefore claimed to provide a reliable index of that disposition to long-term observers. However, as illustrated by the case of spies and sleeper cells, if the benefits are great enough, it will be worth paying substantial costs to repeatedly emit false signals. Moreover, even in cases that do not involve malice aforethought, at the time that an opportunity for substantial gain through defection occurs, past signaling costs are already sunk; hence, from a cost-benefit perspective, the only logical consideration is whether the costs of the reputational damage caused by defection, reduced by the probability of detection, outweigh the benefits of defection. If the benefits of defection are sufficiently large as to outweigh the reputational costs, it is economically rational (calculating over *both* the short and the long term) to defect. Similarly, expenditures in support of a positive reputation are probably characterized by diminishing returns, especially for repeated low-cost actions of similar type, a feature that potentially lowers the threshold for rational defection. In short, cost-benefit considerations suggest that, although many past acts of accurate signaling may well predict future instances when costs are low, there is no inherent strategic impetus for continued consistency when opportunities for substantial gain through Machiavellian manipulation present themselves. This means that, although the mechanisms that underlie much of subjective commitment may well have evolved in part as a result of the benefits of being able to persuade recipients of the validity of signals concerning future action, such mechanisms will often not suffice in this regard in high-stakes situations, that is, precisely those contexts that, from an evolutionary perspective, will often matter most.

A second tenet of Frank's (1987, 1988) position is that the reliability of repeated observations of signals derives in part from the greater cognitive demands of deception relative to honesty. Because sustained pretense is more cognitively taxing than acting in a manner that accurately reflects one's motives, multiple observations over prolonged periods will reveal the reliability of a focal actor's signals, as a manipulative actor will make occasional mistakes that reveal the underlying misrepresentation. While evidence continues to accumulate that deception is indeed more cognitively demanding than honesty (e.g., Walczyk et al., 2009), this argument nonetheless suffers from limitations similar to those that weaken the budgetary reasoning discussed above. First, we can expect cognitive resources to be marshaled in proportion to the importance of the task at hand. In pursuit of substantial benefits through long-term deception, spies and sleeper cells succeed at manipulating others in part through rehearsal and memorization, practices that, though costly, are worth the price given the benefits at issue. Second, all such considerations apply primarily to cases of malice aforethought, and are less relevant to decision making at the time that a new opportunity for substantial gains through defection presents itself. Overall, the observation that deception is sufficiently common as to have apparently selected for evolved mechanisms for detecting cheaters (Cosmides & Tooby, 2005; Verplaetse, Vanneste & Braeckman, 2007) suggests that, over evolutionary time, cognitive constraints have often not been prohibitive when it really matters. In sum, existing signaling accounts of emotions and morality may shed light on these contributors to subjective commitment, but they do not resolve the problem of the fundamentally unreliable nature of signals of subjective commitment.[2]

In contrast to signals associated with subjective commitments, objective commitments are maintained by factors that are external to the individual. As a consequence, observers will often be able to more directly discern the determinants of the focal actor's actions, thus increasing their ability to forecast the focal actor's future behavior. Of course, although the underpinnings of objective commitment are more amenable to inspection by virtue of their status in the external world, signals associated with objective commitment are nevertheless not entirely immune to deceptive manipulation. Focal actors can falsely create the impression that such external factors exist when they do not, often in regard to the purported deployment of commitment devices. For example, Kahn's (1965, p. 11) proposed strategy of winning the drivers' game of chicken by removing the steering wheel and conspicuously throwing it out the window hinges on the truth-value attached by the opponent to the actions witnessed. However, such truth-value is undermined by the fact that these actions are potentially subject to deception, as the thrown wheel may not, in fact, be the steering wheel from the given car, the focal actor may have a secondary means of steering (such as a smaller steering wheel not visible to the opponent), and so on (R. Kurzban, personal comm.). Nonetheless, while both subjective and objective commitment signals are amenable to deception, signals associated with objective commitment can more readily be subjected to systematic scrutiny, thus increasing their reliability.

As a consequence, we may expect objective commitments to have a stronger effect on others' behavior.

A key factor influencing the ability of observers to forecast the behavior of focal actors in situations of objective commitment is the extent of the costs underlying the given commitment. If alternative courses of action have not been eliminated, then signals associated with objective commitments are only as reliable as the size of the costs attending alternative courses of action, as this predicts the likelihood that the focal actor will not subsequently change course—when the costs are low, objective commitment is weak, that is, the actor may subsequently decide that the costs are worth paying, and alter course accordingly. However, once again, by virtue of the fact that the determinants of the costs are external to the focal actor, observers will have greater access to this information than is true in subjective commitment, and hence signals referencing this information will be more reliable than is often the case with regard to subjective commitment.

This does not mean that, in cases in which no deception is involved and the costs of alternative courses of action are high, forecasts based on signals associated with objective commitment are perfect. With the exception of those objective commitments in which alternative courses of action have been irreversibly eliminated, it is possible that external costs and benefits will change before the fulfillment of the course of action, whereafter the focal actor's behavior may then change accordingly. However, particularly when compared with the vicissitudes of emotions and values, many features of the world are quite stable. More importantly, even when they are not stable, changes in them can be observed in ways that changes in internal states cannot. Correspondingly, observers can be expected to have greater confidence in forecasts associated with objective commitment than in those associated with subjective commitment.

Taken together, the above considerations indicate that, although caveats apply, signals associated with subjective commitment will generally be less reliable than signals associated with objective commitment. The focal actor motivated by subjective commitment who stands to benefit from signaling her commitment to others therefore faces the dilemma that, being unreliable, such signals will often likely not have sufficient effect on others' actions. A partial solution is to create an observable commitment device that bolsters subjective commitment, as knowledge of this device can provide some reassurance to observers. For example, our aforementioned demoted alcoholic might document for his employer that he has joined Alcoholics Anonymous. However, because commitment devices of this type merely enhance subjective commitment, signals associated with them still suffer from the problems of reliability intrinsic to subjective commitment itself. Because of the fundamental asymmetry in reliability between subjective and objective commitments, a focal actor motivated by subjective commitment who wishes to substantially increase her ability to influence observers will therefore often be best served by initiating parallel objective commitments through the use of commitment devices.

Consider the problem of forming a stable socio-sexual union (i.e., "marriage," howsoever locally defined). A substantial corpus of literature supports the assertion that, consonant with the central role of reproduction in natural selection, much of contemporary human behavior in this domain reflects the workings of evolved psychological mechanisms (see, e.g., Sefcek et al., 2006, for review). Viewed in evolutionary terms, heterosexual courtship in an open market (i.e., free of ancillary social obligations and constraints) presents a prototypical signaling dilemma of the type described above. For women, a committed male partner affords substantial fitness benefits through provisioning, coparenting, and protection. However, women face a signaling problem: Men who wish to pursue a high-investment strategy run the risk that their partners will be surreptitiously unfaithful, leading them to misallocate their investment by provisioning another man's progeny. A woman who seeks to secure a high-investing partner therefore profits by signaling that she will be faithful to her prospective husband. However, given the benefits to women of securing investment from one man and genes from another (reviewed in Pillsworth & Haselton, 2006), men, in turn, should be skeptical of women's declarations of subjective commitment. Elizabeth Pillsworth and Robert Kurzban (personal comm.) have each proposed that limerence, the form of romantic love motivating sincere courtship (Tennov, 1979), impels the woman to conspicuously and consistently spurn alternative suitors, thereby generating observable objective commitment by narrowing the options available to her.[3] Once a woman has a child by a man, the costs to her of being abandoned by her mate should he suspect infidelity increase. These costs create an objective commitment that enhances the likelihood that the woman will be faithful. With this observable objective commitment in place, the need for limerence diminishes, and, correspondingly, this highly disruptive emotion fades.[4]

Now, consider the same situation from the perspective of a male suitor. Because the obligate biological costs of reproduction are low for men, men have the option of pursuing either a quality-over-quantity strategy (one or a few partners in whom much is invested) or a quantity-over-quality strategy (many partners in whom little is invested). As a consequence, women should be skeptical of men's declarations of subjective commitment in this regard, as some purported dads are likely to be deceptive cads. As in the female case, limerence can motivate men to spurn alternative partners as a commitment device. However, this behavior will generate less commitment, and thus be less informative as a signal, when emitted by a man. A single man can produce a far larger number of children than can a single woman; hence access to female partners is the limiting factor in male reproductive success. Men's pursuit of women will therefore generally be more extensive than women's pursuit of men. As a result, in small groups such as those in which our ancestors lived (see note 3), a limerent woman has the opportunity to spurn a far larger proportion of alternative suitors than does a limerent man, creating a stronger objective commitment, and thus a more reliable signal. However, spurning alternatives is not the only commitment device available to male suitors. Because provisioning loomed large among the benefits that men provided to women in ancestral populations, and hence plays a central role in mating psychology today, one solution available to the male suitor is to provide initial gifts that are sufficiently substantial as to constitute an objective commitment. American folk culture specifies that the man should give the woman a diamond engagement ring and, importantly, that it should cost 25 percent of his annual salary; this is to be followed later by a similarly priced wedding ring. Because the gifts become the property of the recipient (i.e., the man cannot subsequently retrieve the rings),[5] these practices constitute sequential additive objective commitment devices (Sozou & Seymour, 2005; see also Brinig, 1990)—owing to the financial constraints involved, the farther down the path to the altar that the man proceeds, the less feasible it becomes for him to alter course and seek to woo another woman.[6] Lastly, although engagement and wedding rings are a culturally and historically parochial invention, the institution of requiring male suitors (or their families, to whom they are beholden) to provide a substantial up-front payment is not unique to modern nations, occurring in two-thirds of societies (Anderson, 2007).[7] Given the insights that contemporary hunter-gatherers provide into the lifeways of ancestral human populations, it is particularly noteworthy that bridewealth or brideservice (providing labor, rather than wealth, to the bride's family) occur in almost half of extant foraging societies (Apostolou, 2008). In light of substantial cultural variation in the details of such practices, we do not expect the evolved mechanisms underlying male limerence to include any specific practice as part of their innately specified output. Rather, we expect such mechanisms to incline limerent men to find attractive any behavior that involves sacrifices (often concerning resource transfer), particularly those that constrain their options in a conspicuous fashion.[8]

In the above account of limerence, the emotion underlying a subjective commitment leads to behavior that generates an observable objective commitment device, thus creating a signal that allows the targeted party to reliably forecast the focal actor's future actions. We think that this general approach holds considerable promise. For example, whereas existing perspectives on anger in the context of commitment and communication focus on the message that the focal actor is not obeying rational calculations regarding immediate costs and benefits (Hirshleifer, 1987, 2001; Frank, 1988), we suggest that this is only part of the story. Anger is indeed associated with changes in sensitivity to risk, particularly in men (Fessler, Pillsworth & Flamson, 2004). Nevertheless, because risk-taking comes in many forms, engaging in acts that demonstrate this state does not by itself provide a reliable indication that, unless appeased, the focal actor will spitefully inflict costs on the target individual. Displaying evidence of such "temporary irrationality" may thus be a necessary condition for generating a signal that, by virtue of being reliable, can effectively shape the behavior of the target individual, but, by itself, it is not sufficient to do so. Rather, we suggest, as with limerent individuals, the goals of angry individuals are best served when their subjective commitment leads them to act in ways that create objective commitment devices. That anger acts in this fashion is evident in the natural history of violent altercations.

Anger plays a central role in much noninstrumental violence (Fessler, 2010). Altercations of trivial origin often escalate when they take the form of *character contests* wherein individuals attempt to save face in light of perceived challenges or transgressions (Luckenbill, 1977). Particularly for men (who are vastly more violent than women—Daly & Wilson, 1988; Kellerman & Mercy, 1992), a public setting increases the likelihood that such character contests will end in assault (Deibert & Miethe, 2003). In our view, threats and insults uttered in anger can operate as objective commitment devices, as making such statements in public during an altercation creates a situation wherein failing to follow through on the statement entails reputational damage for the speaker and invites additional transgressions against him. Moreover, the reputational costs of both failing to respond to threats and insults and failing to follow through on one's own statements are symmetrical across the two parties in an altercation, a feature that explains the ratcheting escalation (see Felson & Steadman, 1983) of such interactions. Each party is bound by his objective commitment devices to continue pouring fuel on the fire, thus igniting both further subjective commitment by the other party and the corresponding deployment of additional objective commitment devices. The result is often tragic.

6 The Role of Institutions in Commitment

The above discussion of anger underscores the role that reputation can play in objective commitment—some reputations can deter transgressors, whereas others can shape access to opportunities for profitable cooperation. Many classes of public behaviors can thus act as objective commitments or commitment devices, as, once the chosen course of action has been broadcast, deviation from that course can entail reputational costs. However, this is not the only avenue whereby public behaviors can play a role in commitment. When individuals take on a social role associated with culturally defined obligations or responsibilities, failure to properly perform that role may lead not only to lack of inclusion in cooperative ventures but, moreover, to active punishment by other parties. For several reasons, the more public the assumption of responsibilities (and, by extension, the failure to perform them), the more likely such punishment is to occur, and the more severe it is likely to be. First, the more widely the information is disseminated, the greater the likelihood that it will reach prospective punishers. Second, because the costs of punishing for any given punisher are inversely related to the number of punishers, the above feature makes it more likely that any given prospective punisher will punish (see DeScioli & Kurzban, 2009; Boyd, Gintis & Bowles, 2010).

Acts that generate objective commitments by virtue of being public and addressing culturally defined roles are often themselves culturally defined—culture supplies both the roles and the formulaic behaviors whereby one objectively commits to them. For example, in regard to the case mentioned in the introduction, the role of "living martyr," the final stage of preparation for a suicide bombing mission, has been highly codified among Palestinian terrorist organizations, and the testimonial video shot during this phase is likewise culturally dictated, with standardized props and set phrases being employed (see Moghadam, 2003). Performing the prescribed actions that will serve to publicize one's status thus generates powerful objective commitment that bolsters the subjective commitment that originally motivated one's joining the terrorist organization (Sosis & Alcorta, 2008). Importantly, such practices are neither limited to rare behaviors such as suicidal terrorism nor usually intended to publicize the assumption of the role only retrospectively. For example, in the cultures in which they occur, engagement and wedding rings are precursors to a formal marriage ceremony, the components of which are standardized. Ceremonies such as this serve the dual purpose of publicizing the commitments at issue and linking them to a widely shared set of cultural pre- and proscriptions. This linkage increases the likelihood that failure to adhere to cultural strictures will elicit punishment, since locating behavior within a framework of standardized cultural expectations reduces ambiguity as to whether or not a given action is acceptable. In turn, because prospective punishers decide whether or not to punish in part based on the likelihood that others will punish, reducing ambiguity as to the acceptability of any given act increases the likelihood that a given prospective punisher will punish (see DeScioli & Kurzban, 2009; Boyd, Gintis & Bowles, 2010). By formalizing the public acts whereby roles are assumed, culture thus provides actors with powerful objective commitment devices that allow them to reliably signal their future actions to other parties.

Not all societies have formal marriage ceremonies in which there is a public pronouncement of new roles and attendant responsibilities. However, although marriage ceremonies are not universal, it appears that all societies employ institutionalized practices to formalize and publicize commitments in a manner that generates objective commitment. In particular, oaths and similar formal pledges or promises appear to be a human universal (Brown, 1991). Oaths can serve as objective commitment devices via a variety of pathways. Although the degree varies across instances, all oaths and promises are attended by norms of sanctity, such that violating them constitutes a moral failing above and beyond mere dishonesty, inconsistency, or hypocrisy. As such, these acts attach substantial additional reputation- and punishment-derived costs to deviation from the selected course of action. Indeed, oaths often contain within them an invocation of external costs should the oath be violated. A common variant of this is to appeal to supernatural agents or forces as sanctioners, as in "May God strike me dead," or "Cross my heart and hope to die." Supernaturalism is not mandatory, however, and some oaths even contain within them an explicit social contract empowering others to enforce the oath, as in the initiation vow of the Nuestra Familia prison gang in California: "If I go forward, follow me; if I hesitate, push me; if they kill me, avenge me; if I am a traitor, kill me" (quoted in Phelan & Hunt, 1998).

As illustrated by the last example, an extremely common pattern cross-culturally is the use of culturally constituted practices to generate objective commitment to a group. Consonant with the above argument regarding publicity, punishment, and objective commitment, these rites often serve notice of the assumption of a new role. However, their power as commitment mechanisms is often increased through techniques that broadcast the assumed role not only to members of the in-group who serve to enforce the relevant cultural standards, but also to members of out-groups who, by virtue of their own agendas, influence the courses of action available to the focal actor. For example, some of the most powerful rites binding a new member to a group involve permanent body modification. Such advertisements of affiliation can constitute objective commitment devices as, in the event of intergroup hostility or discrimination, emblematic body modification that identifies group identity will elicit hostility from out-group members, thereby casting the initiate's lot with the in-group. Consonant with this thesis, practices such as ritual surgical modification and scarification are more common in groups that engage in intergroup warfare than in peaceful groups or those that engage in intragroup conflict, a pattern explicable in terms of the need for warlike groups to bind their members to the group in order to prevent free riding, obtaining the benefits of the group's aggressivity without paying the (often substantial) costs (Sosis, Kress & Boster, 2007).[9] Similarly, in the United States, prison gangs such as Nuestra Familia are engaged in endemic violent conflict; these gangs employ a graded system of tattooing wherein the more visible the tattoo, the greater the perceived dedication to, and hence status within, the group (Phelan & Hunt, 1998). This is clearly a product of the levels of objective commitment generated by different tattoos, as, for example, individuals tattooed on the face and neck cannot pass as nonmembers of the gang, and hence neutrality is impossible during intergroup conflicts.

7 Commitment and Gene-Culture Coevolution

As previous commentators have noted at length (Hirshleifer, 1987; Frank, 1987, 1988), objective commitment can be used to increase cooperation in a wide variety of domains, as each party can be confident that, because of the increases in the cost of defection entailed by the given objective commitment, the other party will cooperate. Importantly, although people occasionally invent novel objective commitment devices to facilitate cooperation, consonant with their linkage with both culturally defined roles and reputation and punishment, this goal is usually achieved through existing culturally defined practices. That such institutions are frequently available precisely in those contexts in which they are most useful is explicable in terms of cultural group selection. Consider the case of body modification discussed above. Intergroup conflict is an important source of selective pressure in cultural evolution, as groups that can successfully solve the free rider problem can marshal larger and more cohesive combat forces than those that cannot, leading the former to decimate, or conquer and assimilate, the latter (Boyd & Richerson, 2009). Moreover, though dramatic, the case of intergroup conflict is not unique. Because groups characterized by higher rates of cooperation will be more prosperous, more stable, and more competitive than groups characterized by lower rates, cultural group selection can be expected to favor the cultural evolution of a wide variety of institutionalized practices that support cooperation (Boyd & Richerson, 2009). This principle is nicely illustrated in Kanter's (1968) study of utopian communities, in which community longevity is shown to be correlated with both the contribution of all private assets to the common pool and the public renunciation of outside social ties upon initiation. These two measures objectively narrow alternative courses of action for all parties. The magnitude of the cost, together with the group-wide nature of the measure, increase dedication to, and cooperation within, the collective, as all parties know that the others are subject to the same objective commitments.

Noting that longstanding patterns of cultural practice can generate selective pressure for the biological evolution of traits that enhance fitness in such cultural environments, Richerson and Boyd (2001; Boyd & Richerson, 2009) argue that institutions supporting cooperation and collective action have coevolved with psychological propensities that undergird subjective commitment to groups. We concur, noting that many institutions appear exquisitely well designed to bolster subjective commitment and, correspondingly, humans seem markedly susceptible to such practices. Indeed, Kanter's (1968) study also revealed that the longevity of utopian groups is positively related to the presence of ideologies and rituals that involve the diminution of autonomy and surrender to the collective, practices that, in even more extreme form, are found among terrorist organizations (reviewed in Atran, 2003, and Moghadam, 2003), and, in less extreme form, are readily observed in the ceremonies of all major religions, the folk rituals practiced at professional and collegiate sporting events, and so on. Nevertheless, much as we agree with Richerson and Boyd in this regard, as our earlier example of engagement rings implies, we wish to carry their position one step further. We suggest that the psychological features responsible for many kinds of commitment, and not merely commitment to groups, have coevolved with diverse cultural practices that shape commitment. We thus hypothesize that actors are innately prepared to recognize (albeit not always explicitly) the affordances that cultural practices offer as commitment devices, including opportunities to employ institutionalized commitment devices that both add an objective component to subjective commitment and provide other parties with a reliable signal whereby they can forecast the focal actor's future behavior.[10]

The above hypothesis has two entailments, one cognitive-behavioral, the other affective. First, people will seize upon culturally provided means to solve both the personal and the social aspects of commitment problems. For example, Brinig (1990) has compellingly demonstrated that, although the institution of the engagement ring had existed in some form for centuries, the practice only became widespread (and, eventually, highly normative) in the United States following the repeal of breach of promise-to-marry laws in the first half of the twentieth century. These laws had allowed a woman to sue a man for substantial sums if he broke off their engagement, with additional compensation awarded in the event that she had lost her virginity with him during that period (as was fairly common at the time). This legal recourse for women created an objective commitment for men who proposed marriage. As a consequence, women enjoyed increased certainty that proposals would be followed by marriage. When legislatures repealed these laws, both skeptical women and their sincere suitors were left without an institutionalized objective commitment device to solve their cooperation problem. Engagement rings, already known but not widely employed, provided a solution, and a previously rare behavior rapidly became common. Moreover, rings provided affordances absent from breach of promise laws, as wearing the ring signals to a woman's other potential suitors that she is no longer available; accordingly, confidence that his partner will display the ring in public (e.g., because she has elected to size the ring such that it cannot readily be removed) provides additional reassurance to a man that his betrothed will be faithful, that is, the ring can enhance a woman's own efforts at reliable signaling via an objective commitment device. Our point here is that individual commitment psychology and culture work together. On the one hand, cultural practices provide standardized mechanisms for solving cooperative commitment problems, and, on the other hand, such practices are refined and transmitted in part owing to individuals' propensities to recognize the relative utility in this regard of different practices.

The second entailment of our gene-culture coevolutionary account is our claim that humans are unique in that, living within a culturally constituted reality, they experience an intimate feedback between subjective commitment and culturally shaped objective commitment. In our view, participants in cultural practices that generate objective commitment are often deeply moved by those practices precisely because they recognize the objective commitment aspect, and thus understand that their participation provides a reliable signal to others. We expect, for example, that the intensity of the sentiments experienced by a bride and groom at their wedding will increase with the size of the audience and the degree to which the wedding conforms to cultural templates; the same will be true of the allegiance of a suicide bomber to his cause, the devotion of an initiate to his group at an initiation, or the dedication of an oath-taker to the principles espoused in her oath. This is because, as discussed previously, audience size and cultural standardization are determinants of the degree of objective commitment. We propose that the mental mechanisms that generate subjective commitment are sensitive to the extent to which recipients of signals concerning the focal actor's future behavior can accept those signals as reliable—when subjective commitment leads to objective commitment, the focal actor experiences intense emotions in part because she knows that both she and others can be confident that her future path is laid.

In sum, we suggest that, although nonhuman animals likely experience subjective commitment of sorts (Dugatkin, 2001) and certainly recognize the external factors that generate objective commitment, unlike humans, they are not psychologically equipped to recognize how socially transmitted practices can translate subjective commitments into objective commitments, nor are they able to experience an enhancement of subjective commitments through the enactment of cultural commitment devices. Consonant with the highly cooperative nature of our species, humans possess an evolved psychology and a corresponding cultural repertoire that allow us to engage in commitment to a degree unprecedented in evolutionary history.

Acknowledgments

The authors thank Ben Fraser, Kim Sterelny, Richard Joyce, and Brett Calcott for helpful feedback. K. Q. is grateful for the financial support of the Fonds Wetenschappelijk Onderzoek-Vlaanderen: She holds an aspirant doctoral fellowship from the FWO and benefits from a FWO travel grant. During a portion of this project, D. F. was supported by the U.S. Air Force Office of Scientific Research, grant no. FA9550-10-1-0511. D. F., who was subjectively but not objectively committed to writing this essay, is grateful to the editors for their patience and for having given him the opportunity to observe Kim Sterelny in his natural habitat (P. J. O'Reilly's).

Notes

1. Note that, paralleling Elster's (2003, p. 1754) definition of what he terms *precommitment*, our action-based definition of a commitment device excludes internal psychological features, such as emotions, that scholars such as Frank (1988) include as possible commitment devices. We find the latter usage confusing, as it entails splitting the actor into a strategic agent and a subjective agent. The subjective agent's experiences would involve subjective costs or benefits that influence the strategic agent. Rather than enter into a discussion of the plausibility of such splitting, we adopt an approach that simply preserves the unity of the actor.

2. We leave for another day the question of the evolution of emotions and morality, as each topic merits lengthy treatment on its own.

3. Note that the postulated system is only functional if the total pool of possible partners from which the woman selects is fairly small. If the pool is large, then spurning various suitors will not substantially narrow the woman's options, and thus will not reduce the likelihood that she will subsequently select an alternative partner. However, if the pool is small, then each such act removes a nontrivial fraction of the possible alternatives. Recognizing that the human mind evolved in circumstances very different from the present day, although reconstructions of the size of typical ancestral human groups vary, it is unquestionably the case that the scale at issue was vastly smaller than is typical of modern urban societies. Hence, reflecting a situation of evolutionary disequilibrium, the postulated mechanisms have little real signal value in much of the contemporary world.

4. Although not yet subjected to direct testing, this hypothesis is consonant with available findings. In well-nourished urban populations in which the timing of copulation is optimized, time to first conception is approximately six months (Gnoth et al., 2003). This period is likely at least twice as long in hunter-gatherers, who are subject to substantial variation in food availability, and who presumably do not optimize the timing of copulation. Accordingly, in ancestral populations, for fertile couples, first birth would generally have occurred between the end of the second year of marriage and the start of the third year. Correspondingly, Tennov (1979) reports that limerence generally lasts approximately two years. Across 58 societies, divorce rates steadily increase from the first to the fourth years of marriage (Fisher, 1989); data for hunter-gatherer societies are scarcer, but similar patterns appear to apply (see Fisher, 1989, and Blurton Jones et al., 2000), suggesting that this is the critical period during which the relationship must be cemented—indeed, Hadza hunters explicitly state that they will seek a new partner if a child is not born within several years (Marlowe, 2004). Nevertheless, as K. Sterelny (personal comm.) has pointed out, one limitation of this hypothesis is that being rejected as a suitor is not isomorphic with being ruled out as a partner for future extra-pair copulations, and hence, even in small groups akin to those of our ancestors, such behavior on the woman's part does not provide absolute certainty of paternity to the target of her limerence.

5. Men are protected from the possibility that unscrupulous women seeking wealth would procure an engagement ring and then break off the engagement by legal precedents allowing them to retrieve the ring in such cases (see Brinig, 1990).

6. We do not contest prevailing costly signaling explanations (see Bliege Bird & Smith, 2005) of behavior such as this (e.g., that, by virtue of their cost, expensive engagement rings honestly signal a suitor's ability and willingness to invest—Cronk & Dunham, 2007). Our aim here is simply to point out that the same actions can serve as both (i) signals that intrinsically reveal underlying features (as in costly signaling accounts) and (ii) objective commitment devices that increase the reliability of forecasts of future behavior. Much of the time, these two processes will operate in parallel: On the one hand, the greater the relative cost of a signal, the more it is likely to be honest owing to handicap considerations; on the other hand, the greater that relative cost, the more that the signal's production limits alternative courses of action. Indeed, at least under certain simplified assumptions, these two processes are mathematically equivalent (Sozou & Seymour, 2005).

7. Skeptics are often quick to point out that bridewealth is but one version of resource transfer, as dowries also occur. However, dowries are found only in a small fraction of the world's societies (see Anderson, 2007; Apostolou, 2008), and are often associated with social stratification wherein parents are able to move up the social hierarchy by providing inducements for higher-status men to marry their daughters (Anderson, 2007), a very different function from that discussed here. Revealingly, dowry occurs in only 2.44 percent of foraging societies (Apostolou, 2008).

8. Skeptics critical of the notion that natural selection favored adaptations for own mate selection and courtship point to the importance of parents and other family members in marriage decisions in small-scale societies, including hunter-gatherers whose lifeways are thought to resemble those of ancestral humans (Apostolou, 2007, 2008). However, such critiques greatly underestimate a woman's autonomy in such societies, including both strategies that countermand others' efforts in regard to her first marriage, and the diminution of others' influence in regard to subsequent marriages (Pillsworth & Barrett, in prep.).

9. In our reading of the literature on public goods problems in groups vulnerable to free riding, we find that authors often do not explicitly differentiate between (i) costs that are employed as barriers to entry (the theory being that only individuals who are subjectively committed to the aims of the group would be willing to pay them), and (ii) costs that constitute objective commitments, such that their enactment eliminates alternative courses of action or makes them prohibitively expensive. Some authors (e.g., Sosis, Kress & Boster, 2007) emphasize the former, whereas others (e.g., Berman & Laitin, 2008) attend more to the latter. While these have been important contributions, we feel that it is useful to distinguish clearly between signals that constitute sunk costs and signals that constitute objective commitments, as these are not isomorphic.

10. The plausibility of this proposal rests in part on the extent to which precursor adaptations operate proximally using subjective states rather than relying on fixed action patterns. For example, many nonhuman appeasement displays constitute objective commitment devices in that they entail reducing one's immediate fighting capacity through postural changes and the exposure of vulnerable body parts. If, rather than being immutable fixed action patterns, such displays are undergirded by a subjective desire to maximize one's own vulnerability, then any novel circumstance that affords increasing said can be pursued. We are confident that human subjective commitments are flexible in this manner; it is an open question whether the same is true of nonhuman animals, particularly those species of primates from which we are best able to discern the probable attributes of earlier hominid species.

References

Anderson, S. (2007). The economics of dowry and brideprice. *Journal of Economic Perspectives, 21*(4), 151–174.

Apostolou, M. (2007). Sexual selection under parental choice: The role of parents in the evolution of human mating. *Evolution and Human Behavior, 28*(6), 403–409.

Apostolou, M. (2008). Bridewealth and brideservice as instruments of parental choice. *Journal of Social, Evolutionary, and Cultural Psychology, 2*(3), 89–102.

Atran, S. (2003). Genesis of suicide terrorism. *Science, 299*(5612), 1534–1539.

Berman, E., & Laitin, D. D. (2008). Religion, terrorism, and public goods: Testing the club model. *Journal of Public Economics, 92*(10), 1942–1967.

Bliege Bird, R., & Smith, E. (2005). Signaling theory, strategic interaction, and symbolic capital. *Current Anthropology, 46*, 221–248.

Blurton Jones, N. G., Marlowe, F. W., Hawkes, K., & O'Connell, J. F. (2000). Paternal investment and hunter-gatherer divorce rates. In L. Cronk, N. A. Chagnon, & W. Irons (Eds.), *Adaptation and human behavior: An anthropological perspective* (pp. 69–90). Piscataway, NJ: Transaction Publishers.

Boyd, R., Gintis, H., & Bowles, S. (2010). Coordinated punishment of defectors sustains cooperation and can proliferate when rare. *Science, 328*, 617–620.

Boyd, R., & Richerson, P. J. (2009). Culture and the evolution of human cooperation. *Philosophical Transactions of the Royal Society of London, Series B: Biological Sciences, 364*, 3281–3288.

Brinig, M. F. (1990). Rings and promises. *Journal of Law Economics and Organization, 6*(1), 203–215.

Brown, D. E. (1991). *Human universals*. New York: McGraw-Hill.

Cosmides, L., & Tooby, J. (2005). Neurocognitive adaptations designed for social exchange. In D. M. Buss (Ed.), *Evolutionary psychology handbook* (pp. 584–627). New York: Wiley.

Crenshaw, M. (1987). Theories of terrorism: Instrumental and organizational approaches. *Journal of Strategic Studies, 10*(4), 13–31.

Cronk, L., & Dunham, B. (2007). Amounts spent on engagement rings reflect aspects of male and female mate quality. *Human Nature, 18*(4), 329–333.

Daly, M., & Wilson, M. (1988). *Homicide*. New York: Aldine de Gruyter.

Deibert, G. R., & Miethe, T. D. (2003). Character contests and dispute-related offenses. *Deviant Behavior, 24*(3), 245–267.

DeScioli, P., & Kurzban, R. (2009). Mysteries of morality. *Cognition, 112*(2), 281–299.

Dugatkin, L. A. (2001). Subjective commitment in nonhumans: What should we be looking for, and where should we be looking? In R. Nesse (Ed.), *Evolution and the capacity for commitment* (pp. 120–137). New York: Sage Press.

Elster, J. (2003). Don't burn your bridge before you come to it: Some ambiguities and complexities of precommitment. *Texas Law Review, 81*, 1751–1787.

Felson, R. B., & Steadman, H. J. (1983). Situational factors in disputes leading to criminal violence. *Criminology, 21*(1), 59–74.

Fessler, D. M. T. (2002). Windfall and socially distributed willpower: The psychocultural dynamics of Rotating Savings and Credit Associations in a Bengkulu village. *Ethos, 30*, 25–48.

Fessler, D. M. T. (2010). Madmen: An evolutionary perspective on anger and men's violent responses to transgression. In M. Potegal, G. Stemmler & C. D. Spielberger (Eds.), *Handbook of anger: Constituent and concomitant biological, psychological, and social processes* (pp. 361–381). Springer.

Fessler, D. M. T., Pillsworth, E. G., & Flamson, T. J. (2004). Angry men and disgusted women: An evolutionary approach to the influence of emotions on risk taking. *Organizational Behavior and Human Decision Processes, 95*(1), 107–123.

Fisher, H. E. (1989). Evolution of human serial pairbonding. *American Journal of Physical Anthropology, 78*(3), 331–354.

Frank, R. H. (1987). If *Homo economicus* could choose his own utility function, would he want one with a conscience? *American Economic Review, 77*(4), 593–604.

Frank, R. H. (1988). *Passions within reason: The strategic role of the emotions*. New York: Norton.

Frank, R. H. (2001). Cooperation through emotional commitment. In R. M. Nesse (Ed.), *Evolution and the capacity for commitment* (pp. 57–76). New York: Sage Press.

Gauthier, D. (1986). *Morals by agreement*. Oxford: Oxford University Press.

Gnoth, C. D., Godehardt, D., Godehardt, E., Frank-Herrmann, P., & Freundl, G. (2003). Time to pregnancy: Results of the German prospective study and impact on the management of infertility. *Human Reproduction, 18*(9), 1959–1966.

Gonzaga, G., Haselton, M. G., Smurda, J., Davies, M. S., & Poore, J. C. (2008). Love, desire, and the suppression of thoughts of romantic alternatives. *Evolution and Human Behavior, 29*(2), 119–126.

Hirshleifer, J. (1987). On the emotions as guarantors of threats and promises. In J. Dupre (Ed.), *The latest on the best* (pp. 307–326). Cambridge, MA: MIT Press.

Hirshleifer, J. (2001). Game-theoretic interpretations of commitment. In R. M. Nesse (Ed.), *Evolution and the capacity for commitment* (pp. 77–93). New York: Sage Press.

Kahn, H. (1965). *On escalation: Metaphors and scenarios*. New York: Praeger.

Kanter, R. M. (1968). Commitment and social organization: A study of commitment mechanisms in utopian communities. *American Sociological Review, 33*(4), 499–517.

Kellerman, A. A., & Mercy, J. A. (1992). Men, women, and murder: Gender-specific differences in rates of fatal violence and victimization. *Journal of Trauma, 33*(1), 1–5.

Luckenbill, D. F. (1977). Criminal homicide as a situated transaction. *Social Problems, 25*(2), 176–186.

Marlowe, F. W. (2004). Mate preferences among Hadza hunter-gatherers. *Human Nature, 15*(4), 365–376.

Miller, L. (2006). The terrorist mind. I. A psychological and political analysis. *International Journal of Offender Therapy and Comparative Criminology, 50*(2), 121–138.

Moghadam, A. (2003). Palestinian suicide terrorism in the Second Intifada: Motivations and organizational aspects. *Studies in Conflict and Terrorism, 26*(2), 65–92.

Nesse, R. M. (2001). Natural selection and the capacity for subjective commitment. In R. M. Nesse (Ed.), *Evolution and the capacity for commitment* (pp. 1–44). New York: Sage Press.

Phelan, M. P., & Hunt, S. A. (1998). Prison gang members' tattoos as identity work: The visual communication of moral careers. *Symbolic Interaction, 21*(3), 277–298.

Pillsworth, E. G., and Barrett, H. C. (In prep.). Women's subordination and resistance in Shuar marriage: A case for female choice in the evolution of human mating.

Pillsworth, E. G., & Haselton, M. G. (2006). Women's sexual strategies: The evolution of long-term bonds and extrapair sex. *Annual Review of Sex Research, 17*, 59–100.

Richerson, P. J., & Boyd, R. (2001). The evolution of subjective commitment to groups: A tribal instincts hypothesis. In R. M. Nesse (Ed.), *Evolution and the capacity for commitment* (pp. 186–220). New York: Sage Press.

Schelling, T. C. (1960). *The strategy of conflict*. Cambridge, MA: Harvard University Press.

Schelling, T. C. (2001). Commitment: Deliberate versus involuntary. In R. M. Nesse (Ed.), *Evolution and the capacity for commitment* (pp. 48–56). New York: Sage Press.

Sefcek, J. A., Brumbach, B. H., Vasquez, G., & Miller, G. F. (2006). The evolutionary psychology of human mate choice: How ecology, genes, fertility, and fashion influence mating behavior. *Journal of Psychology & Human Sexuality, 18*, 125–135.

Sosis, R., & Alcorta, C. (2008). Militants and martyrs: Evolutionary perspectives on religion and terrorism. In R. Sagarin & T. Taylor (Eds.), *Natural security: A Darwinian approach to a dangerous world* (pp. 105–124). Berkeley: University of California Press.

Sosis, R., Kress, H. C., & Boster, J. S. (2007). Scars for war: Evaluating alternative signaling explanations for cross-cultural variance in ritual costs. *Evolution and Human Behavior, 28*(4), 234–247.

Sozou, P. D., & Seymour, R. M. (2005). Costly but worthless gifts facilitate courtship. *Proceedings of the Royal Society of London, Series B: Biological Sciences, 272*(1575), 1877–1884.

Tennov, D. (1979). *Love and limerence: The experience of being in love*. Chelsea, MI: Scarborough House.

Verplaetse, J., Vanneste, S., & Braeckman, J. (2007). You can judge a book by its cover: The sequel: A kernel of truth in predictive cheating detection. *Evolution and Human Behavior, 28*(4), 260–271.

Walczyk, J. J., Mahoney, K. T., Doverspike, D., & Griffith-Ross, D. A. (2009). Cognitive lie detection: Response time and consistency of answers as cues to deception. *Journal of Business and Psychology, 24*(1), 33–49.

23 Communicative Functions of Shame and Guilt

June P. Tangney, Jeffrey Stuewig, Elizabeth T. Malouf, and Kerstin Youman

Shame and guilt have been variously referred to as self-conscious emotions, moral emotions, secondary or derived emotions, and social emotions. Such varied terminology hints at the rich and varied functions served by shame and guilt among humans. In this chapter, we describe the difference between shame and guilt. We then discuss these varied functions, focusing on shame and guilt as internal processes that bear on self-regulation as well as the communicative functions of shame and guilt in the interpersonal realm.

Conceptualizing Shame and Guilt: Self-Conscious, Moral, Secondary, and Social Emotions

Together with embarrassment and pride, theorists have conceived of shame and guilt as members of a family of "self-conscious" emotions (Lewis, 1990; Fischer & Tangney, 1995) that are evoked by self-reflection and self-evaluation. In the face of transgression or error, the self turns toward the self, evaluating and rendering judgment in reference to standards, rules, and goals. This self-evaluation may be implicit or explicit, consciously experienced or transpiring beneath the radar of our awareness. But importantly, the self is the object of these self-conscious emotions. For example, when good things happen, we may feel a range of positive emotions—joy, happiness, satisfaction, or contentment. But we feel pride in our *own* positive attributes or actions. By the same token, when bad things happen, many different types of negative emotions are possible—sadness, disappointment, frustration, or anger. But feelings of shame and guilt typically arise from a recognition of one's *own* negative attributes or negative behaviors. We feel ashamed of our self, guilty over our behavior, and embarrassed by our pratfalls. Even when we feel shame because of another person's behavior, that person is almost invariably someone with whom we are closely affiliated or identified (e.g., a family member, friend, or colleague closely associated with our self). We experience shame because that person is part of our self-definition.[1]

Shame and guilt are also often cited as two key "moral emotions" because of the presumed role they play in fostering moral behavior and in inhibiting antisocial behavior and aggression. In effect, shame, guilt, embarrassment, and pride function as an emotional moral barometer, providing immediate and salient feedback on our social and moral acceptability. When we sin, transgress, or err, aversive feelings of shame, guilt, or embarrassment are likely to ensue. When we "do the right thing," positive feelings of pride and self-approval are likely to result.

Shame and guilt (and the other "self-conscious" emotions of embarrassment and pride) have been described as "secondary," "derived," or "complex" emotions because they emerge later in development and require several key cognitive abilities (Lewis, 1992; Lewis et al., 1989; Fischer & Tangney, 1995). In contrast to the "basic" emotions (e.g., anger, fear, joy) which emerge very early in life and which are characterized in part by unique, universally recognizable facial expressions, shame, guilt, embarrassment, and pride require the development of a sense of self—a recognition of one's self as separate and distinct from others. In fact, most emotions theorists believe that a recognized self, as distinct from others, is a *prerequisite* for emotions such as embarrassment, shame, guilt, and pride (Lewis, 1992; Stipek, 1995; Stipek, Recchia & McClintic, 1992; Wallbott & Scherer, 1995; see Barrett, 1995, however, for an opposing view). For this reason, very young children (e.g., prior to age 15 months) do not have the cognitive capacity to experience shame and guilt because there is not yet a developed self to be conscious of. Shame and guilt also require the development of a set of standards against which the self (or behavior) is evaluated. Such standards need not be fully internalized (owned by the self as intrinsic values and standards); they may rely heavily on significant others in the social environment. But a sense of what constitutes "good" and "bad," "acceptable" and "unacceptable," "desirable" and "inappropriate" is a precondition for experiences of shame and guilt.

Finally, shame and guilt are part of a family of "social" emotions that arise primarily in interpersonal contexts, evoked by actions that have interpersonal contexts. For example, in a study of both children's and adults' personal shame and guilt narratives, Tangney et al. (1994) and Tangney, Miller et al. (1996) found that the vast majority of shame- and guilt-eliciting events described by respondents occurred in social contexts, involving transgressions or failures known to others. Moreover, as described below, interpersonal concerns of one sort or another are central to the phenomenology of both shame and guilt.

What's the Difference between Shame and Guilt?

Psychologists and laypersons alike tend to use the terms "shame" and "guilt" interchangeably. It is not unusual to hear people refer to "feelings of shame and guilt" or "problems with shame and guilt." In Western societies, people have a tendency to use "guilt" as a nonspecific term to refer to aspects of both emotions. But two decades of research indicate that these are distinct emotions with very different implications for motivation and behavior.

In contrast to earlier anthropological theory (Ausubel, 1955; Benedict, 1946) that conceptualized shame as a "public" emotion arising from public exposure versus guilt as a "private" emotion arising from self-generated pangs of conscience, the weight of empirical evidence supports Helen Block Lewis's (1971) distinction between these often confused emotions. From Lewis's perspective, the crux of the difference between shame and guilt is whether people focus on themselves (their character) or their behavior in response to failures and transgressions. When people feel shame, their focus is on the self ("*I* did that horrible thing"), whereas when people feel guilt, their focus is on a behavior ("I *did* that horrible *thing*"). According to Lewis, this differential focus on self versus behavior gives rise to quite distinct emotional experiences.

Shame is typically the more painful, disruptive emotion because the self, not simply one's behavior, is the object of judgment. When people feel shame about the self, they feel "small," worthless, and powerless. There is the sense that *I* am unworthy, incompetent, or just plain bad. Shamed people also feel exposed. Even though an actual observing audience is not necessary, people in the midst of a shame experience often imagine how their defective self would appear to others. Lewis (1971) described this as a split in self-functioning, where the self is both agent and object of observation and disapproval. As shamed people judge themselves negatively, they become more intensely aware of social evaluation, fearing that others will share their negative judgments. Regarding motivations or "action tendencies," shame often prompts efforts to hide or defend the diminished, defective self (Ketelaar & Au, 2003; Lewis, 1971; Lindsay-Hartz 1984; Tangney, Miller, et al., 1996; Wallbott & Scherer, 1995; Wicker, Payne, & Morgan, 1983).

Guilt is an unpleasant emotion, but it is generally less painful and debilitating than shame. Feelings of guilt involve a negative evaluation of some specific behavior (or failure to act). The failure or transgression is self-relevant, in the sense that the person feels responsible, but it does not carry with it an indictment of the self as a person. In this way, one's sense of identity and self-worth remains intact. When feeling guilt, people focus not on the self but on their *behavior* and its consequences for others. They ruminate over the misdeed, feeling the pain of remorse and regret, wishing they had behaved differently or could somehow undo the bad deed that was done. Regarding action tendencies, whereas shame often motivates hiding, guilt often motivates reparative action—a confession, an apology, or efforts to make amends for the wrongdoing.

This "self versus behavior" distinction between shame and guilt has received impressive empirical support from research using a range of methods—including qualitative case studies (Lewis, 1971; Lindsay-Hartz, 1984; Lindsay-Hartz, de Rivera &

Mascolo, 1995), qualitative analyses of shame and guilt autobiographical narratives (Ferguson, Stegge & Damhuis, 1990; Tangney, 1992; Tangney et al., 1994), quantitative ratings of shame and guilt experiences (e.g., Ferguson, Stegge & Damhuis, 1991; Tangney, 1993; Tangney, Miller et al., 1996; Wallbott & Scherer, 1995; Wicker, Payne & Morgan, 1983), and analyses of participants' counterfactual thinking (Niedenthal, Tangney & Gavanski, 1994). Most recently, Tracy and Robins (2006) used both experimental and correlational methods to demonstrate that internal, stable, uncontrollable self-attributions for failure (i.e., depressogenic attributions) are associated with shame, whereas internal, unstable, controllable (i.e., behavior) attributions are associated with guilt. Together, these studies underscore that shame and guilt are distinct emotional experiences, differing substantially along cognitive, affective, and motivational dimensions.

Shame and Guilt Are Not Equally "Moral" or Adaptive Emotions

Several decades of research suggest that shame and guilt are not equally "moral" or adaptive emotions, in the sense of inhibiting wrongdoing and promoting right-doing, especially vis-à-vis others. On balance, guilt appears to be the more useful emotion, benefiting individuals and their relationships in a variety of ways (Baumeister, Stillwell & Heatherton, 1994; Tangney, 1991; Tangney & Dearing, 2002). Five lines of research illustrate the adaptive functions of guilt, in contrast to the hidden costs of shame (for more comprehensive reviews, see Tangney & Dearing, 2002; Tangney, Stuewig & Mashek, 2007).

Hiding versus Amending

Research consistently shows that shame and guilt lead to contrasting motivations or "action tendencies" (de Hooge, 2008; de Hooge, Zeelenberg & Breugelmans, 2007; Ketelaar & Au, 2003; Lindsay-Hartz, 1984; Tangney, Miller et al., 1996; Wallbott & Scherer, 1995; Wicker et al., 1983). Shame often motivates efforts to deny, hide, or escape the shame-inducing situation. Guilt often motivates reparative action (e.g., confession, apology, efforts to undo the harm done). For example, when people are asked to anonymously describe and rate personal shame and guilt experiences along a number of phenomenological dimensions (Tangney, Miller et al., 1996), their ratings indicate that they feel more compelled to hide from others and less inclined to admit what they had done when feeling shame as opposed to guilt. Thus, although shame is generally the more painful emotion, it is not the more powerful motivator of corrective action. In fact, the pain of shame may be so overwhelming as to be counterproductive (Tangney & Dearing, 2011). Importantly, guilt is not merely a pale version of shame that people can readily shrug off. Rather, it is often sufficient to motivate positive, future-oriented action without overwhelming the self.

Interpersonal Concerns and Other-Oriented Empathy

The situations that give rise to shame and guilt are objectively quite similar in terms of the types of failures and transgressions involved and the degree to which others are aware of the event (Keltner & Buswell, 1996; Tangney, 1992; Tangney & Dearing, 2002; Tangney, Miller et al., 1996; Tracy & Robins, 2006). And they are both "social" emotions in that they most often arise in connection with interpersonal events. Nonetheless, people's interpersonal *concerns* differ, depending on whether they are experiencing shame (about the self) or guilt (about a specific behavior). For example, a study of children's and adults' autobiographical accounts of personal shame and guilt experiences (Tangney et al., 1994) found systematic differences in people's interpersonal focus as they described past failures, misdeeds, and transgressions, depending on whether they were describing shame or guilt events. Among adults, especially, shame experiences were more likely to involve a concern with others' evaluations of the self, whereas guilt experiences were more likely to involve a concern with the effect of one's behavior on others. This difference in "egocentric" versus "other-oriented" interpersonal concerns likely derives from shame's self-focus versus guilt's more specific behavioral focus. A shamed person who is focusing on negative *self*-evaluations would naturally be concerned with others' evaluations of the self, as well. In contrast, a person experiencing guilt is already less self-absorbed (focusing on a negative *behavior* somewhat apart from the self) and thus is more likely to recognize (and become concerned with) the effects of that behavior on others.

Not surprisingly, research consistently indicates that feelings of guilt go hand in hand with other-oriented empathy, whereas, if anything, shame is apt to disrupt people's ability to connect empathically with others. This differential relationship of shame and guilt to empathy is apparent both at the level of emotional dispositions (i.e., shame-proneness and guilt-proneness) and at the level of emotional states (Joireman, 2004; Leith & Baumeister, 1998; Tangney, 1991; Tangney & Dearing, 2002; Tangney et al., 1994).[2]

Externalization of Blame, Anger, and Aggression

Research indicates a robust link between shame and tendencies to externalize blame and anger, again observed at both the dispositional and state levels. Among individuals from all walks of life, proneness to shame is positively correlated with anger, hostility, and the propensity to blame others (Andrews et al., 2000; Bennett, Sullivan & Lewis, 2005; Harper et al., 2005; Harper & Arias, 2004; Paulhus et al., 2004; Stuewig et al., 2010; Tangney, 1990; Tangney, Wagner & Gramzow, 1992; Tangney & Dearing, 2002). In an effort to escape painful feelings of shame, shamed people are inclined to defensively "turn the tables," externalizing blame and anger outward onto whoever is handy as a convenient scapegoat. By doing so, the shamed persons attempt to regain some sense of control and superiority in their lives, but the long-term costs can be steep.

Friends, coworkers, and loved ones may feel confused and alienated by apparently irrational bursts of anger. The link between shame and overt physical aggression, observed in many but not all studies (Tangney, Wagner et al., 1996; for a review, see Tangney, Stuewig & Mashek, 2007), appears to be almost entirely mediated by externalization of blame (Stuewig et al., 2010).

In sharp contrast, guilt-prone individuals are inclined to take responsibility for their transgressions and errors. They do not show the same defensive externalization of blame that has been associated with shame. Externalization of blame has been consistently negatively correlated with guilt at both the state and trait levels (for reviews, see Tangney & Dearing, 2002; Tangney, Stuewig & Mashek, 2007). That is, once they recognize that they have failed or transgressed, their sense of responsibility remains solid. Moreover, guilt-proneness is unrelated to anger—that is, guilt-prone people are just as prone to anger as anyone else. But when angered, guilt-prone individuals are inclined manage their anger constructively (e.g., nonhostile discussion, direct corrective action) and they are *disinclined* toward aggression (Ahmed & Braithwaite, 2004; Lutwak et al., 2001; Paulhus et al., 2004; Tangney, Wagner et al., 1996).

Psychological Symptoms and Substance Abuse

Research consistently shows a link between shame-proneness and all manner of psychological symptoms, including low self-esteem, depression, anxiety, eating disorders, post-traumatic stress disorder (PTSD), suicidal ideation, and substance dependence (for reviews, see Tangney & Dearing, 2002; Tangney, Stuewig & Mashek, 2007). The negative psychological implications of shame are evident across measurement methods, across diverse age groups and across diverse populations.

Although Freud cited guilt (an emotion theoretically rooted in id and/or ego conflicts with the superego) as the culprit behind neurosis and other psychological ills, the empirical findings tell a different story. When guilt is assessed using measures sensitive to Lewis's (1971) distinction between shame about the self versus guilt about a specific behavior, the propensity to experience "shame-free" guilt is essentially unrelated to psychological symptoms. Problems with guilt appear more likely to arise when people have an exaggerated or distorted sense of responsibility for events (Tangney & Dearing, 2002; Zahn-Waxler & Robinson, 1995), or when guilt becomes fused with shame.

Shame and guilt-proneness are also differentially related to substance use and abuse. The propensity to experience shame about the self is positively related to substance use, abuse, and dependence. The propensity to experience guilt about specific behaviors is negatively or negligibly related to substance-use problems. In two studies, adults in recovery programs had lower guilt-prone scores and higher shame-prone scores than individuals in community samples (Meehan et al., 1996; O'Connor et al., 1994). In two samples of undergraduates and one sample of jail inmates (a subset of the current sample), shame-proneness was consistently positively related to both alcohol and drug problems, whereas guilt-proneness was (less consistently) negatively related to such problems (Dearing, Stuewig & Tangney, 2005). In a longitudinal study, shame and guilt proneness in the fifth grade predicted alcohol and drug use as reported at 18 years of age (Kendall, Reinsmith & Tangney, 2002). Children high in shame tended to start drinking earlier than those low in shame and were more likely to later use heroin, uppers, and hallucinogens. Those high in guilt started drinking at a later age than those low in guilt and were less likely to use heroin, with similar trends for marijuana and uppers.

Delinquent, Criminal, and Otherwise "Immoral" Behavior

Because shame and guilt are painful emotions providing negative feedback for wrongdoing, it is often assumed that both motivate individuals to do the right thing. But research tells a different story. There is stronger empirical support for the moral function of guilt as opposed to shame (Stuewig & Tangney, 2007). Among all age groups, guilt-proneness is associated with low levels of antisocial behavior, but there is little evidence for the presumed moral inhibitory functions of shame. If anything, shame-prone individuals have difficulty following the straight and narrow, relative to their peers.

Tibbetts (1997) found that undergraduates' *intention* to drive drunk or shoplift was negatively related to anticipated shame, but not the dispositional tendency to experience shame. Regarding *actual illegal behavior*, in a second study of undergraduates, Tibbetts (2003) found that criminal offending indexed by number of illegal behaviors (including use of marijuana and other illegal drugs) was strongly and consistently negatively related to guilt-proneness whereas results involving shame-proneness were mixed. An overall shame-proneness index, comprising three dispositional measures of shame, was unrelated to illegal behavior, calling into question the often assumed inhibitory function of shame. Similarly, college undergraduates' self-reported moral behaviors (assessed by the Conventional Morality Scale; Tooke & Ickes, 1988) were substantially positively correlated with proneness to guilt, but unrelated to proneness to shame (Tangney, 1994). For example, compared to their less guilt-prone peers, guilt-prone individuals were more likely to endorse such items as "I would not steal something I needed, even if I were sure I could get away with it."

Two prospective studies of community samples have investigated the long-term predictive effects of shame- and guilt-proneness in the domain of criminal behavior. In the first, shame-proneness in the fifth grade was unrelated to either arrests or convictions reported by the participant at age 18. In contrast, guilt-proneness negatively predicted both (Tangney & Dearing, 2002). In the second study, Stuewig and McCloskey (2005) examined whether shame or guilt in early adolescence mediated the relationship between maltreatment in childhood and subsequent delinquency and depression measured in late adolescence. "Guilt-free" shame was unrelated to delinquency using either juvenile court records or self-report of delinquency, while "shame-free" guilt negatively predicted delinquency. Furthermore, guilt-proneness continued to be negatively related to delinquency even when a number of other variables, including symptoms of conduct disorder in childhood and parenting in adolescence, were controlled.

Do studies conducted in community settings involving relatively minor transgressions generalize to more serious offenses among individuals involved in the criminal justice system? We were able to locate only two published studies that examined the behavioral implications of shame and guilt in incarcerated samples.[3] In a small-scale study of incarcerated adolescent male offenders and a community comparison group, shame and guilt proneness only marginally differentiated between groups (Robinson et al., 2007). Consistent with the age-crime curve, however, adolescents from the community sample engaged in antisocial behavior at a fairly high rate. Combining the two samples revealed robust negative correlations between guilt-proneness and self-reported antisocial attitudes and behavior. Shame-proneness was unrelated to these antisocial indicators, with the exception of a *positive* link with aggression and anger. In a much larger German sample of incarcerated adolescents and young adults (ages 14–24), single item shame and guilt ratings assessed within four weeks of incarceration predicted postrelease recidivism (Hosser, Windzio & Greve, 2008). Specifically, shame ratings at the outset of incarceration predicted higher recidivism rates whereas guilt ratings predicted lower recidivism. These findings held when controlling for a host of potentially confounding variables.

Finally, in our own study of "general population" adult male and female jail inmates detained on felony charges (Tangney et al., under review), age, one of the most consistent correlates of crime, was positively correlated with guilt-proneness but unrelated to shame-proneness in this sample of inmates ranging from 18 to 69 years old. Consistent with the age-crime curve (i.e., people's tendency to "age out of" antisocial behavior as they move from late adolescence into adulthood), older inmates evidenced a greater propensity to experience guilt than younger inmates. Additionally, self-control, the backbone of Gottfredson and Hirschi's (1990) general theory of crime, was positively correlated with inmates' propensity to experience guilt and negatively correlated with inmates' proneness to shame. Furthermore, inmates' proneness to guilt was substantially negatively correlated with risk-assessment measures and psychological factors known to predict violent and nonviolent criminal recidivism. Individuals higher in guilt-proneness also self-reported lower levels of antisocial personality, criminogenic cognitions, and scored lower on the PAI Violence Potential Index (VPI). In contrast, inmates' shame-proneness was unrelated to clinician ratings of psychopathy and violence risk, and positively correlated with self-reported antisocial personality and criminogenic cognitions. Finally, inmates' proneness to guilt, assessed shortly after incarceration, was negatively correlated with severity of current charges, prior jail experience, prior felony convictions, and custody level at the jail. In contrast, proneness to shame was unrelated to severity of current charges, prior jail experience, and custody level at the jail. (Proneness to "guilt-free" shame [i.e., the unique variance in shame, unrelated to guilt] was modestly but significantly negatively correlated with serious offense history and prior felony convictions.)

In sum, guilt appears to be a protective factor vis-à-vis severity of crime, involvement in the criminal justice system, and known predictors of recidivism. In contrast, there is little evidence that the propensity to experience shame serves an inhibitory function; in fact, shame-proneness may represent an additional vulnerability among those already involved in the criminal justice system.

Communicative Functions of Shame and Guilt

As fundamentally social beings, we spend much of our life involved in relationships of significance, interacting with people who matter to us. Given such ongoing social interaction, mistakes and transgressions—tactless remarks, unintended slights, betrayals large and small—are inevitable. Feelings of shame and guilt can serve as immediate, painful feedback that we have "done wrong" and that some kind of action is necessary. Equally important, *expressions* of these emotions can serve critical social signaling functions, communicating important messages about our intentions and level of commitment to those whom we've harmed or offended.

Social Signals of Shame and Guilt

The social signals of emotions are most obvious in the case of "basic" emotions because they are paired with universally recognized facial expressions (e.g., anger, disgust, fear, joy). Although the more complex, "self-conscious" emotions such as shame and guilt are not associated with clearly definable, codable facial expressions (Ekman & Rosenberg, 1997; Izard, 1977; Keltner, 1995), nonetheless, shame and guilt can be effectively communicated to others.

Research (Keltner, 1995; Keltner & Buswell, 1996) and clinical observations (Lewis, 1971; Tangney & Dearing, in press) indicate that the social signals of shame lie in the body, not just the face. Shame is most readily recognized by postural cues of submission—hunched shoulders, gaze aversion or downcast eyes. Other more subtle cues signaling the possibility of a shame episode include an abrupt interruption in the flow of conversation accompanied by signs of discomfort or agitation, physical or emotional withdrawal, nervous laughter, stammering, face-touching, lip manipulation, and disproportionate expressions of anger. Relevant to postural cues, recent research has identified a physiological component of the experience of shame—episodes of shame are typically accompanied by elevations in pro-inflammatory cytokines, which prompt muscles to contract into an archetypal posture of submission (Dickerson, Gruenewald & Kemeny, 2004).

In the case of guilt, there are no clear nonverbal expressions—facial, postural, or otherwise (Keltner, 1995; Keltner & Buswell, 1996). Rather, expressions of guilt are communicated exclusively via verbal and behavioral means. People signal their guilt via confessions, apologies, and efforts to repair the harm that was done. A heartfelt "I'm so sorry," timely efforts to replace a valued broken item, fervent and sincere pledges to never again break a promise—these verbal and behavioral expressions communicate a perpetrator's feeling of guilt.

Social Functions of the Communication of Shame

What social functions does the communication of shame serve? One view, based in evolutionary psychology, regards shame as a relatively primitive emotion that served its most adaptive "appeasement" functions in the distant past, in the context of much simpler hierarchical societies, among ancestors whose cognitive processes were less sophisticated. This sociobiological approach put forth by Gilbert (1997) and others (de Waal, 1996; Fessler, 2007; Keltner, 1995; Leary et al., 1992; Leary, Landel & Patton, 1996) emphasizes the appeasement functions of displays of shame, humiliation, and embarrassment among early humans and among nonhuman primates. In brief, displays of shame and embarrassment communicate a subordinate's recognition of offense (deviations from expected patterns of behavior) and his or her submission. Such communication reaffirms the offender's relative rank in a dominance hierarchy and minimizes harmful intragroup aggression. That is, shamelike submissive displays have been shown to defuse anger and aggressive retaliation from dominant peers (de Waal, 1996). Furthermore, the motivation to withdraw—so often a component of the shame experience—may be a useful response, interrupting potentially threatening social interactions, further allowing shamed subordinates to escape imminent threats of attack, and permitting parties to later regroup once the conflict has deescalated. In short, from this perspective, shame evolved as a strategy to limit damage in contexts where the likelihood of aggression was high and the consequences were often life threatening.

Fessler (2007) recently proposed an additional adaptive communicative function of shame arising from an evolutionary perspective. Fessler suggests that as human society has evolved, dominance hierarchies have been replaced by "prestige hierarchies." Whereas in dominance hierarchies, an elevated social position is acquired by threat or force, in prestige hierarchies, individuals are selected to elevated positions *by* the lower rank and file. In other words, elevated positions are *taken* from others in dominance hierarchies; elevated positions are *given* in prestige hierarchies. Fessler believes that as humankind evolved, the "appeasement" functions of shame became less relevant as a means of avoiding bodily injury, since prestige competitions generally do not involve physical aggression.

But he suggests that such signs of appeasement may continue to play an important role in modern society by signaling that one is a trustworthy partner who takes seriously social norms. Modern prestige hierarchies rely heavily on cooperative ventures, which in turn rely on mutual trust. In cooperative ventures, participants risk significant cost by investing time, energy, and/or resources, passing up other opportunities in order to participate in activities that will benefit all involved. The upshot is that there is a high potential for exploitation. For this reason, one's reputation as a trustworthy partner is extremely important. Individuals who transgress, but then express clear signs of shame, may protect their reputation as a trustworthy potential partner who is still "on the same page" as others. In contrast, the reputations of apparently shameless transgressors (e.g., Bernie Madoff) are severely tarnished. They are no longer attractive as trustworthy cooperative partners. As indicated by current research on the differential correlates and motivations associated with shame and guilt, shame may not be the most reliable marker of long-term trustworthy behavior—especially in the wake of transgressions. But it may be preferable to a blatant absence of shame or guilt.

Social Functions of the Communication of Guilt

Humankind has evolved over the millennia not only in terms of physical characteristics, but also in terms of emotional, cognitive, and social complexity. With increasingly complex perspective-taking and attributional abilities, modern human beings have the capacity to distinguish between self and behavior, to take another person's perspective, and to empathize with others' distress. Whereas early moral goals centered on reducing potentially lethal aggression, clarifying social rank, and enhancing conformity to social norms, modern morality centers on the ability to acknowledge one's wrongdoing, accept responsibility, and take reparative action. In this sense, guilt may be the quintessential moral emotion of modern humans (Tangney, 2003).

Baumeister, Stillwell, and Heatherton (1994) identified several "relationship-enhancing functions" that arise from the communication of guilt. First, Baumeister et al. (1994) observed that, in expressing feelings of guilt, people "affirm their social bonds," signaling to one another that the relationship and each other's welfare is important. We feel guilty because we care—an important message of reassurance for those we've hurt or offended. The message—sometimes implicit, sometimes explicit—is that the perpetrator is invested in not repeating the hurtful or offending behavior, and that, despite the rift, trust is warranted. In this sense, Fessler's emphasis on reputation as a trustworthy partner in cooperative relationships seems highly relevant. Our sense is that expressions of guilt (especially when accompanied by apologies and efforts to make reparation) are especially well placed to serve important reputation-repairing functions, perhaps even more so than expressions of shame. The degree to which expressions of guilt may be more effective than expressions of shame in this regard is a matter that remains to be examined empirically.

Second, expressions of guilt can serve to restore equity in a relationship, a function especially important in modern cooperative, communitarian societies as opposed to societies characterized by dominance hierarchies. There are several distinct mechanisms by which guilt may serve this function. Baumeister et al. observe that it is usually the more powerful person in relationships who behaves in a manner that takes advantage of or offends a less powerful partner. Expressions of guilt acknowledging the importance of the relationship (often prompting apologies and concessions) can function to reallocate power and resources, thereby moving the dyad closer to a state of equality. Moreover, interpersonal harm is not necessary to induce feelings of guilt. As Baumeister et al. observe, sometimes positive inequity itself (i.e., getting more than one deserves) is sufficient to engender guilt. Here, too, expressions of guilt (often prompting emotional or tangible concessions) reaffirm a commitment to fairness and equality, bringing the dyad closer to a state of parity.

Third, in the domain of emotion, expressions of guilt can level the affective playing field. As Baumeister et al. point out, in instances of interpersonal harm, the victim is initially the distressed party. (And often, the perpetrator may experience disproportionate positive affect corresponding to the significant benefits accruing from the transgression.) In such cases of affective inequity, experiences and expressions of guilt may serve to "redistribute" emotional distress. When perpetrators subsequently experience and express negative emotions of guilt (spontaneously or in response to guilt-inducing behaviors of the victim), victims are apt to feel better—they feel attended to, cared for, and validated. In effect, negative affect is redistributed, so that the affective experiences of victim and perpetrator are closer in valence. As Baumeister et al. point out, affective similarity enhances communication, empathy, and attraction. Thus, the relationship in the moment in strengthened. (Shame may serve a similar function of redistributing emotional affect, perhaps even more effectively. To the extent, however, that shame fosters denial and externalization of blame and inhibits other-oriented empathy, the advantages of leveling the affective playing field may be outweighed by the other attendant interpersonal costs.)

In sum, feeling guilt leads to positive intrapersonal and interpersonal processes. Expressions of guilt can strengthen relationships in a number of ways, especially in contexts requiring cooperation and interpersonal trust and based on assumptions of equity and fairness.

Notes

1. Further, people may experience self-conscious emotions over a range of objects that are closely linked to the self—by ownership or by affiliation. For example, people may feel shame, embarrassment, or pride over their car, their clothing, or their home. People also feel self-conscious emotions vis-à-vis characteristics and behavior of groups with which they closely identify (e.g., nations, ethnic

groups, favorite soccer teams, alma maters—see Lickel, Schmader & Spanovic, 2007, for a review of group-based shame and guilt). Importantly, in each case, there is a psychologically meaningful link between self and object—the object has been incorporated as part of one's self-definition.

2. Much of the research summarized thus far has focused on emotion *states*—situation-specific experiences of shame and guilt "in the moment." In the realm of moral emotions, researchers are also interested in *dispositional* tendencies to experience shame and guilt in the face of failure or transgression. By definition, shame-prone (or guilt-prone) individuals are more susceptible to state experiences of shame (or guilt), relative to their peers. Notably, shame-prone and guilt-prone people are not thought to walk through life in a constant state of shame or guilt. Rather, when they encounter emotion-relevant situations (e.g., failure or transgression), shame-prone people are inclined to respond with shame about the self, and guilt-prone people are inclined to respond with guilt about a specific behavior. In this way, shame-proneness is conceptually distinct from "internalized shame" defined by Cook (1988) as an "enduring, chronic shame that has become internalized as part of one's identity and which can be most succinctly characterized as a deep sense of inferiority, inadequacy, or deficiency." Internalized shame is thus more akin to low self-esteem, whereas proneness to shame is the propensity to experience episodic shame states in response to failures or transgressions.

Shame-proneness and guilt-proneness are often assessed using scenario-based measures (e.g., the Test of Self-Conscious Affect; Tangney, Wagner & Gramzow, 1989). These measures consist of a series of hypothetical situations that are relatively common (e.g., "You make a mistake at work and find out a coworker is blamed for the error"), followed by brief phenomenological descriptions of shame and guilt reactions, as described in the theoretical, phenomenological, and empirical literature. For the scenario described above, the shame response is "You would keep quiet and avoid the coworker." The guilt response is "You would feel unhappy and eager to correct the situation." People rate their likelihood of responding in each manner indicated. Thus people may endorse both shame and guilt, which can co-occur in a given situation. We favor this scenario-based approach because it is conceptually consistent with our current understanding of guilt as a behavior-specific negative appraisal and affective response experienced in a situational context (for an extended discussion, see Tangney, Youman & Stuewig, 2009). In contrast to global adjective checklists, scenario-based measures provide a means of assessing tendencies to experience guilt about specific behaviors, distinct from shame about the self. A second advantage of the scenario-based approach is that the situation-specific phenomenological descriptions of shame and guilt do not require the respondent to distinguish between the terms "shame" and "guilt."

3. Harris (2003) examined event-specific shame and guilt among drunk driver offenders, following their appearance in court or at a restorative justice conference. In contrast to most extant studies, Harris found no evidence that shame and guilt form distinct factors. This study, however, focused on a unique, homogeneous sample (convicted drunk drivers, many with substance abuse problems) and a single type of transgression. Harris's findings raise the intriguing possibility that individuals with substance abuse problems may not have well-differentiated experiences of shame and guilt. Alternatively, guilt and its attendant empathic focus on the harmed other may be less relevant to transgressions such as drunk driving, which typically do not result in objective physical harm to others. (Only drunk drivers not involved in automobile accidents were selected

for the study.) In short, generalizability is unclear given the unique, homogeneous population (convicted drunk drivers, presumably with substance abuse problems) and the consideration of moral emotional responses to a single type of transgression.

References

Ahmed, E., & Braithwaite, V. (2004). "What, me ashamed?" Shame management and school bullying. *Journal of Research in Crime and Delinquency, 41*(3), 269–294.

Andrews, B., Brewin, C. R., Rose, S., & Kirk, M. (2000). Predicting PTSD symptoms in victims of violent crime: The role of shame, anger, and childhood abuse. *Journal of Abnormal Psychology, 109*, 69–73.

Ausubel, D. P. (1955). Relationships between shame and guilt in the socialization process. *Psychological Review, 62*, 378–390.

Barrett, K. C. (1995). A functionalist approach to shame and guilt. In J. P. Tangney & K. W. Fischer (Eds.), *Self-conscious emotions: The psychology of shame, guilt, embarrassment, and pride* (pp. 25–63). New York: Guilford Press.

Baumeister, R. F., Stillwell, A. M., & Heatherton, T. F. (1994). Guilt: An interpersonal approach. *Psychological Bulletin, 115*, 243–267.

Benedict, R. (1946). *The chrysanthemum and the sword*. Boston: Houghton Mifflin.

Bennett, D. S., Sullivan, M. W., & Lewis, M. (2005). Young children's adjustment as a function of maltreatment, shame, and anger. *Child Maltreatment: Journal of the American Professional Society on the Abuse of Children, 10*(4), 311–323.

Cook, D. R. (1988). The measurement of shame: the Internalized Shame Scale. Paper presented at the Annual Meetings of the American Psychological Association.

Dearing, R. L., Stuewig, J., & Tangney, J. P. (2005). On the importance of distinguishing shame from guilt: Relations to problematic alcohol and drug use. *Addictive Behaviors, 30*, 1392–1404.

Dickerson, S. S., Gruenewald, T. L., & Kemeny, M. E. (2004). When the social self is threatened: Shame, physiology, and health. *Journal of Personality, 72*, 1191–1216.

de Hooge, I. E. (2008). *Moral emotions in decision making: Towards a better understanding of shame and guilt*. Unpublished doctoral dissertation, University of Tilburg, the Netherlands.

de Hooge, I. E., Zeelenberg, M., & Breugelmans, S. M. (2007). Moral sentiments and cooperation: Differential influences of shame and guilt. *Cognition and Emotion, 21*, 1025–1042.

de Waal, F. B. M. (1996). *Good natured: The origins of right and wrong in humans and other animals*. Cambridge, MA: Harvard University Press.

Ekman, P., & Rosenberg, E. L. (1997). *What the face reveals: Basic and applied studies of spontaneous expression using the Facial Action Coding System (FACS)*. New York: Oxford University Press.

Ferguson, T. J., Stegge, H., & Damhuis, I. (1990). Guilt and shame experiences in elementary school-age children. In R. J. Takens (Ed.), *European perspectives in psychology* (Vol. 1, pp. 195–218). New York: Wiley.

Ferguson, T. J., Stegge, H., & Damhuis, I. (1991). Children's understanding of guilt and shame. *Child Development, 62*, 827–839.

Fessler, D. M. T. (2007). From appeasement to conformity: Evolutionary and cultural perspectives on shame, competition, and cooperation. In J. L. Tracy, R. W. Robins & J. P. Tangney (Eds.), *The self-conscious emotions: Theory and research* (pp. 174–193). New York: Guilford Press.

Fischer, K. W., & Tangney, J. P. (1995). Self-conscious emotions and the affect revolution: Framework and introduction. In J. P. Tangney & K. W. Fischer (Eds.), *Self-conscious emotions: Shame, guilt, embarrassment, and pride* (pp. 3–22). New York: Guilford Press.

Gilbert, P. (1997). The evolution of social attractiveness and its role in shame, humiliation, guilt, and therapy. *British Journal of Medical Psychology, 70*, 113–147.

Gottfredson, M. R., & Hirschi, T. (1990). *A general theory of crime*. Stanford, CA: Stanford University Press.

Harper, F. W. K., Austin, A. G., Cercone, J. J., & Arias, I. (2005). The role of shame, anger, and affect regulation in men's perpetration of psychological abuse in dating relationships. *Journal of Interpersonal Violence, 20*, 1648–1662.

Harper, F. W. K., & Arias, I. (2004). The role of shame in predicting adult anger and depressive symptoms among victims of child psychological maltreatment. *Journal of Family Violence, 19*(6), 367–375.

Harris, N. (2003). Reassessing the dimensionality of the moral emotions. *British Journal of Psychology, 94*, 457–473.

Hosser, D., Windzio, M., & Greve, W. (2008). Guilt and shame as predictors of recidivism: A longitudinal study with young prisoners. *Criminal Justice and Behavior, 35*(1), 138–152.

Izard, C. E. (1977). *Human emotions*. New York: Plenum Press.

Joireman, J. (2004). Empathy and the self-absorption paradox II: Self-rumination and self-reflection as mediators between shame, guilt, and empathy. *Self and Identity, 3*, 225–238.

Keltner, D. (1995). Signs of appeasement: Evidence for the distinct displays of embarrassment, amusement, and shame. *Journal of Personality and Social Psychology, 68*, 441–454.

Keltner, D., & Buswell, B. N. (1996). Evidence for the distinctness of embarrassment, shame, and guilt: A study of recalled antecedents and facial expressions of emotion. *Cognition and Emotion, 10*, 155–171.

Kendall, S., Reinsmith, C., & Tangney, J. P. (2002). Implications of childhood shame and guilt or criminal behavior in early adulthood. Paper presented at the meetings of the American Society of Criminology, Chicago.

Ketelaar, T., & Au, W. T. (2003). The effects of feelings of guilt on the behavior of uncooperative individuals in repeated social bargaining games: An affect-as-information interpretation of the role of emotion in social interaction. *Cognition and Emotion, 17,* 429–453.

Leary, M. R., Britt, T. W., Cutlip, W. D., II, & Templeton, J. L. (1992). Social blushing. *Psychological Bulletin, 112,* 446–460.

Leary, M. R., Landel, J. L., & Patton, K. M. (1996). The motivated expression of embarrassment following a self-presentational predicament. *Journal of Personality, 64,* 619–637.

Leith, K. P., & Baumeister, R. F. (1998). Empathy, shame, guilt, and narratives of interpersonal conflicts: Guilt-prone people are better at perspective taking. *Journal of Personality, 66,* 1–37.

Lewis, H. B. (1971). *Shame and guilt in neurosis.* New York: International Universities Press.

Lewis, M. (1990). Thinking and feeling—the elephant's tail. In C. A. Maher, M. Schwebel, & N. S. Fagley (Eds.), *Thinking and problem solving in the developmental process: International perspectives* (pp. 89–110). Hillsdale, NJ: Erlbaum.

Lewis, M. (1992). *Shame: The exposed self.* New York: Free Press.

Lewis, M., Sullivan, M. W., Stanger, C., & Weiss, M. (1989). Self-development and self-conscious emotions. *Child Development, 60,* 146–156.

Lickel, B., Schmader, T., & Spanovic, M. (2007). Group-conscious emotions: The implications of others' wrongdoings for identity and relationships. In J. L. Tracy, R. W. Robins, & J. P. Tangney (Eds.), *The self-conscious emotions: Theory and research* (pp. 351–370). New York: Guilford Press.

Lindsay-Hartz, J. (1984). Contrasting experiences of shame and guilt. *American Behavioral Scientist, 27,* 689–704.

Lindsay-Hartz, J., de Rivera, J., & Mascolo, M. (1995). Differentiating shame and guilt and their effects on motivation. In J. P. Tangney & K. W. Fischer (Eds.), *Self-conscious emotions: Shame, guilt, embarrassment, and pride* (pp. 274–300). New York: Guilford.

Lutwak, N., Panish, J. B., Ferrari, J. R., & Razzino, B. E. (2001). Shame and guilt and their relationship to positive expectations and anger expressiveness. *Adolescence, 36*(144), 641–653.

Meehan, M. A., O'Connor, L. E., Berry, J. W., Weiss, J., Morrison, A., & Acampora, A. (1996). Guilt, shame, and depression in clients in recovery from addiction. *Journal of Psychoactive Drugs, 28,* 125–134.

Niedenthal, P. M., Tangney, J. P., & Gavanski, I. (1994). "If only I weren't" versus "If only I hadn't": Distinguishing shame and guilt in counterfactual thinking. *Journal of Personality and Social Psychology, 67,* 585–595.

O'Connor, L. E., Berry, J. W., Inaba, D., & Weiss, J. (1994). Shame, guilt, and depression in men and women in recovery from addiction. *Journal of Substance Abuse Treatment, 11,* 503–510.

Paulhus, D. L., Robins, R. W., Trzesniewski, K. H., & Tracy, J. L. (2004). Two replicable suppressor situations in personality research. *Multivariate Behavioral Research, 39,* 303–328.

Robinson, R., Roberts, W. L., Strayer, J., & Koopman, R. (2007). Empathy and emotional responsiveness in delinquent and non-delinquent adolescents. *Social Development, 16*(3), 555–579.

Stipek, D. (1995). The development of pride and shame in toddlers. In J. P. Tangney & K. W. Fischer (Eds.), *Self-conscious emotions: Shame, guilt, embarrassment, and pride* (pp. 237–252). New York: Guilford Press.

Stipek, D., Recchia, S., & McClintic, S. (1992). Self-evaluation in young children. *Monographs of the Society for Research in Child Development, 57* (1, Serial No. 226).

Stuewig, J., & McCloskey, L. A. (2005). The relation of child maltreatment to shame and guilt among adolescents: Psychological routes to depression and delinquency. *Child Maltreatment, 10*, 324–336.

Stuewig, J., & Tangney, J. P. (2007). Shame and guilt in antisocial and risky behaviors. In J. L. Tracy, R. W. Robins, & J. P. Tangney (Eds.), *The self-conscious emotions: Theory and research* (pp. 371–388). New York: Guilford Press.

Stuewig, J., Tangney, J. P., Heigel, C., Harty, L., & McCloskey, L. (2010). Shaming, blaming, and maiming: Functional links among the moral emotions, externalization of blame, and aggression. *Journal of Research in Personality, 44*(1), 91–102.

Tangney, J. P. (1990). Assessing individual differences in proneness to shame and guilt: Development of the self-conscious affect and attribution inventory. *Journal of Personality and Social Psychology, 59*, 102–111.

Tangney, J. P. (1991). Moral affect: The good, the bad, and the ugly. *Journal of Personality and Social Psychology, 61*, 598–607.

Tangney, J. P. (1992). Situational determinants of shame and guilt in young adulthood. *Personality and Social Psychology Bulletin, 18*, 199–206.

Tangney, J. P. (1993). Shame and guilt. In C. G. Costello (Ed.), *Symptoms of depression* (pp. 161–180). New York: John Wiley.

Tangney, J. P. (1994). The mixed legacy of the super-ego: Adaptive and maladaptive aspects of shame and guilt. In J. M. Masling & R. F. Bornstein (Eds.), *Empirical perspectives on object relations theory* (pp. 1–28). Washington, DC: American Psychological Association.

Tangney, J. P. (2003). Self-relevant emotions. In M. Leary & J. P. Tangney (Eds.), *Handbook of self and identity* (pp. 384–400). New York: Guilford Press.

Tangney, J. P., & Dearing, R. (2002). *Shame and guilt.* New York: Guilford Press.

Tangney, J. P., & Dearing, R. L. (2011). Working with shame in the therapy hour: Summary and integration. In R. L. Dearing & J. P. Tangney (Eds.), *Shame in the therapy hour* (pp. 375–404). Washington, DC: APA Books.

Tangney, J. P., Marschall, D. E., Rosenberg, K., Barlow, D. H., & Wagner, P. E. (1994). *Children's and adults' autobiographical accounts of shame, guilt and pride experiences: An analysis of situational determinants and interpersonal concerns.* Unpublished manuscript, George Mason University, Fairfax, VA.

Tangney, J. P., Miller, R. S., Flicker, L., & Barlow, D. H. (1996). Are shame, guilt, and embarrassment distinct emotions? *Journal of Personality and Social Psychology, 70*, 1256–1269.

Tangney, J. P., Stuewig, J., Mashek, D., & Hastings, M. (2011). Assessing jail inmates' proneness to shame and guilt: Feeling bad about the behavior or the self? *Criminal Justice and Behavior, 38*, 710–734.

Tangney, J. P., Stuewig, J., & Mashek, D. J. (2007). Moral emotions and moral behavior. *Annual Review of Psychology, 58*, 345–372.

Tangney, J. P., Wagner, P., & Gramzow, R. (1989). *The test of self-conscious affect*. Fairfax, VA: George Mason University.

Tangney, J. P., Wagner, P., & Gramzow, R. (1992). Proneness to guilt, and psychopathology. *Journal of Abnormal Psychology, 101*, 469–478.

Tangney, J. P., Wagner, P. E., Hill-Barlow, D., Marschall, D. E., & Gramzow, R. (1996). Relation of shame and guilt to constructive versus destructive responses to anger across the lifespan. *Journal of Personality and Social Psychology, 70*, 797–809.

Tangney, J. P., Youman, K., & Stuewig, J. (2009). Proneness to shame and proneness to guilt. In M. R. Leary & R. H. Hoyle (Eds.), *Handbook of individual differences in social behavior* (pp. 192–209). New York: Guilford Press.

Tibbetts, S. G. (1997). Shame and rational choice in offending decisions. *Criminal Justice and Behavior, 24*, 234–255.

Tibbetts, S. G. (2003). Self-conscious emotions and criminal offending. *Psychological Reports, 93*, 101–126.

Tooke, W. S., & Ickes, W. (1988). A measure of adherence to conventional morality. *Journal of Social and Clinical Psychology, 6*, 310–334.

Tracy, J. L., & Robins, R. W. (2006). Appraisal antecedents of shame and guilt: Support for a theoretical model. *Personality and Social Psychology Bulletin, 32*(10), 1339–1351.

Wallbott, H. G., & Scherer, K. R. (1995). Cultural determinants in experiencing shame and guilt. In J. P. Tangney & K. W. Fischer (Eds.), *Self-conscious emotions: The psychology of shame, guilt, embarrassment, and pride* (pp. 465–487). New York: Guilford Press.

Wicker, F. W., Payne, G. C., & Morgan, R. D. (1983). Participant descriptions of guilt and shame. *Motivation and Emotion, 7*, 25–39.

Zahn-Waxler, C., & Robinson, J. (1995). Empathy and guilt: Early origins of feelings of responsibility. In J. P. Tangney & K. W. Fischer (Eds.), *Self-conscious emotions: The psychology of shame, guilt, embarrassment, and pride* (pp. 143–173). New York: Guilford Press.

24 Moral Disgust and the Tribal Instincts Hypothesis

Daniel R. Kelly

1 Introduction: Some Interesting Facts about Morality

Psychological research has been discovering a number of puzzling features of morality and moral cognition recently.[1] Zhong and Liljenquist (2006) found that when people are asked to think about an unethical deed or recall one they themselves have committed in the past, issues of *physical cleanliness* become salient. Zhong and Liljenquist cleverly designate this phenomenon the "Macbeth Effect," and it takes some interesting forms. For instance, reading a story describing an immoral deed increased people's desire for products related to cleansing, like shower soap, disinfectants, or antiseptic wipes. Moreover, Zhong and Liljenquist found that cleaning one's hands after describing a past unethical deed actually reduced moral emotions such as guilt and shame—so much so that those who did "wash away their sins" were less likely than other participants to help out a desperate graduate student. Other researchers report similar findings. Schnall and her colleagues explored how issues of cleanliness influence judgments of moral *severity*. In one experiment (Schnall et al., 2008b), feelings of disgust were induced in participants in a number of ways, including having them remember a disgusting experience, having them watch a disgusting video, or having them fill out questionnaires at a filthy desk or in the presence of a foul and unpleasant odor. Schnall and colleagues found that many participants' moral judgments about events described in vignettes were more severe in such disgusting conditions, even though what triggers the feelings of disgust (a bad smell, a disgusting memory) is "extraneous" to the events described in the vignettes, about which the moral judgments are putatively being made. In another (Schnall et al., 2008a), one group of participants was made to physically wash their hands between experiencing disgust and making a series of moral judgments, while a second group was not. Those who had just washed themselves down were significantly less severe in their condemnation of the vignettes they were asked to evaluate.

Another series of experiments has sought to demonstrate a link between morality and worries about contamination—or more precisely, it explored the link between often tacit worries about contamination, moral taint, and *immorality*. Paul Rozin and his colleagues found that many people are slightly reluctant to put on a sweater that once belonged to and was worn by an undescribed stranger, even if it had subsequently been thoroughly cleaned. They also found that people tend to become increasingly reluctant to put on, and in some cases even *touch*, a similarly laundered sweater if they are told that the previous owner had committed some extreme moral violation such as murder. The link between immorality and contamination was made especially evident by the piece of clothing that people were most reluctant to come into contact with, which was treated as the most aversive and contaminated of all: a sweater that once belonged to the ultimate moral monster, Adolf Hitler (Rozin, Lowery & Ebert, 1994; see also Haidt et al., 1997, for discussion).

Coming at the issue of morality from another angle are those exploring prejudicial attitudes that members of one group may harbor about members outside their own group. As is all too familiar, either from anecdotal report or even from common first-hand experience, members of one cultural "in-group" will often consider members of other cultural "out-groups" to be below them in one way or another. In exaggerated cases of this, one group is liable to completely demonize and dehumanize the other, considering them not fully human but merely animal, undeserving of any moral treatment or consideration whatsoever, and somehow tainted and tainting. The Indian caste system is often said be deeply informed by worries about the threat of contamination flowing up from lower to higher castes, or even from those, like Europeans, outside the caste structure itself (see Bouglé, 1971, especially pp. 22–23; see also Shweder et al., 1997, and Rozin et al., 1999). Perhaps the most extreme and well-known example was the Nazi attitude toward Jews. Indeed, anti-Jewish Nazi propaganda flagrantly invoked the imagery and language of purity, contamination, and dehumanization. Hitler's rhetoric portrayed Jews as maggots in a festering abscess, hidden away inside the clean, healthy body of the nation. Experimenters have begun exploring the psychological bases of a variety of kinds of prejudices, and one of the most interesting findings has been that among members of a particular in-group, different emotions are commonly associated with attitudes toward different out-groups (Cottrell & Neuberg, 2005).

Finally, Wheatley and Haidt (2005) have used the emotion of disgust to induce some striking and seemingly irrational moral judgments as well. In one of their most devious experiments, they ask subjects to consider the following vignette:

> Dan is a student council representative at his school. This semester he is in charge of scheduling discussions about academic issues. He often picks topics that appeal to both professors and students in order to stimulate discussion.

All of those given this vignette were asked to rate how morally wrong Dan's behavior is. However, some of the subjects were also hypnotically induced to undergo a brief flash of disgust when they saw the word "often," which occurs near the middle of the vignette. Amazingly, many of those hypnotized subjects judged Dan to be doing something morally wrong. They arrived at this judgment despite the completely innocuous description of Dan's actions. Moreover, they were remarkably persistent in standing by their initial impulses; they upheld their dim opinion of Dan even after admitting they had little or no justification for the judgment. Indeed, when asked to explain themselves, participants ended up saying things like Dan seems like a "popularity seeking snob," and that he "just seems like he's up to something." Most revealing, perhaps, was one subject who bluntly stated "I don't know [why it's wrong], it just is."

The explicit mention of disgust in the last example points to a common thread running through all of these otherwise disparate findings. Indeed, these are just a few examples of recent work that has been exploring a link between disgust and morality that is fascinating, puzzling, and often troubling. In the rest of this chapter, I will offer an account of moral disgust that illuminates this link, and makes sense of some of its more unsettling features. In the next section, I will briefly sketch the tribal instincts hypothesis, a component of gene-culture coevolutionary theory, which posits a set of cognitive mechanisms that allow humans to navigate an important part of their social world. This sketch will provide a theoretic background for the rest of the chapter. In the third section, I will advance and defend my account of the basic disgust response. This account, encapsulated by what I call the Entanglement thesis, holds that disgust has a pair of primary functions, one associated with protecting the digestive system from potentially poisonous foods, and the other geared toward preventing disease and parasitic infection in general. Finally, in the fourth section, I put the pieces together and argue for the Co-opt thesis, which holds disgust acquired several auxiliary functions associated with sociality, tribal instincts, and morality, without losing the features that allowed it to adequately perform its primary functions associated with poisons and diseases. Finally, I show how my account is able to shed light on some of the puzzling features of morality and moral cognition mentioned above, for instance, how concerns with spiritual purity or moral taint can be explained as products of a mismatch between elicitor and response, produced when this emotion was brought to bear on issues that it did not initially evolve to deal with.

2 The Tribal Instincts Hypothesis: Social Norms and Ethnic Boundary Markers

The tribal instincts hypothesis supposes that the human ability to interact and cooperate on the scale of entire tribes and cultures is facilitated by a number of reliably produced impulses, which are at least partly innately structured. Indeed, Richerson and Boyd (1998, 1999) argue that human "ultrasociality" is too complex to be fully explained by appeal to only kinship and reciprocity, the standard resources of evolutionary theorists attempting to explain cooperation, and so these tribal instincts promise to be a crucial part of a complete explanation of distinctively human social structure. The tribal instincts hypothesis itself is derived from a much broader theory concerned with human nature and evolution, namely gene-culture coevolutionary theory, or GCC for short. As its name suggests, both culture and cultural evolution, on the one hand, and genetic and biological evolution, on the other, fall within the scope of GCC. However, it does not treat either of these topics in isolation from the other. Rather, GCC sees cultural and genetic evolution as deeply intertwined in humans and seeks to understand the ways in which culture can and has influenced genetic evolution, and alternatively, how genetic factors have influenced the evolution of cultures (Boyd & Richerson, 1985, 2005a; see also Richerson & Boyd, 2005, for a more accessible overview).

GCC conceives of culture very broadly, as information in the social environment that can be passed from one member to another by social (therefore nongenetic) means. Culture can be transmitted not just through populations but also across generations. Information accumulates as it is passed from one generation to the next, and as such, the entire body of information can be seen as an inheritance system, sharing some important properties with the genetic inheritance system. One factor motivating the tribal instincts hypothesis is the insight that continuous and increasing reliance on the information in this cultural inheritance system imposes new requirements on those who rely on it. As both the volume and import of the information in the cultural inheritance system increase, new selective pressures would be created that favor certain psychological capacities, namely those capacities allowing individuals to easily access and use information stored in the cultural repository.

One important result of humans' extended immersion in culture is that among the many types of socially relevant cues to which humans are sensitive (indications of kinship, hints about others' intentions, etc.) are cues about norms and indicators of what kinds of norms others embrace. That is, humans have become innately disposed to see their social world in tribal terms, and to react accordingly.[2] The enfolding of cultural and natural selective pressures in humans evolution has produced a set of social, tribal "instincts" that are sensitive to particular types of cultural information, namely information that facilitates living within the context of large, cooperative groups or tribes. Into this category falls information about both *social norms* and *ethnic boundaries*.

It is nearly a platitude that human social interactions are regulated by complex systems of norms. GCC sees enormous importance in this platitude, however, and suggests that it is largely these systems of behavior-guiding social norms that make it possible for humans to smoothly interact in large groups. Pairing this evaluation of the importance of social norms with the tribal instinct hypothesis yields the prediction that humans will have a distinctive capacity to cognize those social norms. Recent research suggests that this is, indeed, the case. The details are far from settled, but it appears that the capacity has a number of important features, including the ability to easily acquire and internalize norms from the social environment. Once a norm is internalized, the capacity produces characteristic types of motivation to both comply with the norm and to punish those who violate it (Sripada & Stich, 2007; see also Nichols, 2004).

Whereas social norms help coordinate social interactions within a tribe, ethnic symbols serve to mark the boundaries between different tribes. Such symbols, or ethnic boundary markers, as they are often called in the GCC literature, allow members of the same tribe to identify and selectively interact with each other (McElreath, Boyd & Richerson, 2003; Barth, 1969). Members of the same tribe, in this sense, share a large set of beliefs and values. More importantly, they also share large clusters of social and moral norms. Such beliefs, values, and norms are not themselves immediately visible to the naked eye, however. More directly and easily detectable symbols of various sorts (displayed colors, styles of clothing, use of different dialects, varieties of cuisine, and so forth) often serve to signal information about which norms, values, and beliefs a person holds, and thus which tribe he or she belongs to.

Moreover, there is a sound rationale for displaying such symbols: Coordinated interactions will go much more smoothly and efficiently than noncoordinated interactions, and other things being equal, it is in no one's interest to engage in the latter, more difficult sort of exchanges. Social norms often govern social interactions in such a way that actors who share the same relevant norms will have similar and complementary expectations about the "proper" form of the interactions, practices, and customs in which they might mutually engage. Against this backdrop, ethnic markers perform an important signaling function: They serve as visible cues for the behavioral dispositions and otherwise unobservable values and norms that guide the behavior of others. The perceivable symbolic markers thus allow actors to identify and selectively interact with those who have similar and complementary expectations about the type of interaction in question, and to avoid difficult and inefficient exchanges with those who do not.

Pairing this idea with the tribal instincts hypothesis also yields a prediction about human psychology: Humans will tend to express commitment to their tribe and their tribe's characteristic cluster of norms by displaying ethnic boundary makers, will be predisposed to be sensitive to the ethnic boundary markers displayed by others, and will be motivated and make inferences about whom to socially engage and whom to avoid based on them. Moreover, such capacities can give rise to ethnocentric attitudes in favor of one's own tribe (and perhaps, to a lesser extent, friendly and allied tribes) along with its members, customs, values, and norms. The dark flipside of this is that those same tendencies can all too easily give rise to prejudicial attitudes against other, hostile tribes, and their members, customs, values, and norms. In an eloquent expression of this line of thought, Boyd and Richerson speculate that such attitudes often involve the emotion of disgust:

> [G]roups of people who share distinctive moral norms, particularly norms that govern social interactions, quite likely become ethnically marked. This suggests that ethnocentric judgments easily arise because "we the people" behave properly, while those "others" behave improperly, doing disgusting, immoral things, and showing no remorse for it, either. (Boyd & Richerson, 2005a, p. 101)

3 The Basic Disgust Response and the Entanglement Thesis: Poisons and Parasites

For the moment, set aside the tribal instincts hypothesis and consider the emotion of disgust on its own. A wide range of empirical work has shown that the basic disgust response comprises a diverse but highly coordinated set of elements, including affective, behavioral, and cognitive components (see Ekman, 1992; Rozin, Haidt & McCauley, 2008). Among the most recognizable of the behavioral components is the gape face, the characteristic facial expression associated with disgust. In especially intense episodes, production of the gape face can tip into the retching that it so clearly prefigures. Gapes are also accompanied by a feeling of nausea, an orally based sense of aversion, and a reflex-like withdrawal, the quick physical recoil from the disgusting entity. The basic disgust response includes cognitive components as well, including a more sustained sense of offensiveness and contamination. When some entity is found disgusting, it is considered offensive in a particular way: the thing is repulsive; one does not want to come into physical contact with it; mere physical proximity to the entity often is off-putting, repugnant, barely tolerable. Moreover, while a disgusting entity often captures the attention, even thinking about it is unpleasant. More striking, an entity that is considered disgusting has the ability to transmit its disgustingness to other things it comes into contact with. Those things thus contaminated are thereby considered disgusting, and elicit the same suite of response elements. Together, the operation of these two cognitive components of disgust can quite naturally lead to concerns about cleanliness and desires to purify oneself.

Opposite the basic disgust response are the sorts of things that can trigger it. Here, the data are even more puzzling (see again Rozin, Haidt & McCauley, 2008). Disgust can be elicited by an extraordinarily diverse set of triggers, ranging from the concrete and physical to the abstract and social. On the one hand, some of the most universally disgusting things are closely associated with the body, like spit, feces, blood, and organic decay of all sorts. Disgust is also sensitive to bodily boundaries in a particular way; the emotion enforces a "no reentry" policy. If something was once within or a part of a body, even your own, but then exits or breaks off, it thereby becomes an object of disgust; common examples include saliva, blood, hair, fingernails, and severed limbs. Also disgusting to many is a set of creatures that might be called "creepy crawlies": slugs, roaches, rats, and the like (Davey, Forster & Mayhew, 1993; Webb and Davey, 1993; Ware et al., 1994). Certain types of perfectly edible (i.e., nonpoisonous) food disgust some people as well. Common offenders in this category include cuisine like Brussels sprouts, escargot, caviar, pork rinds, Whoppers, and deep-fried Twinkies. Other common elicitors of disgust are nonstandard sexual behaviors and practices, including most notably incest—though what counts as nonstandard, and, in the case of incest, which kin are off limits, varies from culture to culture. Finally, certain types of social behavior can elicit disgust. Crooked politicians and ambulance-chasing lawyers are emblematic of such behaviors in our own culture, but violating certain social norms, especially those that govern how antecedently disgusting entities are to be dealt with (norms regulating burial rituals, the correct way to prepare food or maintain bodily hygiene, etc.) can also often elicit disgust (Shweder et al., 1997; Rozin et al., 1999; see also Haidt et al., 1997).

Another noteworthy feature of the elicitors is that some appear to be universally and perhaps innately disgusting. These include those elicitors closely linked to the body, as well as incest (Lieberman, Tooby & Cosmides, 2003; Fessler & Navarrete, 2004; cf. Prinz, 2008, for a dissenting view). In the case of other elicitors, however, what is considered disgusting can exhibit patterned variation from culture to culture. For instance, whether it is escargot, caviar, and Brussels sprouts that are typically considered repulsive by the locals, or whether it is pork rinds, Whoppers, and deep-fried Twinkies, will depend on whether you are at a state fair in the United States' rural Midwest or at a posh bistro on Paris's Left Bank. Likewise, while many cultures consider some types of deviant sexual activities not just wrong but also disgusting, exactly which of those activities are deviant in this way varies from culture to culture. In terms of social behaviors, as the relevant norms vary from one group to another, so will the transgressions that are considered disgusting. In extreme cases, the norms and even ideologies of entire opposing social groups can come to disgust as well, for example, conservative ideology can be mildly disgusting to liberals; liberal ideology likewise to conservatives.[3]

Taking a step back and surveying all of these data invites some difficult questions. First and foremost is the simplest one: given (1) this puzzling and variable array of elicitors that (2) evokes a response composed of an equally puzzling cluster of components, *how are all of these things connected*? My aim is to sketch an answer to this question. That answer comes in two parts: the Entanglement thesis and the Co-opt thesis. The rest of this section will be occupied with the former, and the next section will take up in the later.

The Entanglement thesis holds that disgust is a uniquely human kludge.[4] Underlying the basic disgust response are two distinguishable cognitive mechanisms that were once distinct, but have become deeply entangled with each other in modern human beings. Through the evolutionary process of descent with modification, these two mechanisms became more and more functionally integrated with each other, eventually forming the single emotion now recognized as disgust. The character of that human emotion remains informed by the character of those two entangled mechanisms and the adaptive problem each initially (and separately) evolved to solve. However, whereas homologies with similar features and functions to each distinguishable mechanism can be found in primates and other animals, the Entanglement thesis holds that only in humans have these two mechanisms become so tightly intertwined as to form disgust. Thus, the Entanglement thesis also provides an explanation for why this particular emotion is found in human beings, but not other animals (see Rozin, Haidt & McCauley, 2008; Morris, Doe & Godsell, 2007).

One of those two entangled mechanisms is directly linked to digestion. This mechanism initially evolved to regulate food intake and to protect the gut and gastro-intestinal system against substances that are poisonous, toxic, or otherwise harmful when swallowed. It mainly protects against such substances by preventing them from being fully ingested. The mechanism can also produce orally based aversion toward specific types of food, to prevent them being eaten. Indeed, foods that, once fully consumed, induced gut-based distress are often not only expelled, but also generate what have been called acquired taste aversions, so that they are not consumed again (Garcia, Hankins & Rusiniak, 1974; Bernstein, 1999). For shorthand, I will call this the *poison mechanism*.[5]

Returning to the characterization of disgust offered above, certain features of the basic response are easily traced to the poison mechanism and its proprietary adaptive function. In general, the aversion in episodes of disgust is often produced via physiological systems primarily based in the mouth and gut, giving it a strong oral aspect (what Rozin calls a "sense of oral incorporation"). The gape face often precedes and uses many of the same muscle groups as retching, which is how the body expels substances from the gut and mouth. The accompanying feeling of nausea is also useful in preventing ingestion of food in the first place. Finally, culturally local patterns of disgust to certain types of cuisine also indicate the operation of a mechanism dedicated to monitoring food intake.

The other of the two entangled mechanisms that shaped human disgust is linked to disease and parasites. This mechanism originally evolved to protect the entire organism from all forms of pathogenic or parasitic infection. It does this by causing the organism to avoid any close physical proximity to infectious agents, or anything that is likely to be infected and contagious. Since many infectious agents are microbes that cannot themselves be detected by the naked eye, protecting against infection involves avoiding not only visible pathogens and parasites, but also avoiding places, substances, and other organisms that might be harboring them. Unlike the one protecting against poisons, this mechanism is not specific to ingestion, and so obviously has a much larger domain and range of cues to monitor. I'll often refer to this as the *parasite mechanism* in what follows.[6]

Once again returning to the characterization of disgust offered above, certain features of the basic response are easily traced to the parasite mechanism and its proprietary adaptive function. One of these is the reflex-like withdrawal: Quickly recoiling from a disgusting entity instantly decreases its physical proximity. The more cognitive sense of offensiveness can effectively prevent getting close to disgusting entities in the first place, and motivate moving further away from them once they are detected. Finally, the sensitivity to the possibility of contamination, the motivation to cleanse oneself, and the concern about physical purity are all clearly fitted to the adaptive problems linked to infection.

As in the case of the poison mechanism, evidence of the parasite mechanism can also be found in the common elicitors of the emotion. Disgust does appear to have a special link to the mouth (the intensity of an episode of disgust can usually be increased by imagining the offending entity coming into contact with one's mouth or tongue), but its domain is by no means restricted to the oral; the emotion monitors all of the bodily orifices and boundaries (Rozin et al., 1995; Fessler & Haley, 2006). Feces and organic decay are some of the most effective vectors of disease transmission, and are also some of the most potent elicitors of disgust, perhaps universally so (Haidt, McCauley & Rozin, 1994). Finally, phenotypic abnormalities and other reliable indicators of infection in conspecifics are also possible universal and innate elicitors of disgust (Curtis, Aunger & Rabie, 2004).

Another form of evidence in support of the Entanglement thesis comes from developmental psychology. Patterns in ontogeny suggest that components of the disgust response are on different developmental schedules, and emerge at different ages. Indications of mere distaste (not liking the taste of certain foods, but not reacting to them with disgust) and the ability to make and react to gape faces are present within the first year of life (Bandura, 1992). Other components of the mature disgust response such as contamination sensitivity, however, do not emerge until much later. Some researchers mark the usual appearance of full-blown disgust as late as 4 to 8 years (Rozin, Fallon & Augustoni-Ziskind, 1985), while others put it earlier, around 2.5 to 3 years (Siegal & Share, 1990). Whichever turns out to be closer to the truth, both estimates place components of the response linked to the parasite mechanism as emerging significantly later than those linked to the poison mechanism.

While the Entanglement thesis makes sense of much of the extant data and proximate structure of disgust, it also raises questions about the ultimate explanation of the emotion and the sorts of selective pressures that drove the poison and the parasite mechanisms together. Though evolutionary accounts of mental structure are notoriously difficulty to confirm or falsify, several factors would have been instrumental in creating what we now recognize as disgust in modern humans. First, there must have been a nontrivial degree of functional overlap between the two systems; food is a major vector of disease transmission, and so the parasite mechanism would have antecedently been sensitive to food-related cues, making the two mechanisms easily poised to become entangled by the right kind of tweak here or there. A major, related tweak, however, might have come when ancestral humans began to consume more and more meat, procured either by advances in hunting or active scavenging. The increase in meat consumption marks an important difference between humans and their primate cousins and has been linked to increased brain size (see Aiello & Wheeler, 1995, Sterelny, 2003). This is relevant to the evolution of disgust because ancestral humans did not have a long history of scavenging, and so would not have been able to simply rely on a scavenger's powerful gastro-intestinal system to neutralize the dangers in the larger quantities of meat they were developing the wherewithal to acquire. This situation is just the sort that would have pressured the parasite mechanism to become even more sensitive to potential foods, thus creating the kind of selection that would have pushed that mechanism to simply merge with the poison mechanism.

This line of thought also suggests why researchers fail to find anything fitting the description of disgust in other mammals like dogs, who *are* long-time scavengers and have evolved the type of iron-clad gut to accommodate it, or in other primate species, who's evolutionary past *did not* include the transition to a meat-intensive diet the way humans' did. As such, an ultimate explanation along these lines is compatible with the claim that disgust is a uniquely human emotion, while at the same time being compatible with there being unentangled homologies of its component parts present in a range of other species.[7]

Finally, it is worth nothing that despite the differences in ontogenetic and evolutionary history, the Entanglement thesis holds that the poison and parasite mechanisms have merged in mature, modern humans. Once the emotion of disgust is fully developed in an individual, the many components of the response come as a package; they are thereafter produced together with lawlike regularity, forming what philosophers of science sometimes call a "nomological cluster" (Boyd, 1991). In the case of disgust, this means that any elicitor will reliably produce those clustered components, both those linked to the evolutionary problem of food regulation (reflexive production of gape face, nausea, sense of oral incorporation) and those linked to pathogen and parasite avoidance (quick withdrawal, sense of offensiveness, sensitivity to contamination, and the desire to cleanse or purify). While the intensity or vividness of different episodes of disgust will obviously vary from the mild to the extreme, elicitors of all sorts trigger this full nomological cluster, from creepy crawlies, to bodily fluids, to the relevant types of moral transgressions.

4 Moral Disgust and the Co-opt Thesis

Recall the question posed earlier. Given, on the one hand, the puzzling cluster of affective, behavioral, and cognitive components that make up the disgust response, and on the other hand, the equally puzzling array of elicitors that trigger the response, how are all of these things connected? The first part of my answer was largely (though not exclusively) focused on the character of the response itself. The second part of my answer is the Co-opt thesis, which assumes that the Entanglement thesis is correct and embeds it within the context of GGC and the tribal instincts hypothesis.[8] It takes the set of elicitors as its point of departure.

The Co-opt thesis holds that as humans became more reliant on social groups and the cultural information they provided, basic disgust was co-opted by the emerging tribal instincts to help perform a variety of novel functions that arose in conjunction with this increased sociality. In doing so, disgust's most characteristic features, features that initially evolved to solve adaptive problems linked to poisons and parasites, were brought to bear on those new functions in the social domain. Moreover, it is exactly this *imperfect fit* between the basic disgust response and many of those social functions it was later co-opted to perform that gives rise to the sorts of puzzling results turning up in the recent research on moral cognition. In short, some of the more troubling features of moral judgments discussed in the first section can be understood as cognitive by-products, generated by the mismatch between "unanticipated" problems and the kludgy solution disgust helps provide.

As an example of co-optation, consider the gape face discussed above. Gaping utilizes most of the same facial muscular movements as retching, the physical act that it sometimes precedes and accompanies. According to the Entanglement thesis, as the poison and parasite mechanisms fused, however, that facial expression acquired a new purpose: It was recruited to send signals to conspecifics. Generally speaking, the importance of communication increased in tandem with human sociality and the propensity to live in larger and larger groups. As the significance of communication rose, so too did the need for perspicuous signals that could cleanly transmit important information. Faces and different facial expressions in general could already convey a rich assortment of information, and on my view, the gape face was co-opted to perform a similar signaling role. It can warn others, including small children, against eating something known to be toxic or poisonous. The gape also expresses a sort of "Warning! Biohazard!" message, useful for cautioning others to avoid nearby pathogens or contaminated areas. In being co-opted, the gape went from merely preceding the actual expulsion of substances from the mouth to acting as a warning sign. Moreover, once this broader signaling function was acquired, gapes were able to transmit other socially relevant information as well, including information related to the other functions disgust acquired in the social and moral arenas.[9] For instance, expressions of disgust, together with the types of things that elicit them—disgust toward specific types of food, like deep-fried Twinkies or escargot, or toward particular behaviors and social practices, like driving a gas-guzzling SUV or smoking cigarettes—can themselves act as ethnic boundary markers, signaling information about group membership and hence about which values and norms a person is committed to.[10]

For more insight into the character of the auxiliary functions that disgust has acquired, recall GCC and the tribal instinct hypothesis. GCC maintains that one factor that greatly contributes to the human ability to cooperate and coordinate on such a large scale (compared, e.g., to other primates and most other animals) is that human social interactions are governed by a complex set of norms. Common sense and anecdotal evidence are supported by recent research showing that disgust is indeed operative in a number of different types of these social norms. In these cases, the emotion provides the types of intrinsic motivation mentioned above, including motivation to comply with the norm in question, to avoid the actions they prohibit, and to punish or direct punitive attitudes at transgressors of the norm. Indeed, disgust has been shown to play such roles in a number of different types of norms, including the rules of table etiquette (Nichols 2002a,b, 2004), taboos restricting the consumption of meat (Fessler & Navarrete, 2003), and taboos against incest (Lieberman, Tooby & Cosmides, 2003; Fessler & Navarrete, 2004).

More generally, the anthropologist Richard Shweder and his colleagues have called attention to an entire class of norms that follow the logic of disgust, which they call purity norms (Shweder et al., 1997; Haidt et al., 1997; Rozin et al., 1999). As their name suggests, purity norms are often understood as regulating issues of purity, not only guarding the sanctity of the physical body, but also protecting the soul from contamination and spiritual defilement. Indeed, purity norms are often distinguished from other classes of norms, such as harm norms or fairness norms, in that transgressions of purity norms usually do not result in direct physical harm or the inequitable treatment of any person.[11] More traditional or religious cultures often see transgressors of a purity norm as defiling themselves by disrespecting the sacredness of God (or the gods), or by violating the divine order. Purity norms are not completely absent from largely secular cultures, however; their presence is just not as central to the social structure or prevailing moral code. They are often given a different justification in secular cultures, as well: Transgressions of purity norms are usually conceived of as "crimes against nature" or violations of the natural order. According to Shweder, norms fitting this description regulate a range of issues, such as the proper foods to eat, when it is admissible to eat them, and often the proper way to prepare them; the details of sexual activities and even sleeping arrangements among family members; proper attire in a variety of settings, especially ritual and religious settings; the proper way to deal with organic materials, like corpses, blood, feces, and so forth; and how to interact with members of other social groups, particularly how to avoid being polluted by members of lower classes or castes. In addition to the obvious themes of purification and contamination, preliminary research supports the idea that the character of purity norms is heavily influenced by the emotion of disgust (see especially Rozin et al., 1999).

Also confirming commonsense suspicions are recent neuroimaging experiments that link the disgust response to prejudices and ethnic membership. This research shows disgust to be operative in sustaining a class of biases and prejudicial attitudes toward those in particular out-groups or tribes.[12] As was mentioned above, distinct emotions are often associated with the different types of attitudes directed at different out-groups and their members (Cottrell & Neuberg, 2005). Particularly interesting (if not completely surprising) is the demonstration that disgust is often the emotion linked to the most extreme prejudices, directed at members of the lowliest, most vilified, and dehumanized ethnicities (Harris & Fiske, 2006).

Finally, the pieces are ready to be put together. On the one hand, GCC provides details about a number of relatively novel adaptive problems that arise in the wake of increased human sociality and reliance on cultural information, and posits a set of tribal instincts that evolved to help deal with them. On the other hand, we have basic disgust, an emotion that appears to have been cobbled together from parts that originally and separately evolved to deal with poisons and parasites, but which also appears to be acting as an important component of certain tribal instincts. The Co-opt thesis offers an explanation: At some point, the poison and parasite mechanisms that make up basic disgust were co-opted to perform a variety of novel *auxiliary* functions unrelated to either poisons or parasites. Furthermore, the Co-opt thesis maintains that in performing those novel functions linked to social norms and monitoring ethnic boundaries, the full nomological cluster of components that make up the basic disgust response is brought to bear on those social functions, from the more cognitively complex sensitivity to contamination, to the gape face and physical recoil, to the more visceral feelings of nausea and repulsion. Moreover, those social behaviors and social attitudes driven by disgust will be also informed, perhaps implicitly, by the components of the emotion, such as worries about contamination and feelings of revulsion, that initially evolved in response to adaptive problems associated with the avoidance of toxic foods and diseases.

As such, the Co-opt hypothesis opens up the possibility of explaining persistent concerns about contamination and purity in moral affairs to be understandable but misplaced. Such concerns are revealed as by-products of the imperfect fit between the character of the disgust response and the new function it has been co-opted to perform in conjunction with human tribal instincts. The social norms that recruit disgust appear to require, most basically, some kind of avoidance and aversion motivation. In co-opting disgust in particular, the activities proscribed by those norms, as well as those actors who transgress them, are not *simply* avoided and found aversive. Rather, they are also subliminally infused with a very specific kind of offensiveness, are often considered tainted and contaminating, so much so that they can induce a desire to cleanse or purify oneself. The same type of explanation applies to tribal instincts that monitor ethnic boundaries and their symbolic markers. According to GCC, what is needed is motivation to avoid members of other tribes who have internalized different social norms, in order to avoid uncoordinated (and perhaps hostile) exchanges. When disgust is the emotion co-opted to provide that motivation, however, along with it come attendant components like contamination sensitivity, offensiveness, visceral aversion—the full nomological cluster of the disgust response.

Also troubling can be the way in which feelings of disgust can induce judgments that are remarkably persistent, and clung to dogmatically. Remember Dan, the "popularity-seeking snob":

> Dan is a student council representative at his school. This semester he is in charge of scheduling discussions about academic issues. He often picks topics that appeal to both professors and students in order to stimulate discussion.

Those hypnotized to feel disgust at the word "often" maintained their initial judgment that Dan was doing something morally wrong even when they were unable to provide any supporting reasons. The vividness and visceral power of the emotion could lead people to remain doggedly committed to other attitudes and norms that involve disgust, even if those attitudes and norms can be shown to be unjustified or rationally unfounded (cf. Haidt, Bjorklund & Murphy, ms.).

5 Conclusion

Although the view of disgust advanced here is primarily descriptive and explanatory, it is tempting to think that it has normative implications, and that the account provides the materials to construct an argument concerning what sort of moral authority ought to be accorded the emotion, and what role feelings of disgust deserve in moral deliberation and debate. Though I believe this suspicion is correct, here is not the place to spell out and defend such an argument (though see Kelly, 2011, ch. 5). Rather, I will conclude by pointing out that whatever its conclusion, an argument of this sort will begin from facts about the nature of the emotion itself, and be much informed by the rarified perspective of evolutionary theory. This may be surprising to certain philosophical sensibilities; such facts and evolutionary considerations are often thought to be morally neutral, morally irrelevant, or even corrosive to a moral outlook. I doubt such a view is sustainable, however, especially in this case. At the very least, in calling attention to the imperfect fit between this cognitive system that initially evolved to deal with poisons and parasites, on the one hand, and the social dynamics it was later co-opted to help navigate, on the other, the Entanglement and Co-opt theses are able to expose concerns with things like moral taint and spiritual purity as baseless projections even as they explain their source and prevalence.

Acknowledgments

My thanks to Brett Calcott, Christine Clavien, Ben Fraser, Richard Joyce, and Kim Sterelny for comments on previous drafts of this chapter, as well as the participants in the 2008 Minds and Societies conference at UQAM and the participants in the 2008 Emotion and Commitment conference at ANU for helpful comments on presentations of material.

Notes

1. In what follows, I'll call attention only to those findings that are relevant to my argument. For more encompassing overviews of recent work in empirical moral psychology, see Doris & Stich (2005, 2006) and Doris et al. (2010).

2. A note on terminology: the name "tribal instincts hypothesis" I am taking from gene culture coevolutionary theorists like Boyd and Richerson and their collaborators. However, in what follows, I will focus primarily on only two specific types of "tribal instincts," namely those psychological capacities associated with social norms and those associated with ethnic boundary markers. I take no stand on whether there are other, distinct tribal instincts beyond these two. Moreover, when I use the term "tribe" I will be guided by the usage in that literature, and so mean, roughly, groups of several hundred people who may not be bound together by relations of kinship or histories of direct cooperative interaction, but who are culturally homologous in that they share beliefs and values, and most importantly, embrace a common cluster of norms and display common ethnic markings. For some discussion of the troublesome terminology, see Gil-White (2006).

I should note that the literature on gene culture coevolutionary theory is not only complex, but large and growing. My discussion here will of necessity be somewhat cursory, but I hope to give the flavor of the outlook and highlight those elements most important to my aim of illuminating the nature of what is often called "moral disgust."

3. This need not imply that *every* time someone describes himself as "disgusted" by something he actually is disgusted, that he is correctly reporting the full activation of his disgust system. Surely sometimes claiming to be disgusted by, for instance, a particular type of campaign finance reform can be merely one way to verbally express fervent disagreement or outrage. However, there is reason to believe that at least sometimes, social and moral issues can genuinely activate the disgust system, and thus produce (mild) episodes of disgust, complete with the full suite of components. See Chapman et al. (2009); Rozin, Haidt, and Fincher (2009); for a skeptical view see Bloom (2004); for discussion see Kelly (2011), especially ch. 4.

4. The term "kludge" is taken from engineering and computer science, where it is usually used to refer to a clumsy, piecemeal, or inelegant solution to a problem, or the clumsy, piecemeal, inelegant device used to solve the problem. Kludges are often gerrymandered, constructed to fix problems that were themselves unanticipated. The clumsiness stems from the fact that kludges are often constructed out of whatever parts are available when the problem unexpectedly arises. Those initially unrelated parts are cobbled together to construct the kludge, and in the process often put to new uses. Given that natural selection is a blind, tinkering process that operates without the benefit of foresight, the term also comes in handy in evolutionary explanations, though obviously in such contexts the metaphorical use of terms like "unexpected" or "unanticipated" needs to be taken with a grain of salt.

5. Previous theorists have emphasized this feature of disgust. For example, Griffiths (1997) follows Darwin (1872) in casting the emotion as a food rejection system, and much of Rozin's work on disgust stresses its oral character (see Rozin, Haidt & McCauley, 2008, and citations therein).

6. Psychologists have previously noted this aspect of disgust as well. Steven Pinker (1997), for instance, calls disgust "intuitive microbiology," and Curtis, Aunger, and Rabie (2004) present an impressive body of evidence in support of the claim that disgust is an "evolved response to threats of infectious disease." Kurzban and Leary (2001) cite "parasite avoidance" behavior to explain a certain type of social stigmatization, though the link to disgust is less explicit in their discussion.

7. Another factor that could have helped push the poison and parasite mechanisms toward entanglement has to do with the increased need for a perspicuous signal that early humans could have used to socially transmit (perhaps parochial) information about disease- and parasite-related issues in the local environment. The gape face, associated with retching and the poison mechanism, provided a good candidate. Selection for an effective signaling system would have generated pressure to link the gape with the parasite system, which could have also contributed to the integration of the two systems. Moreover, there would be little incentive for deception given the relevant adaptive problems, especially those having to do with contagious disease, and so no clear selective pressure militating *against* the evolution of a clear, trustworthy signaling system. See the third chapter of Kelly (2011) for more detailed discussion.

8. Although my presentation of the Co-opt thesis assumes the Entanglement thesis is correct, the two are logically distinct. Others may consistently subscribe to something similar in spirit to the Co-opt thesis, e.g., the idea that disgust initially evolved to do something other than the functions it now performs in human moral psychology, while rejecting the Entanglement thesis, and replacing it with an alternative account of what, exactly, disgust initially evolved to do.

9. It will be useful to make terminological caveat, in order to forestall a number of (interesting, open) philosophic questions about the definition of "morality," and thus the proper domain of the moral—which judgments, mechanisms, or roles are "really" about moral issues and which are not. (For discussion of these issues, see Nado, Kelly & Stich, 2009; Kelly & Stich, 2007; Nichols, 2004.) In the text, "moral disgust" will be used in the same way it is used in the empirical psychological literature, namely to capture those roles that disgust systematically plays in social life. These include its role in guiding social coordination, motivating behavior in potentially cooperative situations, influencing interactions between individuals and between groups of people, and influencing judgments about similar matters.

10. It is worth pointing out that on the explanation offered here, expressions of disgust have acquired a role in signaling commitment, but it is commitment to a group and the set of norms that bind it together. This is an importantly different type of commitment than that often associated with emotional expression, according to which social emotions like guilt and anger are fundamentally commitment devices that evolved to help navigate the Machiavellian vicissitudes of deception, credibility, and defection (for the classical statement of this view, see Frank, 1988; see also Pinker, 1997). For more discussion of the similarities and differences between these two

types of commitment and the different explanations of cooperation they are associated with, see Richerson and Boyd (2001) and Boyd and Richerson (2005b).

11. I do not take these categories—harm norms, purity norms, fairness norms—to delineate categorically distinct and disjoint sets, but rather poles on a continuum. One complicated and contentious issue is the role of perceived harm in various norms and norm transgressions, including so-called purity norms; see Turiel, Killen, and Helwig (1987); Haidt, Koller, and Dias (1993); Kelly et al. (2007); Sousa, Holbrook, and Piazza (2009); and Stich, Fessler, and Kelly (2009).

12. In using the term "tribes" here, I do not mean to suggest that today's social networks are still structured along anything approximating tribal lines. However, this is completely compatible with the possibility that important components of the human psychological system devoted to social cognition still *see* social interactions in largely tribal terms, and conceptualize and attempt to navigate those interactions accordingly. Indeed, this is one way to understand the tribal instinct hypothesis: Many of the mechanisms underlying social cognition, even in contemporary human beings, originally evolved to allow living in tribal-sized groups, and as such they are sensitive to the types of observable cues that are likely to convey information relevant to which tribe someone is a member of, and which cluster of norms they embrace. Moreover, those mechanisms are wont to process that information, make inferences, and form intentions in ways that are well suited to a tribal existence, even if the inferences and intentions they produce are not always optimal or even efficient in the context of our modern social institutions.

References

Aiello, L., & Wheeler, P. (1995). The Expensive Tissue hypothesis: The brain and the digestive system in human and primate evolution. *Current Anthropology, 36*, 199–221.

Bandura, A. (1992). Social cognitive theory of social referencing. In S. Feinman (Ed.), *Social referencing and the social construction of reality in infancy* (pp. 175–208). New York: Plenum.

Barth, F. (Ed.). (1969). *Ethnic groups and boundaries: The social organization of cultural differences*. Boston: Little, Brown.

Bernstein, I. (1999). Taste aversion learning: A contemporary perspective. *Nutrition, 15*(3), 229–234.

Bloom, P. (2004). *Descartes' baby*. New York: Basic Books.

Bouglé, C. (1971). *Essays on the caste system* (Pocock, D. F., Trans.). Cambridge: Cambridge University Press.

Boyd, R. (1991). Realism, anti-foundationalism, and the enthusiasm for natural kinds. *Philosophical Studies, 61*, 127–148.

Boyd, R., & Richerson, P. J. (1985). *Culture and the evolutionary process*. Chicago: University of Chicago Press.

Boyd, R., & Richerson, P. (2005a). *The origin and evolution of cultures*. New York: Oxford University Press.

Boyd, R., & Richerson, P. (2005b). Solving the puzzle of human cooperation. In S. Levinson (Ed.), *Evolution and culture* (pp. 105–132). Cambridge, MA: MIT Press.

Chapman, H., Kim, D., Susskind, J., & Anderson, A. (2009). In bad taste: Evidence for the oral origins of moral disgust. *Science, 323*, 1222–1226.

Cottrell, C. A., & Neuberg, S. L. (2005). Different emotional reactions to different groups: A sociofunctional threat-based approach to "prejudice." *Journal of Personality and Social Psychology, 88*, 770–789.

Curtis, V., Aunger, R., & Rabie, T. (2004). Evidence that disgust evolved to protect from risk of disease. *Proceedings of the Royal Society of London, Series B: Biological Sciences, 271*(4), S131–S133.

Darwin, C. (1872). *The expressions of emotions in man and animals*. New York: Philosophical Library.

Davey, G., Forster, L., & Mayhew, G. (1993). Familial resemblances in disgust sensitivity and animal phobias. *Behaviour Research and Therapy, 31*(1), 41–50.

Doris, J., & Stich, S. (2005). As a matter of fact: Empirical perspectives on ethics. In F. Jackson & M. Smith (Eds.), *The Oxford handbook of contemporary philosophy* (pp. 114–152). Oxford: Oxford University Press.

Doris, J., & Stich, S. 2006. Moral psychology: Empirical approaches. In Edward N. Zalta (Ed.), *The Stanford Encyclopedia of Philosophy* (summer 2006 edition), http://plato.stanford.edu/archives/sum2006/entries/moral-psych-emp/.

Doris, J., & The Moral Psychology Research Group (Ed.). (2010). *The Oxford handbook of moral psychology*. New York: Oxford University Press.

Ekman, P. (1992). An argument for basic emotions. *Cognition and Emotion, 6*, 169–200.

Fessler, D., & Haley, K. (2006). Guarding the perimeter: The outside–inside dichotomy in disgust and bodily experience. *Cognition and Emotion, 20*(1), 3–19.

Fessler, D., & Navarrete, C. (2003). Meat is good to taboo: Dietary proscriptions as a product of the interaction of psychological mechanisms and social processes. *Journal of Cognition and Culture, 3*(1), 1–40.

Fessler, D., & Navarrete, C. (2004). Third-party attitudes toward sibling incest: Evidence for Westermarck's hypothesis. *Evolution and Human Behavior, 25*, 277–294.

Frank, R. (1988). *Passions within reason: The strategic role of the emotions*. New York: W. W. Norton.

Garcia, J., Hankins, W., & Rusiniak, K. (1974). Behavioral regulation of the milieu interne in man and rat. *Science, 185*, 824–831.

Gil-White, F. (2001). Are ethnic groups biological "species" to the human brain? *Current Anthropology, 42*(4), 515–554.

Gil-White, F. (2006). The study of ethnicity and nationalism needs better categories: Clearing up the confusions that result from blurring analytic and lay concepts. *Journal of Bioeconomics, 7*, 239–270.

Griffiths, P. (1997). *What the emotions really are.* Chicago: University of Chicago Press.

Haidt, J., Bjorklund, F., & Murphy, S. (Unpublished ms.). Moral dumbfounding: When intuition finds no reason. University of Virginia.

Haidt, J., Koller, S., & Dias, M. (1993). Affect, culture, and morality, or is it wrong to eat your dog? *Journal of Personality and Social Psychology, 65*(4), 613–628.

Haidt, J., McCauley, C., & Rozin, P. (1994). Individual differences in sensitivity to disgust: A scale sampling seven domains of disgust elicitors. *Personality and Individual Differences, 16,* 701–713.

Haidt, J., Rozin, P., McCauley, C., & Imada, S. (1997). Body, psyche, and culture: The relationship between disgust and morality. *Psychology and Developing Societies, 9,* 107–131.

Harris, L. T., & Fiske, S. T. (2006). Dehumanizing the lowest of the low: Neuro-imaging responses to extreme outgroups. *Psychological Science, 14*(10), 847–853.

Kelly, D. (2011). *Yuck! The nature and moral significance of disgust.* Cambridge, MA: MIT Press.

Kelly, D., & Stich, S. 2007. Two theories about the cognitive architecture underlying morality. In P. Carruthers, S. Laurence & S. Stich (Eds.), *Innateness and the structure of the mind: Foundations and the future* (pp. 348–366). New York: Oxford University Press.

Kelly, D., Stich, S., Haley, K., Eng, S., & Fessler, D. (2007). Harm, affect, and the moral/conventional distinction. *Mind & Language, 22*(2), 117–131.

Kurzban, R., & Leary, M. (2001). Evolutionary origins of stigmatization: The functions of social exclusion. *Psychological Bulletin, 127*(2), 187–208.

Lieberman, D., Tooby, J., & Cosmides, L. (2003). Does morality have a biological basis? An empirical test of the factors governing moral sentiments relating to incest. *Proceedings of the Royal Society of London, Series B: Biological Sciences, 270,* 819–826.

McElreath, R., Boyd, R., & Richerson, P. (2003). Shared norms can lead to the evolution of ethnic markers. *Current Anthropology, 44*(1), 123–129.

Morris, P., Doe, C., & Godsell, E. (2007). Secondary emotions in non-primate species? Behavioral reports and subjective claims by animal owners. *Cognition and Emotion, 22*(1), 3–20.

Nado, J., Kelly, D., & Stich, S. (2009). Moral judgment. In J. Symons & P. Calvo (Eds.), *The Routledge companion to the philosophy of psychology* (pp. 621–633). New York: Routledge.

Nichols, S. (2002a). Norms with feeling: Towards a psychological account of moral judgment. *Cognition, 84,* 221–236.

Nichols, S. (2002b). On the genealogy of norms: A case for the role of emotion in cultural evolution. *Philosophy of Science, 69,* 234–255.

Nichols, S. (2004). *Sentimental rules: On the natural foundations of moral judgment.* New York: Oxford University Press.

Pinker, S. (1997). *How the mind works.* New York: W. W. Norton.

Prinz, J. (2008). *The emotional construction of morals*. Oxford: Oxford University Press.

Richerson, P., & Boyd, R. (1998). The evolution of human ultra-sociality. In I. Eibl-Eibisfeldt & F. Salter (Eds.), *Ideology, warfare, and indoctrinability* (pp. 71–95). Oxford: Oxford University Press.

Richerson, P., & Boyd, R. (1999). Complex societies: The evolutionary origins of a crude superorganism. *Human Nature, 10*, 253–289.

Richerson, P., & Boyd, R. 2001. The evolution of subjective commitment to groups: A tribal instincts hypothesis. In R. Nesse (Ed.), *The evolution and the capacity for commitment* (pp. 186–220). New York: Russell Sage.

Richerson, P., & Boyd, R. (2005). *Not by genes alone*. Chicago: University of Chicago Press.

Rozin, P., Fallon, A., & Augustoni-Ziskind, M. (1985). The child's conception of food: The development of contamination sensitivity to "disgusting" substances. *Developmental Psychology, 21*, 1075–1079.

Rozin, P., Haidt, J., & Fincher, K. (2009). From moral to oral. *Science, 323*, 1179–1180.

Rozin, P., Haidt, J., & McCauley, C. (2008). Disgust. In M. Lewis & J. M. Haviland-Jones (Eds.), *Handbook of emotions* (3rd Ed., pp. 757–776). New York: Guilford Press.

Rozin, P., Haidt, J., McCauley, C., & Imada, S. (1997). Disgust: Preadaptation and the cultural evolution of a food-based emotion. In H. M. Macbeth (Ed.), *Food preferences and taste: Continuity and change* (pp. 65–82). Providence, RI: Berghahn Books.

Rozin, P., Hammer, L., Oster, H., Horowitz, T., & Marmora, V. (1986). The child's conception of food: Differentiation of categories of rejected substances in the 16 months to 5 year age range. *Appetite, 7*, 141–151.

Rozin, P., Lowery, L., & Ebert, R. (1994). Varieties of disgust faces and the structure of disgust. *Journal of Personality and Social Psychology, 66*(5), 870–881.

Rozin, P., Lowery, L., Imada, S., & Haidt, J. (1999). The CAD triad hypothesis: A mapping between three moral emotions (contempt, anger, disgust) and three moral codes (community, autonomy, divinity). *Journal of Personality and Social Psychology, 76*(4), 574–586.

Rozin, P., Millman, L., & Nemeroff, C. (1986). Operation of the laws of sympathetic magic in disgust and other domains. *Journal of Personality and Social Psychology, 50*, 703–712.

Rozin, P., Nemeroff, C., Horowitz, M., Gordon, B., & Voet, W. (1995). The borders of the self: Contamination sensitivity and potency of the body apertures and other body parts. *Journal of Research in Personality, 29*, 318–340.

Schnall, S., Benton, J., & Harvey, S. (2008a). With a clean conscience: Cleanliness reduces the severity of moral judgments. *Psychological Science, 19*, 1219–1222.

Schnall, S., Haidt, J., Clore, G., & Jordan, A. (2008b). Disgust as embodied moral judgment. *Personality and Social Psychology Bulletin, 34*, 1096–1109.

Shweder, R., Much, N., Mahapatra, M., & Park, L. (1997). The "big three" of morality (autonomy, community, and divinity), and the "big three" explanations of suffering. In A. Brandt & P. Rozin (Eds.), *Morality + health* (pp. 119–169). New York: Routledge.

Siegal, M., & Share, D. (1990). Contamination sensitivity in young children. *Developmental Psychology, 26*(3), 455–458.

Sousa, P., Holbrook, C., & Piazza, J. (2009). The morality of harm. *Cognition, 113*, 80–92.

Sripada, C., & Stich, S. (2007). A framework for the psychology of norms. In P. Carruthers, S. Laurence & S. Stich (Eds.), *The innate mind: Culture and cognition* (pp. 280–301). New York: Oxford University Press.

Sterelny, K. (2003). *Thought in a hostile world*. New York: Blackwell.

Stich, S., Fessler, D., & Kelly, D. (2009). On the morality of harm: A response to Sousa, Holbrook, and Piazza. *Cognition, 113*(1), 93–97.

Turiel, E., Killen, M., & Helwig, C. 1987. Morality: Its structure, functions, and vagaries. In J. Kagan & S. Lamb (Eds.), *The emergence of morality in young children* (pp. 155–245). Chicago: University of Chicago Press.

Ware, J., Jain, K., Burgess, I., & Davey, G. (1994). Disease-avoidance model: Factor analysis of common animal fears. *Behaviour Research and Therapy, 32*(1), 57–63.

Webb, K., & Davey, G. (1993). Disgust sensitivity and fear of animals: Effect of exposure to violent and revulsive material. *Anxiety, Coping and Stress, 5*, 329–335.

Wheatley, T., & Haidt, J. (2005). Hypnotic disgust makes moral judgments more severe. *Psychological Science, 16*(10), 780–784.

Zhong, C., & Liljenquist, K. (2006). Washing away your sins: Threatened morality and physical cleansing. *Science, 313*(5792), 1451–1452.

25 Evolution, Motivation, and Moral Beliefs

Matteo Mameli

1 Moral Beliefs and Moral Demands

According to John Mackie, someone who makes a moral judgment is making a judgment that "involves a call for action, or for the refraining from action, and one that is absolute, not contingent upon any desire, or preference or policy or choice, his own or anyone else's" (Mackie, 1977, p. 33). In his book *The Evolution of Morality*, Richard Joyce (2006) develops Mackie's suggestion and gives an interesting characterization of the mental states expressed by moral judgments. I will call such mental states *moral beliefs*. Whether moral beliefs are actually *beliefs* is controversial. Noncognitivists deny that moral judgments—unlike other kinds of judgments—express belief-like states (Gibbard, 1990; Blackburn, 1998). According to Joyce, moral beliefs are beliefs. More specifically, they are beliefs about the existence of practical demands that are, in his terminology, *inescapable* and *authoritative*.

A practical demand is inescapable if it applies to a person irrespective of her ends (desires or interests). Institutional practical demands, such as rules of etiquette, can be and often are inescapable in this sense. Even someone who does not care about etiquette can transgress against its demands. The rule "You ought not to speak with your mouth full" is supposed to apply to everyone, even to those who have no desire whose satisfaction depends on following the rule and—independently of their current desires perhaps—no interest served by following the rule. Moral beliefs purport to be about practical demands that are not just inescapable but also authoritative. They purport to be about demands that a person—any person—"would be irrational in ignoring, or at least . . . has a reason of genuine deliberative weight to comply" (Joyce, 2006, p. 62).[1]

Institutional demands *apply* to individuals irrespective of their ends, but they do not *bind* individuals irrespective of their ends. Moral beliefs purport to be about practical demands that have this special kind of *normative authority*: They bind individuals irrespective of their desires and interests. Unlike the case of institutional demands, opting out is not an available option, even for those who have interests that would be best served by ignoring the practical demands of morality and for those who have desires whose satisfaction depends on acting against morality's demands. Of course, as Joyce points out, the authority of morality—as he conceives of it—may be an illusion. It is not clear whether there is room in nature for practical demands with inescapable authority. Such practical demands may not exist. But leaving aside questions about their existence, what matters for the purposes at hand is that, on this characterization, moral beliefs *purport* to be about practical demands with inescapable authority.

In discussing Turiel's work on the moral–conventional distinction (Turiel, 1998), Joyce claims that moral beliefs purport to be about practical demands that are *authority independent*. This is a sense of "authority" different from the one just introduced. If Turiel is right, humans across all cultures are able to distinguish from a young age between rules whose normative force derives from their having been issued or approved by an authority (a parent, a teacher, an institution, including the state) and rules whose normative force does not depend on what anyone might think or do. Moral beliefs purport to be about rules of the second kind. In other words, moral beliefs purport to be about practical demands whose authority in Joyce's sense (normative/binding force) does not derive from any authority in Turiel's sense (any person or institution, no matter how powerful). We can call "moral demands" the practical demands that, on this account, moral beliefs purport to be about. Moral demands, if they exist, are *inescapable* and have *automatic* or *underived* normative authority.

In this essay, I will use Joyce's characterization as a starting point for exploring various views about the nature of moral beliefs and moral concepts, including views that are very different from Joyce's. Here is how I will proceed. In section 2, I will discuss Joyce's hypothesis about the evolution of the ability to form moral beliefs and the idea that these beliefs played an important role in the motivation of fitness-enhancing cooperative behaviors. In section 3, I will argue that the motivational power of moral beliefs is domain general: When we form moral beliefs in favor of noncooperative behaviors, their motivational power is not diminished because of their being about noncooperative behaviors. In section 4, I will discuss two hypotheses about the source of the motivational power of moral beliefs. The first hypothesis—which fits well with Joyce's characterization, though he does not seem to endorse it—is that moral beliefs derive their motivational power from a (possibly implicit and subconscious) desire to avoid irrationality. The second hypothesis is that the motivational power of moral beliefs derives from connections between them and emotions like guilt and moral disapproval. In section 5, I will explore the hypothesis that moral beliefs are *intrinsically* motivating. On this view, to have a moral belief is—among other things—to have a mental state connected in specific ways with emotions like guilt and moral disapproval. In sections 6 and 7, I will explain what moral concepts must be like if moral beliefs are intrinsically motivating beliefs. In section 8,

I will compare the kind of moral concepts discussed in the previous two sections—intrinsically motivating moral concepts—with the kind of (parasitic) moral concepts that, according to some ways of interpreting the empirical evidence, are found in psychopaths. In section 9, I will go back to the two hypotheses discussed in section 4 and, in the light of what said in the intervening sections, draw some tentative conclusions.

2 Moral Beliefs and Evolution

Joyce (2006, ch. 4) argues that the ability to form moral beliefs—belief-like states that purport to be about practical demands with inescapable and underived normative authority—is a universal feature of human beings. According to him, it is found in all cultures that we know of, both present and past, and in all normal individuals. He also argues that this ability is a genetic Darwinian adaptation. The ability is common in our species because in the evolutionary past of our lineage there was genetically heritable variation with respect to the possession of this ability and, moreover, possession of the ability conferred higher genetic fitness. How did the ability to form moral beliefs benefit our ancestors' genetic fitness? Joyce gives an interesting and complex answer. I will ignore the subtleties and focus instead on what I take to be the core of his answer.

Joyce argues as follows. From the point of view of genetic fitness, cooperating was crucially important for our ancestors in a very broad range of circumstances. But, motivationally speaking, cooperation often poses problems. The benefits of cooperation are often long term, indirect, and more generally, for a variety of reasons, difficult for agents to detect and to compute. Because of this, those in our lineage who had moral beliefs in favor of (certain kinds of) cooperative behaviors had higher fitness (on average). Thinking of cooperation in moral terms generates a robust and reliable motivation to cooperate. Believing that cooperation is morally required—demanded in an inescapable and automatically authoritative way—makes one more likely to cooperate.

On this view, moral beliefs are not the only source of cooperative motivations. Other sources exist. The emotion of sympathy is one example: Concern for the well-being of others can motivate altruistic behavior. Selfish prudential reasoning is another example: One may deliberate that on a particular occasion cooperation is the utility-maximizing strategy. But, according to Joyce, moral beliefs are a more robust and reliable source of cooperative motivations than these other ones. More specifically, moral beliefs are a source of cooperative motivations that can work in cases where (i) cooperating is extremely important for genetic fitness and (ii) the other sources of motivation are unlikely to be effective. It is because of this that, on his view, the ability to form moral beliefs was genetically selected for.

Is this hypothesis correct? The mental states we express when we make moral judgments seem to have important motivational powers. Moreover, these mental states often, though not always, seem to be about practical demands in favor of cooperative behaviors. These two facts make plausible the hypothesis that the ability to form such mental states was selected (in part perhaps) because of their ability to motivate cooperative behaviors. But one can agree that moral beliefs are motivationally powerful, and agree that they were selected because they helped motivate fitness-enhancing cooperative behaviors, while denying Joyce's characterization of their content and also denying that moral beliefs are actually beliefs.

I will not discuss noncognitivism here. But even if we agree that moral beliefs are actually beliefs, it is far from clear whether Joyce's characterization captures their content correctly, especially if we assume—as Joyce does—that mental states of this kind are present in virtually all adult humans. Different bits of evidence seem to point in different directions. Consider, for example, the view that atheism leads to moral nihilism. It is not easy to explain the popularity of this view if we assume that believing that someone is immoral is believing that he is (in a special sort of way) irrational.[2] But consider also the study by Nucci and Turiel (1993), which indicates that religious people do *not* in general think of the authority of morality as deriving from God's authority. Young teenagers belonging to conservative religious groups were asked whether a variety of actions would be wrong if God thought they were okay. For actions that are normally classified as immoral (e.g., hitting others, stealing), the teenagers in general said that the actions would be wrong even if God had said nothing negative about them.

Perhaps people have an incoherent conception of the authority of morality, which is seen as both derived and underived from God's authority.[3] Or perhaps people's moral beliefs have a Joyce-style content, though people's explicit views—their metabeliefs—mistakenly characterize the authority of morality as dependent on God's authority. Perhaps different people—or even the same people at different stages of their lives—conceive of the authority of morality in different ways. While this issue is extremely interesting and important, I cannot discuss it any further here. In what follows, I will just assume that, insofar as a *coherent* characterization of the authority of morality can be given, this coincides with Joyce's characterization.

One thing to notice in this context is that even if we established that the ability to form Joyce-style moral beliefs is universal, there would still be room for exploring evolutionary hypotheses different from the one put forward by Joyce, such as by-product hypotheses and cultural evolution hypotheses. A by-product hypothesis would say that the ability to form moral beliefs is a by-product of features of the human mind genetically selected for other purposes, which is what, for example, some cognitive anthropologists claim about the ability to form religious beliefs (e.g., Atran,

2004).[4] A cultural evolution hypothesis would say that the ability to form moral beliefs is transmitted, at least in part, through social learning (e.g., the belief that some practical demands are inescapable and automatically authoritative may be transmitted in this way) and that the ability, for some reasons, perhaps reasons that have to do with the motivational power of moral beliefs, has (had) high cultural fitness. I will not discuss the plausibility of these alternative explanations here. Instead, I will focus on one of the presuppositions of Joyce's hypothesis: the idea that moral beliefs are motivationally powerful. What is the best way to make sense of this?

3 Motivational Power: Domain Generality

First of all, notice that, while Joyce focuses on motivations to behave cooperatively, his argument can be generalized to other kinds of behaviors. Some behaviors are fitness enhancing. Some of these are cooperative behaviors, while others are not. If thinking about these behaviors as morally required—believing that there are inescapable and automatically authoritative practical demands in their favor—provides a robust motivation to perform these behaviors, then thinking of these behaviors as morally required is fitness enhancing. This is true independently of whether the behaviors in question are cooperative or not. If, say, killing the members of other groups is fitness enhancing, and if the members of other groups have a different religion, then thinking of the killing of people who do not believe in your gods as morally demanded is fitness enhancing.

More generally, and independently of any hypotheses about their evolutionary origins, it seems that the motivational power of moral beliefs is not domain specific, in the sense that it is not confined to moral beliefs in favor of cooperative behaviors. There are after all many moral norms in many cultures that have nothing to do with cooperation, at least not in any obvious sense: norms of purity are an example. Of course, it may well be the case that, for evolutionary reasons, we humans are more likely to form moral beliefs in favor of (certain kinds of) cooperative behaviors rather than in favor of noncooperative behaviors. The evolutionary scenario would be one where it is advantageous to "moralize" certain kinds of cooperative behaviors, and so humans evolved both the ability to "moralize" and the tendency to "moralize" cooperative behaviors more than other behaviors. Once available, this source of motivational power could also be applied to practical demands that have nothing to do with cooperation.

Whether we are more likely to form moral beliefs in favor of cooperation than in favor of antisocial behaviors or behaviors that have nothing to do with cooperation is obviously an interesting and difficult empirical question. Independently of the answer to that question, though, it seems that, as a matter of psychological fact, when we form moral beliefs in favor of noncooperative behaviors, their motivational power is not diminished because of their being about noncooperative behaviors. Where does this motivational power come from?

4 Motivational Power: Rationality and Emotions

Where does the motivational power of moral beliefs come from? If Joyce's characterization is correct, judging that—say—keeping promises is morally required is believing that that there is an inescapable and automatically authoritative practical demand in favor of keeping promises. If someone, call him Charles, believes in the existence of this practical demand, he believes that breaking promises is irrational, and that it is so independently of one's ends (desires, interests). How can this belief motivate Charles to do anything? This belief can motivate Charles in the presence of a desire not to be irrational. This desire need not be explicit or conscious; it can be implicit and subconscious. If Charles does not want (perhaps implicitly, subconsciously) to be irrational, and he believes that breaking promises is irrational, he will be motivated to keep promises in order to avoid irrationality.

Interestingly, Joyce does not appeal to a desire to avoid irrationality when he gives his account of the motivational power of moral beliefs. He never mentions such a desire. Instead, he appeals to connections between moral beliefs and certain emotions. There are various ways of conceiving of such connections. I will now provide my own version of this hypothesis.

When a person believes he has violated a moral norm—when he has a moral belief *M* and also believes that he has behaved in a way that goes against the practical demand that *M* is about—he feels a self-directed negative emotion. This emotion is what in ordinary language we often call "guilt." This feeling of guilt often results in efforts to make amends. In addition, when a person believes that someone else has transgressed against a moral norm, he feels a negative emotion toward the transgressors. This other-directed emotion is sometimes called "moral disapproval," and it often gives rise to a disposition to punish the transgressor.

On this view, the disposition to feel guilt when a moral demand is disregarded and people's strong motivation to avoid guilt—due to its negative valence—result in this: When someone believes that something ought (morally) to be done, he is motivated to do it, and when he believes that something ought (morally) not to be done, he is motivated not to do it. The mechanism could be similar to the one described by Damasio (1994; see also Mameli, 2004). If someone believes that an action ought not to be performed, when he imagines himself performing the action, he will feel guilt, or something motivationally equivalent to it. He will experience a specific negative emotional reaction at the thought of performing the imagined action. Because of this, the potential action will become negatively emotionally marked and the person will be motivated not to perform it. So, for example, if Charles believes that keeping promises is morally required, when he considers breaking a specific promise, he will have a negatively valenced emotional reaction. He will feel bad at the thought of breaking the promise, and this will motivate him not to break the promise. *Mutatis mutandis* for the case of someone who believes that an action ought (morally) to be performed. The motivation to behave in accordance with one's own moral beliefs is brought about by the unpleasantness of guilt.

In an interesting sense, on this account, the motivational power of a moral belief can be said to extend *beyond* the person who has the belief. If Charles believes that Tim has violated a moral norm—that is, Charles has a moral belief *M* about a practical demand *m* and also believes that Tim has violated *m*—then Charles will feel moral disapproval toward Tim and may be motivated to express this moral disapproval and punish Tim. Punishment is of course unpleasant, and so is being morally disapproved of, which can be seen as a particular kind of punishment. Insofar as Tim is motivated to avoid being punished by Charles and to avoid Charles's moral disapproval, Tim will also be motivated to behave in accordance with Charles's moral belief. Again, the mechanism could be similar to the one described by Damasio. When Tim considers acting in a way that goes against a moral belief that Tim knows Charles to have, Tim will experience a negative emotion at the thought that Charles will feel moral disapproval toward him. The potential action will become negatively emotionally marked, and Tim will thereby be motivated not to perform the action. So, for example, if Charles believes that keeping promises is morally required, and Tim knows this, Tim's fear of Charles's moral disapproval will motivate Tim to keep promises.[5]

Let us go back to the alternative view mentioned above, according to which the motivational power of moral beliefs derives from a desire to be rational. Can this view account for the long reach of the motivational power of moral beliefs? If Charles desires to be rational and believes that breaking promises is—irrespective of one's ends and interests—irrational, Charles will be motivated to keep promises. But will Charles be motivated to criticize or punish others if they do not keep promises? If Charles does not just desire that he himself avoid irrationality but also desires *everyone* to avoid irrationality, he will be motivated to do something that those who do not keep promises will find unpleasant, such as criticizing or punishing them. So, if the motivational power of moral beliefs has a long reach, this way of explaining the motivational power needs to posit not just a desire to avoid irrationality for oneself but a desire (implicit, subconscious) that *everyone* avoid irrationality.

The idea that the motivational power of moral beliefs has a long reach, whatever its explanation, fits well with the hypothesis that moral beliefs—or more precisely the ability to form them—evolved in order to make cooperative behaviors more likely to be performed, at least in certain circumstances. In general, cooperative behaviors are fitness enhancing only if the individuals with whom one interacts are likely to be disposed to cooperate. It is normally bad for one's fitness to cooperate with individuals that do not cooperate, and cooperation can evolve if—on average—cooperators tend to benefit other cooperators more than they benefit noncooperators, though strict reciprocation may not be necessary (e.g., Godfrey-Smith, 2009).

However, what we said about the domain-general nature of the motivational power of moral beliefs still stands. The ability to form moral beliefs may have evolved in order to make certain cooperative behaviors more likely to be performed *both by the individual who has the belief and by those individuals with whom he interacts*. But the motivational power of a moral belief is independent of its specific content. A firm belief that a noncooperative behavior is morally required has the same motivational power—*both internal and external*—as a firm belief that a cooperative behavior is morally required.

Let me take stock. I have introduced two hypotheses about the motivational power of moral beliefs:

(1) The motivational power of moral beliefs derives from a (possibly subconscious and implicit, and perhaps also generalized) desire to avoid irrationality.
(2) The motivational power of moral beliefs is due to connections between such beliefs and the emotions of guilt and moral disapproval.

These hypotheses can also be combined:

(3) The motivational power of moral beliefs derives *both* from a desire to avoid irrationality *and* from connections between such beliefs and the emotions of guilt and moral disapproval.

This third hypothesis can be developed in various ways. One could theorize that these two sources of motivational power of moral beliefs are independent of each other, or that they are not. If they are not, one option is that the connections between moral beliefs and guilt/moral disapproval depend (causally or conceptually) on the desire to avoid irrationality. Another option is that the reverse is true.

According to the evolutionary hypothesis sketched in section 2, the evolution—by natural selection—of the ability to form moral beliefs is due to the motivational power of such mental states. In this context, an account of this motivational power is crucially important. A complete evolutionary account of moral beliefs will have to say something about the evolution of what gives (or gave) such moral beliefs their motivational power. If the remarks made in this section are on the right track, such an evolutionary account may have to refer to the evolution of a desire to avoid irrationality or to the evolution of certain kinds of emotional dispositions, or to both.

In the final section, I will return to the idea that a desire to avoid irrationality may play a role in explaining the motivational power of the mental states we express when we make moral judgments. In the following sections, instead, I will develop the view that the motivational power of these mental states is due to certain emotions. One possible way to make sense of this view is this: For a mental state to *be* a moral belief it must have the relevant connections to certain emotions. That is, certain emotional dispositions are part of what it takes to have a moral belief. On this account, moral beliefs are *intrinsically* motivating, and thereby they are—among other things, perhaps—motivational states.

As said above, despite the characterization he gives of moral beliefs, Joyce thinks that the motivational power of moral beliefs is due to their connection to emotions like guilt and moral disapproval, and not to a desire to avoid irrationality. According to him, though, such connections are not an essential feature of moral beliefs: Moral beliefs are *not* intrinsically motivating and the connections between such mental states and the emotional dispositions mentioned above are just reliable but *contingent* causal links. Below, I will try to determine whether any sense can be made of the view that moral beliefs are intrinsically motivating.

5 Intrinsically Motivating Beliefs

If moral beliefs are not actually beliefs, then they can be intrinsically motivating simply by being straightforward motivational states. But if moral beliefs are beliefs, things are more complicated. Intrinsically motivating beliefs are sometimes called "besires" in the philosophical literature and they are often discussed in the context of the Humean Theory of Motivation, which is thought by some to imply that such states cannot exist (e.g., Smith, 1994). These mental states are called besires because they are beliefs that, in addition to having a belief-like content, also have a desire-like content, which gives them their motivational power. There is no room here to discuss all the arguments that have been presented for and against the Humean Theory and the possibility of besires. I will just sketch an account according to which mental states that have both a belief-like content and a desire-like content are not only possible but also common, both in humans and in other animals. A detailed defense of this account from Humean objections will have to be left for some other occasion.

Here is the account. A mental state with a belief-like content is a mental state with the function to track the obtaining of some specific state of affairs. For example, a mental state with the belief-like content *that there is Cannonau wine in the cellar* is a mental state with the function to track the presence of Cannonau in the cellar. In contrast, a mental state with a desire-like content is a mental state with the function to bring about some specific state of affairs. For example, a mental state with a desire-like content *that there be Vermentino wine in the fridge* is a mental state with the function to bring about the presence of Vermentino in the fridge.[6]

Some mental states have a belief-like content and no desire-like content. Standard beliefs are of this kind. They are motivationally inert and can only motivate action in conjunction with a desire-like state. So, for example, the belief that there is Cannonau in the cellar can motivate action only in conjunction with some appropriate desire, such as the desire to drink Cannonau. Some mental states have a desire-like content and no belief-like content. We can suppose that standard desires, like the one just mentioned, are of this kind. But there are also mental states with both a desire-like content and a belief-like content. An example is an emotion like fear.

Let me illustrate this with a little story. Sarah is afraid of spiders. She suddenly sees a spider a few meters away from her and goes into a state of fear. This mental state of hers has two functions: to signal the presence of a potential threat in Sarah's environment and to motivate Sarah to find a way to avoid the threat. It has both a belief-like content (*that that spider is dangerous*) and a desire-like content (*that I avoid contact with that spider*). So, Sarah's fear is a mental state with a belief-like content that, because of its desire-like content, is intrinsically motivating. No other mental state is needed for Sarah to be motivated to act.

Notice that, in a sense, the two contents—the two functions—of this mental state of Sarah's can be pulled apart modally.[7] That is, it is possible to have a mental state that has the same belief-like content as Sarah's mental state but does not have the same desire-like content, or has no desire-like content at all. Consider Sarah's friend, Rita, who is not afraid of spiders. Rita is a spider expert. She knows a great deal about them and is fond of them. Sarah draws Rita's attention to the spider she has just spotted. Thanks to her expertise, Rita immediately realizes that it belongs to a very poisonous species. But Rita is not in the least scared. She has dealt with spiders of this species many times in the past. Despite this, she forms the belief that that spider is dangerous; as said, she knows that the spider is poisonous. Her belief that the spider is dangerous has the same belief-like content as Sarah's fear, but unlike Sarah's fear it is motivationally inert. It is only in conjunction with a distinct and independent desire to avoid contact with dangerous spiders that Rita's belief can motivate Rita to act.

Both Sarah and Rita are motivated to avoid contact with the spider. In Rita's case, the motivation is due to a mental state with the function to track the obtaining of a certain state of affairs and a *distinct* mental state with the function to bring about a certain state of affairs. In contrast, in Sarah's case the motivation is due to single mental state with both the function to track a certain state of affairs and the function to bring about a certain (different) state of affairs. But notice that the fact that the two functions of Sarah's fear can be pulled apart modally does not mean that Sarah's fear is nothing but the conjunction of a belief and a desire. If fear were reducible to a contingent conjunction of a belief and a desire there would be no psychological differences between Sarah and Rita. But there *are* important psychological differences between the two women. These are the psychological differences that we refer to when we say that Sarah is afraid and Rita is not.

When I say that Sarah's fear and Rita's belief have the same tracking function and thereby the same belief-like content, I am using a notion of "broad"—as opposed to "narrow"—content. Fear refers to dangerousness (it has the function to track potential threats) *differently* from the way the concept [DANGEROUS] that Rita deploys in her belief does. Fear and the concept [DANGEROUS] are two different *modes of presentation* of the same property (dangerousness), just like, for example, the concept [WATER] and the concept [H_2O] are two different modes of presentation of the same chemical compound. Fear is an intrinsically motivating way of mentally representing dangerousness, whereas the concept [DANGEROUS] is not. More precisely, fear has the function to track dangerousness in an intrinsically motivating way, whereas [DANGEROUS] does not.

Being a certain kind of motivational state is part of what it takes for a mental state to be a state of fear. The same is not true of beliefs like the one attributed to Rita. So, it is only when we think about a particular kind of broad content—the sorts of events that they have the function to track—that we can say that Sarah's fear and Rita's belief have the same belief-like content. When we think about "thinner" ways of classifying contents, and in particular when we think about the "cognitive value" of mental states—the specific way in which mental states track the events they have the function to track—we have to say that Sarah's fear and Rita's belief have different contents.

Let us go back to moral beliefs. Here is one possible way to make sense of the view that moral beliefs are actually beliefs and that, at the same time, the relation between them and certain emotional dispositions is not contingent. Moral beliefs have a belief-like content in that they have the function to track the obtaining of certain states of affairs. More precisely—and here I am borrowing Joyce's characterization—they have the function to track the existence of practical demands with inescapable automatic authority. In addition, moral beliefs have the function to bring about certain states of affairs. More precisely, a moral belief about a practical demand *m* has the function to bring it about that one behaves in accordance with *m*, and it has the function to do this by bringing it about that one feels guilt at the thought of acting in ways that constitute violations of *m*. So, for example, Charles's belief that keeping promises is morally required has a belief-like content: that there is an inescapable and automatically authoritative practical demand for keeping promises. Moreover, Charles's belief has the function to motivate Charles to keep promises.

Given our discussion about the long reach of the motivational power of moral beliefs, we can also say that Charles's belief has the function to motivate *everyone* to keep promises, not just Charles. A moral belief *M* of a person *P* about a practical demand *m* has the function to bring it about that *everyone* (not just *P*) is motivated to behave in accordance with *m*. It has the function to do this by causing *P* to feel guilt at the thought of acting in ways that constitute a violation of *m* and by causing *P* to feel moral disapproval toward *anyone* who behaves in ways that violate *m*. So, on this view, Charles's belief that keeping promises is morally required has the function to motivate every individual, not just Charles, to keep promises.

Just like for fear, the tracking function and the motivating function of a moral belief can be pulled apart modally. But this does not mean that a moral belief is the contingent conjunction of a motivationally inert belief and a standard desire. In this very important respect, according to this view, moral beliefs are like emotional states, even though in other respects—for example, in the way they are formed, or in the way they interact with other mental states—they may be more similar to motivationally inert beliefs.

6 The Open Question Argument

Moral beliefs are formed by deploying moral concepts, in conjunction with other concepts. Nonmoral beliefs are formed by deploying (exclusively) nonmoral concepts. In this section, I will try to explain what moral concepts must be like if the view sketched in the previous section is right. I will start with the so-called open question argument, not because this argument works, but because understanding how this argument *fails* can help us understand the nature of moral concepts.

The open question argument is an argument against moral naturalism, that is, against the view that there are moral facts and that they are natural facts. Here is one version of the argument:[8]

OQ1 According to moral naturalists, for some natural property *N*, being morally required is identical with *N*, and so for every *x*, *x* is morally required if and only if *x* is *N*.
OQ2 But for any choice of *N*, it is possible that a person who possesses the concept [MORALLY REQUIRED] may be sure that something is *N* while being unsure whether it is morally required.
OQ3 Hence, for any choice of *N*, the concept [MORALLY REQUIRED] does not refer to property *N* and therefore being morally required and *N* are different properties. That is, the moral naturalist is wrong.

The argument is not valid. This can be seen by examining this other argument:

W1 According to current scientific theories, being water and being H_2O are the same property, and so for every *x*, *x* is water if and only if *x* is H_2O.
W2 But a person who has the concept [WATER] may be sure that something is water while being unsure whether it is H_2O. (Consider, for example, a person who has not yet discovered or learned that water is H_2O.)
W3 Hence, the concept [WATER] does not refer to H_2O and therefore water and H_2O are different things. That is, current scientific theories on water are wrong.

From W2, it follows is that the concept [WATER] and the concept [H_2O] have different cognitive values. But two concepts with different cognitive values may corefer. They may be different modes of presentation *of the same property*. That is, two concepts may have the function to track the same substance, but they may do this in different ways. So, W2 does not threaten current scientific theories, and the inference from W2 to W3 is invalid.

Similarly, from OQ2—the existence of an open question—it follows that the concept [MORALLY REQUIRED] and the nonmoral concept [N] have different cognitive values, and this is true for every choice of natural property *N*. (The nonmoral concept [N] is simply a concept that refers to the natural property *N* and that can be expressed using nonmoral language.) Even if we agree that OQ2 is true, it is still possible that, for some natural property *N*, the concepts [MORALLY REQUIRED] and [N] corefer. So, it is still possible that, for some natural property *N*, the property of being morally required is *N*. Some version of moral naturalism may be correct. The inference from OQ2 to OQ3 is invalid.[9]

Is OQ2 true though? If the account of moral beliefs sketched in the previous section is correct, moral concepts differ in cognitive value from nonmoral concepts of natural properties. In fact, in this context, the difference between natural and nonnatural properties is irrelevant. Consider this:

*OQ2** For any choice of nonnatural property *N* (e.g., being desired by God), it is be possible that a person who possesses the concept [MORALLY REQUIRED] may be sure that something is *N* while being unsure whether it is morally required.

On the account given above, both OQ2 and OQ2* are true. Moral concepts differ in cognitive value (in narrow content) from nonmoral concepts, be they nonmoral concepts of natural properties or nonmoral concepts of nonnatural properties. What does this difference in cognitive value amount to? If the analogy with fear works, one of the differences—though not the only difference—is that moral concepts are intrinsically motivating ways of mentally representing, whereas nonmoral concepts (at least in general) are not.

7 Intrinsically Motivating Moral Concepts

Consider the following scenario. Peter has a strong disposition to apply his concept [MORALLY REQUIRED] to all and only those actions that he thinks maximize total aggregate happiness. This disposition is due to a firm belief that being morally required and maximizing happiness are the same property, the belief that [MORALLY REQUIRED = HAPPINESS MAXIMIZING].[10] John, in contrast, has a strong disposition to apply his concept [MORALLY REQUIRED] to all and only those actions that he thinks maximize group cohesion. This disposition is due to a firm belief that being morally required and maximizing group cohesion are the same property, the belief that [MORALLY REQUIRED = GROUP COHESION MAXIMIZING]. Is Peter's concept [MORALLY REQUIRED] the same as John's concept? Do these concepts have the same cognitive value?

The maximization of happiness and the maximization of group cohesion do not always coincide. Suppose, just for the sake of argument, that the capital punishment of terrorists maximizes group cohesion but not collective happiness, and that both Peter and John know this. If Peter and John had concepts with different cognitive value, when they ask themselves "Is capital punishment for pedophiles morally required?" they would be asking different questions and they would be entertaining different thoughts. If so, when they debate about whether capital punishment for pedophiles is morally required, they would be talking past each other. But they are not talking past each other. They genuinely disagree.

Peter and John have concepts with the same cognitive value, despite differences in their dispositions to apply the concepts to particular actions. Peter and John are in a situation similar to that of two individuals who share the concept [WATER] but disagree on what water is: one thinks that [WATER] refers to H_2O and the other one thinks that [WATER] refers to XYZ, where H_2O and XYZ are two different chemical compounds. But while it is relatively easy to understand in what sense two individuals who disagree about the chemical nature of water may share the concept [WATER]—they both think of water as the colorless, odorless substance that fills rivers and seas, that people drink when they are thirsty, and so on—in what sense do Peter and John share the concept [MORALLY REQUIRED]?

If the account sketched in the two previous sections is correct, Peter and John agree on the nature of moral demands. They both think of moral demands as having inescapable automatic authority. Moreover, they both have a disposition to feel guilt when they disregard or consider disregarding (what they take to be) a moral demand. Because of this they are both motivated to act in accordance with (those practical demands that they take to be) moral demands. Also, they both have a disposition to feel moral disapproval toward those who disregard (those practical demands that they take to be) moral demands; and in this way they may be able to motivate others to act in accordance with (those practical demands that they, Peter and John, take to be) moral demands.

Questions about the concept [MORALLY REQUIRED] are questions about beliefs of the form [*x* IS MORALLY REQUIRED]. On the view explored here, both in Peter and in John, beliefs of this form have the function to track practical demands with inescapable automatic authority. Moreover, in both of them, mental states of this kind have the function to bring it about that everyone behaves in accordance with the relevant practical demand, and they have the function to do this by generating guilt and moral disapproval in the appropriate contexts. That is, in both Peter and John, beliefs of this form are intrinsically motivating ways of tracking practical demands with inescapable automatic authority. It is in this sense that Peter and John can be said to have concepts with the same cognitive value.

On this account, having certain kinds of emotional dispositions is part of what it takes to possess the concept [MORALLY REQUIRED]: this concept is an *intrinsically motivating mode of presentation*. Both Peter and John have the relevant dispositions even though, as said, they apply this concept to different actions—happiness-maximizing actions in the case of Peter and cohesion-maximizing actions in the case of John—and thereby have the relevant emotional dispositions toward different actions.

Consider Peter. Even if he has a firm belief that being morally required and being happiness maximizing are the same property—that the concepts [MORALLY REQUIRED] and [HAPPINESS MAXIMIZING] corefer—he can still entertain the thought that the two modes of presentation do not corefer. In the same way, a person who believes that water is H_2O can entertain the thought that the concepts [WATER] and [H_2O] do not corefer. Peter can, for example, entertain the thought that, contrary to what he believes, actions that maximize group cohesion—rather than happiness—are morally required. He can entertain the thought that [MORALLY REQUIRED] and [COHESION MAXIMIZING] corefer, and if he were to believe this thought to be true, he would be disposed to act in ways that he regards as group-cohesion maximizing. It is because Peter can entertain this thought that he can understand John when the two of them debate about moral matters, just like we can understand the claims of a heretical scientist who argues that, contrary to current scientific orthodoxy, the stuff that people call "water" is not in fact H_2O but some other chemical compound. *Mutatis mutandis* for John.

Now compare Peter and John with two other characters, Ben and Will. Just like the first two characters, Ben and Will have the concept [MORALLY REQUIRED]. They can entertain the thought that [MORALLY REQUIRED] and [HAPPINESS MAXIMIZING] corefer, but they believe this thought to be false. And they can entertain the thought that [MORALLY REQUIRED] and [COHESION MAXIMIZING] corefer, but they believe this thought to be false too. Ben believes these thoughts to be false because he believes the property of being morally required to be identical with the property of being desired by God: he believes that [MORALLY REQUIRED = DESIRED BY GOD]. Facts about God's desires are supernatural facts, so Ben can be said to be a supernaturalist. In contrast, Will believes that there are no practical demands of the moral sort. He is a moral nihilist and thinks that the concept [MORALLY REQUIRED] has an empty extension.

Let us introduce yet another character, Mark. Just like the other guys, Mark has the concept [MORALLY REQUIRED]. Mark applies this concept to some actions and not to others. For example, Mark believes that helping one's friends when they are in need is morally required, that giving some money to charity every now and then is morally required, that preserving the planet for future generations is morally required, that keeping promises is morally required, and so on. But unlike Peter, John, and Ben, this new character does not have a nondisjunctive nonmoral mode of presentation for referring to all and only the actions that he considers morally required. That is, Mark does not have a nondisjunctive nonmoral concept [ND] (such as [HAPPINESS MAXIMIZING], [COHESION MAXIMIZING], or [DESIRED BY GOD]) such that he can be said to believe that [MORALLY REQUIRED = ND]. Unlike the other guys, when asked which actions he believes to be morally required, Mark can only answer by providing a (possibly very long) list. That is, if asked with what property the property of being morally required is identical, he can only characterize this property in a disjunctive way.

Even though this is empirical issue, it is plausible to think that most people—and especially people who have not thought much and in a reflective and systematic way about their moral views and about the way they apply the concept [MORALLY REQUIRED]—are like Mark in this respect. People like Peter, John, and Ben seem to be relatively rare in the general population. And so, of course, are people like Will.

Peter, John, Ben, Will, and Mark have the moral concept [MORALLY REQUIRED]—and this also means, on the view examined here, that they all have the right kinds of emotional *dispositions*—but they disagree on which actions are morally required and on whether there actually are any morally required actions. That is, they disagree on which practical demands are demands of the moral sort and on whether there actually are any practical demands of the moral sort—and given the way they think about practical demands of the moral sort, their emotional dispositions get triggered by different events and in different occasions.

These five characters can all understand each other in the sense that for any property *P*—be it natural or supernatural—they can all entertain the thought that the concepts [MORALLY REQUIRED] and [P] corefer, where [P] is a nonmoral concept referring to *P*. More generally, for any property *P*, be it natural or supernatural, it is possible in principle to find someone who possesses the moral concept [MORALLY REQUIRED] and does not believe that [MORALLY REQUIRED] and [P] corefer. This person may be sure that something is *P* while being unsure, or in fact while denying, that it is morally required. It is in this sense that, on this account, OQ2 and OQ2* are true.

8 Parasitic Moral Concepts

Let us now introduce Pat. Let us suppose Pat does not possess the concept [MORALLY REQUIRED] that the characters mentioned so far possess. If you ask him whether he believes that, say, keeping promises is morally required, he answers "yes" and he is sincere. But, by giving this answer, he is not expressing the mental states that the others express when they give the same answer. Pat does not think of keeping promises as something for which there is a practical demand that is inescapable and automatically authoritative. Moreover, he does not have the disposition to feel guilt when he disregards or considers disregarding this practical demand, and he does not have the disposition to feel moral disapproval toward those who disregard this practical demand.

Pat is not motivated to keep promises—at least not by the belief that he expresses when he says "Keeping promises is morally required," when such belief is taken on its own. This belief of Pat's is motivationally inert.

When Pat says "Keeping promises is morally required," he is simply expressing his belief that people around him call certain things "morally required." The concept Pat expresses by the words "morally required"—let us call it the concept [MORALLY REQUIRED*]—is not the concept [MORALLY REQUIRED] that the other characters mentioned so far have. Pat's concept is simply a way of tracking the property of being classified as morally required by other people, people who do have the concept [MORALLY REQUIRED]. It is a motivationally inert way of tracking this property.

Pat's beliefs of the form [*x* IS MORALLY REQUIRED*] do not motivate behavior. They do not have the function to bring about specific behaviors. So, for example, Pat's belief that [KEEPING PROMISES IS MORALLY REQUIRED*]—that is, the belief that keeping promises is something that other people classify as morally required—does not *by itself* motivate Pat to keep promises or, for that matter, to do anything. This belief can motivate behavior only in conjunction with distinct and specific desires. For example, Pat may desire not to be punished, and he may have learned that doing things that others classify as morally required often results in punishment.

Compare Pat with Will, the nihilist. Will possesses the intrinsically motivating concept [MORALLY REQUIRED] but he does not apply it to anything, because he believes the concept to have an empty extension. If Will and Pat were to break a promise, they would not feel guilt, but for different reasons. Will would not feel guilt because he believes that his concept [MORALLY REQUIRED] does not apply to anything, and so in particular he believes it does not apply to keeping promises. Will has the emotional dispositions required to possess [MORALLY REQUIRED] but such dispositions are never triggered in him, because he never deploys the concept in belief. In contrast, Pat would not feel guilt because, despite the fact that he is happy to apply his concept [MORALLY REQUIRED*] to keeping promises, his concept has no intrinsic connection with emotions like guilt. Neither Will nor Pat has the belief that [KEEPING PROMISES IS MORALLY REQUIRED], but in the case of Pat this is simply due to the fact that he does not have the concept [MORALLY REQUIRED].

Compare now Pat with Mark. Suppose that these two characters belong to the same cultural group. They live in the same country, speak the same language, and so on. Suppose moreover that Mark's moral beliefs are representative of the moral beliefs that are common in that cultural group. That is, for any action *A*, Mark believes that [*A* IS MORALLY REQUIRED] if and only if most people in that cultural group believe that [*A* IS MORALLY REQUIRED]. Suppose also that Pat knows what actions most people in his cultural group classify as morally required. That is, for any action *A*, Pat believes that [*A* IS MORALLY REQUIRED*] if and only if most people in his cultural group believe that [*A* IS MORALLY REQUIRED]. Thus, for any action *A*, Pat believes that [*A* IS MORALLY REQUIRED*] if and only if Mark believes that [*A* IS MORALLY REQUIRED]. Given that the function of Pat's concept [MORALLY REQUIRED*] is to track the actions that most people—that is, people like Mark—classify as morally required, we can say that Pat's concept is *parasitic on* the concept of people like Mark.

The topic is extremely controversial, but there seems to be some evidence for the view that psychopaths are like Pat (Blair, 1995; 1997; Nichols, 2004). Psychopaths know what actions are classified as morally required by others, and they are happy to call those actions "morally required." The concept they express when they call something "morally required" seems to be parasitic on the moral concept possessed by other people. The psychopaths' concept does not by itself have any motivational powers. Psychopaths do not feel guilt when they fail to act or consider not acting in ways that they are happy to call "morally required." As a result of this, psychopaths are not motivated to behave in ways that they call "morally required," at least not *directly*. Insofar as they are motivated to perform such actions, this is due to a distinct desire to avoid punishment and their knowledge that not doing things that others consider morally required often results in punishment. If this is right, psychopaths have beliefs of the form [*x* IS MORALLY REQUIRED*] but they do not have beliefs of the form [*x* IS MORALLY REQUIRED]. They possess the concept [MORALLY REQUIRED*] but not the concept [MORALLY REQUIRED].[11]

What is important for the purposes at hand is that the concepts [MORALLY REQUIRED] and [MORALLY REQUIRED*] are different and that they differ in *two* important ways. First of all, [MORALLY REQUIRED] has a motivating function whereas [MORALLY REQUIRED*] does not. Second, the two concepts have two different tracking functions: [MORALLY REQUIRED] has the function to track practical demands with inescapable automatic authority, whereas [MORALLY REQUIRED*] has the function to track practical demands that are *believed* by others to have inescapable automatic authority. To illustrate this point, suppose Will is right in thinking that there are no practical demands with inescapable automatic authority, despite the fact that some practical demands are widely believed to have inescapable automatic authority. In this case, all beliefs of the form [*x* IS MORALLY REQUIRED] would be false, while some beliefs of the form [*x* IS MORALLY REQUIRED*] would be true. Ironically, on this view, if Will is right, psychopaths can express true beliefs by saying things like "Keeping promises is morally required," while nonpsychopaths cannot.

9 Other Kinds of Moral Concepts

If the concepts [MORALLY REQUIRED] and [MORALLY REQUIRED*] differ in two ways, there is room for two concepts each of which differs only in *one* way from the concept [MORALLY REQUIRED]:

(i) a concept with the same tracking function as [MORALLY REQUIRED] but without the motivating function that this concept has; let us call this concept [MORALLY REQUIRED^T], where T stands for "tracking function only."
(ii) a concept with the same motivating function as [MORALLY REQUIRED] but without the tracking function that this concept has; let us call this concept [MORALLY REQUIRED^M], where M stands for "motivating function only."

[MORALLY REQUIRED] and [MORALLY REQUIRED^T] are, in some respects, like fear and the concept [DANGEROUS]. They have the same tracking function but the second of each pair does not have a motivating function. What exactly is the tracking function that [MORALLY REQUIRED] and [MORALLY REQUIRED^T] have in common? What does it mean to say that a mental state has the function to track practical demands with inescapable automatic authority?

If Joyce is right, to say that an action is morally required is to say that it would be irrational for any person, independently of her specific ends, not to perform that action. Thus, having the function to track practical demands with inescapable automatic authority is having the function to track practical demands that it would be irrational for any person, independently of her specific ends, to ignore. On this view, having the belief that [*A* IS MORALLY REQUIRED^T] is having the motivationally inert belief that [NOT DOING *A* IS IRRATIONAL INDEPENDENTLY OF ONE'S DESIRES AND INTERESTS].[12]

As we saw in section 4, this belief cannot motivate action *on its own*, but it can certainly do so in the presence of a desire to be rational, or—which is the same—in the presence of a desire to avoid irrationality. So, for example, if one desires to avoid irrationality, and one believes that not keeping promises would be irrational, one is motivated to keep promises. By itself, the belief that [KEEPING PROMISES IS MORALLY REQUIRED^T] is motivationally inert, but in the presence of a desire to avoid irrationality it acquires motivational "oomph." As said above, this desire does not need to be explicit and conscious. It can be implicit and subconscious.

Both the belief that [*A* IS MORALLY REQUIRED^T] and the belief that [*A* IS MORALLY REQUIRED] mentally represent that not doing action *A* would be irrational, and that it would be irrational for any person, independently of her specific desires and interests. But the belief that [*A* IS MORALLY REQUIRED] mentally represents this in an intrinsically motivating way, just like fear represents dangerousness in an intrinsically motivating way. In contrast, the belief that [*A* IS MORALLY REQUIRED^T] represents it in a way that is not intrinsically motivating, just like the concept [DANGEROUS] represents dangerousness. Because of their shared tracking function, though, both beliefs can motivate action *in the presence of a desire to be rational*. The belief that [*A* IS MORALLY REQUIRED] retains some of its motivational power even in a person who does not desire to avoid irrationality, whereas the belief that [*A* IS MORALLY REQUIRED^T] does not. The reason why the belief that [*A* IS MORALLY REQUIRED] retains some of its motivational power in the absence of this desire is that possession of the concept [MORALLY REQUIRED] comes with the possession of certain dispositions, dispositions to feel negatively valenced emotions (guilt and moral disapproval) in certain circumstances.

In contrast with the two concepts just discussed, the concept [MORALLY REQUIRED^{M}] does not have a tracking function and so cannot be used to token beliefs. A mental state formed by applying [MORALLY REQUIRED^{M}] to, for example, [KEEPING PROMISES] is not a belief and does not have a belief-like content. This mental state only has a desire-like content: *that everyone keep their promises*. It has the function to bring it about that everyone keeps their promises and it can fulfill this function by causing guilt and moral disapproval in the appropriate circumstances. The presence or absence of a desire to avoid irrationality has no effect on this mental state.

The concept [MORALLY REQUIRED^{T}] has motivational power only in the presence of a desire to avoid irrationality. The concept [MORALLY REQUIRED^{M}] derives its motivational power exclusively from emotional dispositions. The concept [MORALLY REQUIRED] combines these two sources of motivational power. One hypothesis is that the mental states we express when we make moral judgments are beliefs of the form [*x* IS MORALLY REQUIRED^{T}] and that they have motivational power only insofar as there is a desire to avoid irrationality. A second hypothesis is that these mental states are not beliefs but rather straightforward motivational states involving the deployment of [MORALLY REQUIRED^{M}]. A third hypothesis is that these mental states are intrinsically motivating beliefs of the form [*x* IS MORALLY REQUIRED], with two associated sources of motivational power.

The first hypothesis looks implausible, at least *prima facie*, in that it seems to imply that guilt and moral disapproval play no role in explaining the motivational power of moral beliefs.[13] The second option is certainly interesting, though many will find it unsatisfactory because of its noncognitivist nature. The third hypothesis is also interesting, though many questions about it still need to be addressed.[14]

I will conclude with some speculations about the development and evolution of [MORALLY REQUIRED], assuming—at least for exploratory purposes—that this is the concept that all adult nonpsychopathic humans express when they say that something is morally required. One possible—though not very plausible—view is that a desire to avoid irrationality is present in human infants and plays a crucial role in the developmental acquisition of [MORALLY REQUIRED], and perhaps also in the acquisition of the emotional dispositions that the concept involves. It could be the case, for example, that guilt at the thought of not doing something morally required is the developmental outcome of, among other things, a desire to avoid rational criticism. Another—in some ways much more plausible—view is that the emotional dispositions from which [MORALLY REQUIRED] derives some of its motivational power *precede* the acquisition of this concept and play an important role in such acquisition. Perhaps it is the concept [MORALLY REQUIRED^{M}], the one with a motivating function but no tracking function, that first emerges in development.[15] Only later does this concept acquire a tracking function—the function to track practical demands that it would be irrational to disregard irrespective of one's ends—and thereby is transformed into the concept [MORALLY REQUIRED]. After this transformation has taken place, the desire to avoid irrationality becomes able to *contribute* to the motivational power of the concept, though the concept also has some motivational power that comes directly from its connection with preexisting emotional dispositions.

This second view fits well with an interesting evolutionary hypothesis according to which humans first evolved the ability to form mental states (without a belief-like content) of the form [*x* IS MORALLY REQUIRED^{M}] and only later evolved—either genetically or culturally—the ability to form *beliefs* of the form [*x* IS MORALLY REQUIRED].

Needless to say, all these hypotheses will have to be explored further on some other occasion.

Acknowledgments

I would like to thank Richard Joyce, Kim Sterelny, Brett Calcott, and Ben Fraser for their useful comments on a previous draft. I would also like to thank David Papineau for helpful discussions on the topics of this essay.

Notes

1. "A person who has good reason to be rude (e.g., to speak with his mouth full in order to stop a friend from eating a wasp) is still violating etiquette, but he has done the right thing, and in fact he has no reason to do otherwise. Only people who have interests that will be served by following etiquette (i.e., most of us, most of the time) have any genuine reason to comply. It is often thought that morality requires something stronger and more authoritative. . . . A prescription may inescapably *apply* to a person without her having any reason to comply; if, however, the prescription brings reason for compliance with it ('automatically,' as it were), let us say that it has *authority*" (Joyce, 2006, p. 193). For the distinction between inescapable rules that are authoritative and inescapable rules that are not authoritative—or between strong and weak categorical imperatives—see Foot (1972) and ch. 2 of Joyce (2001).

2. Kim Sterelny drew my attention to this point.

3. On a draft of this essay, Joyce commented:

I think it's pretty clear that mere inescapability doesn't capture all that the folk want from morality—not when one really reflects on how wimpy inescapable norms can be. The folk definitely want *something* more from morality, though they may be pretty inchoate as to what. In any case, let's call the something-more "authority." Then there's a question of whether this authority can be elucidated and, if so, how. Saying that immorality is "contrary to reason" is one hypothesis of how we might understand the authority. Perhaps it's not the best way, but I've yet to hear of a better. I suppose saying that it's "contrary to God's will" might be another. Perhaps in the end the "authority" that the folk want is just mysterious and forever inchoate. (So much the better for the error theorist if that's true.)

4. An exaptation hypothesis is also possible. The ability to form moral beliefs may be the result of capacities that were initially selected for other purposes and that, at some point in our evolutionary history, started working together in a specific way and started being selected because of their resulting in the tokening of moral beliefs. (Thanks to Ben Fraser for pointing this out.)

5. It is true that, once he knows about Charles's moral beliefs, Tim may be motivated to cover up his *m*-violations rather than to behave in accordance with *m*. Much will depend on what Tim can gain by violating *m* and on the risks and costs involved in being caught violating *m*. Much will also depend on whether Tim shares Charles's moral belief or not. If Tim thinks Charles's moral views are wrong, he may resent attempts at being punished for violating *m* and may try to counterpunish Charles. But if Charles is a more powerful individual than Tim, and Tim knows this, Tim may still be motivated to behave in accordance with Charles's moral belief in order to avoid Charles's disapproval and punishment, even if Tim does not share Charles's moral views.

6. For some examples of theories that try to give an account of the content of mental states in terms of the function of mental states, see Millikan (1993), Papineau (1995), Dretske (1997). The relevant notion of function is a teleological one, that is, one that leaves room for the possibility of malfunctioning. I am not committed to the details of any of these theories, nor am I committed to an evolutionary account of the content-determining teleofunction of mental states. I actually find developmental accounts of such functions more promising, though this is irrelevant for the purposes at hand. All that matters here is that *some* teleological notion of function can be used to give an account of mental content. Teleological notions of function are attractive in the context of discussions about the nature of mental content because they allow for misrepresentation to be conceived of as a specific form of malfunctioning.

7. Those who embrace the Humean Theory of Motivation often argue for the view that besires are impossible in this way: Since belief-like contents and desire-like contents can always be pulled apart modally, any state with two such contents is nothing but a conjunction of a belief and a desire (Smith, 1994; Joyce, 2001).

8. The original version of the argument was developed by Moore (1903). Arguably, Moore's version—with its focus on "definability"—is not particularly transparent and is not useful for the discussion in this essay.

9. According to one kind of moral naturalism, the identity between moral properties and natural properties can be known a priori, whereas according another kind of moral naturalism, the identity can only be known a posteriori, just like the identity between water and H_2O. It should be clear from what I have said that the open question argument cannot be used against a posteriori naturalism. Can it at least be used against a priori naturalism? No, it cannot. Consider the concept [293] and the concept [((((333 × 3) + 856)/5) – 78)]. These two concepts both refer to the number 293, and the fact that they corefer can be known a priori: by doing the calculations. Yet, before having done the calculations, one may be sure *that there are 293 pins in a box* without being sure *that there are ((((333 × 3) + 856)/5) – 78) pins in the same box*. The two concepts have different cognitive values. They are two modes of presentation of the same mathematical object, and that this is so can be established a priori. But because of their differences in cognitive value—including, importantly, differences in structure—the fact that the two concepts corefer is *not*

obvious. The same could be true, for some natural property *N*, of the concept [MORALLY REQUIRED] and the concept [N]. So neither a posteriori naturalism nor a priori naturalism seems to be incompatible with OQ2.

10. This belief may be implicit or explicit, and Peter may have formed it a priori or a posteriori. All these distinctions are irrelevant here.

11. It would be also interesting to think about the moral concepts of individuals with acquired sociopathy, i.e., individuals who suffer brain damage and as a result seem no longer motivated to do what they believe they morally ought to do (see Roskies, 2003).

12. One thing to note is that the concept [MORALLY REQUIRED^T] has a complex structure. This concept is constituted by other, simpler concepts, each of which has its own tracking function. This is important for someone who wants to combine a teleofunctional account of content with Joyce's view that there actually are no practical demands with inescapable automatic authority, i.e., the view that no action is such that it would be irrational not to do that action no matter what one's desires and interests are. On a teleological theory of content, only complex concepts—concepts constituted by other concepts—can have empty extensions. Primitive concepts cannot. A concept can have the function to track something that does not exist only if its tracking function derives by composition from the function of simpler concepts, each of which has the function to track something that actually exists. So, for example, the concept [UNICORN] has a tracking function despite the fact that unicorns do not exist, but it has a tracking function only because [UNICORN] is simply the complex concept [HORSE WITH A HORN], and the tracking function of this concept derives by composition from the tracking function of concepts like [HORSE], [HORN], etc. In a nutshell: If there are no practical demands with inescapable automatic authority, how can there be a mental representation that has the function to track these practical demands, unless this mental representation derives its tracking function from its constitutive mental representations, each of which has the function to track something that actually exists?

13. But see my remark in section 4 about the possibility that a desire to avoid irrationality and the disposition to feel guilt and moral disapproval are connected in some interesting way.

14. Notice that, as suggested by the discussion in section 2, these hypotheses are not exhaustive.

15. Some of the evidence discussed by Nichols (2004) could be used in support of this view.

References

Atran, S. (2004). *In gods we trust*. Oxford: Oxford University Press.

Blackburn, S. (1998). *Ruling passions*. Oxford: Oxford University Press.

Blair, R. J. (1995). A cognitive developmental approach to morality: Investigating the psychopath. *Cognition*, *57*, 1–29.

Blair, R. J. (1997). Moral reasoning and the child with psychopathic tendencies. *Personality and Individual Differences*, *26*, 731–739.

Damasio, A. (1994). *Descartes' error*. New York: Avon.

Dretske, F. (1997). *Naturalizing the mind*. Cambridge, MA: MIT Press.

Foot, P. (1972). Morality as a system of hypothetical imperatives. *Philosophical Review, 81*(3), 305–316.

Gibbard, A. (1990). *Wise choices, apt feelings*. Cambridge, MA: Harvard University Press.

Godfrey-Smith, P. (2009). *Darwinian populations and natural selection*. Oxford: Oxford University Press.

Joyce, R. (2001). *The myth of morality*. Cambridge: Cambridge University Press.

Joyce, R. (2006). *The evolution of morality*. Cambridge, MA: MIT Press.

Mameli, M. (2004). The role of emotions in ecological and practical rationality. In D. Evans & P. Cruse (Eds.), *Emotion, evolution, and rationality* (pp. 158–178). Oxford: Oxford University Press.

Mackie, J. (1977). *Ethics: Inventing right and wrong*. London: Penguin.

Millikan, R. (1993). *White queen psychology and other essays for Alice*. Cambridge, MA: MIT Press.

Moore, G. E. (1903). *Principia ethica*. Cambridge: Cambridge University Press.

Nichols, S. (2004). *Sentimental rules*. Oxford: Oxford University Press.

Nucci, L., & Turiel, E. (1993). God's word, religious rules, and their relation to Christian and Jewish children's concepts of morality. *Child Development, 74*(5), 1475–1491.

Papineau, D. (1995). *Philosophical naturalism*. Malden: Blackwell.

Roskies, A. (2003). Are ethical judgments intrinsically motivational? Lessons from "acquired sociopathy." *Philosophical Psychology, 16*(1), 51–66.

Smith, M. (1994). *The moral problem*. Malden: Blackwell.

Turiel, E. (1998). The development of morality. In W. Damon (Ed.-in-Chief) & N. Eisenberg (Vol. Ed.), *Social, emotional, and personality development: Handbook of child psychology* (5th Ed., Vol. 3, pp. 863–932). New York: Wiley.

26 The Many Moral Nativisms

Richard Joyce

Introduction

John Stuart Mill's opinion that "moral feelings are not innate, but acquired" (Mill, 1861, p. 527) was, in the estimation of Charles Darwin, destined to be judged as "a most serious blemish" on that moral philosopher's future reputation (Darwin, 1879/2004, p. 121). But Darwin's prophesy has so far proved incorrect; Mill's opinion on the matter has hardly been commented upon, let alone decried. Indeed, the whole question of the origin of human morality received remarkably little discussion in the century or so after Darwin's *Descent*.[1] The last two decades, however, have seen the question placed back on the agenda. The emergence of *fin de siècle* evolutionary psychology—and in particular its pioneers' decision to focus on the moralistic trait of "cheater detection" as their favorite case study (see Cosmides & Tooby, 1992)—has prompted burgeoning debate about moral nativism. While this debate has yet to mature, and though one of its striking characteristics is a tendency for claims to be pressed (both for and against) with a confidence disproportionate to available evidence, we nevertheless might reasonably hope for genuine progress in the foreseeable future. Before that progress can occur, however, we need to understand the hypothesis. Currently there are a number of points of significant imprecision in the debate over moral nativism that often pass unnoticed and which lead to seemingly opposed factions speaking at crossed purposes. I think it is fair to say that we are at present in the same state that William Darwin (in a letter to his father) attributed to Mill: of being "rather in a muddle on the whole subject."[2]

In previous works I have advocated moral nativism (Joyce, 2006a,b)—though I did so provisionally and cautiously; my objective was concerned more with clarification than all-out endorsement. Advocating moral nativism is not my intention in this chapter; my goal here is principally diagnostic. I will highlight three places where the nativist–nonnativist debate fragments in such a way that it ceases to be clear what the hypothesis is that is under dispute. In two of the three problematic places, the options for reinstating precision are reasonably well defined, so my conclusion is that disputants simply need to take care to specify which understanding of the hypothesis is under discussion. In the third case, however, my attitude is rather more pessimistic. Here, it seems, we find at the heart of the debate an inchoate concept—that of *moral judgment*—regarding which the options for precisification are not well understood, and for which any stipulative specificity appears more of a misleading distortion than a welcome clarification. One possible consequence of this is that on some legitimate conceptions of *moral judgment* moral nativism is true, but on other equally legitimate conceptions moral nativism is false. And if there is no satisfactory way of deciding among these conceptions, then the debate over moral nativism is undecidable—not just in the sense that we lack decisive data, but in the sense that there is really no fact of the matter.

The First Node of Imprecision: Innateness

In its crudest form, moral nativism is the view that human morality is innate. What might be meant by "human morality" is a question that will occupy much of this chapter, but first our attention should pause on what is meant by "innate." Some participants in the debate over moral nativism know what they mean by "innate," but many employ an intuitive folk notion that doesn't withstand critical scrutiny. Of those that do have a clear view of what they mean, not all mean the same thing.

The folk notion of innateness is a blend of several subclusters of ideas. One such group of ideas pertains to a trait's being present at birth, to its being not learned, to its being determined by genes rather than environment, to its being developmentally robust in the face of environmental variation. Another idea central to innateness is the Darwinian notion of a trait's existing because it was selected for by the process of natural selection—that is, of a trait's being an adaptation. Another is the essentialist idea of a trait's being species-typical: present in all members of the species or at least in all "normal" members. (For diagnosis and discussion of such options, see Griffiths, 2002; Mameli & Bateson, 2007; Mameli, 2008.)

These ideas are not all equally scientifically respectable, and, more to the point, they are far from coextensional. Down's syndrome is present at birth, genetically influenced, and developmentally robust, but it is not an adaptation. The possession of a certain stone-knapping technique may satisfy the criteria for being an adaptation (it may be transmitted from parent to offspring and may owe its existence to the fact that it enhanced reproductive fitness), but is neither nonlearned nor developmentally robust. And so forth. Hence we must reject the common unexamined presupposition that these phenomena more or less come together and thus can be treated as facets of a single "cluster" concept. In light of the way these disparate ideas get lumped together, Matteo Mameli (2008) disparagingly refers to *innateness* as a "clutter" concept.

In the literature on moral nativism, two conceptions of innateness are most conspicuous: an evolutionary conception and a developmental conception. A typical statement of the evolutionary conception comes from Jesse Prinz, who sums up moral nativism as the claim that "morality is an evolved capacity" (Prinz, 2009, p. 168). I have myself described moral nativism as the view that "morality (under some specification) . . . is to be explained by reference to a genotype having granted ancestors reproductive advantage" (Joyce 2006a, p. 2). On this view, moral nativism is the claim that morality is a Darwinian adaptation.

Standing in contrast to this is the developmental conception, according to which the emergence of the trait is buffered against variation in the developmental environment (Ariew, 1996, 1999). Chandra Sripada and Stephen Stich use such a conception when they write that "we can consider a normative rule to be innate if various genetic and developmental factors make it the case that the rule would emerge . . . in a wide range of environmental conditions" (Sripada & Stich, 2006, p. 299).

These two conceptions of innateness are by no means coextensional. The trait of morality might be a specific adaptation but may nonetheless require particular structured environmental inputs in order to become manifest. If such inputs were reliably available in the environment in which morality evolved, then there would be no selective pressure to make the developmental emergence of morality robust in the face of environmental perturbation. The reverse is also true: Human morality may be developmentally canalized while not being an adaptation. Some of the well-known ways by which traits may become canalized without being adaptations—genetic drift, mutation, genetic disease—are admittedly far-fetched in the case of morality. However, one way is entirely plausible: that morality is a by-product of other adaptations. (This possibility shall be discussed later.)

Clearly, this introduces potential confusion into the debate over moral nativism, for it allows that moral nativism may be true in one respect but false in another. Even when advocates of a particular view are conscientious in articulating which thesis they mean to defend or attack, casual readers may miss the qualification. For example, in his paper "Moral nativism: A sceptical response," Kim Sterelny is careful to explain that he is skeptical of the *developmental* nativist thesis. He allows that "there is a plausible . . . case for the idea that moral cognition is an adaptation," but adds that "even if that is right, it does not follow that this capacity is innate" (Sterelny, 2010, p. 280). If such comments are overlooked, however, then one might gain the impression that Sterelny is in the same camp as other opponents of moral nativism when in fact these others are skeptical of the *adaptational* nativist thesis. More worryingly, one may gain the erroneous impression that Sterelny is in the opposing camp to someone like myself, who has advocated the adaptational nativist thesis, when it is entirely possible that we agree on everything of substance.

We have seen that in assessing the thesis of moral nativism possible misunderstandings lurk around the term "innate." Yet the possible misunderstandings surrounding the term referring to the trait in question—"human morality"—are even greater. In subsequent sections I will tease this matter apart into two further particular points of imprecision, but first I will introduce the general problem via a discussion of altruism and Darwin's views on moral nativism. My reason for doing so is as follows. I want to demonstrate that the trait in question, human morality, is difficult to define with any precision—that it admits of more liberal and more strict characterizations. To illustrate this, it is useful to begin with another trait, altruism, that is not a million miles from morality but which is pretty clearly not the same thing. Identifying the difference between altruism and morality forces us to ask what exactly the trait of "morality" is. Darwin's own views are worth discussing here because he begins with prosocial attitudes, like altruism, which he then supplements with further psychological traits in order to achieve something that, he believes, deserves the label "the moral sense." (Darwin, obviously, is focused on evolutionary rather than developmental emergence.) This transition from nonmoral organism to moral organism is exactly what we are interested in. But Darwin's efforts also exemplify the difficulty and obscurity of the task—the fact that it is radically unclear what an adequate account of the transition from the nonmoral to the moral would have to involve. There is, I wish ultimately to argue, no single answer to this question.

From Altruism to Darwin

It is standard to distinguish two forms of altruism: psychological and evolutionary. An action is psychologically altruistic if and only if it is motivated by an ultimate desire for the well-being of some other organism. A behavioral trait is evolutionarily altruistic if and only if it benefits another at some cost to the individual, where benefits and costs are understood in terms of reproductive fitness. (It must be added that the trait has been selected for *because* it benefits another, otherwise one ends up counting as altruistic such things as a sea turtle's drive to lay its eggs on the beach, which makes its hatchlings such easy prey for seagulls.) The former is an articulation of a vernacular notion, whereas the latter is very much a term of art.

The extensive literature ostensibly concerning the "evolution of altruism" often fudges this important distinction, and, indeed, frequently concerns neither. Consider the so-called altruistic behavior of bees. It is surely not psychologically altruistic (since bees simply lack the motivational prerequisites), but nor is it obviously evolutionarily altruistic: William Hamilton's breakthrough work on kin selection (Hamilton, 1964) demonstrated how the individual bee who dies to save her nest-mates is in fact advancing her own inclusive fitness. Or consider the reciprocal grooming behavior of primates (see Schino & Aureli, 2010). If the explanation of primate A's tendency to take the time and effort to groom primate B is that this increases the probability of A's being groomed in return, then in performing this behavior A is reproductively better off (eventually) than if it did not. (See West, Griffin & Gardner, 2007.) Similar considerations pertaining to hunting lions, mobbing birds, meerkats on sentry duty, and so on will also reveal neither psychological nor evolutionary altruism. For this reason, it is best to call such behaviors simply "cooperation" (leaving this an intuitive term), which then allows the questions of whether these cooperative behaviors are also instances of psychological altruism or evolutionary altruism to be substantive inquiries. (For discussion of how true evolutionary altruism is possible, see Sober, 1988; Sober & Wilson, 1998.)

Without pausing to investigate the details of how much cooperation in nature really is evolutionarily altruistic, one can at least safely say that cooperation often turns out to be evolutionarily selfish, in the sense that the cooperative behavior ultimately enhances the actor's reproductive fitness better than not cooperating. The temptation that it is crucial to resist is thinking that this evolutionary selfishness has any bearing on psychological selfishness. Organisms that do not have psychological states at all, such as plants, may be evolutionarily selfish or altruistic. To satisfy the prerequisites for being *psychologically* altruistic or selfish, a creature must be able to have ultimate motives concerning others' or their own welfare, which requires them to have the concepts of *other* and *self*. The only creatures for which we can be confident of the satisfaction of these prerequisites are humans.

I will take it as obvious that the mere fact that a behavioral trait is to be explained by reference to *evolutionary* altruism is insufficient to make the introduction of talk of "morality" appropriate. A plant may have evolutionarily altruistic traits, but the plant neither makes moral judgments nor is a suitable subject of our moral appraisals. But it is not so obvious that there is no connection between *psychological* altruism and morality, so this requires some discussion. It is particularly important here because a plausible case can be made that psychological altruism in humans is innate; hence this may have direct implications for the prospects of moral nativism.

The details of the argument for nativism concerning psychological altruism need not delay us; a sketch will suffice for present purposes. The argument concerns *evolutionary* nativism rather than developmental nativism, and has been advocated by Elliott Sober (2000). Sober's principal opponent is the psychological egoist, who holds that all human actions are performed with the ultimate motive of benefiting the actor. Given that natural selection has clearly forged humans to be cooperative in certain ways—at the very least, caring for our offspring—Sober wonders what kinds of psychological mechanisms would be likely to be favored to govern these cooperative tendencies. Assuming that it is adaptive to come to the aid of one's children when they are in distress, for example, what is the better psychological setup? On the one hand, we can envisage a parent motivated to provide aid simply because he loves his daughter—he cares directly for her in such a way that a perceived threat to her welfare directly prompts action. On the other hand, we can imagine the egoistic parent: moved via a combination of the belief that his daughter's suffering has a negative effect on his own welfare plus his love for himself. One might plausibly claim that the former mechanism is more reliable and less complicated—and thus, *ceteris paribus*, more adaptive—than the latter. By analogy, a person prompted to withdraw her fingers from a flame *by pain* seems moved by a more reliable and less complicated process than a person who forms a belief about the bodily damage caused by fire and calculates the costs and benefits of action versus nonaction. This argument may not be without problems (see Stich, 2007), but here my intention is not to evaluate or endorse the argument, but rather to examine what would follow—or, more precisely, what would *not* follow—if it were sound.

We have seen that to be psychologically altruistic a creature needs to be fairly cognitively sophisticated, but it doesn't follow that the creature is therefore capable of making moral judgments. This truism is potentially muddied by the fact that the only clear-cut case of a species capable of psychological altruism (and selfishness) is also the only clear-cut case of a species of which we speak in moral terms: namely, humans. Still, the conceptual distinction does not seem difficult to discern. One can imagine members of a cognitively sophisticated social species, motivated by love and altruistic tendencies toward their fellows, but who fail to "moralize" these feelings—who are, in fact, constitutionally incapable of making a moral judgment. Such creatures have powerful desires to see their loved ones flourish, but cannot conceive of actions satisfying those desires as morally right or obligatory.

It might be conceded that these imaginary creatures don't make moral judgments but maintained that they are at least morally *praiseworthy* (that is, that they warrant *our* moral judgment). But upon reflection even this is unclear. After all, altruistic motives can prompt someone to act in a morally despicable manner. Consider a mother who genuinely adores her child, and who poisons all the other children at the sports day so her child can win. In any case, it seems misguided to identify moral nativism with the claim that the trait of *being morally praiseworthy* is innate. Such a proposal would lead straight into a metaethical quagmire from which the debate is unlikely ever to emerge. We are not primarily interested in the question of at what point, either in evolution or development, humans become *morally admirable*; we are interested in at what point they become capable of *making moral judgments*. Popular discussions of moral nativism with headlines like "Are we born to be good?" or "The moral animal" or "Chimps display morality" (etc.) blur this basic distinction, and in doing so spread more misunderstanding than illumination.

Once we focus nativism on the question of *moral judgment*, it becomes clear that we are asking about something different from (or perhaps more than) psychological altruism. None of this is to deny that the emergence of psychological altruism (both evolutionarily and developmentally) might be a crucial precursor to moral judgment; I'm not claiming that someone with an interest in moral nativism should dismiss all discussion of the emergence of psychological altruism as irrelevant. My claim is simply that moral judgment is not the same thing as altruism, and that establishing nativism about altruism does not establish moral nativism. Though this much seems assured, the natural further question of what exactly *is* required for moral judgment is much harder to answer.

As a way of illustrating this problem, I turn now to Darwin's views on the matter. Darwin undertakes the task of supplementing prosocial emotions (like altruism) with further psychological capacities in an attempt to "build" a human moral sense. Drawing attention to the difficulties inherent in this project is one of the goals of this chapter, so sketching his attempt is a useful exercise.

Darwin is no psychological egoist. He writes:

> With respect to the impulse which leads certain animals to associate together, and to aid one another in many ways, we may infer that in most cases they are impelled by the same sense of satisfaction or pleasure which they experience in performing other instinctive actions. . . . In many instances, however, it is probable that instincts are persistently followed from the mere force of inheritance, without the stimulus of either pleasure or pain. . . . Hence the common assumption that men must be impelled to every action by experiencing some pleasure or pain may be erroneous. (Darwin, 1879/2004, p. 128)

Darwin speaks frequently of the "social instincts" of animals—which include such affections as sympathy, love, and pleasure in the company of one's fellows—instincts that promote cooperative behavior among animals in many circumstances. But he is adamant that they do not suffice for a moral sense: "I fully subscribe to the judgment of those writers who maintain that of all the differences between man and the lower animals, the moral sense or conscience is by far the most important." He goes on:

> [A]ny animal whatever, endowed with well-marked social instincts . . . would inevitably acquire a moral sense or conscience, as soon as its intellectual powers had become as well, or nearly as well developed, as in man. (Ibid., pp. 120–121)[3]

What are these "intellectual powers"? First of all, Darwin thinks, one needs a good memory, in order to recall those times in the past when one has failed to act cooperatively and (as a result of one's social instincts) felt dissatisfaction. One needs to recall that the benefits gained from failing to cooperate (i.e., the profits of defection) were fleeting. Second, the emergence of language allows that "the wishes of the community could be expressed, [and] the common opinion how each member ought to act for the public good, would naturally become . . . the guide to action" (ibid., p. 122). Lastly, one needs the capacity to form *habits* of acting for the good of one's fellows.

This might be interpreted as an argument for moral nativism (of the adaptational variety), but on another interpretation Darwin thinks of the moral sense as a kind of "spandrel" derived from faculties that evolved for other purposes.[4] In fact, he is explicitly undecided on the matter. Referring just to the social instincts, he writes that it is "impossible to decide in many cases whether certain social instincts have been acquired through natural selection, or are the indirect result of other instincts and faculties" (ibid., p. 130). We will return to this distinction later; currently what interests me is how, precisely, the moral sense is supposed to emerge from these elements. My strategy will be to grant Darwin these ingredients and attempt to motivate doubt that we have enough to warrant the label "a moral sense."

Darwin certainly has plenty of persuasive things to say about the evolution of the social instincts; on this topic he is squarely in his "comfort zone." But his explanation of how certain "intellectual powers" get married to those instincts, resulting in a moral sense, is considerably sketchier and less convincing. The latter two ingredients listed in his initial presentation—language and habit—hardly get a further mention. (It is, besides, unclear whether he thinks of these two traits as *necessary* for a moral sense.) It is the role of *memory* that he mentions repeatedly and evidently judges of paramount importance. But the case is underdescribed at best.

Consider a creature brimming with altruistic sentiment for its conspecifics. I argued above that this alone does not suffice for a moral judgment. The creature needn't think that it *ought* to help its fellows; it doesn't think of failure to help as *prohibited*; it doesn't think that such failures warrant punishment or disapproval, or that helping merits praise. It simply *wants* to help. Yet suppose that occasionally the creature experiences temptations to do otherwise, since there are other competing instincts operative in its psyche. When this creature succumbs to such temptations, it enjoys the satisfaction of the tempting outcome (whatever it may be) and yet also feels bad because of the frustration of its natural desire to cooperate. Let us stipulate that the creature's instincts are such that the pleasures achieved at the expense of cooperation tend to be short lived. Let us now grant it the intellectual powers both to realize and to remember this fact. Thus, when temptation arises, the creature is able to deliberate along the lines of: "Well, that sure looks enticing, but I remember how rotten I felt last time I succumbed to temptation, so I'll cooperate." So now we have a creature with self-control in favor of cooperative behavior.

But where does the *moral judgment* emerge in this process? Acting cooperatively is still, essentially, just something that this creature *wants* to do. Compare a monkey that is often tempted to climb its favorite tree using the dangling outer branches, but who, through trial and error, comes to learn that it is safer to ascend by the trunk. When faced with the temptation to dart up the dangly branches the monkey may pause and recall the bruises of earlier decisions. So now we have a creature with self-control in favor of climbing a tree via the trunk. But do we credit the monkey with anything like the judgment that climbing the outer branches is a *transgression*? If it does climb by the outer branches without mishap, we can imagine it thinking "That was a bit stupid, but, phew, I got away with it!" Where would be the *guilt*? Where would be the thought that it *deserves punishment* for its crime? Why would it take an interest in punishing other monkeys that exhibit foolish climbing habits?

In the case of the first creature whose instincts are in favor of cooperation, we need also to factor in the reactions of its conspecifics, but I don't believe that this alleviates the puzzle. The conspecifics *don't like it* when the individual defects on some cooperative enterprise, and we can imagine that their disappointment and anger is something that our individual will take into account. It controls itself by remembering how bad its failures to cooperate made it feel in the past, and when those failures are accompanied by its fellows expressing their anger with (say) violence and ostracization, then self-control will be all the easier since the negative repercussions of such failures will be even worse. Thus the influence of the conspecifics will certainly significantly strengthen the process of self-control, but it in no obvious way brings about a change in *kind* in the sorts of judgments and attitudes that we attribute to the individual.

It appears, therefore, that one can identify elements that seem important to moral judgment—such as the ideas of transgression, guilt, and desert—for which Darwin's hypothesis does not account. In assessing this matter one needs to be wary of projecting one's own "moralizing" thoughts onto the imaginary characters involved. It is difficult to cleanly imagine someone simply *not wanting* to perform noncooperative actions (in part because she recognizes that other parties *don't want* her to) without positing the seemingly innocuous extra assumption that she also judges that she *ought not* perform those actions. It is natural for us to assume that as our imaginary creature forms the habit of acting cooperatively, surely at some point it "internalizes the norm": its *expectation* of negative outcomes morphs naturally into the thought that such outcomes are *warranted*; its *desire* for its fellows' welfare gradually begets the judgment that acting for their welfare is *desirable*; it moves from habitually not wanting something to judging it *prohibited*; and so forth. But assuming that this transition occurs naturally is exactly what we must not do in this context, for how such a transition occurs is precisely what is under scrutiny.

Darwin brings the discussion to the edge of "moralization," but it is not obvious that he succeeds in crossing the conceptual gap. Perhaps the ingredients he provides suffice for a thin notion of moral judgment, but there is a richer folk conception whose evolutionary emergence remains mysterious. As we shall see, the same can be said of some modern participants in this debate: They provide ingredients that may be adequate to account for moral judgment in some attenuated sense but which fail to explain important components of a robust conception of moral judgment. Thus the debate founders not merely through lack of empirical data, but through an absence of any single phenomenon uniquely deserving of the name "moral judgment." Before discussing this matter further, however, I should like to note another source of confusion about the nature of the trait whose origin is under discussion.

The Second Node of Imprecision: Content versus Concept

It is important to distinguish between moral concepts and moral judgments. Let us say that a complete paradigm moral judgment consists of the application of a moral concept, like *moral wrongness*, to a general subject, like *incest*, or to a particular subject, like *John and Mary's incestuous relation*.[5] Given this framework, we can identify another way in which moral nativist hypotheses may vary.

One version of moral nativism will allow that certain complete moral judgments are innate. There is certainly nothing to be said in favor of the claim that complete moral judgments concerning *particulars* are innate. For example, to hold that the judgment "John and Mary's incestuous relation is morally wrong" is an adaptation would involve accepting that our ancestors somehow knew about the individuals John and Mary, formed a moral opinion about what they got up to, and that this opinion enhanced reproductive fitness. Given that a great many of our moral judgments do concern particulars, nativism about complete judgments is going to be utterly implausible for a great many of our moral judgments. Even for those moral judgments that take universals as subjects, nativism concerning the complete moral judgment is feasible only when the subject is something that was present in the environment of evolutionary adaptiveness (the EEA). One may, for example, countenance nativism for "Incest is wrong," but nativism for "Shoplifting is wrong" is a nonstarter.

Another version of nativism eschews any commitment to complete judgments being innate and prefers the image of a moral faculty as a "toolkit" of moral *concepts*, with the individual's socialization process as the sole determinant of to which subjects these concepts get attached. Thus, according to this hypothesis (expressed in simplistic terms), a concept like *moral wrongness* is innate, and one social environment may lead the individual to apply the concept to *incest*, another environment may lead the child to apply it to *John and Mary's incestuous relationship* but not to *Ptolemy and Cleopatra's incestuous relationship*, while yet another may lead the child not to apply the concept to any incestuous relationship.

These two nativist positions represent extremes, between which lie a variety of hypotheses. Some allow that a few broad abstract moral principles are innate but that the environment sets the parameters of how these create specific moral judgments (Hauser, 2006). Some allow that content is learned but that the moral sense comes "prepared" to latch on to certain domains more easily than others (see Haidt & Joseph, 2004; Sripada, 2008).[6]

Even with some options in the moral nativism spectrum sketched in so heavy-handed a manner, we have seen enough to recognize that evidence favoring one version of moral nativism will not favor another. As a way of illustrating the muddle that ensues, I will examine the debate over moral universals.

In fact, even if all parties were in complete concurrence regarding which trait is under scrutiny, the place of universals in the debate over nativism would be far from straightforward. The tempting assumption that if a trait is innate then we can expect to find it manifest everywhere must be rejected. If one is focused on developmental innateness, then many innate traits are not universal (e.g., Down's syndrome, eye color, lactose tolerance). If one is discussing adaptational innateness, then innate traits may well require substantial environmental input—input that may have been reliably present in the EEA but is absent, patchy, or distorted in the modern environment. I intend to put these important complications aside, however, in order to focus on another simpler point about universals. So for the sake of argument let us allow the assumption that innate traits will reliably emerge and thus tend toward universality. The question is: For what kind of universals should we be looking? And the answer is: It depends which version of moral nativism is under scrutiny.

In one of a series of papers arguing against moral nativism, Prinz discusses three possible moral universals: *don't harm innocent people*; *respect and obey authorities*; and *incest is prohibited* (Prinz, 2009; see also Prinz, 2008a,b, 2013). He carefully examines historical and anthropological evidence in an attempt to find counterexamples to the claim of universality for each, thus discrediting moral nativism. But the limitations of this strategy should by now be clear: Many moral nativisms will not hold that such complete moral judgments are innate.[7]

This is not to say that Prinz's efforts are wasted. Certain versions of moral nativism may well claim that precisely these three complete moral judgments are innate, and I share Prinz's determination to reject such views. Prinz, moreover, knows that he is challenging only one form of moral nativism. He is aware of the kind of toolkit moral nativism mentioned earlier—which holds no complete moral judgment to be innate but rather postulates innate moral concepts. Prinz labels this kind of moral nativism "minimal" (Prinz, 2009) and "weak" (Prinz, 2013). I confess to finding this labeling system unfortunate, since it allows the antinativist to proceed by first refuting the "strong" versions of moral nativism (the kinds that were never terribly plausible in the first place), thus giving the impression of the moral nativist retreating to an ever weaker position in a desperate bid to defend his or her hypothesis. The rhetorical narrative this suggests is inaccurate and is exasperating to anyone who begins with a desire to defend toolkit moral nativism while agreeing wholeheartedly that there is little to be said in favor of the more content-complete versions of nativism.

If we are investigating evidence for and against universality, and have the more plausible toolkit kind of moral nativism in mind, then we should be examining whether any cultures lack moral judgments altogether. If one culture thinks that incest is morally acceptable while another judges it repugnant, this is no counterexample to universality, for both cultures are still evaluating the world in moral terms. All too often the debate has revolved around the question of whether "moral universals" exist, but if I am correct then this is misguided; what we should be investigating is whether *having a system of moral judgments* is a human universal. And while it is not my intention on this occasion to press the case in favor of this latter hypothesis, it is reasonable to suppose that the prospects of its being true are far better than the likelihood of finding moral universals.

Prinz is certainly unable to provide a counterexample. At one point, he mentions the Ik group of Uganda, famously described by anthropologist Colin Turnbull (1972) as a "vicious people" with "sadistic customs." We now know that Turnbull's account of the Ik was flawed in numerous ways (see Heine, 1985; Knight, 1994), but even if that were not so, the "viciousness" of which he spoke is compatible with the Ik having a moral system—one that might seem blighted and alien to us, but a moral system nonetheless. Indeed, when, several years later, the Ik elders heard of how Turnbull had portrayed them to the world, they were angry that he had "spoilt" their reputation, and threatened to make him "eat his own faeces" if he ever showed his face again (Heine, 1985, p. 3). To the extent that they thought that Turnbull *deserved* this unenviable fate, the Ik proved themselves capable of wielding a moral concept.

Prinz doesn't seriously think that the Ik lack any moral system. When he squarely addresses the toolkit version of moral nativism, he admits "I certainly don't know of any exceptions to this claim" (Prinz, 2008a, p. 386). This concession forces a change of tactic in his pursuit of the nonnativist agenda: He moves from trying to provide counterexamples to universality and instead sets out to demonstrate that an appeal to nativism is not required to explain moral judgment; he endeavors to provide an empiricist explanation of the (possibly universal) phenomenon. In doing so, he aims to discredit a focal argument in favor of moral nativism: the poverty of the stimulus (POS) argument. According to this argument, the capacities evident in moral cognition are acquired in a manner that far outstrips the information that is available in the learning environment. The structure of the argument comes, of course, from the debate over nativist explanations of human linguistic abilities (see Chomsky, 1967, 1987/1990), where the POS argument is widely judged to be triumphant in establishing some form of nativism.[8] It is not my intention here to evaluate the prospects of a moral POS argument, but rather point out how progress gets confounded by distinct theoretical options being conflated.

One obvious way of countering a POS argument is to show that the stimulus is in fact a great deal less impoverished than one might have thought. Thus moral nonnativists are eager to point out how rich is the moral learning environment of the child. Shaun Nichols reminds us that "the child is exposed to *lots* of admonitions and instruction in the normative domain. Parents and teachers are constantly telling kids what shouldn't be done" (Nichols, 2005, p. 358). Sterelny makes a similar observation:

> The narrative life of a community—the stock of stories, songs, myths and tales to which children are exposed—is full of information about the actions to be admired and to be deplored. Young children's stories include many moral fables: stories of virtue, of right action and motivation rewarded; of vice punished. So their narrative world is richly populated with moral examples. (Sterelny, 2010, p. 289)

This is all undeniable. The child's moral world is richly structured, and the explicit moral instruction is coordinated and unrelenting.

It is not sufficient, however, simply to remark upon the wealth of the moral stimulus in a general way. We need to decide which version of moral nativism is under discussion, for this determines what kind of moral task it is whose acquisition process is under scrutiny. If our interest is in toolkit moral nativism, then focusing on how children acquire complete moral judgments is misleading; rather, our attention should be on how children acquire their basic moral conceptual tools. If this is the target trait, then wondering how children acquire the belief that shoplifting is wrong (say) would be a distraction (for I'm sure all parties can agree that they are taught it by adults); instead we should be wondering about how children acquire the concept of *moral wrongness* in the first place. Is the environment rich enough to provide them with *that*?

This is a crucial disambiguation to make before assessing the prospects of any moral POS argument, yet it still leaves progress hampered by a serious conceptual imprecision, for one is still left wondering "What is a moral judgment?" The possibility remains that moral nativism may be more plausible with certain conceptions than others. This is discernible in antinativist attempts to oppose the moral POS argument, as the following short review will demonstrate.

Some Antinativist Hypotheses

The opponent of moral nativism will usually try to account for the human trait of making moral judgments by calling attention to other psychological traits that evolved or develop for other purposes. Often moral judgment is described as a by-product or "spandrel" of these other traits. I will sketch a few antinativist views in order to give a flavor of the approach.

Prinz attempts to account for the evolutionary emergence of moral judgment from a cluster of other evolved faculties, each of which has a more general role. At the center of his argument is the view that moral judgments are emotional responses.[9] In one paper (Prinz, 2009), he proposes to construct a moral response out of emotions that are not distinctively moral: anger and sadness. We feel sad in many circumstances, but when we feel sad at having transgressed against a norm, Prinz argues, then the sadness is called "guilt." "Guilt is an accidental byproduct of sadness" (ibid., p. 183). In other works, Prinz develops a somewhat different empiricist hypothesis. In his 2008a he mentions not only the nonmoral emotions, but some additional traits: *meta-emotions* (emotions directed at our own emotions or at others' emotions), *perspective taking* (allowing for third-party concern), and other *non-moral preferences* (e.g., the "social instincts" which were Darwin's starting point). The important point is that these are all general cognitive skills; thus, if moral judgment is a natural by-product of these traits, the moral nativist would be defeated.

The view that emotion has a central role in moral judgment is also at the heart of Nichols's attempt to provide an empiricist account of the origin of moral judgment (Nichols, 2005). Nichols allows that "rule nativism" is plausible, where the rules in question are non-hypothetical. "There is no obvious story about how the empiricist learner might come to acknowledge *nonhypothetical imperatives*" (ibid., p. 357). He correctly argues that morality is but a proper subset of nonhypothetical rule systems, citing etiquette and institutional rules (e.g., of a gentlemen's club) as involving nonmoral but nonhypothetical imperatives (following Philippa Foot, 1972). A key question, then, is what is distinctive about *moral* nonhypothetical imperatives. Nichols's answer starts by noting the distinctive subject matter of morality—namely, that it pertains to harm.[10] Given this characterization of morality, the second ingredient in Nichols's hypothesis is an innate affective mechanism that responds to suffering in others. This emotional response imbues a certain subset of nonhypothetical imperatives with a particular flavor (call it "moral"), picking them out as salient, resonant, and memorable. Nichols concludes:

> [B]oth of the mechanisms that I've suggested contribute to moral judgment might well be adaptations. However, it is distinctly less plausible that the capacity for core moral judgment itself is an adaptation. It's more likely that core moral judgment emerges as a kind of byproduct of (*inter alia*) the innate affective and innate rule comprehension mechanisms. (Nichols, 2005, p. 369)

Another antinativist argument comes from Sterelny, though, as noted earlier, he is focused more on the developmental trajectory than the evolutionary emergence of the trait. Like both Prinz and Nichols, Sterelny holds that one of the key psychological ingredients in a nonnativist explanation of moral judgment is emotion. He appears willing to endorse nativist hypotheses for emotional contagion, for sensitivity to interactions involving harm, and for the emotions associated with "reciprocation, sympathy, empathy, disgust, and esteem" (Sterelny, 2010, p. 293). He argues at length that moral learning is largely a matter of generalizing from exemplars—which explains why moral intuitions can be fast and automatic—and also stresses that this would not mark moral learning as unusual (i.e., the faculties involved in prototype-comparison learning are general mechanisms). Sterelny further persuasively emphasizes the extremely rich and structured nature of the moral learning environment, arguing that the "parental generation engineers the informational environment in which the next generation develops, thus guaranteeing the development of moral competence" (ibid., p. 294). Sterelny concludes that moral norms "are grafted on top of our dispositions to respond emotionally" (p. 292), that moral cognition "is a natural development of our existing emotional, intellectual and social repertoire" (p. 293), and that moral cognition "develops from an interaction between emotions, exemplar-guided intuitions and explicit principles" (p. 293).

Clearly, it is beyond the ambitions of this chapter to attempt to analyze or refute these proposals in detail; I aim to make a more general point. First, I will pursue the same strategy as was deployed earlier against Darwin: taking the ingredients offered and questioning whether they suffice for making a moral judgment. My ultimate goal, however, is not to declare that all such arguments simply fail, but rather to argue that varying conceptions of moral judgment are in play.

Consider, first, Prinz's argument that guilt is just sadness directed at having transgressed against a norm. There appear to be important components of full-blooded guilt that remain unaccounted for. Sadness predicts social withdrawal, whereas guilt (unlike shame) urges reparative action (Tangney & Fischer, 1995; Tangney, this vol.). Extreme sadness cripples a person's capacity to engage in everyday activities, whereas guilt, even acute guilt, is a burden that a person can usually shoulder while getting on with things. Even the manifestation of weeping that we associate with sadness we do not associate so readily with guilt (which is not to deny that guilt can cause a person to cry[11]). Indeed, language itself should be a giveaway here. We do have words for some special instances of sadness defined according to their object. "Grief," for example, denotes sadness directed at the loss of someone or something dear to us. Notice that just as we can say "I feel grief about Fred's death," we can say "I feel really sad about Fred's death," and no one will bat an eye-lid. But compare the huge difference between saying "I feel guilty about having committed that crime" and "I feel sad about having committed that crime."

Consider, second, Prinz's argument that attempts to build moral judgment out of nonmoral emotions (e.g., blame, which includes "other-directed emotions, such as anger, contempt, disgust, resentment, and indignation" [Prinz 2008a, pp. 368–369]) combined with metaemotions, third-party concern, and abstract ideas.[12] As a way of testing the adequacy of this empiricist hypothesis, let us imagine someone who satisfies all these components for one of the other-directed emotions that Prinz mentions: disgust. Suppose Ernie sees Bert vomit and feels disgust. Perhaps Ernie feels embarrassed at this response, or perhaps he is pleased with it; in either case, he manifests metaemotions. When Ernie thinks about some distant other person vomiting, he finds this idea pretty disgusting too; hence the emotion can be directed at third parties. Ernie is also capable of forming abstract ideas, so even the thought of vomit in some abstract sense makes him feel queasy.

It is clear that Ernie is pretty unhappy about Bert's vomiting, but it is considerably less clear that he has made a full-blooded moral judgment about it. We apparently need not credit him with the ideas that vomiting is wrong, that Bert has transgressed, or that vomiters deserve reprimand (or that nonvomiters deserve praise). These, it will be noticed, are distinctly *cognitive* elements that are lacking in Prinz's account. If our conception of moral judgment privileges such cognitive elements, then Prinz's project must be deemed inadequate.

According to Nichols, core moral judgments concern harm prohibitions that are lent resonance and prominence by an innate affective program. One might also want to insist that a key element of moral norms (as opposed to other kinds of nonhypothetical norms) is that they have a special kind of practical authority. Foot, for example, discusses the Kantian idea that to transgress against a moral imperative is irrational, whereas transgressions against etiquette need not be. Elsewhere, I have followed John Mackie (1977) in suggesting that moral imperatives are conceptually "non-institutional" whereas those of etiquette are not (see Joyce, 2001, 2011). Nichols doesn't deny this extra authority with which morality is imbued, but he argues that it comes into the picture later: as a consequence of the affective resonance of this class of norms. He writes that "the affective response seems to play a major role in determining the strength of one's normative commitments. . . . [T]he affect-backed norms are treated as having justifications that go beyond the conventional" (Nichols, 2004, p. 159).

But the nature of this connection remains puzzling. It can be granted that emotionally charged norms may be more memorable and seem more important. Yet it does not obviously follow that such resonant norms must also be accorded a stronger binding quality, that they will seem to hold independently of any institutional backing, that they will appear to require no further justification, or that one will be tempted to treat their violation as a form of irrationality. If affectively underwritten norms happen to produce this air of practical authority, then this is a phenomenon requiring explanation. Until such an explanation is offered, then to the extent that one's conception of a moral judgment makes central this idea of special practical *authority*, Nichols's empiricist hypothesis doesn't pass muster.

The ingredients offered by Sterelny suffice for a social creature who is sensitive to harm situations, who feels empathy for his fellows, who generalizes from exemplars, for whom departures from the cooperative order are memorable and salient, and who, as a consequence, operates extremely well in his social world. But where is the morality? The language Sterelny uses does seem to acknowledge that there is at least some important element of morality that is *more* than the joint exercise of these capacities, for he writes of moral norms "developing from" and being "grafted on top of" these capacities. This seems correct, for it appears no great feat of the imagination to envisage a social creature who enjoys the traits allowed by Sterelny but who is nevertheless constitutionally incapable of making moral judgments concerning an action's meriting punishment, a norm's having convention-transcending practical authority, or even an outcome's being desirable (as opposed to being desired). It is, in other words, not hard to imagine a creature who enjoys all Sterelny's ingredients but for whom full-blooded moral cognition does not simply "develop." Hence, if one's conception of moral judgment privileges such cognitive accomplishments, then what is required is an explanation for why and how it does develop from these ingredients in the normal human case.

The Third Node of Imprecision: Moral Judgment

From this review of some antinativist hypotheses, a pattern has emerged. Antinativists tend to understand moral judgment in term of emotional traits which, they think, have more general psychological roles and thus are unlikely to count as mechanisms dedicated to the production of moral judgment. However, the ingredients they offer appear to leave certain more cognitive elements of moral judgment unaccounted for. Though I am tempted by the hard-nosed response of insisting that these cognitive components are *essential* to moral judgment and thus that these antinativist arguments fail, my considered stance is more pluralistic.[13]

I suggest that the notion of *moral judgment* is sufficiently pliable as to allow different legitimate precisifications. A less demanding conception can be built largely out of emotional resources. To the extent that the less demanding conception might feel unsatisfying, in that it leaves certain cognitive elements of moral judgment unaccounted for, we must recognize the existence of a more demanding conception.[14] It is not a matter of there being two or more concepts; it's a matter of there being competing precisifications of the same somewhat indeterminate concept. A liberal conception will count as moral judgments items that the strict conception will not. And even for a paradigm moral judgment about which there is no doubt, the competing conceptions will disagree regarding the criteria in virtue of which the item counts as a moral judgment. It's not a matter of our not knowing which is the correct conception (because we lack data); it's that there is no unique fact of the matter.

A similar view has been expressed in the useful comparison case of the human language faculty. Marc Hauser, Noam Chomsky, and W. Tecumseh Fitch—recognizing that "the word 'language' has highly divergent meanings in different contexts and disciplines" (Hauser, Chomsky & Fitch, 2002, p. 1570)—distinguish between a faculty of language in a broad sense and in a narrow sense. The former, they hypothesize, consists largely if not entirely of capacities that humans share with other animals, whereas the latter (which is basically the capacity for linguistic recursion) is a uniquely human trait.

But whereas Hauser, Chomsky, and Fitch do an admirable job of delineating the various skills and capacities involved in the two senses of "language faculty," I feel somewhat pessimistic that the same can be done for "moral faculty," for here, it seems to me, matters are considerably more nebulous. The three examples of nonnativists described above—Prinz, Nichols, and Sterelny—hardly present a univocal picture of what a liberal conception of moral judgment might look like. They do, very broadly, all think that emotions are terribly important, but beyond this, three noticeably different views are articulated. To the extent that my own views have represented the advocacy of an opposing more cognitivist position, I haven't denied the importance of emotions but have maintained that cognitive components are vital too (cognitive components, that is, for which the antinativist proposals do not succeed in accounting). Yet if asked to characterize the crucial cognitive elements of the more demanding conception, I have nothing so simple and distinct as "recursion" to say. Rather, I will point to aspects of moral judgment like *desert, transgression, practical authority* (etc.), and declare (A) that these are *cognitions* (e.g., judging that *X* deserves punishment is not something one just "feels"), and (B) that emotional resources alone do not suffice to account for them. But the answer lacks precision (though is no less reasonable for that): The list of cognitions is worryingly open-ended (note the "etc."), and, moreover, not one of the items listed is easily defined. The literature regarding with what kind of "practical authority" our moral norms are invested, for example, stretches back to the ancient Greeks and continues unabated.

Perhaps my pessimism is premature, and distinct senses of "moral judgment" can be delineated with a reasonable amount of specificity. Or perhaps my doubt will be borne out, and the whole concept will remain inchoate and ill defined. In either case, what is evident is that it is a mistake to choose one particular characterization of "moral judgment" and declare it to be the true and unique deserver of that name. I have argued elsewhere (Joyce, 2012) that this kind of indeterminacy may span the difference between metaethical cognitivism and noncognitivism, and also the difference between moral realism and moral skepticism. In other words, there may be some legitimate precisification of the concept *moral rightness* (for example) according to which rightness is a real property of certain actions; but there may be other equally legitimate precisifications according to which no such property exists anywhere. How might this sort of indeterminacy affect the debate over moral nativism?

It is possible (and not unlikely) that on *any* precisification of "moral judgment" (and on any disambiguation of "innate"), moral nativism is false. But it is also possible that moral nativism is true for certain precifisications and false for others. Certainly the *plausibility* of various pronativist and antinativist arguments varies according to different conceptions of the target trait. For example, if one is concerned with questions of universality, then the less demanding our conception of a moral judgment, the more likely it is that we will find evidence of universality, since, as a truistic rule of thumb, X+Y is going to occur more often than X+Y+Z. On these grounds, Stich objects that the rich conception of moral judgment that I offered (in Joyce, 2006a) spells problems for moral nativism: "For if moral judgment requires *all of that*, what reason is there to think that people in cultures very different from ours *make* moral judgments?" (Stich, 2008, p. 233).[15] If this is correct, then (roughly) richly construed moral judgments are less likely to be universal, thus favoring the nonnativist case (various aforementioned complications with universality aside).

A number of opponents of moral nativism allow that some kind of *normative* nativism might be true. Earlier we saw Nichols accept nativism about non-hypothetical norms. Edouard Machery and Ron Mallon (2010) also accept the plausibility of nativism about normative cognition ("that is, the capacity to grasp norms and to make normative judgments" [p. 4])—where nativism is understood in evolutionary terms.

What they insist upon is that *moral* judgment is but a proper subset of the normative, and there is no evidence for any psychological adaptations dedicated to moral thinking in particular. While it cannot be reasonably denied that the category of the normative is larger than the category of the moral, it should also be noted that *how much larger* depends on what conception of the moral one endorses. A demanding conception will make the moral a smaller subset of the normative; a less demanding conception will yield a larger subset. The larger the subset, however, the more plausibility there is to the claim that it is in fact *moral* judgment that is the distinct adaptation, while the human capacity to make nonmoral normative judgments is a case of aspects of a biological adaptation being coopted for new uses.[16] (This position will be strengthened if we have a plausible hypothesis about why moral judgment in particular might have been adaptive to our ancestors while lacking a hypothesis about why normative judgments in general might have been adaptive.) Thus, again, a less demanding conception of moral judgment might be more amenable to a nativist explanation than a more demanding one.

On the other hand, POS arguments seem to cut the other way. If a thin moral judgment can be constructed out of evolutionarily preexisting mechanisms, then heaping more demands on the conception of moral judgment ("thickening" it) lowers the probability that these mechanisms will remain sufficient to the explanatory task. Again, speaking roughly: Richly construed moral judgments will need more mechanisms to explain them; and the more mechanisms to which one must appeal, the more likely it is that at some point one will need to appeal to a *dedicated* mechanism, thus favoring the nativist case. In this chapter, I haven't attempted the difficult task of arguing that a POS-style argument is plausible even for a demanding conception of moral judgment (though I admit to some sympathy with the project); my objective is simply to draw attention to the fact that the plausibility of the argument may vary according to how the target trait is drawn.

Conclusion

The upshot is that both moral nativism and moral nonnativism may be perfectly defensible positions, and may remain so even when all data are in. This, I predict, will not be a popular conclusion—philosophers and scientists alike prefer their truths tidier—but it is surely worthwhile to diagnose, in advance, those points of conceptual imprecision that may confound future debate.

Acknowledgments

Thanks to my coeditors for valuable feedback. Some of the passages in this chapter concerned with psychological altruism are lifted more or less verbatim from my entry on "altruism and biology" for the *International Encyclopedia of Ethics*.

Notes

1. Of course, one would have little trouble assembling a list of books and articles from 1880 to 1980 (say) that would appear to counter this claim (Edvard Westermarck's works in particular spring to mind); but I would maintain that this list—though superficially impressive if gathered in an endnote—still constitutes "remarkably little attention" for a century's worth of intellectual labor on the topic.

2. Darwin Archives: DAR88.76–77. Charles had evidently asked William to read and summarize Mill's *Utilitarianism* for him while he (Charles) was preparing the second edition of *Descent*. Given that the point of this delegation of labor was to discern Mill's views on the origin of the moral sense, I cannot resist remarking that it was William who had many years earlier been the subject of his father's article "A biographical sketch of an infant," and whose "first sign of moral sense" was allegedly observed at just over a year old (Darwin, 1877, p. 291).

3. Darwin uses "moral sense" and "conscience" seemingly interchangeably. One interesting implication is that he sees the moral sense primarily in terms of *self-directed* moral evaluations—for that is what a conscience is. It seems to me, moreover, that this gives license to assume that when Darwin talks of a "moral sense" it is a faculty of making moral *judgments* that is under discussion. While I am aware that there is some room for debate about this assumption, here I'm willing to forgo argument and treat it as a simplifying supposition.

4. In previous work I have interpreted Darwin as a moral nativist; I now think that this is not straightforward.

5. This statement may seem metaethically question-begging and also surprising in light of other claims I have just made, so a couple of quick explanations are called for. First, at this stage of the discussion I don't intend this notion of "applying a concept" to be theoretically deep; thus my claim is meant to be metaethically neutral. I take it that the locution "applying a concept" is something that even the modern noncognitivist will seek to accommodate. Simon Blackburn's quasi-realist program sets out to "earn the right" to such realist-sounding talk but from an anti-realist position that eschews any genuine metaphysical commitment to such entities. (See Blackburn, 1993, 1998.) Second, given the emerging worries about the indeterminacy surrounding the notion of *moral judgment*, one may wonder on what grounds I can confidently make such an assertion. The answer is that even if there are thinner and richer explications available of the notion of *moral judgment*—such that the former counts certain things as moral judgments that the latter will not—nevertheless, there is surely a class of paradigm instances of moral judgments about which all parties will agree. Of these paradigms, though disagreement may remain concerning in virtue of what they count as moral judgments, it hardly follows that we can say nothing about their characteristics. The statement to which this note is appended is intended to be just such a platitudinous description.

6. Note my avoidance of speaking of "innate moral *knowledge*"—an unnecessary practice that seems to beg several large questions. Moral nativists who seemingly lack such qualms include Sue Dwyer (Dwyer, 2009; Dwyer, Huebner & Hauser, 2010) and John Mikhail (2008).

7. Another potential problem is that Prinz sets out to investigate the existence of *cultural* universals, whereas if nativism did imply universality, we should be examining evidence of *psychological* (i.e., individualistic) universals. For the sake of argument I'll play along with the focus on cultural universals. See Buller (2006, pp. 457–458), for critical discussion.

8. For powerful criticism of this orthodoxy, see Cowie (1999).

9. Or so Prinz claims when he's summarizing his view, but the more detailed presentation is rather more complicated. First, it turns out that having emotions is just the "standard" way to assess things morally (Prinz, 2007, p. 42). Second, moral judgments are linked by Prinz not directly to emotions but to *sentiments*—where a sentiment is a disposition to have an emotion (ibid., p. 84). Thus Prinz has at least two "escape routes" should evidence come forward of moral judgments made with no emotional arousal. For further criticism of Prinz's view, see Joyce (2009).

10. Nichols is aware of moral norms that have nothing obvious to do with harm (concerning, e.g., cleaning the toilet with the national flag), but he states that "it is plausible that judgments about harm-based violations constitute an important core of moral judgment" (Nichols, 2004, p. 7).

11. Yet when one pictures guilt prompting tears, it is natural to picture the scene as one where the transgressor is confronted and accused. By contrast, we have no trouble imagining the tears of sadness falling in private.

12. Prinz (2013) adds the capacity for abstraction to his list of general mechanisms that account for moral judgment.

13. In the past I have offered a fairly detailed description of what I take moral judgments to be, involving strong cognitive elements (Joyce, 2006a, ch. 2). This characterization has been criticized as being nonmandatory (see Machery & Mallon, 2010), and, indeed, Stich finds it necessary to speak of "Joyce-style moral judgments" (Stich, 2008, p. 234).

14. I should point out that in the interests both of simple expression and playing along with an entrenched dialectic, I am drawing a line between "emotions" and "cognitions" in the orthodox ham-fisted manner. Of course, the real distinction is nuanced and complicated. I should also say something to clarify the relation (or lack thereof) between the view under discussion and the literature on the neuroscience of moral judgment, in which the question of emotions versus cognitions looms large. Joshua Greene argues that some moral judgments (deontological ones) stem from emotional arousal whereas others (consequentialist judgments) flow from rational faculties. (See Greene et al., 2001.) Be that as it may, the deontological judgments that are prompted by emotional responses still, in my book, involve obvious cognitive elements. For example, judging that someone has an inalienable right to something (for which consequentialist considerations are irrelevant) involves the deployment of the hefty abstract concept *inalienable right*. Similarly, Jon Haidt's (2001) work may show that moral judgments are little more than post hoc rationalizations of knee-jerk emotional responses, but this should not be confused with the claim that moral judgments are nothing more than emotional responses. In sum, though Greene and Haidt (and others) underline the central role of emotion in moral judgment, they need not be interpreted as proponents of a less demanding conception of moral judgment.

15. Machery and Mallon make the same point: "[Joyce's] claim is substantive and provocative precisely because of the rich characterization of moral judgments that he offers" (2010, p. 21).

16. I am making a debatable background assumption here: that if trait *T* has adaptive function *Fa*, then, for whatever processes make possible "co-opting" *T* for new functions *Fb*, *Fc*., etc., it will be *prima facie* more probable that these processes will have co-opted *T* for fewer new functions than for more new functions. Assessing such a principle would be a complicated task; here I leave it at an intuitive level.

References

Ariew, A. (1996). Innateness and canalization. *Philosophy of Science, 63*(suppl.), S19–S27.

Ariew, A. (1999). Innateness is canalization: In defense of a developmental account of innateness. In V. G. Hardcastle (Ed.), *Where biology meets psychology: Philosophical essays* (pp. 117–138). Cambridge, MA: MIT Press.

Blackburn, S. (1993). *Essays in quasi-realism.* Oxford: Oxford University Press.

Blackburn, S. (1998). *Ruling passions.* Oxford: Oxford University Press.

Buller, D. (2006). *Adapting minds.* Cambridge, MA: MIT Press.

Chomsky, N. (1967). Recent contributions to the theory of innate ideas. *Synthese, 17*, 2–11.

Chomsky, N. (1987/1990). On the nature, use, and acquisition of language. In W. Lycan (Ed.), *Mind and cognition* (pp. 627–646). Oxford: Blackwell.

Cosmides, L., & Tooby, J. (1992). Cognitive adaptations for social exchange. In J. H. Barkow, L. Cosmides, & J. Tooby (Eds.), *The adapted mind* (pp. 163–228). Oxford: Oxford University Press.

Cowie, F. (1999). *What's within: Nativism reconsidered.* Oxford: Oxford University Press.

Darwin, C. [1879] (2004). *The descent of man.* London: Penguin Books.

Darwin, C. (1877). A biographical sketch of an infant. *Mind, 2*, 285–294.

Dwyer, S. (2009). Moral dumbfounding and the linguistic analogy: Methodological implications for the study of moral judgment. *Mind & Language, 24*, 274–296.

Dwyer, S., Huebner, B., & Hauser, M. (2010). The linguistic analogy: Motivations, results, speculations. *Topics in Cognitive Science, 2*, 486–510.

Foot, P. (1972). Morality as a system of hypothetical imperatives. *Philosophical Review, 81*, 305–316.

Greene, J., Sommerville, R., Nystrom, L., Darley, J., & Cohen, J. (2001). An fMRI investigation of emotional engagement in moral judgment. *Science, 293*, 2105–2108.

Griffiths, P. (2002). What is innateness? *Monist, 85*, 70–85.

Haidt, J. (2001). The emotional dog and its rational tail: A social intuitionist approach to moral judgment. *Psychological Review, 108*, 814–834.

Haidt, J., & Joseph, C. (2004). Intuitive ethics: How innately prepared intuitions generate culturally variable virtues. *Daedalus, 133*, 55–66.

Hamilton, W. (1964). The genetical evolution of social behavior, I and II. *Journal of Theoretical Biology, 7*, 1–52.

Hauser, M. (2006). *Moral minds*. New York: HarperCollins.

Hauser, M., Chomsky, N., & Fitch, W. T. (2002). The faculty of language: What is it, who has it, and how did it evolve? *Science, 298*, 1569–1579.

Heine, B. (1985). The mountain people: Some notes on the Ik of north-eastern Uganda. *Africa: Journal of the International African Institute, 55*, 3–16.

Joyce, R. (2001). *The myth of morality*. Cambridge: Cambridge University Press.

Joyce, R. (2006a). *The evolution of morality*. Cambridge, MA: MIT Press.

Joyce, R. (2006b). Is human morality innate? In P. Carruthers, S. Laurence, & S. Stich (Eds.), *The innate mind: Culture and cognition* (pp. 257–279). New York: Oxford University Press.

Joyce, R. (2009). Review of Jesse Prinz's The emotional construction of morals. *Mind, 118*, 508–518.

Joyce, R. (2011). The error in "The error in the error theory." *Australasian Journal of Philosophy, 89*, 519–534.

Joyce, R. (2012). Metaethical pluralism: How both moral naturalism and moral skepticism may be permissible positions. In S. Nuccetelli & G. Seay (Eds.), *Ethical naturalism: Current debates* (pp. 89–109). Cambridge: Cambridge University Press.

Knight, J. (1994). The mountain people as tribal mirror. *Anthropology Today, 10*, 1–3.

Machery, E., & Mallon, R. (2010). The evolution of morality. In J. Doris, G. Harman, S. Nichols, et al. (Eds.), *The moral psychology handbook* (pp. 3–46). Oxford: Oxford University Press.

Mackie, J. (1977). *Ethics: Inventing right and wrong*. New York: Penguin.

Mameli, M. (2008). On innateness: The clutter hypothesis and the cluster hypothesis. *Journal of Philosophy, 105*, 719–737.

Mameli, M., & Bateson, P. (2007). The innate and the acquired: Useful clusters or a residual distinction from folk biology? *Developmental Psychobiology, 49*, 818–831.

Mikhail, J. (2008). The poverty of the moral stimulus. In W. Sinnott-Armstrong (Ed.), *Moral psychology*, Vol. 1: *The evolution of morality: Adaptations and innateness* (pp. 353–360). Cambridge, MA: MIT Press.

Mill, J. S. (1861). Utilitarianism, chapter 3. In *Fraser's Magazine*, November, 525–534.

Nichols, S. (2004). *Sentimental rules: On the natural foundations of moral judgment*. New York: Oxford University Press.

Nichols, S. (2005). Innateness and moral psychology. In P. Carruthers, S. Laurence, & S. Stich (Eds.), *The innate mind: Structure and contents* (pp. 353–430). New York: Oxford University Press.

Prinz, J. (2007). *The emotional construction of morals.* Oxford: Oxford University Press.

Prinz, J. 2008a. Is morality innate? In W. Sinnott-Armstrong (Ed.), *Moral psychology*, Vol. 1: *The evolution of morality: Adaptations and innateness* (pp. 367–406). Cambridge, MA: MIT Press.

Prinz, J. 2008b. Resisting the linguistic analogy: A commentary on Hauser, Young, and Cushman. In W. Sinnott-Armstrong (Ed.), *Moral psychology*, Vol. 2: *The cognitive science of morality: Intuition and diversity* (pp. 157–170). Cambridge, MA: MIT Press.

Prinz, J. (2009). Against moral nativism. In D. Murphy & M. Bishop (Eds.), *Stich and his critics* (pp. 167–189). Malden: Blackwell.

Prinz, J. (Forthcoming). Where do morals come from?—a plea for a cultural approach. In M. Christen, M. Huppenbauer, C. Tanner & C. van Schaik (Eds.), *Empirically informed ethics*. New York: Springer.

Schino, G., & Aureli, F. (2010). The relative roles of kinship and reciprocity in explaining primate altruism. *Ecology Letters, 13*, 45–50.

Sober, E. (1988). What is evolutionary altruism? In M. Matthen & B. Linsky (Eds.), *New essays on philosophy and biology* (*Canadian Journal of Philosophy* suppl. vol. 14), 75–99.

Sober, E. (2000). Psychological egoism. In H. LaFollette (Ed.), *The Blackwell guide to ethical theory* (pp. 129–148). Oxford: Blackwell.

Sober, E., & Wilson, D. S. (1998). *Unto others: The evolution and psychology of unselfish behavior.* Cambridge, MA: Harvard University Press.

Sripada, C. S. 2008. Nativism and moral psychology: Three models of the innate structure that shapes the contents of moral norms. In W. Sinnott-Armstrong (Ed.), *Moral psychology*, Vol. 1: *The evolution of morality: Adaptations and innateness* (pp. 319–343). Cambridge, MA: MIT Press.

Sripada, C., & Stich, S. (2006). A framework for the psychology of norms. In P. Carruthers, S. Laurence, & S. Stich (Eds.), *The innate mind: Culture and cognition* (pp. 280–301). New York: Oxford University Press.

Sterelny, K. (2010). Moral nativism: A sceptical response. *Mind & Language, 25*, 279–297.

Stich, S. (2007). Evolution, altruism, and cognitive architecture: A critique of Sober and Wilson's argument for psychological altruism. *Biology and Philosophy, 22*, 267–281.

Stich, S. (2008). Some questions about The Evolution of Morality. *Philosophy and Phenomenological Research, 77*, 228–236.

Tangney, J., & Fischer, K. (Eds.). (1995). *Self-conscious emotions: The psychology of shame, guilt, embarrassment, and pride.* New York: Guilford Press.

Turnbull, C. (1972). *The mountain people.* New York: Simon & Schuster.

West, S., Griffin, A., & Gardner, A. (2007). Social semantics: Altruism, cooperation, mutualism, strong reciprocity, and group selection. *European Society for Evolutionary Biology, 20*, 415–432.

Contributors

Sam Brown
Oxford University

Brett Calcott
Arizona State University

Maciek Chudek
University of British Columbia

Andrew Cockburn
Australian National University

Fiery Cushman
Harvard University

Tanya Elliot
Santa Fe Institute

Doug Erwin
Santa Fe Institute; Smithsonian Institution

Daniel M. T. Fessler
University of California, Los Angeles

Jessica C. Flack
Santa Fe Institute

Ben Fraser
Australian National University

Herbert Gintis
Santa Fe Institute; Central European University

Deborah M. Gordon
Stanford University

Adam G. Hart
University of Gloucestershire

Joe Henrich
University of British Columbia

Katja Heubel
University of Helsinki

Cecilia Heyes
Oxford University

Simon M. Huttegger
University of California, Irvine

Richard Joyce
Victoria University of Wellington

Daniel R. Kelly
Purdue University

Hanna Kokko
Australian National University

David C. Krakauer
Santa Fe Institute

Elizabeth T. Malouf
George Mason University

Matteo Mameli
King's College London

Hugo Mercier
University of Neuchâtel

Ronald Noë
University of Strasbourg

Haim Ofek
Binghamton University

Katinka Quintelier
Ghent University

Livio Riboli-Sasco
Paris Descartes University

Don Ross
University of Cape Town

Paul Seabright
Toulouse School of Economics

Nicholas Shea
Oxford University

Brian Skyrms
University of California, Irvine

Kim Sterelny
Australian National University; Victoria University of Wellington

Jeffrey Stuewig
George Mason University

François Taddei
Paris Descartes University

June P. Tangney
George Mason University

Bernhard Voelkl
Humboldt University of Berlin

Felix Warneken
Harvard University

Kerstin Youman
George Mason University

Wanying Zhao
University of British Columbia

Index

Printed in the United States
by Baker & Taylor Publisher Services